The chemical economy

A guide to the technology and economics of the chemical industry

B. G. REUBEN, M.A., B.Sc., D.Phil.
and
M. L. BURSTALL, M.A., B.Sc., D.Phil.

LONGMAN

Longman
1724-1974

LONGMAN GROUP LIMITED
London
Associated companies, branches and representatives throughout the world

© B. G. Reuben and M. L. Burstall 1973

All rights reserved. No part of this publication may be reproduced, stored in a retrieval system, or transmitted in any form or by any means, electronic, mechanical, photocopying, recording, or otherwise, without the prior permission of the Copyright owner.

First published 1973

ISBN 0 582 46307 6

Library of Congress Catalog Card Number: 73–85210

*Made and printed in Great Britain by
William Clowes & Sons, Limited, London, Beccles and Colchester*

To Teresa

Preface

How do we put chemistry to work for us? This is an important question. The chemical industry is now the fourth largest industry in the United Kingdom and occupies a similar position in the economies of most advanced countries. Forty years ago four qualified chemists went into teaching and pure research for every one who entered industry; in 1965 four went into industry for every one who entered university teaching. Furthermore, by the age of 35 half of all chemists have ceased to practise and have become administrators. If the image of a chemist in 1900 was of a dedicated scientist immured in his laboratory, the image today may well be that of a marketing man clinching a deal for 10 000 tons of acetone over a business lunch.

In spite of these changes the education provided for chemists has scarcely altered. Industrialists grumble incessantly about the unsuitable training their recruits have received and students complain that their courses lack relevance. Everyone pays lip-service to university–industry cooperation but little is done. In 1964 we were asked to design a course to bridge this gap which was to be taken as an option by Chemistry students at the new University of Surrey. The course has flourished and is now taken by students from several departments. We were unable to find any good general books to recommend so we have written this one. It is also the book which we would both have liked to have read before entering industry ourselves.

We have tried to place the chemical industry in its social and economic perspective. We have taken a broad view because we feel that it is important to see both the plastic wood and the polypropylene trees. We have concentrated on trying to make sense of the industry from the viewpoint of the man at the operating level. Thus we have dealt with microeconomics rather than macroeconomics because the chemist is likely to be involved with the former but merely a victim of the latter. Similarly we have assumed the social and economic system of the Western World as it now is, partly because the modern chemical industry is a product of that world and partly because we have experience of no other. Our economic theory is meant to be an approximate description of what happens and is not intended ideologically.

Our book is directed primarily at chemists and assumes a GCE Advanced Level or first year university knowledge of chemistry, although large sections should be comprehensible without this. We have assumed no previous knowledge of chemical engineering or economics and have therefore provided an introduction to the concepts and vocabulary of these subjects which those who already possess such background should omit. We admit to having treated them somewhat cursorily. Our aim, however, is to help the non-specialist to see how the chemical industry works rather than to look good in the eyes of professionals who are always reluctant to believe that the eternal verities of their subjects can be encompassed in less than one thousand closely-printed pages.

We are uneasily aware that in an interdisciplinary book of this length we are bound to have made a number of gaffes. We apologize for these in advance and would be genuinely grateful to have our attention drawn to them.

Of the many people who have made this book possible we would especially like to thank Dr Teresa Poole, Professor John Salmon, Professor Colin Robinson, Dr John Lamb, and our other colleagues at the University of Surrey, Mr Jack Wickings of BP Chemicals International, Stirling and Co., Stock and Share brokers, and our families and friends whose experience of the birth pangs, albeit at second hand, must have been even more traumatic than our own.

Notes and references

It is tempting for anyone in the academic world to append to every publication a detailed list of primary sources referred to by a mass of numbers in the text. The method is scholarly but cumbersome, and inappropriate in a book of this kind. On the other hand, a general bibliography leaves the reader without any indication of the sources of individual bits of an author's information. We have therefore followed each chapter with an annotated bibliography designed to introduce readers to the literature, to show which books we found useful and to give the sources of our more important facts. A general bibliography is given in the first section – '*How to use this book*'.

Anyone who is interested in the development of the ideas in this preface should consult B. G. Reuben and M. L. Burstall, 'Objectives in the Teaching of Chemistry', *Chem. and Ind.*, 1968, 1794; M. L. Burstall, 'The Education of Industrial Scientists' in *Scientists in British Industry*, ed. S. G. Cotgrove and G. Walter, Bath University Press, 1968, and B. G. Reuben, 'Humane Education for the Industrial Chemist', *New Scientist*, 5 November, 1970.

The drift of scientists to managerial jobs was measured in *The Survey of Professional Scientists, 1968*, Ministry of Technology and Council for Science and Technology, HMSO, London, 1970, which showed that after the age of 35, more than half the professional scientists in each age group were in managerial posts.

Contents

Preface		vii
How to use this book		**1**
0.1	Structure and layout	1
0.2	Definitions	1
0.3	General bibliography of chemical technology	2
	0.3.1 Bibliography	2
	0.3.2 Encyclopaedia	2
	0.3.3 Books	2
	0.3.4 Journals	3
	0.3.4.1 *European Chemical News*	3
	0.3.4.2 British	3
	0.3.4.3 United States of America	3
	0.3.4.4 European	3
	0.3.4.5 Asian	4
	0.3.5 Annuals	4
0.4	Statistics of the chemical industry	4
	0.4.1 Statistics	5
	0.4.1.1 International	5
	0.4.1.2 Great Britain	5
	0.4.1.3 United States of America	5
	0.4.1.4 Europe	6
	0.4.1.5 Japan	6
0.5	Organizations	6
0.6	Prices and costs	7
0.7	Units	7

Part I   The growth of the chemical industry

1	**The old chemical industry from 1800 onwards**	**11**
1.1	Early days	11
1.2	The market for chemicals	12
	1.2.1 Soap-making	12
	1.2.2 Glass	12
	1.2.3 Cotton textiles	12

x   Contents

		1.2.4 Paper-making	12
		1.2.5 Agriculture	13
		1.2.6 Explosives	13
	1.3	The Leblanc process	13
	1.4	The golden age of the alkali trade	13
	1.5	Solvay versus Leblanc	14
	1.6	Crisis and stagnation	15
	1.7	Dyestuffs and the rise of Germany	16
	1.8	Deutschland über Alles	17
	1.9	The impact of war, 1914–18	18
		1.9.1 Germany	18
		1.9.2 Britain	18
		1.9.3 The USA	19
	1.10	The Post First World War scene	19
	1.11	The formation of ICI	20
	1.12	The slump and after	21
	Notes and references		22

2	The new chemical industry from 1820 onwards		25
	2.1	The polymer revolution	25
	2.2	Rubber	25
	2.3	Cellulosics	28
	2.4	The early days of synthetic resins	29
	2.5	The science of large molecules	30
	2.6	Three countries in the 1930s	30
		2.6.1 Germany	31
		2.6.2 USA	32
		2.6.3 Great Britain	32
	2.7	The impact of war: 1939–45	32
	2.8	From coal to oil	33
	2.9	Polymers triumphant	34
		2.9.1 The rise of Japan	34
		2.9.2 The natural history of a polymer	35
	Notes and references		38

Part II   Economic aspects of the chemical industry

3	Introduction to microeconomics		43
	3.1	Problems of the economic system	44
	3.2	Money	45
	3.3	Utility	45
		3.3.1 Changes of price and income	48
	3.4	Demand	50
		3.4.1 Elasticity of demand	51

			Contents	xi

	3.5	Production		53
		3.5.1 Diminishing returns		54
		3.5.2 Economies of scale		56
		3.5.3 Alternative production possibilities		58
	3.6	Supply		60
		3.6.1 Long-term balance of supply and demand		63
	3.7	Competition		64
	3.8	International trade		67
		3.8.1 Comparative advantage		68
		3.8.2 Import duties on chemicals		69
		3.8.3 Chemical exports and imports		69
	Notes and references			72

4	**The decision to produce**		73
	4.1 The demand schedule		73
	4.2 Costs and marginal costs		76
	4.3 Profits		76
	4.4 Fixed and variable costs		79
	4.5 Response of a firm to varying prices		82
	4.6 Finding the resources		83
	4.7 Limited liability		86
	Notes and references		87

5	**How do you know that you are making money?**		88
	5.1 Accounts and accountants		88
	5.2 The balance sheet		88
	5.3 The profit and loss account		91
	5.4 Measures of profitability		92
	5.5 Measures of financial stability		93
	5.6 The impact of inflation		94
	5.7 The choice of projects		95
		5.7.1 The Discounted Cash Flow method	95
		5.7.2 How to use the DCF method	96
		5.7.3 Implications of the DCF method	98
		5.7.4 Alternatives to the DCF method	100
	Notes and references		100

6	**The world chemical industry**		101
	6.1 The UK chemical industry		101
		6.1.1 The national income	101
		6.1.2 The chemical industry and the UK economy	107
		6.1.3 The nature of the UK chemical industry	109
		6.1.4 The performance of the UK chemical industry	112
		6.1.5 Individual companies	114

xii *Contents*

	6.1.5.1	Imperial Chemical Industries	114
	6.1.5.2	British Oxygen Company	117
	6.1.5.3	BP (Chemicals) International	117
	6.1.5.4	Glaxo Group	118
	6.1.5.5	Albright and Wilson	118
	6.1.5.6	Fisons	119
	6.1.5.7	Wellcome Foundation	119
	6.1.5.8	Rio-Tinto Zinc Corporation	119
	6.1.5.9	Beecham Group	119
	6.1.5.10	Laporte Industries	120
	6.1.5.11	Royal Dutch/Shell	120
	6.1.5.12	Unilever	120

6.2 The chemical industry in the developed world — 121
    6.2.1 Individual countries — 126
        6.2.1.1 United States of America — 127
        6.2.1.2 Germany — 132
        6.2.1.3 France — 133
        6.2.1.4 Japan — 133
        6.2.1.5 Switzerland — 134
        6.2.1.6 Italy — 135
        6.2.1.7 Netherlands and Belgium — 135
        6.2.1.8 Union of Soviet Socialist Republics — 135
        6.2.1.9 Eastern Europe — 136

6.3 The chemical industry in developing countries — 136
    6.3.1 Differences between under-developed countries — 138
    6.3.2 Small size of local markets — 140
    6.3.3 Labour — 141
    6.3.4 Choice of product — 142
    6.3.5 Choice of process — 143
    6.3.6 Raising the capital — 146
    6.3.7 Conclusions — 147

6.4 Industries of special importance to developing countries — 148
    6.4.1 Desalination — 148
        6.4.1.1 Solar evaporation — 149
        6.4.1.2 Multiple effect distillation — 149
        6.4.1.3 Flash distillation — 149
        6.4.1.4 Freezing processes — 150
        6.4.1.5 Electrodialysis — 150
        6.4.1.6 Ion exchange — 151
        6.4.1.7 Reverse osmosis — 151
        6.4.1.8 Comparison of methods — 152
    6.4.2 Semi-synthetic protein — 153
        6.4.2.1 Yeast protein from petroleum — 153
        6.4.2.2 Textured vegetable protein — 153
        6.4.2.3 Fungal protein — 154

Notes and references — 154

Contents xiii

Part III The products of the chemical industry

7 Raw materials : petroleum — 159
   7.1 The petroleum industry — 159
   7.2 Supply and demand — 159
      7.2.1 The exporters — 159
      7.2.2 The importers — 160
      7.2.3 The USA — 161
      7.2.4 The Soviet bloc — 162
   7.3 The composition of petroleum — 162
      7.3.1 Hydrocarbons — 162
      7.3.2 Oxygen, nitrogen and sulphur compounds — 163
      7.3.3 Resins and asphalts — 163
   7.4 Refining — 164
   7.5 Cracking and reforming — 165
   7.6 Alkenes from petroleum — 166
      7.6.1 General considerations — 166
      7.6.2 Chemistry — 167
      7.6.3 Technology — 170
      7.6.4 Economics — 171
   7.7 Acetylene from petroleum — 173
   7.8 Aromatics from petroleum — 175
   7.9 Hydrogen from petroleum — 178
   Notes and references — 179

8 Raw materials : natural gas, coal, fermentation — 182
   8.1 Petroleum's competitors — 182
   8.2 Natural gas — 182
   8.3 Mineral hydrocarbons — 184
   8.4 Coal — 184
      8.4.1 Structure — 184
      8.4.2 Production — 185
      8.4.3 Carbonization — 187
      8.4.4 Acetylene via carbide — 189
      8.4.5 Water-gas and the Fischer–Tropsch process — 190
      8.4.6 Hydrogenation — 191
   8.5 Fermentation — 191
      8.5.1 Nature and scope — 191
      8.5.2 Technology — 191
      8.5.3 Applications — 194
   Notes and references — 194

9 Heavy organic chemicals — 196
   9.1 Ethylene — 196
      9.1.1 Ethylene oxide — 199
          9.1.1.1 Ethylene glycol — 201
      9.1.2 Ethyl chloride — 201

	9.1.3	Ethyl alcohol		202
	9.1.4	Acetaldehyde		203
9.2	Propylene			204
	9.2.1	Isopropyl alcohol		206
		9.2.1.1	Acetone	207
	9.2.2	Propylene oxide		209
		9.2.2.1	Propylene glycol	210
		9.2.2.2	Polypropylene glycol and other polyether polyols	210
	9.2.3	*n*-Butyl and isobutyl alcohols		211
	9.2.4	2-Ethylhexyl alcohol		211
	9.2.5	Propylene tetramer		212
9.3	Butenes and butadiene			212
9.4	Acetic acid			213
	9.4.1	Ketene and diketene		214
	9.4.2	Acetic anhydride		215
9.5	Aromatics			216
	9.5.1	Benzene		216
		9.5.1.1	Phenol	218
	9.5.2	Toluene		224
	9.5.3	Xylenes		225
	9.5.4	Naphthalene		226
9.6	Acetylene			227
Notes and references				227

10 Industrial polymers — 229

10.1	Properties and molecular structure			230
10.2	Thermoplastics (excluding PVC)			235
	10.2.1	Polyethylene		236
	10.2.2	Polypropylene		238
	10.2.3	Polystyrene		239
	10.2.4	Polymethyl methacrylate		242
	10.2.5	Polyvinylidene chloride		244
	10.2.6	Polycarbonates		244
	10.2.7	Polyacetals		244
	10.2.8	Polytetrafluoroethylene		245
10.3	Polyvinyl chloride, plasticizers and other additives			245
	10.3.1	Manufacture of vinyl chloride		247
	10.3.2	End-uses		249
	10.3.3	Stabilizers for PVC		250
	10.3.4	Plasticizers		251
		10.3.4.1	General purpose PVC plasticizers and their molecular structure	253
		10.3.4.2	Specialized phthalate plasticizers	255
		10.3.4.3	Phosphate ester plasticizers	256
		10.3.4.4	Epoxidized oil and polymeric plasticizers	256
		10.3.4.5	Plasticizers based on aliphatic acids	256
		10.3.4.6	Internal and external plasticizers	257

				Contents	xv

		10.3.5	Extenders		257
		10.3.6	Fillers		258
	10.4	Thermosetting plastics			258
		10.4.1	Phenoplasts		259
			10.4.1.1	Alkaline catalysed P/F resins	259
			10.4.1.2	Acid catalysed P/F resins	261
		10.4.2	Aminoplasts		262
			10.4.2.1	Urea–formaldehyde resins	262
			10.4.2.2	Melamine–formaldehyde resins	264
		10.4.3	Economic aspects of phenoplasts and aminoplasts		266
		10.4.4	Unsaturated polyester resins		266
		10.4.5	Polyurethanes		267
		10.4.6	Epoxy resins		269
	10.5	Paints			271
		10.5.1	Polyvinyl acetate and the water-based emulsion paint		274
		10.5.2	Acrylics		277
		10.5.3	Alkyd resins		279
		10.5.4	Nitrocellulose		280
		10.5.5	Other resins		280
		10.5.6	Titanium dioxide		281
	10.6	Rubbers			282
		10.6.1	Natural rubber		284
		10.6.2	Synthetic polyisoprene		284
		10.6.3	Styrene–butadiene		285
		10.6.4	Butyl rubber		286
		10.6.5	Polybutadiene rubbers		286
		10.6.6	Nitrile rubbers		286
		10.6.7	Polychloroprene rubber		287
		10.6.8	Ethylene–propylene rubbers		287
		10.6.9	Miscellaneous rubbers		288
	10.7	Textile fibres			288
		10.7.1	Cellulosics		293
		10.7.2	Nylon		295
		10.7.3	Polyesters		298
		10.7.4	Acrylonitrile and acrylic fibres		299
	Notes and references				301

11	Soaps and detergents				305
	11.1	The soap trade			305
	11.2	Surface-active agents and their properties			305
	11.3	Detergency			310
		11.3.1	Phenomenology		310
		11.3.2	Theories		311
		11.3.3	Formulations		313

11 (continued)

- 11.4 Anionic surfactants — 315
  - 11.4.1 Soap — 315
  - 11.4.2 Synthetic anionics — 317
- 11.5 Cationic surfactants — 318
- 11.6 Non-ionic surfactants — 318
- 11.7 Effluent problems and the detergent industry — 319
- 11.8 Marketing — 320
- 11.9 The economics of soap — 321
- Notes and references — 321

12 Dyes and pigments — 324

- 12.1 The colour industry — 324
- 12.2 Colour and constitution — 325
- 12.3 Dyes and fibres — 328
  - 12.3.1 Acid and basic dyes — 328
  - 12.3.2 Dyes and cellulose — 331
  - 12.3.3 Dyes and synthetic fibres — 334
  - 12.3.4 Fibre-reactive dyes — 335
- 12.4 Miscellaneous applications of dyes — 336
- 12.5 Pigments — 336
- 12.6 Production and trends — 338
- Notes and references — 338

13 Pharmaceuticals — 340

- 13.1 Sources of drugs — 344
- 13.2 Characteristics of the pharmaceutical industry — 345
- 13.3 Aspirin (acetylsalicylic acid) — 350
- 13.4 Sulphonamides — 352
- 13.5 Barbiturates — 354
- 13.6 Antibiotics — 354
  - 13.6.1 Penicillin — 355
  - 13.6.2 Streptomyces antibiotics — 358
  - 13.6.3 Tetracyclines — 358
- 13.7 Psychotropic drugs — 359
- Notes and references — 363

14 Heavy inorganic chemicals — 365

- 14.1 The nitrogen industry — 367
  - 14.1.1 The Haber process — 369
  - 14.1.2 'Synthesis' or 'make-up' gas — 372
  - 14.1.3 Nitric acid — 376
  - 14.1.4 Ammonium sulphate — 377
  - 14.1.5 Ammonium nitrate — 377
  - 14.1.6 Urea — 378
  - 14.1.7 Methanol and formaldehyde — 378
- 14.2 Sulphuric acid — 379
  - 14.2.1 The lead chamber process — 380

	14.2.2	The contact process	383
	14.2.3	Sulphur dioxide production	384
14.3	The alkali industry		385
	14.3.1	Electrolysis of brine	387
	14.3.2	Bleaching powder and calcium hypochlorite	390
	14.3.3	Sodium hypochlorite	390
	14.3.4	Sodium carbonate, bicarbonate and sesquicarbonate	390
	14.3.5	Chlorine – the Deacon process	390
14.4	The phosphorus industry		391
	14.4.1	Superphosphate fertilizers	392
	14.4.2	Phosphoric acid and phosphates	392
14.5	The borax industry		393
14.6	Potash		395
	14.6.1	Extraction from sylvinite by hot leaching	396
	14.6.2	Extraction from sylvinite by flotation	397
	14.6.3	Potash from the Dead Sea (via carnallite)	397
14.7	Hydrogen peroxide		398
Notes and references			399

Part IV Large scale chemistry

15 The scaling-up of laboratory experiments — 403

15.1	Economic factors in scaling-up		403
	15.1.1	Yields	403
	15.1.2	Planning	404
15.2	Technical factors in scaling-up		404
15.3	Reactor design and optimization of reaction conditions		407
	15.3.1	Batch and continuous reactors	407
	15.3.2	Reactors for heterogeneous reactions	409
15.4	Unit operations		409
	15.4.1	Fluid dynamics	410
	15.4.2	Solids handling	410
	15.4.3	Heat transfer	410
		15.4.3.1 Conduction	411
		15.4.3.2 Convection	411
		15.4.3.3 Radiation	412
		15.4.3.4 Heat exchangers	412
	15.4.4	Mass transfer and distillation	416
		15.4.4.1 Application of distillation theory	417
		15.4.4.2 Variation of reflux ratio	424
		15.4.4.3 Optimization of reflux ratio	424
		15.4.4.4 Types of fractionating column	426
15.5	Fire and explosion hazards		429
15.6	Toxicity and related hazards		429
15.7	Corrosion and materials of construction		429
15.8	Instrumentation and control		431

	15.9	The scaled-up plant	431
	Notes and references		434

16 Unit processes — 436

16.1	Oxidation		436
16.2	Hydrogenation		437
16.3	The OXO process		439
16.4	Esterification		440
16.5	Hydrolysis and hydration		441
16.6	Alkylation		443
16.7	Amination		446
16.8	Nitration		447
16.9	Sulphation and sulphonation		449
16.10	Halogenation		449
	16.10.1	Fluorination	449
	16.10.2	Chlorination	450
	16.10.3	Oxychlorination	451
	16.10.4	Bromination	452
	16.10.5	Iodination	452
16.11	Polymerization		453
	16.11.1	Condensation polymerization	453
	16.11.2	Addition polymerization	454
	16.11.3	Free radical polymerization	455
	16.11.4	Cationic polymerization	456
	16.11.5	Anionic polymerization	456
	16.11.6	Bulk polymerization	457
	16.11.7	Solution polymerization	457
	16.11.8	Emulsion polymerization	457
	16.11.9	Suspension polymerization	458
16.12	Processing and fabrication of plastics		458
	16.12.1	Processing of plastics	458
	16.12.2	Fabrication of plastics	459
16.13	Other unit processes		465
Notes and references			465

17 The effluent society — 467

17.1	Historical background		467
17.2	Domestic sewage treatment		468
	17.2.1	Screening	471
	17.2.2	Sedimentation	471
	17.2.3	Treatment of clear effluent	472
	17.2.4	Sludge treatment	474
17.3	Industrial effluents		475
	17.3.1	Effluents from electroplating	475
	17.3.2	Woollen industry wastes	476
	17.3.3	Petrochemical effluents	477

			Contents	xix

	17.4	Air pollution		477
		17.4.1 Cigarette smoke		479
		17.4.2 Domestic and industrial fuel burning		481
		17.4.3 Automobile exhaust fumes		484
	17.5	Long-term pollution of the biosphere		489
		17.5.1 Medium-scale effects		489
		17.5.2 The 'greenhouse' effect		489
		17.5.3 Insecticides		490
		17.5.4 High level jet aircraft flights		491
	17.6	Conservation and regeneration		491
	Notes and references			492

Part V **The future**

18	**And will the trees grow up to the sky?**			**497**
	18.1	Yesterday's tomorrows		497
	18.2	Five versions of the future		497
		18.2.1 The trees *will* grow up to the sky		497
		18.2.2 Doom is just around the corner		498
		18.2.3 Rich world, poor world		499
		18.2.4 Like, there's been a change of values, man		500
		18.2.5 A generation of difficulties		501
	18.3	But what's in it for chemistry?		501
		18.3.1 The logistic curve and problems of forecasting		502
		18.3.2 The British chemical industry		507
			18.3.2.1 Britain in the 1970s	507
			18.3.2.2 The chemical industry in the 1970s	508
			18.3.2.3 The future of polymers	508
			18.3.2.4 The future of pharmaceuticals	510
			18.3.2.5 The future of the old chemical industry	510
			18.3.2.6 The environmental problems of the industry	510
			18.3.2.7 Is the golden age over?	511
		18.3.3 The world chemical industry		511
		18.3.4 New technologies and their impact		512
	Notes and references			514

Compound index	**515**
Subject index	**523**

How to use this book

0.1 Structure and layout

The book is divided into five parts: Part I (ch. 1–2) deals with the development of the chemical industry; Part II (ch. 3–6) with the economics of the firm, accounting and the position of the industry in the national and world economies; Part III (ch. 7–14) with the major sectors of the industry; Part IV (ch. 15–17) with the problem of large-scale chemistry; and Part V (ch. 18) with the future. Certain chapters are intended as elementary introductions to particular topics for readers with a scientific background. They need not be read by those familiar with such matters: thus economists should omit chapters 3–5, and chemical engineers omit chapters 15–16.

Each chapter is divided into numbered sections which have been further subdivided. For example chapter 10 deals with industrial polymers; section 10.3 with polyvinylchloride, plasticizers and additives, subsection 10.3.4 with plasticizers, and sub-sub-section 10.3.4.1 with general purpose PVC plasticizers. Cross references are indicated by this numbering system. Thus in section 7.6.4. one reads: 'A consequence of the DCF method of estimating profitability (section 5.7.3) has been to make executives acutely aware of the drawbacks as well as the advantages of the very large single-stream plant.'

The note 'section 5.7.3' refers the reader to section 5.7.3 which discusses the implications of the DCF method. Tables and figures are similarly numbered: the first table in chapter 1 is Table 1.1, the first figure 1.1 and so on.

0.2 Definitions

The definition of the chemical industry used in this book is essentially that used in the United Nations Standard International Trade Classification (SITC). The chemical industry forms section 5 of the SITC and comprises chemical elements and compounds; mineral tar and crude chemicals from coal, petroleum and natural gas; dyeing, tanning and colouring materials; medicinal and pharmaceuticals products; essential oils and perfume materials; polishing and cleansing preparations; manufactured fertilizers; explosives and pyrotechnic products; plastic materials, regenerated cellulose and artificial resins; and chemical materials and products not elsewhere classified. As is customary, synthetic rubber, chemical products used in photography, and photographic film, all of which fall into other sections of the SITC are included as part of the chemical industry. This definition of the industry corresponds to that implied by sections 6 and 7 of the *Brussels Nomenclature for the Classification of Goods in Customs Tariffs*, 4th edition, Brussels, 1972.

It should be noted that the SITC definition often differs in detail from that used by individual countries. In the UK, for example, imports and exports of chemicals are classi-

2 *How to use this book*

fied according to the SITC rules, but production statistics are recorded in the categories of the UK Standard Classification (SIC) 1968. The differences between Order V of the SIC – chemical and allied industries – and the chemical industry as defined above are small but appreciable; accordingly where the SIC definition is used we refer to 'chemical and allied industries'. Other variations are noted where appropriate.

In certain cases it is convenient to use a more restricted definition of the chemical industry. Where we have done so, the definition used is indicated in the notes.

0.3 General bibliography of chemical technology

At the end of each chapter there is a short critical bibliography which indicates useful books, papers and journals for the particular field in question. Certain sources of information are of general significance and are discussed here. When cited in chapter bibliographies they are referred to by name only: thus Kirk–Othmer means Kirk–Othmer. *Encyclopaedia of Chemical Technology*, 2nd ed., New York: Interscience, 1962–71.

0.3.1 Bibliography

The bibliography *Literature of Chemical Technology*, J. F. Smith (Ed.), American Chemical Society: Advances in Chemistry Series No. 78, Washington D.C., 1968, is a concise guide to books, journals and conference proceedings up to 1966–67. It is rather uncritical and definitely stronger on technology than on economics.

0.3.2 Encyclopaedia

Undoubtedly the most useful single work in the whole field is Kirk–Othmer. *Encyclopaedia of Chemical Technology*, 2nd ed., 22 vols. and supplementary volume, New York: Interscience, 1962–71, which gives comprehensive, authoritative and fully referenced coverage of virtually every subject. Copiously illustrated and most attractively produced, it is a genuine pleasure to use. It should be noted, however, that it is naturally biased towards American practice and that its discussions of economic factors are almost always qualitative in nature.

0.3.3 Books

R. N. Shreve. *Chemical Process Industries*, 3rd ed., New York, McGraw-Hill, 1967, is a classic work now rather out of date. It is strong on heavy inorganic chemicals but weak on petrochemicals and polymers.
F. A. Henglein. *Chemical Technology*, London, Pergamon, 1969, describes the German chemical industry of the early 1950s in enormous detail.
Much more useful is W. L. Faith, D. B. Keyes and R. L. Clark. *Industrial Chemicals*, 3rd ed., New York, Wiley, 1965, which describes the manufacture of 137 common industrial chemicals with brief details of markets and economic aspects.
A. V. Hahn. *The Petrochemicals Industry – Production and Markets*, New York, McGraw-Hill, 1970, covers not only the main petroleum-based chemicals from hydrogen onwards but also their principle derivatives; it includes a wealth of detail about the manufacturing costs, markets and trends prevailing in the USA in the mid-1960s.

0.3.4 Journals

0.3.4.1. European Chemical News In a class by itself is this weekly, (IPC, International Press, London; quarterly index; cited as *ECN*) which reports technical, economic and financial news of the West European chemical industry with briefer coverage of Eastern Europe, Japan and the USA. Frank, lively and remarkably comprehensive it is indispensable reading for the chemical technologist. Particularly useful are the numerous articles on process economics and the periodic supplements on special subjects. Topics covered recently in the latter include the chemical industries of Belgium (3.3.72), France (26.11.71), Austria (2.7.71), and Italy (25.12.70), synthetic fibres (29.10.71), petrochemical development areas (30.7.71) and PVC processes (29.1.71). Regular new plant and large plant supplements are published in the last issues for February and September respectively. All numbers include the current European and American prices of about 230 common chemicals and details of Dutch patent applications of chemical significance.

0.3.4.2 British: Chemical Age (Benn Group, London; quarterly index) is a weekly covering the same ground as *ECN* in a much more sedate way and from a more UK-centred standpoint. Its most valuable feature is the annual survey of the 200 biggest chemical companies, which appears in the last issue for July. Michael Hyde, quondam editor of *Chemical Age*, now produces his own fortnightly chemical industry newsheet, *Chemical Insight*, thin, expensive, and very well informed. *Chemistry and Industry* (fortnightly; published by the Society of Chemical Industry; annual index) emphasizes chemistry and chemical engineering but is a favourite forum for high-level thinking by senior managers. The most useful of the chemical engineering journals is *Chemical and Process Engineering* (monthly; published by Morgan–Grampian Ltd., London; annual index; cited as *CPE*) which publishes detailed process surveys with full cost data from time to time; all issues contain indices of productivity and plant construction costs for the UK chemical industry.

0.3.4.3 United States of America US journals benefit greatly from the frankness of American industry and contain far more hard information than those elsewhere. *Chemical and Engineering News* (Washington D.C., American Chemical Society; annual index) is a glossy well-written weekly news-magazine which covers both chemistry and the chemical industry; an annual supplement on the chemical industry containing the financial results of individual companies and much other information is published each September. The same organization also publishes *Chemical Technology*, a lively monthly review with a technical rather than an economic bias. *The Chemical Marketing Reporter* (formerly the *Oil, Paint and Drug Reporter*; New York, Schnell Publishing Co., weekly) is the commercial newspaper of the industry and the indispensible source of current US prices. The most useful of the chemical engineering journals are *Chemical Engineering* (fortnightly, New York, McGraw-Hill; annual index) and *Chemical Engineering Progress* (monthly, New York, American Institute of Chemical Engineers, annual index). Both publish frequent articles on markets, trends and costs; the former includes a widely used plant cost index. In its more restricted field *Hydrocarbon Processing* (monthly, Houston, Texas, Gulf Publishing; annual index) is of outstanding value, providing authoritative coverage of petroleum refining and petrochemicals. All these journals have a strongly US-centred approach and this must be borne in mind when using the information which they provide.

0.3.4.4 European *ECN* is the best source of information but certain of the national journals are useful. *Chemische Industrie* (in German; monthly, Verlag Handelsblatt, Dusseldorf; biannual index) is the best of the German magazines; it includes a plant cost index. A quarterly English-language version is available. *Chimie et Industrie-Génie Chimique* (in French; twice-monthly, Presse Documentaires, Paris; annual index) is a review and news magazine giving good coverage of France. *L'Industrie Chimique* (in French, Paris, Devernay; no index) is a similar but more staid monthly. *Industrie Chimique Belge* (in French and Flemish; monthly, Fédération des Industries Chimique de Belge; no index) contains an excellent news section.

0.3.4.5 Asian Two excellent English-language reviews are *Chemical Age of India* (monthly, Technical Press Publications, Bombay; annual index) which contains periodic process surveys, and the glossy Japanese monthly *Chemical Economy and Engineering Review* (Chemical Economy Research Institute, Tokyo; annual index).

0.3.5 Annuals

The best annual review of technological aspects is *Reports on Progress in Applied Chemistry*, (Society of Chemical Industry, London) which contains short but fully referenced accounts of all major topics. More specialized fields are covered in rotation. Although the major emphasis is on products and processes, cost studies are often indicated.

The economic aspects of the industry are best studied in *The Chemical Industry*, a report published by the Organization for Economic Co-operation and Development (OECD), Paris and cited as OECD: *Annual Report*. The progress of the chemical and related industries in the OECD countries – Western Europe, North America and Japan – is discussed in detail with a wealth of tabulated data. Occasional supplements are devoted to topics: that to the report of 1968–69 analyses productivity trends for the period 1958–67.

An excellent occasional report on the UK industry is published by Stirling and Co., Stock and Share Brokers, London.

0.4 Statistics of the chemical industry

In this book we have tried to avoid vague generalizations and we have therefore included a considerable amount of statistical data. The bulk of this information comes from generally available official sources but some consists of estimates by informed observers. Sources are indicated in the notes to the individual chapters.

The official statistics should be used with care. Although most advanced nations gather a great deal of information about their respective chemical industries these data are often less easy to use than might at first appear. They may be incomplete since they are usually collected by surveys of the larger firms. The systems of classification used vary from country to country and are periodically changed. In almost all nations information which could help one firm at the expense of another is suppressed: thus, if a chemical is produced by only two companies, total production and sales figures are not given. In the USA, there are many manufacturers and they are legally required to supply the government with production statistics; consequently there are well-documented trade statistics stretching back for many years. In other countries there are fewer manufacturers and no such legal requirement exists. Hence figures are more difficult to obtain. Nonetheless, both the quality and quantity of market data have improved dramatically during the past ten years and still continue to do so.

0.4.1 Statistics

0.4.1.1 International: The *United Nations Statistical Yearbook* (New York, United Nations, last issue 1970) gives world-wide production statistics on a country-by-country basis for major inorganic chemicals, nitrogenous fertilizers, plastics and resins, synthetic fibres and synthetic rubber. The *United Nations Monthly Bulletin of Statistics* (New York, United Nations) contains monthly production figures for many but not all countries. World trade in chemicals is covered in considerable detail in the *World Trade Annual* (New York, Walker and Co. for the United Nations). The OECD *Annual Report* includes details of production, sales and capital investment in all major branches of the chemical industry by each OECD nation. Production statistics for many individual large-tonnage chemicals are also given. A particularly useful feature is the comprehensive coverage of international trade in particular products.

0.4.1.2 Great Britain An excellent guide to British official statistics is *A List of Principal Statistical Series and Publications*, Central Statistical Office (CSO): Studies in Official Statistics No. 20, London, HMSO, 1972. The Central Statistical Office quarterly *Statistical News* should be studied for current developments.

Complete and very detailed production and sales statistics are given in the reports on the quinquennial *Census of Production*, Department of Trade and Industry (DTI), London, HMSO. The last complete census was in 1968: a preliminary report appeared in 1969 and the detailed industry reports between 1970 and 1972. Quarterly reports on some but not all aspects of the chemical industry appear in the *Business Monitor* series, DTI, London, HMSO. Of especial interest are those on general chemicals, organic chemicals, synthetic resins and plastic materials, paint, man-made fibres, rubber, pharmaceuticals, soaps and detergents, pesticides and colours (pigments). Notable omissions are fertilizers and dyestuffs, both of which are covered in the *Monthly Digest of Statistics*, CSO, London, HMSO. The DTI weekly *Trade and Industry* publishes a quarterly review of the industry as a whole which contains overall figures of sales, prices, employment and capital expenditure. Imports and exports are best studied in the *Annual Statement of Trade of the United Kingdom* or in the less detailed *Monthly Trade Statistics of the United Kingdom* (DTI, London, HMSO).

Two useful compilations of official information are the *Annual Abstract of Statistics* (CSO, London, HMSO) and the Chemical Industry Association's *UK Chemical Industry Statistical Handbook*, 2nd ed., London, 1970.

0.4.1.3 United States of America The best guide to US official statistics is *Statistical Services of the US Government*, Executive Office of the President: Bureau of the Budget, Washington, DC, 1968. *Guide to US Government Serials and Periodicals* Vol. 2, by J. L. Andriot, Mclean, Va: Documents Index, 1971, is extremely comprehensive but bulky and expensive.

A *Census of Manufactures* is taken by the Bureau of the Census, Department of Commerce, Washington DC, every five years; the latest to be published is that for 1967. The same agency produces an *Annual Survey of Manufactures*, which gives data on sales, value added, employment and capital expenditure in all major manufacturing industries, and a monthly summary *Current Industrial Reports*. The US Tariff Commission, Washington DC, is an independent agency which produces extremely detailed reports on production and sales of organic chemicals; its principal publications are the annual *Synthetic Organic Chemicals: US Production and Sales* and the monthly *Preliminary Reports on US Production of Selected Synthetic Organic Chemicals* and *US Production and Sales of Plastic and Resin Materials*.

A convenient handbook is the *Statistical Abstract of the USA*, Washington DC, Department of Commerce, Bureau of the Census, published annually.

0.4.1.4 Europe An indispensable bibliography is *European Chemical Market Research Sources*, Zug (Switzerland), Noyes Development S.A., 1969; for general information about European countries see the DTI publications, *Sources of Statistics: the European Economic Community* and *The European Free Trade Area*, London, HMSO, 1971. The most convenient source of information is the *OECD Annual Report*, which summarizes data obtained from the individual countries on a common basis of classification.

Statistics of the West German chemical industry are given in *Industrie und Handwerke: Reihe 3 – Industrielle Produktion*, monthly and annual, Wiesbaden, Statistisches Bundesamt, or, more conveniently, in the biennial *Chemiewirtschaft in Zahlen*, Frankfurt-am-Main, Verbund der Chemischen Industrie. Data on the French industry are given in the *Annuaire de Statistique Industrielle*, Paris, Bureau Central de Statistique Industrielle, and in the *Rapport Annuel* of the Union des Industries Chimiques, Paris. The best source for Italy is the annual *Compendio Statistico* of the Associazone National dell'Industria Chimica, Milan. For Belgium there is a two-volume work in English, *The Belgian Chemical Industry*, Brussels, Business Economics Library, 1969–70; current statistics are to be found in the *Rapport Annuel* of the Fédération des Industries Chimiques de Belge. The Netherlands are covered in the annual *Produkt Statistieken – Chemische Industrie*, The Hague, Centraal Bureau voor de Statistiek.

0.4.1.5 Japan The language problem presents a formidable barrier. The OECD *Annual Report* contains much information derived from official sources. Probably the best source of current information is the *Industrial Statistics Monthly*, published by the Ministry of International Trade and Industry (MITI), Tokyo, which includes production series for about 80 major chemical products, and is in both Japanese and English.

0.5 Organizations

The professional associations and learned societies are mainly concerned with scientific and technical questions but some have sections devoted to industrial and commercial matters. In Britain the Chemical Society – Royal Institute of Chemistry has a rapidly developing industrial division, which promises to rival the well-established Society of Chemical Industry and Institute of Chemical Engineers. A number of the American Chemical Societies' 26 divisions similarly have an industrial orientation, especially the Division of Chemical Marketing and Economics.

There are also several market research associations which organize conferences and symposia of interest to chemical technologists. The Industrial Market Research Association (IMRA), Lichfield, Staffs., has a flourishing chemicals section. The European Association for Industrial Market Research (EVAF), London, is more specifically European; its chemical section, the European Chemical Market Research Association (ECMRA) is particularly active.

0.6 Prices and costs

Prices quoted are taken from *European Chemical News*, *Chemical Age*, and *Chemical Marketing Reporter*, and refer to the first quarter of 1972. It must be emphasized that they relate to specific grades and quantities and do not generally include local taxes. Discounts for quantity, special arrangements for particular customers, and long-term contracts between supplier and consumer are common in the chemical industry, and in consequence the actual prices paid may be substantially lower than those given. Prices are usually quoted c.i.f. (cost, insurance, freight) or f.o.b. (free on board) at a specified point. Thus c.i.f. Liverpool means the material will be delivered to Liverpool at the seller's cost and risk, whereas f.o.b. New York means the material will be delivered on board ship at New York but after that it is the customer's problem.

The manufacturing costs cited are for the purpose of illustration and are only approximate. They are taken from the technical literature – references are given at the ends of the appropriate chapters – and have as far as possible been updated to 1970. Capital charges comprise depreciation and profit. The plant costs quoted are on a battery limits basis, i.e. it is assumed that electricity, gas, steam and water are bought in. The costs of the corresponding green field site plants, which provide their own utilities, are 30–50% higher. The lifetime of a plant is taken as 10 years, depreciation is charged by the straight line method, and its scrap value is assumed to be zero (section 5.3). The pre-tax cash flow (profit + depreciation) corresponds to an internal rate of return of 20% (section 5.7). Annual maintenance and overhead costs are taken to be 6% of the initial investment. Fixed charges per ton of product are calculated on the assumption that the plant is operating at maximum capacity.

Variable costs comprise feedstock, chemicals, utilities and labour. Feedstock and chemical costs are taken from the literature or from the journals cited above. Utility costs are taken to be as follows:

	USA	Western Europe
Fuel (¢/10^6 Btu)	21	45
Electricity (¢/kWh)	0·7	1·1
Steam (¢/1 000 lb)	65	110
Process Water (¢/1 000 US gallons)	25	30
Cooling Water (¢/1 000 US gallons)	2·5	3·0

Labour and supervision are costed at £1 per man-hour in Western Europe and $5 per man-hour in the USA.

At the time of writing the £ is floating and its ultimate value relative to the US $ is uncertain. Throughout this book £1 is taken to be $2.40 which seems to be the most probable point of equilibrium.

Where costs are given they are presented graphically using bar charts with a common shading scheme, shown on p. 8.

Units

Several different systems of units are used in the chemical industry. The so-called British units are used in the USA and in Great Britain, while almost all other nations use the 'traditional' (cgs) metric system. Britain is adopting the use of the highly inconvenient 'Systeme Internationale' (SI or MKS) metric system, so it seems probable that for the

8   *How to use this book*

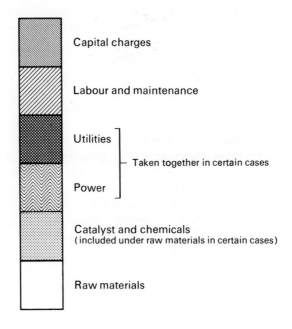

next few years all three systems will be used concurrently.

In this book the main unit of *weight* is the metric ton (tonne) of 1 000 kg = 2 205 lb. = 0·98 long tons = 1·102 short tons; unless otherwise noted 'ton' means metric ton. In the USA, chemical production is normally given in pounds to avoid confusion between the different types of ton; one million pounds = 453·5 metric tons.

The most common metric units of *volume* are the cubic metre and the litre (= 10^{-3} m³). In Britain and the USA the cubic foot (= 0·0283 m³) and the US and Imperial gallon (one US gallon = 0·833 Imperial gallons = 0·00378 m³) are also widely used.

Pressures are given in atmospheres of 1·033 kg/cm²; the latter unit, otherwise the hectopieze (hpz), is the usual metric unit of pressure as the recommended SI unit, the Newton/m² (= 10^{-5} hpz), is inconveniently small. In Britain and the USA the pound/in² (p.s.i. = 0·0703 hpz) is commonly used.

Temperatures are given in degrees centigrade. *Heat* is expressed in kilocalories in metric nations, although the kilojoule (kJ = 0·239 kcal) is preferred in the SI system. The British thermal unit (Btu = 0·252 kcal = 1·055 kJ) and even more the therm (= 100 000 Btu) are still widely used in Britain and the USA.

Part I **The growth of the chemical industry**

1 The old chemical industry from 1800 onwards

1.1 Early days

In a sense there has always been a chemical industry. For many centuries men have wanted soaps, dyes, medicines, love-philtres and poisons; such things were traditionally made on a small – often a domestic – scale and were of little general significance. The history of the chemical industry as we know it begins with the industrial revolution. Between 1780 and 1840 Great Britain changed from a predominantly agricultural to a predominantly industrial country; by 1900 the other West European and North American nations had followed suit. Population, production and consumption rose rapidly (Table 1.1). This unparalleled growth of industry was made possible by a series of technological innovations which removed existing barriers to development. The chemical industry originated in successful attempts to solve certain of these technical problems. From the beginning it was a science-based industry: its progress depended to a large extent on the application of chemical theory to industrial ends. The nature of these ends was determined by the nature of the demand for chemicals.

Table 1.1 The growth of the UK economy and the demand for chemicals 1800–1935

	1801	1841	1881	1907	1924	1935
Population (m.)	15·9	26·7	34·9	43·7	44·9	46·9
Gross National product (£m.)	232	452	1 051	1 995	4 121	4 516
Index of industrial production (1935 = 100)	4	13	40	61	76	100
Consumption of raw cotton ('000 tons)	24	195	638	886	611	563
Production of soap ('000 tons)	22	70	215	386	485	520
Production of glass ('000 tons)	21	34	n.a.	n.a.	n.a.	800
Production of paper ('000 tons)	16	42	n.a.	819	1 150	1 930
Production of major chemicals ('000 tons):						
Sulphuric acid (100%)	4	150?	780*	1 050	920	940
Alkali (Na_2CO_3 + $NaHCO_3$ + $NaOH$)	—	100?	480*	600	1 050	1 360
Ammonium sulphate (N content)	—	—	8*	56	88	97
Bleaching powder	0·1	10	132*	109	66	55
Superphosphate	—	—	n.a.	605	304	445
Synthetic dyestuffs	—	—	2	7	19	28

n.a. = not available.   * 1880 figure.

1.2 The market for chemicals

Throughout the nineteenth and early twentieth centuries the chemical industry remained small – as late as 1935 it accounted for only 2–3% of the output of British manufacturing industry – and relatively specialized. The demand was for a limited range of mainly inorganic products and the prosperity of the industry depended crucially on that of a few major customers at home and abroad. Six industries were of key importance.

1.2.1 Soap-making

The manufacture of soap by the alkaline hydrolysis of animal and vegetable fats has been known since medieval times. The rapid growth in population and rising standards of cleanliness – between 1800 and 1900 the *per capita* consumption of soap roughly quadrupled – resulted in a remarkable expansion of the soap industry and therefore in the demand for sodium hydroxide.

1.2.2 Glass

Glass-making by the fusion of sand (silica) and sodium carbonate was another old-established trade which, like soap-making, grew with the population. The industry profited notably from the housing boom of 1870–1910, while much glass was also used to make bottles and jars.

1.2.3 Cotton textiles

The manufacture of cotton textiles was the greatest single industry of nineteenth century Britain, accounting in 1907 for 6% of the output of British industry and 25% by value of all exports. The chemical industry had helped to make this possible: Charles Tennant's discovery of bleaching powder in 1794 provided a cheap, safe and rapid bleach for cotton. Dyeing also benefited from the application of chemistry. The introduction of auxiliary agents in the late eighteenth century made possible the use of new animal and vegetable materials, while Perkin's discovery of mauve in 1856 was followed rapidly by a host of synthetic dyes of unparalleled brilliance and fastness.

1.2.4 Paper-making

The nineteenth century was an age of increasing literacy; it was even more an era of growth in the supply of reading matter. The mass circulation newspaper made its appearance: in 1831 the average monthly issue of all British newspapers was 3 240 000, by 1881 the corresponding figure was 135 m. More books were published and more copies of each were sold. Posters, wrapped goods and toilet paper became commonplace. Paper was traditionally made from rags, and these were no longer in sufficient supply. Manufacturers began to use other materials – esparto grass, and then, after 1870, wood pulp. The dark resinous lignin was removed by treatment with sodium hydroxide or calcium bisulphite, and the resulting cellulose pulp bleached with calcium hypochlorite.

1.2.5 Agriculture

Agricultural output increased at a greater rate than population. New techniques increased productivity. Farmers in areas where agriculture was intensive rather than extensive began to realize the advantages of chemical products in maintaining soil fertility. As a result of the work of Liebig, Lawes and others, it was known by 1840 that the necessary soil nutrients are nitrogen, phosphorus and potassium, and that these must be supplied in readily metabolized forms. Potassium was normally added as the carbonate or sulphate, and phosphorus as superphosphate, a soluble material obtained by treating phosphate rocks with sulphuric acid, a process introduced by Lawes in 1843. The first source of fixed nitrogen to be used was guano, fossilized bird-dung from the West coast of South America. This was later displaced by sodium nitrate from the rainless deserts of Northern Chile and ammonium sulphate derived from ammonia formed as a by-product of town-gas manufacture.

1.2.6 Explosives

The engineers who sank the mines and built the railways, docks and canals so characteristic of the period needed immense quantities of reliable and effective explosives. The traditional black powder, a mixture of charcoal, sulphur and potassium nitrate, was quite inadequate and after 1867 was rapidly replaced by the various combinations of nitroglycerine invented by Alfred Nobel. In 1889 these discoveries were further exploited in the introduction of the smokeless propellants Cordite and Ballistite which were to be of such importance in twentieth century warfare.

1.3 The Leblanc process

The initial demand for chemicals was thus a demand for bleaching powder, for sulphuric acid, and, above all, for alkali – indeed contemporaries often referred to the chemical industry as the alkali trade. The heart of the nineteenth century industry was the Leblanc process for the production of sodium carbonate. Sodium chloride was heated with strong sulphuric acid to give sodium sulphate (salt cake) and hydrogen chloride. The salt cake was broken up, mixed with coal and limestone, and heated at red heat to yield a mixture of sodium carbonate and calcium sulphide (black ash). Countercurrent extraction of the black ash with water gave a solution of sodium carbonate which was concentrated and decolorized to yield the pure solid. Sodium carbonate could be readily converted to sodium hydroxide by treatment with calcium hydroxide (slaked lime); until the introduction of mild steel drums in 1860 this operation was carried out by customers as caustic soda was too corrosive to be shipped. The chemical reactions involved in the process are:

$$2NaCl + H_2SO_4 \longrightarrow Na_2SO_4 + 2HCl$$

$$Na_2SO_4 + CaCO_3 + 2C \longrightarrow Na_2CO_3 + CaS + 2CO_2$$

$$Na_2CO_3 + Ca(OH)_2 \longrightarrow 2NaOH + CaCO_3$$

1.4 The golden age of the alkali trade

The Leblanc process, invented in 1784, was introduced into Britain as early as 1800 but

14 *The old chemical industry from 1800 onwards*

made little headway until the repeal of the excise duty on salt in 1824. It then developed extremely rapidly. A batch process, it was inefficient in its use of both materials and labour (Table 1.2), and gave rise to most formidable effluent problems; on the other hand the process equipment was easy to design and cheap to build. Moreover the hydrogen chloride produced was a valuable source of chlorine and so of bleaching powder. From 1860 the Leblanc process was invariably combined with the manufacture of

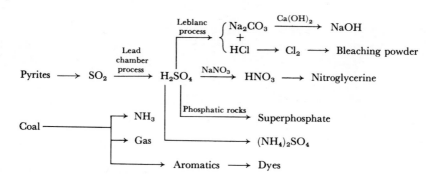

Figure 1.1 The technical structure of the UK chemical industry in 1875.

bleaching powder and of sulphuric acid by the traditional lead chamber process (Fig. 1.1). Demand continued buoyant. British customers continued to expand their operations and a large export trade to Europe and the United States developed. Foreign competition was as yet negligible. By 1878 British production of alkali and bleaching materials had risen to 580 000 tons of which over half was exported. Britain unquestionably dominated the world trade in chemicals.

1.5 Solvay v. Leblanc

British pre-eminence was, however, to prove short-lived. After 1880 the output of alkalis grew only slowly; alkali production, 475 000 tons in 1878, had reached only 700 000 tons by 1913. Exports actually declined, from an annual average of 330 000 tons in 1880–84 to one of 220 000 tons in 1908–13. Bleaching powder was hit even harder, production falling from 150 000 tons in 1890 to 107 000 tons in 1907. A part of this decline was inevitable. British industry as a whole grew slowly and erratically from 1880 until the First World War and so also, in consequence, did the domestic consumption of chemicals. The world demand for alkalis and other staple chemicals remained high, but the British export trade suffered severely from the growth of foreign – especially German – competition and from the general movement towards high import tariffs in Europe and North America. The industry was moreover suffering from technological obsolescence. From 1880 the Leblanc manufacturers came under rapidly increasing pressure from the Solvay ammonia-soda process. In this process a cold ammoniacal salt solution is saturated with carbon dioxide to precipitate the sparingly soluble sodium bicarbonate which is then decomposed by heat to sodium carbonate and carbon dioxide. The ammonium chloride simultaneously produced is treated with slaked lime to regenerate ammonia:

$$CaCO_3 \longrightarrow CaO + CO_2$$

$$CaO + H_2O \longrightarrow Ca(OH)_2$$

$$NaCl + NH_4HCO_3 \longrightarrow NaHCO_3\downarrow + NH_4Cl$$

$$2NaHCO_3 \longrightarrow Na_2CO_3 + CO_2 + H_2O$$

$$2NH_4Cl + Ca(OH)_2 \longrightarrow 2NH_3\uparrow + CaCl_2 + H_2O$$

The Solvay process was the first continuous process to be used in the chemical industry. It was relatively economical in both raw materials and labour but expensive in capital cost (Table 1.2) and, by contemporary standards, required a high degree of skill in plant design. Solvay built his first commercial plant in 1865 and in 1872 licensed John Brunner and Ludwig Mond to operate his process in Britain. Brunner and Mond made their first soda in 1874 and were immediately successful. By 1885 they accounted for 20% of British alkali production and the Leblanc manufacturers were seriously worried men.

Table 1.2 Brunner's estimate of the cost of a ton of sodium carbonate by the Leblanc and Solvay processes

Leblanc process at Widnes	£	s	d	Projected Solvay process at Winsford	£	s	d
Pyrites, $17\frac{1}{2}$ cwt. at 35/0 per ton	1	10	8	Ammonium sulphate, 202 lb. at £20 per ton	1	16	0
Saltpetre, 56 lb. at £14 per ton		7	0	Brine, 40 cwt. at 9d. per ton		1	6
Salt, 29 cwt. at 8/6 per ton		12	4	Limestone, 44 cwt. at 6/6 per ton		14	4
Limestone, 35 cwt. at 6/6 per ton		11	4	Coal, 39 cwt. at 8/10 per ton		17	3
Coal, 88 cwt. at 8/0 per ton	1	15	2	Coke, $5\frac{1}{2}$ cwt. at 18/0 per ton		5	0
Total materials	4	16	6	*Total materials*	3	14	1
Wages	1	12	0	Wages		15	0
Salary, rent and rates	1	5	0	Salary, rent and rates		14	8
Insurance		14	0	Insurance		11	0
Packaging		18	0	Packaging	1	2	0
Grand total	9	1	6	Royalty to Solvay at 8/0 per ton		8	0
				Grand total	7	4	9
Capital cost of 60 tons per week plant	10 000	0	0	Capital cost of 60 tons per week plant	13 500	0	0

1.6 Crisis and stagnation

It was now clear that the Leblanc process was no longer a competitive method of making alkali alone. In effect the Leblanc manufacturers had become makers of bleaching powder with alkali as a mere by-product. In this situation they attempted to safeguard their interests by amalgamation. The United Alkali Company was formed in 1890 to acquire the assets of the Leblanc manufacturers and to improve their profitability by a policy of rationalization and defensive research. Almost at once it was faced with a supreme crisis: the advent of the electrolytic route to chlorine and sodium hydroxide. Unfortunately for the Company its research director, Ferdinand Hurter, was emotionally committed to the Leblanc process and advised the Board against buying the rights to the Castner electrolytic process. The results of this decision were disastrous. The Company rapidly found itself in a position where it had insufficient cash flow to modernize; saved only by international market-sharing agreements it staggered on, through a capital reconstruction in 1913 and the final extinction of the Leblanc process in 1921, to be casually swallowed by ICI in 1926. In contrast Brunner, Mond and Co. went from strength to strength and by the eve of the First World War accounted for some 90% of British alkali production. It remained however a one-product company and played no part in the development of the new catalytic and electrolytic processes that were currently revolutionizing the chemical industry of Europe and North America. The rationalization of the alkali trade remained incomplete. In a situation where total sales increased only slowly the major producers found it more profitable to limit production and share out markets by agreements which were 'favourable to the strong but acceptable to the weak' rather than to push matters to the limit. By the early years of the twentieth century competition in the world alkali market had been limited sufficiently to guarantee at least a living of sorts for all existing producers. Such arrangements were to become general in the chemical industry of the interwar period.

1.7 Dyestuffs and the rise of Germany

Among Britain's competitors Germany was by far the most formidable. After 1870 the German chemical industry expanded with phenomenal speed. It benefited from its late start: the Leblanc process was never important and electrolytic techniques were introduced rapidly and on a large scale. The most dynamic part of the industry was, however, that concerned with dyestuffs. Synthetic dyes were a British invention – mauve, the first commercial example, was discovered by W. H. Perkin in 1856 – and until the early 1870s Britain continued to dominate this area. Subsequently Germany pulled so far ahead that by 1913 German companies were producing 80% of the world output of dyestuffs. German excellence in applied research played a major part in this triumph. Dyestuff development was a field in which technical change was both rapid and continuous and where, for the first time, the application of chemical theory was at a premium. The German universities were able to supply large numbers of research-trained organic chemists and educated managers, whose efforts combined to create the modern system of industrial research. The first great success was the commercial synthesis of alizarin in 1868; the azo and sulphur dyes followed in the 1870s and 1880s; and at the turn of the century, after the expenditure of over £1 m. in research and development, a satisfactory synthetic indigo process was introduced. The British dye industry on the other hand remained in a state of arrested development. The British educational system was inadequate: few graduated in the sciences and training in research was unknown. British management was backward and complacent. The patent situation was unhelpful: a foreign patentee could register his invention in Britain without attempting to work it, so preventing its manufacture. The major consumers of dyestuffs, the textile manufacturers, were only concerned to obtain the best dyes at the lowest prices. By 1913 British dyestuff production was equivalent to only 20% of consumption and consisted mainly of the cheaper and simpler colours.

1.8 Deutschland über Alles

The long-term significance of the dyestuff industry became apparent at the turn of the century when the leading German manufacturers – Hoechst, Bayer, AGFA, Badische Anilin und Soda Fabrik (Badische or BASF) – began to use their research expertise and their large financial resources to diversify their interests. Synthetic pharmaceuticals were an early and obvious development – Aspirin was introduced by Bayer in 1898, and Salvarsan, the first effective antisyphilitic agent, by Hoechst in 1910. They were soon followed by photographic chemicals and film, rubber additives, and synthetic polymers. BASF concentrated on the development of new processes in inorganic chemistry. In the 1890s it pioneered the use of the contact process for sulphuric acid and in the new century initiated studies of direct nitrogen fixation. After a prolonged flirtation with the arc process it was decided in 1909 to concentrate on the development of the Haber process: this involves the synthesis of ammonia by the direct combination of nitrogen and hydrogen at 600°C and 200 atm pressure. The Haber process was the first modern chemical process: the first to use gases at high temperatures and pressures, the first to require specialized plant, the first to be genuinely capital-intensive. The development of the full-scale process was therefore exceptionally difficult and over £1 m. was spent before the

Table 1.3 British, German and US chemical production in the early twentieth century

	Britain 1907			Germany 1913	USA 1914
Total sales	£25 m.			£120 m.	£77 m.
Imports	12			22	n.a.
Exports	13			50	n.a.
	Production	*Imports*	*Exports*	*Production*	*Production*
Sulphuric acid ('000 tons 100% acid)	1 050	4	4	1 700	
Alkali ('000 tons)	600	17	285	460	1 095
Ammonium sulphate ('000 tons N)	56	—	49	108	20
Superphosphate ('000 tons)	605	11	115	1 800	2 900
Synthetic dyes ('000 tons)	7	15	3	140	3
Coal tar distillation products (£m.)	3.5	—	1.4	n.a.	n.a.
Pharmaceuticals (£m.)	4.4	1.7	1.7	n.a.	21

first plant came on stream in September 1913. No other nation could match these achievements. On the eve of the First World War the German chemical industry had established a qualitative and quantitative superiority which was without parallel. A comparison with Britain (Table 1.3) illustrates the magnitude of the German lead. Britain was a producer and exporter of simple basic chemicals, an importer of expensive and sophisticated ones. Germany was strong all round but excelled in the production of new and complex compounds. The British industry remained in the nineteenth century while the German industry moved further and further into the twentieth.

18 *The old chemical industry from 1800 onwards*

1.9 The impact of war, 1914–18

1.9.1 Germany

The situation changed dramatically on the outbreak of the First World War. From August 1914 Germany was subject to a strict blockade and foreign trade ceased. With remarkable speed the dyestuff industry switched to the manufacture of explosives, poison gases, pharmaceuticals and even synthetic rubber. The major problem was the supply of nitrogen compounds. In 1913 Germany consumed 225 000 tons of fixed nitrogen, 85–90% in fertilizers and 10–15% in industrial uses. Production, mainly as by-product ammonium sulphate, amounted to 119 000 tons, the balance being imported as sodium nitrate (Table 1.4). The war created a tremendous demand for explosives and therefore for nitric acid at the very moment when supplies were abruptly reduced. The Haber process provided a solution to the problem. By 1918 German production of fixed nitrogen amounted to 185 000 tons, approximately half being ammonium sulphate and half nitric acid obtained by the catalytic oxidation of ammonia. But for this effort Germany would have run out of food and explosives in 1916 and the war would have ended.

Table 1.4 German production and consumption of nitrogen compounds, 1913–18 ('000 tons N)

	1913	1914	1915	1916	1917	1918
Production						
By-product sulphate	108	95	69	68	57	54
Cyanamide process	10	13	15	34	37	36
Haber process	1	6	12	43	75	95
Total	119	114	96	145	169	185
Consumption						
Agriculture	200	\multicolumn{2}{c}{Uncertain}	73	80	92
Industry	25			72	89	93

1.9.2 Britain

The British chemical industry was notably revived by the War. As in Germany new patterns of demand caused major changes in structure, but in Britain these were of a different nature. Unlike Germany, Britain was able to supply the military demand for nitrogen compounds by existing routes and, until the submarine crisis of 1916–17, serious interest in the Haber process was slow to develop. On the other hand the outbreak of war resulted in acute shortage of dyestuffs which could only be overcome by an immediate expansion of the British dye industry. This was successfully achieved: between 1913 and 1919 the output of dyestuffs increased by 400% and by the latter year production was quantitatively if not qualitatively adequate for the nation's needs. Even more important was the change in political attitudes to the organic chemical industry. The wartime shortage of dyes left a deep impression. A major official report of 1918 cited dyestuffs as an example of 'those industries which we have described as "key" or "pivotal" which should be maintained in this country at all hazards and at any expense'. The wartime government was already deeply involved in dyes: in order to strengthen the industry it had acquired control of the two leading manufacturers which were amalgamated in 1919 to form the British Dyestuffs Corporation, in which the Treasury had a large share holding. The 'key industry' concept was felt to require still further departures from *laissez-faire* and free trade. From 1921 dyestuffs were protected from foreign competition by a system of import licensing, and synthetic organic, fermentation and fine chemicals by an import duty of $33\frac{1}{3}\%$. Both measures were introduced as temporary expedients; both were to last until well after the Second World War.

1.9.3 The USA

In 1914 the American chemical industry was probably the second largest in the world. Negligible as late as 1880 it had since grown at phenomenal speed thanks to the growth of the US economy as a whole and to the exclusion of most chemical imports by the high-tariff policy favoured by the politically dominant Republican party. Like the British industry it specialized in inorganic chemicals; dyestuff production was small and most dyes and intermediates were imported from Germany. Unlike their British counterparts, however, American managers were commercially aggressive and technically adventurous, and were already investing heavily in research and development. Although the USA remained at peace until 1917, the outbreak of war caused a wild boom in the American chemical industry. The output of explosives soared and, after a severe crisis in 1914–15, so did that of dyestuffs. The expansion of the explosives industry forced the development of new routes to industrial solvents. In the Weizmann process, corn was fermented with the organism *Bacillus Clostridium Acetobutylicum* to give butanol and acetone, the latter of vital importance for the gelatinization of high explosives (e.g. cordite) and the production of aeroplane dopes. The process provided its discoverer with access to British government circles and led indirectly to the Balfour Declaration, but though discovered in Britain, it was mainly used in America. Even more important in the long run was Carleton Ellis' development of ways to convert oil refinery waste gases to ethanol, *iso*propanol and acetone. These achievements were due to the private chemical companies. Even after 1917 the attitude of the American government was friendly but non-committal: it distributed the patents and other assets of German-owned firms to their American rivals but refused to invest directly in the industry. As in Britain however, the importance of the chemical industry had been recognized and it was seen as worthy of the highest degree of protection; this it duly received from the Fordney-McCumber Act of 1922, which introduced the American Selling Price tariff system and virtually excluded all imports for the rest of the interwar era.

1.10 The Post First World War scene

Isolated from foreign competition the American industry was to grow rapidly in the 1920s. Other countries faced an uncertain future. In Britain a short boom in 1919–20 was followed by a sharp depression in 1921 and for the rest of the decade growth was slow (cf. Fig. 1.2). The traditional customers marked time or even declined while new markets were slow to appear. Much of the increase in the production of chemicals in the interwar era was merely due to import substitution. The export market was fraught with difficulties. The wartime dislocation of trade had forced all advanced countries to expand their own chemical industries and when peace returned overcapacity came with it. The situation was particularly severe in dyestuffs (world capacity in 1924 was 284 000 tons while output was only 154 000 tons); and in fertilizers – because of the forced growth of the Haber process. International competition was correspondingly intense. Germany remained the dominant force in the world chemical industry. The German dyestuff firms, associated since 1916, amalgamated in 1925 to form IG Farben. This was not only the largest chemical company in the world but also the one which most embodied the dynamic tendencies of the industry. No British – indeed no other European – company could hope to compete on equal terms with the new giant.

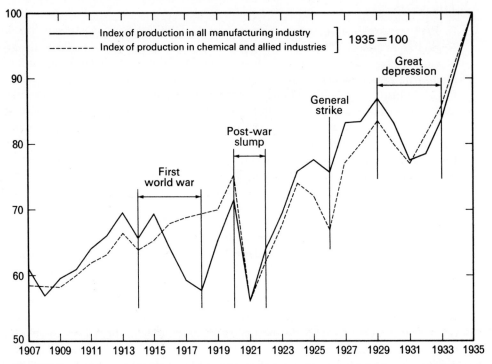

Figure 1.2 Production in British manufacturing industry and in the chemical and allied industries 1907–35.

1.11 The formation of ICI

In the early post-First World War period the fortunes of the major British manufacturers varied markedly. Acquisitions during and immediately after the war had left Brunner, Mond the dominant firm in inorganic chemicals, and in the euphoria of 1919 it decided to introduce the Haber process to Britain. This was a surprising – and a most expensive – departure for a firm with no previous tradition of large-scale innovation. The Ministry of Munitions nitrogen fixation projects, including a site at Billingham, County Durham, were taken over, the necessary process information was bought or stolen, and commercial production was achieved in early 1924. By 1930 the Billingham plant was producing 60 000 tons of fixed nitrogen but, at a cost of £8 m., it had severely strained the resources of Brunner, Mond and its successor. In contrast Nobel Industries, a wartime fusion of the British explosives manufacturers, had diversified successfully into consumer goods as various as paints, leather goods and zip fasteners. British Dyestuffs was less fortunate: import licensing notwithstanding, overexpansion, quarrels between the directors, and severe competition in export markets combined to bring it close to bankruptcy in 1921 and to make it largely unprofitable thereafter. The situation of the British companies was fundamentally unstable. To survive they would have to combine, or to come to terms with IG Farben – or both. After much manoeuvring, and the near amalgamation of Brunner, Mond with IG and Allied and Chemical and Dye of the USA, Brunner, Mond, Nobel Industries, British Dyestuffs and United Alkali, combined in 1926 to form Imperial Chemical Industries. The new company had assets of £57 m. and annual sales in 1927 of £25–30 m., of which 62% came from alkalis and heavy chemicals, 8% from dyestuffs, 20% from explosives and 10% from metals and minor activities. With large financial resources it was able to spend freely on research and development – by the late 1930s the research budget was approximately £1 m. – and to take a long view of the advantages of innovation. Sheer size, moreover, gave it powerful advantages in international commercial diplomacy. In other respects the new company was to prove a disappointment. The process of rationalizing its overlapping facilities was inordinately prolonged and indeed was not completed until 1964. More important, the strategic direction lacked imagination. Far too much money and effort was devoted to the development of such existing fields as nitrogenous fertilizers; far too little to the growth areas of the future, to pesticides, to pharmaceuticals, above all to synthetic polymers. For most of its early years ICI was to be, as a hostile critic remarked, a company 'which depends ... on making comparatively few products extremely well and selling them at prices ... fixed by international agreement'.

1.12 The slump and after

The world-wide depression of the early 1930s hit the chemical industry hard (cf. Fig. 1.2) but recovery was rapid. By 1933 production was back to 1929 levels and it increased steadily thereafter. International market-sharing agreements helped to mitigate the horrors of overcapacity: ICI was a leading member of the nitrogen cartels arranged in 1929–32 and joined the dyestuffs cartel in 1931. The industry also benefited from the general policy of protection instituted by the National government of 1931–35. The existing system of tariffs and licensees was made permanent and then greatly extended. In this cosy environment the chemical industry grew and even flourished. The 1935 Census of Production revealed (Table 1.5) the progress made since the war. The country was now self-sufficient in most important chemicals. The dyestuff industry was fully competitive; five of the seven major innovations of the interwar period had been introduced by British firms. High pressure technology had been introduced on the largest scale. At a cost of £19 m. ICI had built plant capable of producing 130 000 tons of fixed nitrogen; for a further £8 m. a large synthetic gasoline installation was added in 1933–35. Commercially a disaster – the combination of high cost and low profitability played havoc with the company's cash flow – these plants nevertheless marked the coming of age of British chemical engineering. However, the industry still bore resemblances to the German – or even the British – industry of prewar days: prosperity still depended on supplying the same range of chemicals to the same customers. Few foresaw the revolution to come.

Table 1.5 The British chemical industry in 1907 and in 1935

	1907			1935		
	Total	Imports	Exports	Total	Imports	Exports
Sales						
Value (£m.)	25	12	13	70	9	18
Breakdown of output (%)						
Heavy inorganic chemicals	21	0	13	16	0	15
Fertilizers	32	21	31	10	0	9
Synthetic dyestuffs	2	14	2	7	11	8
Coal tar distillation products	14	0	11	11	0	5
Other organic chemicals	17	14	14	4	7	4
Pharmaceuticals, etc.				28	20	16
Other	14	51	31	24	62	43
Volume						
Sulphuric acid	50	4	4	940	3	0
Alkalis ('000 tons)	600	17	285	1 360	3	342
Ammonium sulphate ('000 tons N)	56	0	49	97	0	53
Superphosphate ('000 tons)	605	11	115	445	30	23
Synthetic dyestuffs ('000 tons)	7	16	3	28	2	5
Synthetic and fermentation chemicals ('000 tons)	n.a.	n.a.	n.a.	83	9	2

Notes and references

General By far the best histories of this period are L. F. Haber, *The Chemical Industry in the 19th Century*, and *The Chemical Industry 1900–1930*, Clarendon Press, Oxford, 1958 and 1971.

Nineteenth-century production figures are discussed by Haber, *passim*. For British production 1900–40 the best sources are the *Census of Production* reports, which refer to 1907, 1924, 1930 and 1935; that for 1907 is much less detailed than the later ones. There are many corresponding reports on American industry from 1880 onwards; the US Tariff Commission has produced an annual report on chemical production since 1917.

Definitions In all countries the definition of the chemical industry is surprisingly difficult because of the large number of boundaries involved. During this period the British definition of 'chemical and allied industries' comprised (*a*) chemicals, dyestuffs and drugs, (*b*) fertilizers, disinfectants and glue, (*c*) soap, candles and perfumery, (*d*) paint, colour and varnish, (*e*) seed crushing, (*f*) oil and tallow, (*g*) petroleum, (*h*) explosives, ammunition and fireworks, (*i*) starch and polishes, (*j*) matches and (*k*) ink, gum and typewriter requisites. Following Haber the 'chemical industry proper' (Tables 1.3, 1.5) is taken as groups (*a*) and (*b*) only.

Section 1.1 For the relationship between science, invention and economic growth see J. Schmookler, *Invention and Economic Growth*, Cambridge University Press, 1967. The scientific origins of the chemical industry are described in A. E. Musson and F. Robinson, *Science and Technology in the Industrial Revolution*, Manchester University Press, 1970.

Section 1.2 Much of the information in this section is taken from B. R. Mitchell and P. Deane, *An Abstract of British Historical Statistics*, Cambridge University Press, 1963.

Section 1.2.1 Soap consumption rose from 3·6 to 17·4 lb. per head in the UK according to C. Wilson, *The History of Unilever*, vol. 1, Cassell, London, 1954 and A. E. Musson, *Enterprise in Soap and Chemicals*, Manchester University Press, 1965.

Section 1.2.2 Average annual production of new houses was 69 530 in 1860–69, 93 350 in 1870–79, 79 500 in 1880–89, 105 700 in 1890–99 and 130 500 in 1900–09.

Section 1.2.3 See P. Deane and W. A. Cole, *British Economic Growth 1688–1959*, Cambridge University Press, 1967.

Section 1.2.4 Figures from C. M. Cipolla, *Literacy and Development in the West*, Penguin Books, Harmondsworth, 1969, p. 107.

Pulping with sodium hydroxide was introduced in 1851; it gave a dark pulp which needed extensive bleaching. The Tilghmann process, using calcium bisulphite to produce a light pulp, appeared in 1880.

Section 1.2.5 UK imports of guano reached a peak in 1870 and then fell rapidly. Ammonium sulphate production began in 1870 and was the major source of fixed nitrogen by 1890.

Section 1.2.6 These were dynamite (75% nitroglycerine, 25% kieselguhr) introduced 1866, blasting gelatine (92% nitroglycerine, 8% nitrocellulose), introduced 1875, and ballistite (10% nitroglycerine, 90% nitrocellulose), introduced 1887.

Section 1.3 For a succinct account of the process see Haber, *19th Century*, pp. 252–55. The main waste product was calcium sulphide (tank waste) which had to be dumped on land specially bought for the purpose.

Section 1.4 Figures from Haber, *19th Century*, pp. 59, 214–15. Chlorine was made by the Weldon process:

$$MnO_2 + 4HCl \longrightarrow MnCl_2 + Cl_2 + 2H_2O$$

$$MnCl_2 + Ca(OH)_2 + \tfrac{1}{2}O_2 \longrightarrow MnO_2 + CaCl_2 + H_2O$$

or the Deacon–Hurter vapour-phase process:

$$2HCl + \tfrac{1}{2}O_2 \xrightarrow[300°C]{CuCl_2} Cl_2 + H_2O$$

Section 1.5 The rise of Brunner, Mond is described in W. J. Reader, *ICI – A History. Volume 1: The Forerunners 1870–1926*, Oxford University Press, 1971, which is strong on cartels and company diplomacy but less satisfactory on technological developments. Table 1.2 is from Haber, *19th Century*, p. 101.

Section 1.6 Reader, op. cit., pp. 90–124, is the best account; the quotation is from p. 110.

Section 1.7 German sulphuric acid production rose from 103 000 tons in 1875 to 1·3 m. tons in 1904; soda production increased from 43 000 tons in 1878 to 325 000 tons in 1901 (Haber, *19th Century*, pp. 121–8). The dyestuff industry is well covered in J. J. Beer, *The Emergence of the German Dye Industry*, Illinois University Press, Urbana, 1959, and in Haber, pp. 128–36, 170–80.

Section 1.8 The definitive account of the development of the Haber process is given in Haber, *1900–1930*, pp. 84–103.

Section 1.9.1 In Table 1.4 production figures are from Haber, *1900–1930*, p. 200; consumption figures are estimated from the data given by H. A. Curtis and F. A. Ernst, *The*

Nitrogen Survey. Part 4: *The Nitrogen Situation in the European Countries*, US Department of Commerce, Washington DC, 1924.

Section 1.9.2   The khaki dye used for British army uniforms was imported from Germany; the Kitchener armies therefore took the field in clothing which ranged in colour from olive to mustard yellow. The quotation is from the *Report of the Committee on Commercial and Industrial Policy after the War*, Cd. 9035, HMSO, London, 1918, p. 52. For the dyestuffs companies and their problems see Reader, op. cit., pp. 258–81. I have been privileged to read Dr Haber's unpublished paper on *Government and Chemical Industry 1914–1939* which gives a detailed account of the protective measures of the interwar era.

Section 1.9.3   Volume 3 of W. J. Haynes, *The American Chemical Industry*, 6 vols., Van Nostrand, New York, 1945, gives a detailed account of the US industry between 1914 and 1922.

Section 1.10   Figure 1.2 is compiled from the data of London and Cambridge Economic Service, *Key Statistics of the British Economy 1900–1964*, Times Publishing Co., London, 1966. The Censuses of Production suggest that the chemical industry proper grew rather faster than the chemical and allied industries.

Section 1.11   For the postwar fortunes of these companies and the negotiations which led to the formation of ICI, see Reader, op. cit., pp. 317–466. The product mix of ICI in 1927 was given by Sir Paul Chambers, *ICI Magazine*, **46**, No. 341, p. 41. The quotation is from H. Levinstein, *J. Soc. Chem. Ind.*, 1931, **50**, 251.

Section 1.12   For British progress in dyestuffs see H. W. Richardson, *Scottish Journal of Political Economy*, 1962, **9**, 110. In Britain (Haber, *Government and Chemical Industry*, unpublished) and in Germany (T. P. Hughes, *Past and Present*, 1969, **44**, 106) coal hydrogenation was initially developed as a means of maximizing the use of the hydrogen plants of the Haber process. It was subsequently heavily subsidized in both countries for strategic reasons as it was quite uneconomic.

2 The new chemical industry from 1820 onwards

2.1 The polymer revolution

Since the end of the Second World War the developed nations have enjoyed unbroken prosperity. Industrial production, national wealth and international trade have risen rapidly and continuously. Economic growth has involved structural change. Some industries – coal, cotton textiles, agriculture – have declined relatively or even absolutely, while others – motor cars, aircraft, electronic equipment – have sharply increased in importance. The chemical industry belongs to the latter group; in all countries it has expanded at an altogether exceptional speed. In Britain, for example, while the output of manufacturing industry as a whole rose by 163% between 1935 and 1968, that of the chemical and allied industries rose by 467%. To use another measure, the chemical industry proper accounted for 3·3% of manufacturing income in 1935 and for 6·3% in 1968. Why has this happened? Many of the traditional markets for chemicals (section 6.1) have increased at rates little more – and often less – than those of industry in general, but as Table 2.1 shows, there has been an amazing growth in the use of synthetic organic polymers as plastics, as fibres, and as elastomers. Thus since 1935 the British output of dyestuffs has increased by 17%, that of nitrogen fertilizers by 700%, and that of synthetic polymers by 7 000%. The manufacture of synthetic polymers is now the most important single activity of the chemical industry. In Britain, as in the world as a whole, the unprecedented growth of the industry during the past 30 years has been accompanied by fundamental changes in its nature.

2.2 Rubber

The origins of the polymer industry are to be found in nineteenth century attempts to exploit the natural polymers rubber and cellulose. Rubber, obtained from the Para rubber tree *Hevea Brasiliensis*, was introduced into Europe in 1740 by the French astronomer de la Condamine. For many years it remained a curiosity, used only as an eraser, until in 1823 Charles Mackintosh invented a method of waterproofing cotton with a solution of rubber in naphtha. The mackintosh won a certain acceptance despite notable drawbacks: native rubber is mechanically weak and markedly thermoplastic in nature. It is soft and tacky when warm but rigid when chilled. In 1839 Charles Goodyear, an American inventor, discovered how to overcome these practical limitations. When heated with sulphur, rubber becomes harder, tougher and less sensitive to temperature changes; moreover, the extent to which the properties of rubber are affected by this process of vulcanization depends on the amount of sulphur incorporated, and a whole range of widely differing materials can be readily obtained. In the half-century which followed Goodyear's invention, rubber came to be used for industrial products – springs, washers

Table 2.1 The chemical industry in Britain, the Common Market, USA and Japan 1950–69

	Britain				Common Market				USA				Japan			
	1950	1958	1963	1969	1950	1958	1963	1969	1950	1958	1963	1969	1950	1958	1963	1969
Gross domestic product at constant prices	73	85	100	119	not available		100		65	82	100	129*	41	59	100	195
Index of industrial production	70	84	100	123	39	71	100	142	60	76	100	139	27	47	100	222
Index of chemical production	48	71	100	155	24	56	100	188	43	67	100	162	26	52	100	230
Production of traditional chemicals (m. tons):																
Sulphuric acid (100%)	1·8	2·3	2·9	3·3	5·3	8·6	10·5	14·7	11·8	14·5	18·9	26·0	2·0	3·8	5·0	6·7
Na_2CO_3 (100%)	n.a.	1·1	1·2	1·5	1·8	2·1	2·9	n.a.‡	3·6	3·9	4·2	7·3	0·2	0·4	0·6	1·3
NaOH (100%)	n.a.	0·8	0·9	1·0	0·7	1·4	2·3	4·7†	2·3	3·6	5·3	8·0	0·2	0·6	1·1	2·0
N Fertilizers (as N)	0·3	0·4	0·5	0·8	1·1	2·6	3·4	5·5	1·0	2·1	3·5	6·8	0·4	0·9	1·1	2·1
P Fertilizers (as P_2O_5)	0·4	0·4	0·4	0·5	1·5	2·5	3·3	4·0	1·9	2·4	3·3	n.a.	n.a.	0·3	0·5	0·8
Synthetic dyestuffs ('000 tons)	35	29	37	43	62	61	105	138	n.a.	64	109	113	n.a.	27	47	66
Production of synthetic polymers ('000 tons):																
Thermoplastic resins	50	211	488	883*	125	533	1 670	4 278*	374	1 210	2 751	5 675*	n.a.	364	680	2 521*
Thermosetting resins	105	142	211	355*		428	909	1 889*	365	630	1 056	1 621*			370	887*
Synthetic rubbers	nil	11	105	273	nil	43	391	950	484	1 071	1 634	2 286	nil	nil	103	526
Synthetic fibres	4	30	105	292	5	73	298	781*	55	222	524	1 560	1	46	239	806

n.a. = not available.
* 1968 Figure.
† Excluding Belgium and the Netherlands
‡ Excluding Belgium.

and matting – as well as for waterproof clothing. Appropriate new fabrication methods were developed: it was moulded, calendered into sheets, and extruded as tube. Yet it remained a specialized product; as late as 1910 annual world consumption was only 100 000 tons. Not only was the demand for rubber limited but the only source of supply

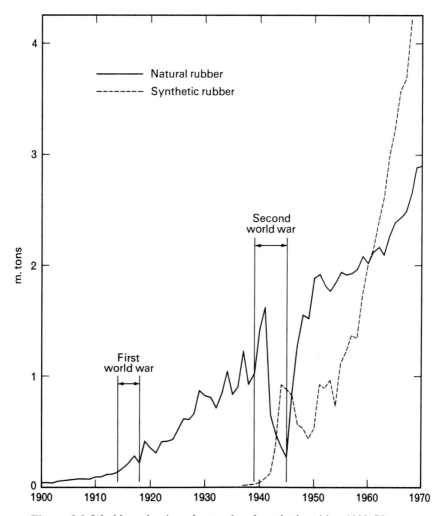

Figure 2.1 World production of natural and synthetic rubber 1900–70.

was the jungles of Brazil and the Congo, where it was collected under atrocious conditions. The situation began to change in the late nineteenth century. An enterprising Englishman, H. A. Wickham, smuggled 70 000 rubber-tree seeds out of Brazil in 1876. After germination at Kew, the seedlings were sent to Ceylon and Malaya for systematic cultivation. Plantation rubber, cheap and of good quality, appeared in the 90s and steadily displaced the wild variety. Almost simultaneously, the advent of the motor-car created a demand for rubber which was to carry world consumption to 1 400 000 tons by 1939 (Fig. 2.1). Rubber had arrived.

28   *The new chemical industry from 1820 onwards*

2.3   Cellulosics

The history of cellulose derivatives is more complex. Cellulose itself is insoluble and infusible, but it may be readily esterified with both organic and inorganic acids to yield products with profoundly modified properties. As early as 1845 Schönbein nitrated cellulose to give the powerful but unpredictable explosive gun-cotton, later used as a propellant. The nitrocelluloses are tough horn-like substances, soluble in ketones and esters and readily softened on mixture with a wide range of organic compounds. After a number of abortive attempts had been made to exploit this behaviour, J. W. Hyatt

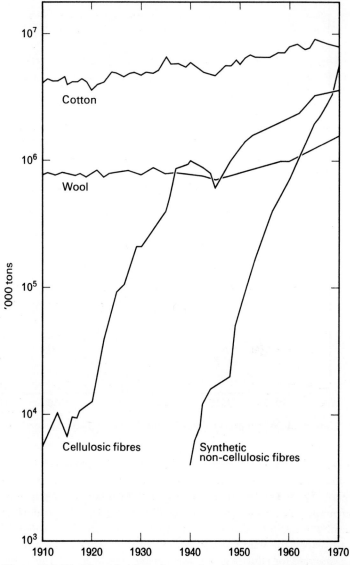

Figure 2.2 World production of natural and man-made fibres 1910–70.

discovered in 1869 that the addition of 25% of camphor to nitrocellulose produced a strong, easily worked thermoplastic material. That the new substance, Celluloid, was appallingly inflammable seemed unimportant and it was rapidly adopted for the manufacture of such novelties as knife-handles, detachable collars and photographic film. Cellulose derivatives also provided the first man-made fibres. In 1892 Cross and Bevan found that the treatment of cellulose with carbon disulphide and alkali yielded a soluble xanthate (viscose). When the latter is injected through fine nozzles into an acid bath, cellulose is regenerated in the form of smooth threads (rayon) which have most of the properties of silk at a fraction of its cost. The long-established silk-weaving firm of Courtaulds introduced the viscose process in 1905 with immediate success. Used mainly in women's wear, rayon was cheap, attractive and classless. With rising living standards, world production grew rapidly in the inter-war era and had reached 485 000 tons by 1938 (cf. Fig. 2.2). The last of the cellulose derivatives to attain commercial importance was cellulose acetate. Introduced on a small scale in 1905 as a less inflammable alternative to celluloid, it proved to be ideally suited to doping the wood and canvas airframes of the First World War and production was greatly expanded. After the armistice of 1918 the manufacturers, led by British Celanese, a most adventurous firm, were forced to diversify in order to avoid bankruptcy. Cellulose acetate was developed as a fibre (acetate rayon) and as a thermoplastic for injection-moulding and extrusion as film. Between the wars the production of both cellulosic fibres and cellulosic plastics developed extensively in all industrial countries and technicians became familiar with the problems involved in the utilization of synthetic fibres and thermoplastic resins. This knowledge was to be of crucial importance when purely synthetic materials appeared in the late 1930s.

2.4   The early days of synthetic resins

While the rubber and cellulose industries were developing, organic chemists had frequently observed the formation of synthetic polymers. These products aroused irritation rather than interest: uncrystallizable syrups or infusible solids, they defied the standard techniques of organic chemistry and were automatically discarded, mentioned only in footnotes as 'substances of unknown structure'. Commercial interest was similarly lacking. Unlike celluloid, the synthetic materials had unpredictable properties; even worse, they were obtained from compounds which at that time were no more than laboratory curiosities. Significantly, the first synthetic polymers to attain industrial importance were those derived from phenol and formaldehyde. These were already commercially available in the late nineteenth century. The reaction between phenol and formaldehyde to give resinous products was first observed in 1891 and at the turn of the century attempts were made to develop these materials as substitutes for the natural resin, shellac. Success came only when Baekeland made a systematic study of the effects of pH and of the phenol–formaldehyde ratio on the course of the reaction. When less than one mole of formaldehyde was used for each mole of phenol the product was a thermoplastic resin, but when more than one mole was used the material obtained could be smoothly converted to a hard infusible solid by the action of heat and alkali (section 10.41). The hardening or curing process required accurate temperature control and the application of pressure to suppress bubbling. Baekeland formed the General Bakelite Corporation in 1909 and began to market compression mouldings, insulating varnishes and laminated sheet for the motor and electrical manufacturers. A new industry had been born.

2.5 The science of large molecules

As yet, polymer chemistry remained an empirical science. There was no understanding of the factors which determine the formation and properties of large molecules. The situation at this time was curious. The very successes of classical organic chemistry had led to the implicit acceptance of several quite erroneous assumptions: that a pure compound must be a gas, a liquid or a crystalline solid; that it must be composed of identical molecules and that these molecules are quite small. The peculiar properties of high polymers — in particular the very large molecular weights indicated by measurements in solution — were held to mean that they were aggregates of small molecules held together by vaguely defined secondary valence forces. Thus the rubber molecule was thought to be a cyclic dimer of isoprene, many of these molecules being combined into large micellar units by the action of the 'partial valencies' arising from their high unsaturation. Like the Freudian view of personality, this approach explained everything, predicted nothing, and provided a convenient and socially acceptable way to dispose of awkward problems. In a series of investigations which began in 1918, Hermann Staudinger attacked such stereotyped thinking and vigorously championed the macromolecular hypothesis. He showed that high polymers are not only colloidal in all solvents but that their solutions, unlike those of genuine micellar colloids, exhibit no ageing phenomena. Moreover, polystyrene and its fully hydrogenated derivative have approximately the same molecular weight in benzene, so showing that 'partial valencies' arising from unsaturation play no part in their behaviour. Particularly significant were the results obtained from the study of the polyoxymethylenes, formed by the polymerization of formaldehyde. Acetylation of these materials gave the easily fractionated compounds $CH_3CO_2CH_2-(OCH_2O)_n-CH_2OCOCH_3$ which were fully characterized by degradation, end-group assay and solution properties. The melting points and boiling points of these compounds increase continuously with increasing molecular weight and their solubilities diminish. Hence the polyoxymethylenes themselves must be composed of linear chains of the type $HOCH_2-(OCH_2O)_n-CH_2OH$ which are of varying lengths. Similar investigations of polyesters, polyamides and polyalkenes by Staudinger himself, by Kienle, and, above all, by W. H. Carothers subsequently led to the formulation of the functionality rules (section 10.1) which both define the conditions necessary for polymer formation and predict the nature of the polymer formed. Finally, the application of X-ray crystallography in the 1920s revealed — after much confusion over the interpretation of the results — that polymers may be amorphous, wholly crystalline or partly crystalline, according to the extent to which the individual molecules are able to form regular arrays.

2.6 Three countries in the 1930s

By 1930 the study of polymers had changed from a black art into a reasonably comprehensible branch of chemistry. Polymer technology benefited greatly. The new science led directly to new products, such as nylon and terylene, and to a better understanding of those already in existence. The large majority of the commercially important synthetic polymers of today first reached production in the inter-war period (Table 2.2). World production was, however, low by modern standards, reaching perhaps 300 000 tons in 1938. Synthetic polymers were used in paints, for electrical insulation, for small mouldings, and for a hundred special applications where their unique properties would justify their high cost. In Britain, for example, the phenolic resin manufacturers rode to prosperity on the radio boom of the 1930s and the resulting demand for control knobs, valve sockets, insulating lacquers and adhesives. The experiences of the major chemical manufacturers in this new field make an instructive contrast.

Table 2.2 Dates of discovery and first commercial production of major synthetic polymers

Product	Discovery	First commercial production			Firm responsible
		Britain	USA	Germany	
Thermosetting products					
Phenol–formaldehyde	1891	1910	1909	1910	General Bakelite
Urea–formaldehyde	1884	1928	1929	1929	British Cyanide
Melamine–formaldehyde		1938	1939	1935	Henkel
Alkyd resins	1863	1929	1926	1927	General Electric
Epoxy resins	1891	1955	1948		Du Pont
Thermoplastic products					
Polyethylene (l.d.)	1933	1938	1941	1944	ICI
Polypropylene	1954	1959	1957	1957	Montecatini
Polystyrene	1835	1950	1933	1930	IG Farben
Polyvinyl chloride	1871	1942	1933	1931	IG Farben
Polyvinyl acetate	1911	1949	1928	1928	Union Carbide
Nylon	1932	1941	1940	1943	Du Pont
Terylene (Dacron)	1941	1949	1950		ICI
Polyacrylonitrile	1922	1959	1948	1952	Du Pont
Elastomers					
Poly-(butadiene–styrene)		1958	1941	1935	IG Farben
Poly-(butadiene–acrylonitrile)		1961	1939	1937	IG Farben
Poly-(2-chlorobutadiene)		1960	1931		Du Pont
Poly-(*iso*butene–butadiene)			1937		Esso

2.6.1 Germany

IG Farben and its predecessors had been actively interested in synthetic polymers since the early years of the century. They had indeed produced small amounts of polyvinyl acetate adhesives and a poly-2,3-dimethylbutadiene rubber in 1917–18. After the war the research and development effort was greatly expanded. With dyestuffs a static field, the directorate correctly decided that their future lay with synthetic polymers and acted accordingly. Between 1925 and 1939 the company spent between 5% and 10% of its turnover on research, of which a large proportion was devoted to polymers. The technical dominance achieved by IG Farben was extraordinary: it accounted for 17% of all patents on plastic materials taken out between 1931 and 1945, and for 30% of those taken out by large firms. Even more remarkable, in a lifetime of only 20 years, it produced 25% of all the major advances made before 1955. The Nazi emphasis on the self-contained economy encouraged the commercial development of synthetic materials and by 1939 the German chemical industry was producing, albeit on a relatively small scale, a wider range of polymers than that of any other nation.

2.6.2 USA

The leading role in America was taken by the Du Pont Company. An old-established explosives manufacturer, it expanded into dyestuffs during the First World War and into cellulose products immediately afterwards. The successful introduction of fast-drying nitrocellulose lacquers in 1923 and of the Dulux phenolic resin paints in 1927 indicated the commercial potential of polymers and led to the decision to undertake a major programme of fundamental research in this field. A team directed by W. H. Carothers began work in 1928 and achieved spectacular success. A series of papers both elucidated the chemistry of the polyamides, polyesters and polyacetylenes, and led to the discovery of chloroprene, the first satisfactory synthetic elastomer, and of nylon, the first synthetic fibre. Few investments have been so richly rewarded in the long run, but the transition from the laboratory to commercial production was both slow and expensive. The first fibres were drawn in 1933, but 8 years and 50 m. dollars were required to bring nylon to large-scale production. Although by 1940 America was manufacturing 120 000 tons of synthetic polymers, the new thermoplastics accounted for only 7% of the total.

2.6.3 Great Britain

By comparison Britain lagged behind. Although phenolic resins had been produced in Britain as early as 1911, the very excellence of the British rubber and metal-working industries discouraged the development of synthetic materials, while the large chemical companies conspicuously lacked the foresight and energy of Du Pont and IG Farben. Preoccupied with the Haber process, ICI showed only sporadic interest in synthetic polymers before 1935; Courtaulds, a sleepy family firm, were content with rayon; and the Distillers Company only entered the field on the eve of the Second World War. The British plastics industry owed its development more to small single-minded firms such as the British Cyanide Company (later British Industrial Plastics) and British Xylonite Ltd., who, however, lacked the funds to undertake large-scale research. In 1938 British production was no more than 25 000 tons and consisted almost entirely of the familiar phenol–formaldehyde and urea–formaldehyde thermosetting resins.

2.7 The impact of war: 1939–45

For the synthetic polymer manufacturers the Second World War was an unmixed blessing. Mechanized warfare required tanks, aircraft, motor transport, radios, radar equipment; polymers could be used for tyres, electrical insulation, tow-ropes, parachutes, cockpit covers. They replaced scarce materials on the home front; thus polyvinyl chloride was used for raincoats, curtains and shoes. Throughout the war Germany was absolutely dependent on synthetic materials. Production of synthetic rubber, mainly of the styrene–butadiene type, rose from 3 150 tons in 1937 to 116 600 tons in 1943, by which time the annual output of other synthetic polymers had also topped the 100 000 ton level. Nylon and polyethylene were independently developed and brought to the production stage. In the Second World War, as in the First, Germany owed much to her chemical industry. Britain and the USA were initially less hard-pressed and the main demand in these countries was for such specialities as polyethylene, polymethyl methacrylate and nylon for the electrical and aircraft industries. The loss of Malaya and its rubber plantations in 1942 caused a major crisis. At a cost of 700 m. dollars the major US chemical companies built plant capable of producing 1 m. tons of styrene–butadiene rubber per year, from which the needs of the Allies were met. British progress, though quantitatively more modest, was substantial. Synthetic rubber was manufactured only experimentally, but polyvinyl chloride was introduced and extensively used as a substitute. By 1945 the production and utilization of a wide range of synthetic polymers was firmly established in all three countries.

2.8 From coal to oil

There remained a major obstacle to the growth of the industry. Synthetic polymers – in particular the highly desirable thermoplastic addition polymers – remained comparatively expensive because they were made from high-cost starting materials. The methods used in the production of basic organic compounds had serious limitations. Until the First World War the chemical industry was mainly concerned with aromatic compounds which were obtained from the by-products of gas-works and coking ovens. Such aliphatic compounds as were needed were made by fermentation or wood distillation. Before addition polymers could be introduced, methods of making the necessary unsaturated monomers had to be found. Two routes to such compounds were exploited in the 1930s: the dehydration of alcohols produced by fermentation, and the addition reactions of carbide-derived acetylene. Neither was cheap (Chapter 8) but both were practicable, and the carbide–acetylene route was to remain the mainstay of the European polymer industry until the mid-50s. Another source of alkenes already existed: petroleum. Cracking processes had been developed in the USA to convert the higher boiling fractions of petroleum into gasoline and it was known that these also produced large quantities of C_2–C_4 compounds (Chapter 7). The American chemical industry moved towards the use of hydrocarbon-based organic chemicals in the 1930s, but European manufacturers, bound to policies of national self-sufficiency, failed to follow suit until after the Second World War. Since then, however, petroleum has become the universal raw material. The British experience is typical: the starting-point for 11% of all organic chemical production in 1949, it was used for 87% in 1968 (Table 2.3). Several factors

Table 2.3 British organic chemical production 1949–68 ('000 tons)

Origin	1949	1955	1959	1963	1968
From coal					
by destructive distillation	172	271	359	460	
from carbide via acetylene	69	108	133	175	388
via synthesis gas	54	79	93	80	
By fermentation	157	138	79	30	n.a.
From petroleum	44	285	586	1 160	2 664
Total	496	881	1 250	1 905	3 052

have combined to make petroleum peculiarly attractive in the post-war era. Throughout the 50s and most of the 60s it remained cheap while the price of coal increased steadily because of its high labour content. The post-war trend towards refining petroleum in consumer rather than producer countries made hydrocarbon feedstocks for cracking processes freely available at low prices. The thermal cracking process itself shows remarkable economies of scale: it has been estimated that ethylene from a 30 000 ton per year cracker costs £80 per ton, whereas that from a 300 000 ton per year cracker costs only £30 per ton. As petrochemicals became available polymer prices fell and consumption grew. The relationship between polymers and petrochemicals is symbiotic: the needs of the polymer manufacturers have forced the growth of petrochemicals but petrochemicals have made possible the growth of the polymer industry.

2.9 Polymers triumphant

Since 1950 the use of synthetic polymers has grown at an explosive rate. World production, perhaps 1 m. tons in 1948, reached 27 m. tons in 1969. Yet in many ways the commercial development of polymers in this period has represented no more than the extrapolation of trends already clearly visible in 1940. Few new large-tonnage materials have appeared since the Second World War – terylene, polypropylene and polyacrylonitrile are the main exceptions – and, despite immense expenditure on research, polymer technology has changed in degree rather than in kind. The growth of production and the development of markets have followed a broadly similar pattern in all industrial countries (cf. Fig. 2.3). The combination of great versatility and steadily falling prices has led to a dramatic increase in the consumption of thermoplastic addition polymers. The thermosetting synthetics have grown more steadily but cellulosic materials have stagnated since the mid-60s. The use of natural polymers has increased only slowly and in some areas has actually declined. The applications of synthetic polymers are similar in most advanced countries. Thus of the major thermoplastics, low-density polyethylene is used mainly for film and high-density polyethylene for mouldings, packaging being the most important outlet for both materials. Polypropylene goes into fibres and industrial mouldings and polystyrene into packaging, toys and holloware. Polyvinyl chloride, the most versatile of polymers (section 10.3), has a correspondingly large range of uses: as sheet and film for packaging and for the motor industry, as flooring and as rigid pipe and sheet for the building trade, and in footware, luggage and records. In all these fields synthetic polymers have supplemented and often displaced the traditional materials. They are light, tough, durable and attractive in appearance. They are uniquely cheap and easy to fabricate. For what they are their price is low. From both a technical and a commercial point of view they are for many purposes the ideal materials.

2.9.1 The rise of Japan

The pattern of growth has been the same in the major industrial nations but the scale of growth has varied greatly. Until 1955 the pioneers – Western Germany, the USA and Britain – dominated the industry through their patents and their accumulated know-how. At this time they accounted for as much as 75% of world production and 85% of the world trade in synthetic polymers. As patents expired and technical expertise became more widely diffused other countries entered the field – the Netherlands, Italy, above all

Japan. With the age of discovery over, the prerequisites for success were low production costs, aggressive technical development and a flourishing domestic market. The extent to which polymers are used is intimately connected with economic growth: *per capita* consumption has risen most rapidly in those countries which have experienced the greatest general progress (Tables 2.1, 2.4). As Japan has emerged as the third industrial nation

Table 2.4 *Per capita* consumption of plastics in industrial nations* (kg per head)

	1950	1955	1960	1961	1962	1963	1964	1965	1966	1967	1968	1969
USA	6·4	8·0	10·7	17·0	18·1	19·1	20·8	24·6	28·5	28·7	32·5	34·7
Britain	2·5	5·2	9·1	9·5	10·4	11·5	13·9	15·1	15·5	17·3	20·4	21·7
West Germany	1·9	5·8	15·0	15·5	21·4	19·5	24·5	26·7	28·5	31·6	40·4	49·9
France	0·9	2·8	7·5	8·3	9·5	10·8	12·5	14·1	17·2	18·9	21·6	26·8
Italy	0·6	1·7	5·0	6·7	8·6	10·2	11·0	11·8	14·7	17·7	19·3	23·5
Netherlands	1·2	3·5	9·1	9·8	10·5	10·6	14·7	16·6	16·5	18·5	18·8	23·6
Japan	0·2	1·4	3·8	8·4	10·6	10·8	13·6	14·2	17·0	23·4	29·0	32·8

* Including all plastics whether cellulosic or synthetic but excluding elastomers and fibres

of the world, so the Japanese production and consumption of polymers has become second only to that of the USA. Conversely the decline of Britain is reflected in the slow growth of British polymer consumption; twenty years of relative stagnation have seen Britain fall from second to seventh place in the league table of advanced nations. Just as the consumption of sulphuric acid was the measure of economic activity in the nineteenth century so the consumption of synthetic resins is that for today.

2.9.2 The natural history of a polymer

It is instructive to consider the progress of a single thermoplastic polymer. Low-density polyethylene was accidentally discovered in 1933 by Gibson and Fawcett of ICI General Chemicals Division. Full-scale production began in 1939 with a 50 ton per year plant supplied with ethylene derived from fermentation alcohol. The polymer was used exclusively as an insulator for high-frequency electric cables and, with the wartime development of radar, production had risen to 1 000 tons per year by 1945. The end of the war forced ICI to develop new outlets for polyethylene. It was already known to be peculiarly suited to injection- and blow-moulding and to extrusion as film; by 1950 it was being used for laboratory bottles, toys, washing-up bowls and luxury packaging. Production had reached only 2 000 tons per year, however, and it remained an expensive speciality used only where its particular properties were at a premium. The turning point was the change in 1952 from ethanol-based ethylene at perhaps £250 per ton to petroleum-based ethylene at about £90 per ton. The price of polyethylene dropped immediately and sharply and has continued to fall (cf. Fig. 2.4). Economies of scale in the production of both ethylene and polyethylene made price reductions possible; competition and overcapacity made them necessary. As prices fell market penetration increased. The main growth area of the 1950s was in mouldings, polyethylene rapidly ousting metals in such applications as household goods and toys; in the 1960s, however, the more rigid high-density polyethylene and polypropylene were increasingly preferred

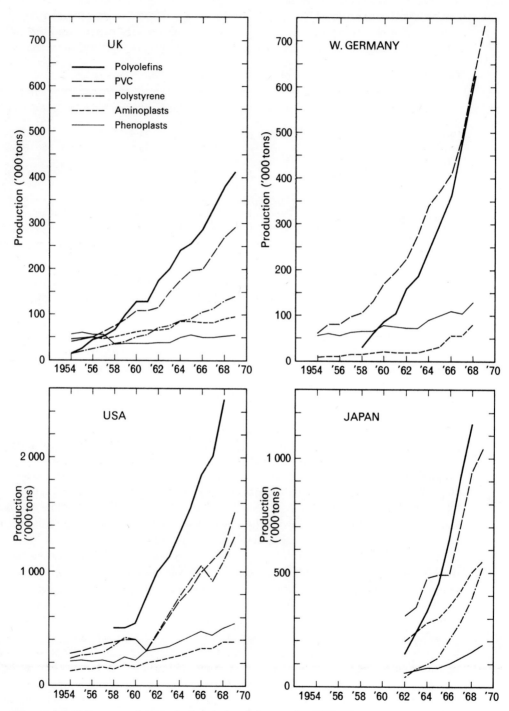

Figure 2.3 Polymer production in four industrial countries 1954–69.

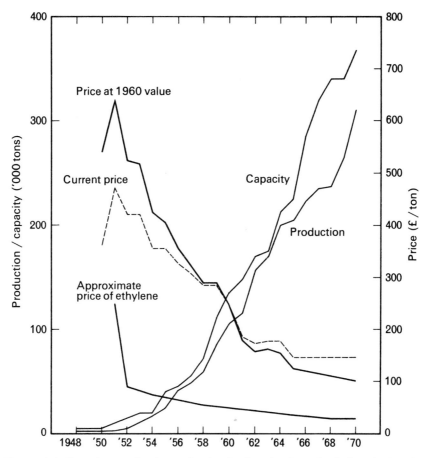

Figure 2.4 Capacity, production and price for low-density polyethylene in Britain 1948–70.

Table 2.5 Production and end-uses of low-density polyethylene in Britain, 1948–68

	1948	1954	1958	1962	1964	1966	1968
Production ('000 tons)	1	17	68	155	202	222	237
Imports ('000 tons)	negl.	negl.	5	15	17	13	43
Exports ('000 tons)	negl.	7	37	67	85	69	75
Consumption ('000 tons)	1	10	36	103	134	166	205
End uses (%)							
Wire and cable	90	30	18	14	10	10	9
Film and sheet		20	30	34	46	49	54
Injection moulding		30	31	25	20	18	15
Blow moulding	10	0	4	12	12	11	10
Other		20	17	15	12	12	12
World production ('000 tons)	30	200	600	1 200	1 800	3 000	4 000

for mouldings, and film for packaging has become the main outlet for the low-density material (cf. Table 2.5). Simultaneously the manufacture of polyethylene became an international business. To further the war effort Du Pont and Union Carbide were licensed by ICI in 1940 and US production rapidly exceeded that of Britain, but until 1955 the output elsewhere remained small. Since then free access to polyethylene technology and the rapidly growing world economy have taken production from 200 000 to 5 m. tons and made it the most widely used of plastics. In 30 years it has changed from an expensive curiosity to a large-volume commodity as familiar as sulphuric acid.

Notes and references

General There is no comprehensive treatment of this subject. The most useful books are by Morris Kaufmann, *The First Century of Plastics*, Institute of Plastics, London, 1963, and *The History of PVC*, Maclaren and Sons, London, 1969. The latter covers more ground than the title suggests and contains many references. Much information is scattered through general histories of the industry, notably W. Haynes, *The American Chemical Industry*, Van Nostrand, New York, 1945–54, vols. 3–6. D. W. F. Hardie and J. Davidson Pratt, *A History of the Modern British Chemical Industry*, Pergamon, Oxford, 1966, otherwise disappointing, has a little on British pioneers.

The only reliable production statistics for the period before 1940 are those of the US Tariff Commission, which, of course, refer to the USA alone. Since the Second World War a variety of government and international agencies have collected some information; the quality and detail of their reports improves markedly after 1955. The most convenient collection of production, consumption and trade data for the major non-Communist countries is the OECD *Annual Report*. The 1965–66 and 1969–70 issues contain statistical supplements covering the years 1953–64 and 1958–68 respectively. Most governments produce information about the polymer industry of their own country (cf. Chapter 10), much of which appears in more digestible form in the trade press. Thus *Modern Plastics* has published a detailed annual survey of the US industry in its January issue since 1946, and *British Plastics* has done the same for the British industry since 1953. The supplements to and special features in *European Chemical News* (1962–) are a mine of information for European developments in the 1960s.

Section 2.1 The British comparisons are based on the *Census of Production*, 1935 and 1968, taking the chemical industry proper to be chemicals, drugs and dyes in both cases, with the addition of fertilizers, etc., in 1935 and of fertilizers, pesticides and disinfectants, photographic materials, and synthetic resins and rubbers in 1968 (cf. Chapter 1, note on definitions). Table 2.1 is based mainly on data given in the OECD *Annual Report 1969–1970* and in various issues of the *United Nations Statistical Yearbook*.

Section 2.2 Figure 2.1 is based on the data of the International Rubber Study Group: *Rubber Statistical Bulletin*, 1971, **25** (10), Table 56, and of various issues of the *United Nations Statistical Yearbook*.

Section 2.3 The best account of the origins and development of viscose rayon is in D. C. Coleman, *Courtaulds*, Oxford University Press, 1969, vol. 2, pp. 1–75 and 171–204, which also contains glances at cellulose acetate and at British Celanese Ltd. Figure 2.2 is based on the data of *Textile Organon*, January 1962, and of the *United Nations Statistical Yearbook 1969*.

Section 2.4 Baekeland described his own researches in *Industrial and Engineering Chemistry*,

1910, **1**, 149–61 and 545–49. There is a complete *catalogue raisonné* of early papers and patents on phenolic resins in Carleton Ellis, *The Chemistry of Synthetic Resins*, Chemical Catalog Co., New York, 1935, vol. 1, pp. 277–90. Baekeland thought that his products contained seven phenolic residues connected by oxygen bridges.

Section 2.5 There is an excellent short review of the development of polymer chemistry in P. J. Florey, *Principles of Polymer Chemistry*, Cornell University Press, Ithaca, New York, 1953, pp. 4–28, which contains almost all the major references. For those who read German, Staudinger, *Arbeitserrinerungen*, Heidelberg, 1961, gives a vivid account of the controversies of the 1920s; some translated extracts have been published in *Education in Chemistry*, 1971, **8**, 63–67.

Section 2.6 C. Freeman, *The Plastics Industry: a Comparative Study of Research and Innovation*, *National Institute Quarterly Review*, November 1963, pp. 22–62 is by far the best analysis of the period 1930–55. For early German work see Kaufmann, *PVC* and *Haber: The Chemical Industry 1900–1930*. Carother's work is best studied in his *Collected Works*, ed. H. Mark and G. S. Whitby, Interscience, New York, 1940. Early British developments are described by Hardie and Davidson Pratt and, less glowingly, by Haber, *Government and Chemical Industry*. Table 2.2 is based on that of Freeman, loc. cit.

Section 2.7 German wartime production is discussed by J. M. Debell, W. C. Goggin and W. E. Gloor, *German Plastics Practice*, Debell and Richardson, Springfield, Mass., 1946. Synthetic rubber is reviewed by F. A. Howard, *Buna Rubber*, Van Nostrand, New York, 1947. British wartime developments are described in *British Plastics*, 1945, **17**, 230–50, and 1946, **18**, 344–46.

Section 2.8 For the origins of petroleum chemistry see Carleton Ellis, *The Chemistry of Petroleum Derivatives*, Reinhold, New York, 1934. The early work was concerned with the manufacture of alcohols and other solvents from petroleum-derived alkenes. For the British experience see H. M. Stanley, *The Petroleum Chemicals Industry*, Royal Institute of Chemistry Lecture Series, No. 4, 1963, from which Table 2.3 is derived, and H. P. Hodge, 'The Petroleum Chemical Industry in the UK', *Institute of Petroleum Review*, 1962, **16**, 114–21.

Section 2.9 Figure 2.3 is based on the data of Freeman and of the OECD *Annual Report 1969–70*, Supplement: Production Statistics for Selected Products 1958–68.

Section 2.10 Table 2.4 is based on the estimates given annually for the OECD countries in the *Annual Report*. A large majority of polymer production is consumed in the country of origin: in 1969 exports made up 8% of US production, 15% of Japanese production, 22% of Common Market production (taking the Common Market as a single unit) and 32% of British production. With the exception of Britain no major consumer of polymers imports more than a small minority of its needs: imports were less than 2% of consumption in Japan and the USA, 7% in the Common Market and 27% in Britain.

Section 2.11 The discovery of low-density polyethylene is described by R. O. Gibson in the *Royal Institute of Chemistry Lecture Series* No. 1, 1963, and by J. A. Allen, *Studies in Innovation in the Steel and Chemical Industries*, Manchester University Press, 1967, pp. 7–36. The commercial development of polyolefins is currently under study by Mr T. J. Russell of Oxford Polytechnic, who most kindly made available the data used in Table 2.5 and Fig. 2.4. It should be noted that the price of ethylene given is notional since most ethylene is captive production.

Part II Economic aspects of the chemical industry

3 Introduction to microeconomics

The chemical industry exists not to make chemicals but to make money. In any study of it we must ask, for example, what profit Imperial Chemical Industries makes from manufacturing chlorine and soda by the electrolysis of brine. In the academic world this would be a meaningless or even a heretical question. To pure scientists, science is an end in itself. It is a part – perhaps the most impressive part – of modern culture; our knowledge of aromatic silicon compounds is philosophically equivalent to, say, the novels of D. H. Lawrence. Academic research is undertaken to push forward the frontiers of knowledge and to enhance the reputations of its practitioners. It also provides applied scientists with a more coherent model of the universe which might eventually lead to discoveries which raise living standards or cure disease – 'Better things for better living through chemistry' as Dupont puts it.

These factors are irrelevant to the manufacture of chlorine. ICI manufactures chlorine because it can sell it – and use it itself to convert to other materials, for example polyvinyl chloride, which it can also sell. The money which ICI receives from these sales goes to pay for the raw materials and services used in chlorine manufacture, the wages and salaries of the people involved in production and sales, and the cost of construction and maintenance of the chlorine producing plant. Any money left over is the profit and after it has been taxed and some of it set aside for future enterprises, the rest is distributed to shareholders. Shareholders invest in a company in the expectation of growth in its activities and profits and in the amount of dividends paid to them; this in turn will cause the market value of the company's shares to rise. If these hopes are not fulfilled and seem unlikely to be so for a long period, then some shareholders will sell some or all of their holdings, the market price will fall and the company's ability to raise fresh capital from its shareholders to boost its activities or expand may be hindered. Thus it behoves ICI to make as large a profit as possible within certain limits set by public opinion and the law. If chlorine manufacture does not run at a profit and ICI can see no way in which it can be made to do so in the foreseeable future, then it will cease to manufacture chlorine. It is irrelevant that ICI has an elegant method for producing chlorine, that chlorine production provides employment or that chlorine is useful to the people who buy it. If chlorine is that important, ICI would say, then purchasers should be prepared to pay enough for it to give us our profit, and if avoiding unemployment is so important to the government then they should offer us taxation relief or a subsidy to turn an unprofitable operation into a profitable one.

In countries with centrally planned economies (USSR, Poland, Cuba etc.) and in nationalized industries in capitalist countries, the government may be regarded as the sole shareholder. Accounting systems may be carefully planned to avoid considerations of profit and loss and unprofitable activities may be supported for political reasons. Such activities, however, will have to be subsidized by profitable ones so that the question of profits is of importance even to the most Marxist of regimes.

In Part II of this book, we shall discuss the economic basis of buying and selling (demand and supply) and the factors governing individual decisions of firms and people, particularly in the chemical industry.

3.1 Problems of the economic system

The desires of human beings are unlimited. No person and no community ever has everything it wants. Once a man's need for food and shelter has been satisfied, he will turn his mind to cars and hi-fi equipment. When he possesses more cars than he can ride in or records than he can play, his desires will turn to emotional satisfactions such as power or prestige. Even the anchorite seeks religious satisfactions, consumes food and is in the market for hair shirts. Even a saint desires the ultimate good and may be prepared to work to bring the Kingdom of God a little nearer.

The desires of a society are satisfied by means of its resources. Resources are defined as the agents or factors of production and consist of land, labour and capital (section 3.5). Though the wants of a society are infinite, its resources are strictly finite, and thus *all* wants and desires cannot be satisfied. A man can buy a car *or* a boat, a community can opt for better roads *or* better schools, or perhaps improve slightly both roads and schools. Individually and collectively, therefore, a choice has to be made between which desires shall and which shall not be satisfied. Frequently, choices must be made between highly desirable ends. Money can be spent on better hospitals or on medical research, on higher family allowances or on higher old age pensions. The choice of which of a society's desires shall be satisfied with the resources available is the economic problem which faces the society. Equally, the individual has to decide which desires to satisfy with his much more limited resources.

Different societies make different choices. The Nazis opted for guns rather than butter; in Egypt under the Pharaohs pyramids had priority over better living conditions for the pyramid builders. Even the most humane and progressive societies are often accused of spending too much on luxuries now instead of investing for the future. The economic system is the mechanism by which this choice is made. It determines how the limited resources of a society may be used to the best advantage in terms of the beliefs and social structure of the society.

The economic system has to make three fundamental decisions: 1. What commodities shall be produced and in what quantities? The term 'commodities' includes goods, such as cars or loaves of bread, and services such as medical attention or entertainment. 2. How shall commodities be produced? That is, who is going to do the producing, by what method and using which resources? Is furniture, for example, to be produced in the cottage of a Staffordshire craftsman or in a factory in High Wycombe? 3. For whom are the commodities to be produced? The Pharaohs derived the benefit, such as it was, from building pyramids, though they did not themselves do the work. In a society, the share-out of the national cake is something which may be decided by collective bargaining, by force, negotiation, tradition and countless other factors.

In a dictatorship, the major decisions are taken by one man with the power to enforce them. In an extreme *laissez-faire* capitalist system, decisions are taken automatically by a system of prices, markets, profits and losses in which every citizen tries to make as much money as possible, and the devil take the hindmost. In a mixed economy, such as exists nowadays in the USA or UK, the government steps in and modifies the working of the price system. For example, an enterprising cigarette manufacturer would not be permitted to market marijuana cigarettes, though a demand exists for them and the enterprise would be commercially profitable. Similarly, the Victoria line on the London Underground was costed on the basis of social benefits, such as the value of the time saved, to the people using it and not on purely commercial considerations.

3.2 Money

If we are to take decisions between various courses of action we must have a yardstick by which to compare a ton of bread with a ton of polyethylene with a ton of steel. This yardstick is called money.

Money has two functions – it acts as a medium of exchange and as a yardstick by which values can be expressed. Every time a man buys a pint of beer with money instead of bartering for it part of whatever he himself produces, he demonstrates the former role. The latter may be exemplified by the man who is paid by credit transfer into a bank account and who shops using a credit card and cheque book. At the end of the month, what he has spent is deducted from his salary and the balance credited to his account. We can imagine an extension of this system by which no medium of exchange (i.e. monetary tokens) would be required, but we would still need a system of value to decide how many suits and cars the man could buy in exchange for his salary.

It seems callous to measure aesthetic or humanitarian needs in terms of money but no alternative has yet been devised. Where profit is not a yardstick, for example in the choice between a new hospital and a new school, it is customary to refer to cost effectiveness, but ultimately these comparisons are made in money terms. We are still forced to take decisions as to how much we are prepared to spend to save a single life on the roads, build a kidney machine or buy a seat at a concert, and to weigh these satisfactions against alternative ways of spending the same money.

3.3 Utility

Why does a person buy goods? At a trivial level he buys to satisfy his needs and desires; having received a wage as a result of this role as a producer, he takes over the role of consumer and spends the wage to buy the things he needs and, which, if there were no division of labour, he would have to produce himself. As much for an individual as for society, however, there are the questions of how much to buy and of what.

Initially a man needs food. The value (or utility or satisfaction) to him of a single loaf of bread will be extremely high, say 50p, since it will save him from hunger. The utility of two loaves will be less than twice this amount and four loaves will have less than twice the utility of two loaves. Ultimately a point will be reached at which the man has no need or desire for any subsequent loaves and in an extreme case may be prepared to pay someone to come and take them away (i.e. they have a negative utility). This is shown in Fig. 3.1. We can also plot the *extra* satisfaction which a consumer obtains from the purchase of each additional loaf. For example, if the value of one loaf is 50p and of two loaves 80p, then the extra satisfaction obtained from the second loaf is 30p. This quantity is called the marginal utility of the second loaf. The marginal utility of bread corresponding

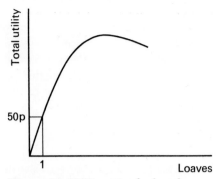
Figure 3.1 Utility curve for bread.

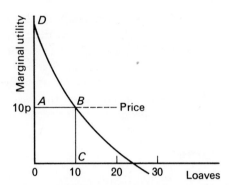
Figure 3.2 Marginal utility curve for bread.

to the utility shown in Fig. 3.1 is shown in Fig. 3.2. Mathematicians will recognize that the marginal utility curve is approximately the differential of the utility curve.

In economics generally, if y is a function of x, then the amount by which y increases for unit increase in x is called the marginal y. In the limit, when x is so large that a unit increase changes it infinitesimally, the marginal y becomes equal to the differential dy/dx. For the situations with which we are concerned, this relationship may be taken as generally true.

If the price of bread is 10p per loaf, the man will go on buying bread until the next loaf he buys is worth less than 10p to him. From Fig. 3.2 it is clear that in this case he will buy ten loaves. In general, a man will go on buying a commodity until the marginal utility of it to him (or the marginal satisfaction he gets from it) becomes equal to the price.

On Fig. 3.2 the outlay on the ten loaves is clearly equal to the area $ABCO$ but the value obtained by the consumer is the area $EBCO$ and the consumer has an excess of satisfaction over outlay equal to the area BEA. This notional 'surplus' of the consumer may be thought of as reason for trade, and when the consumer ceases to receive this surplus, he stops buying.

This analysis deals with only one commodity – bread. But man does not live by bread alone, and he might well decide to buy some wine to go with it. Wine will have total and marginal utility curves similar to those for bread and the consumer will cease to buy when the marginal utility equals the price. He now has a choice however. He can buy bread or wine or some of each. We can construct for him a consumer's indifference curve S_2 (Fig. 3.3), so called because the consumer is indifferent to any combination of food and wine represented by a point on the curve, and gets the same total satisfaction from any. At point A with 16 bottles he is so intoxicated he does not mind the absence of solid food and at point B with 75 loaves he does without drink altogether. At point C he is equally happy with 25 loaves and 5 bottles of wine. At point D he is getting less satisfaction out of his diet than he would get from any point on the curve ACB and at point E he gets more. These therefore lie on other satisfaction curves, S_1 to S_3 represented by the faint lines. The curves are concave because in general a person is glad to give up quite a lot of a plentiful commodity in return for a little of one which is scarce. On curve ACB, near point A, the consumer is prepared to give up a bottle of wine for a single loaf of bread, whereas

Utility 47

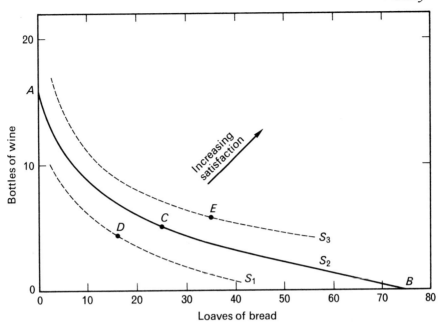

Figure 3.3 Consumer's indifference curve.

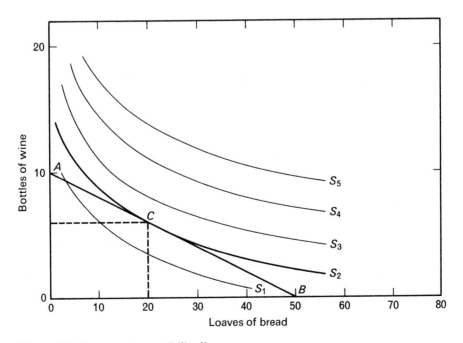

Figure 3.4 Consumption possibility line.

at point B, 10 loaves would be given up for a solitary bottle of wine, and this tends to correspond to people's observed behaviour.

Figure 3.4 shows a series of curves of indifference like those in Fig. 3.3. Suppose our hypothetical consumer has £5 to spend on food and drink and that wine costs 50p per bottle and bread 10p per loaf. If he spends it all on wine he will be at point A on the graph, and if he spends it all on bread he will be at point B. Any other point on the line AB can be seen to represent the same total expenditure on different combinations of food and wine. This is a consumption possibility line. If he wishes to spend £5 the consumer can occupy any point on this line. Which one will he choose to occupy? Clearly the one which affords him the greatest satisfaction, in this case the point C, at which AB touches but does not cross the indifference curve. (If AB crossed an indifference curve, then it would be possible to move to a higher one.)

At the point C, the consumer is buying 20 loaves and 6 bottles; the slope of the consumption possibility curve AB is $-\frac{1}{5}$ (if it is expressed in terms of money) and therefore the slope of the indifference curve at C must also be $-\frac{1}{5}$. It can be seen that the consumer will buy bread and wine in such amounts that the ratio of the marginal utilities of an extra loaf and an extra bottle is equal to the ratio of their prices. That is, if he were to spend an infinitesimally small amount of extra money on either bread or wine, either purchase would afford him equal satisfaction. This can be expressed as 'the marginal utility of money is the same whether it is spent on bread or wine'.

If we wished, we could combine the marginal utility curve for bread in Fig. 3.3 with the similar one for wine and the indifference curves of Fig. 3.4 and obtain a three dimensional graph with surfaces of indifference. A consumer with a fixed income confronted with a given set of prices will adjust his purchases and come to rest at a point on these surfaces defined by the following considerations:

1. The marginal utilities of both bread and wine are equal to their prices.
2. The ratio of the marginal utilities of bread and wine is equal to the ratio of their prices; the marginal value of money is the same whatever it is spent on.

These two principles can be generalized to apply to all commodities, not just bread and wine. Why is this so? Because if the last loaf of bread did not afford the consumer as much satisfaction as the last bottle of wine, then it would pay him to switch some of his money to buying more wine and less bread.

3.3.1 Changes of price and income

Suppose the consumer's income increases by 50%; he can then afford to spend more on bread and wine and will have a different consumption possibility curve. This will be parallel to the earlier one (because its slope depends on the ratio of prices) but lies further from the origin – see Fig. 3.5. It will be noticed that the new balance of bread and wine is different from the old one, 10 bottles and 25 loaves have replaced 6 and 20 and most of the extra money has been spent on wine rather than bread.

A change in the price of bread, on the other hand, is reflected by a change in the slope of the consumption possibility line. In Fig. 3.6 the price of bread has doubled and the consumption possibility line has rotated from ABC to $AB'C'$ tangential to a new, lower satisfaction indifference curve and giving a new equilibrium position of B': 10 bottles and 25 loaves have been altered to approximately 11 bottles and 10 loaves.

Utility 49

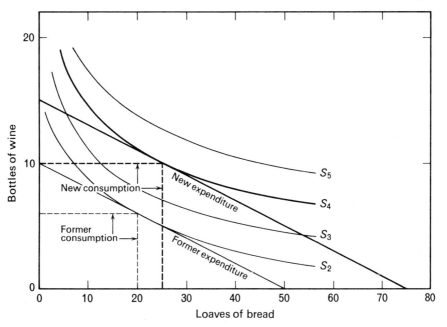

Figure 3.5 Increase in consumer's income.

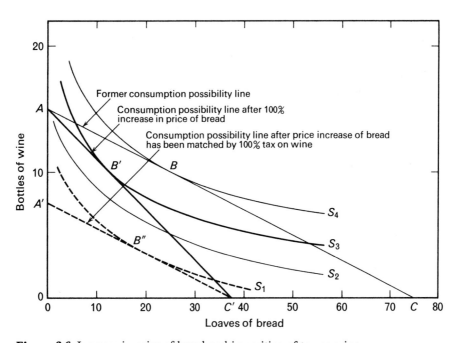

Figure 3.6 Increase in price of bread and imposition of tax on wine.

50 *Introduction to microeconomics*

If the government became alarmed by the drop in bread consumption, it could either subsidize bread to bring the line back to *ABC* (but this would cost money) or put a 100% tax on wine, thus doubling its price and (depending on the shape of the indifference curves) possibly increasing bread consumption but generally lowering total consumption. The line $A'B''C'$ illustrates this and the equilibrium point B'' corresponds to 4 bottles and $17\frac{1}{2}$ loaves. In the end, a small tax on wine might be used to provide a subsidy on bread if pressures on the government made it appear desirable for the bread industry to be supported.

3.4 Demand

We have considered above the problems of demand by a single consumer with a single set of preferences. In fact, each consumer will have a different series of indifference curves – one man may be a teetotaller, another may be overweight and forbidden to eat bread – but it is possible nonetheless to construct utility, marginal utility and indifference curves for whole populations. They look very similar to the ones already considered

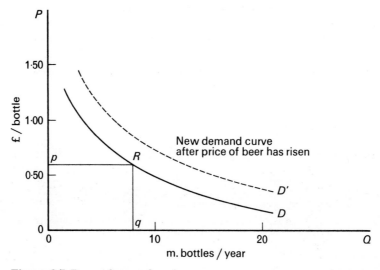

Figure 3.7 Demand curve for wine.

though the numbers concerned are much larger. As the people who construct them are concerned with what the population will buy, it is more usual to construct so-called demand curves which show what quantity q of a commodity the population will buy at any given price p. This is the definition of demand. The demand curve is equivalent to the marginal utility curve for the individual consumer in that it shows for any given price the point at which consumers as a whole will stop buying. Figure 3.7 gives an example.

At a price p, the quantity sold will be q and the amount of money paid known as the revenue will be $p \times q$, i.e. the area of the rectangle $pRqO$.

The amount of a commodity, for example wine, which is bought will depend on the following factors:

(a) Its price.
(b) The prices of other commodities which can reasonably be substituted for it. For example bread is also a food and if it were very cheap money might be spent on it that would otherwise be spent on wine, but it is only a partial substitute. Cheap beer on the other hand could be a practically complete substitute for wine. It is worth noting that certain products complement each other rather than competing. Cheap butter would cut sales of margarine but would increase sales of bread.
(c) The income of the buyers.
(d) The number of buyers.
(e) Their scale of preferences, i.e. the complex of indifference curves that express their attitudes. The aim of advertising is to influence a consumer's preference, i.e. to alter the shapes of his indifference curves.

The last four of these are constant for any given demand curve. If any of them alters, new curves must be plotted. Like marginal utility curves, demand curves customarily slope downwards, i.e. the higher the price the less people buy. There are exceptions to this rule – mink coats and diamonds tend to sell because of rather than in spite of their price; shares on the stock exchange often sell better if their price rises because buyers extrapolate this trend and expect that it will continue to rise.

Furthermore, demand curves tend to move further away from the origin as the income of buyers rises, but again, this is not always true. Demand for both bread and margarine tends to drop once incomes have risen above a certain level. Substances for which the demand drops as standards of living rise are known as inferior commodities.

3.4.1 Elasticity of demand

It is important for any firm to know how responsive the demand for a product will be to a change in price. Faced with the demand curve in Fig. 3.8, a firm might do well to cut its price from p to p' since the revenue would increase from the $pRqO$ to the area $p'R'q'O$. In Fig. 3.9, however, the revenue would decrease and any cut would obviously be foolish.

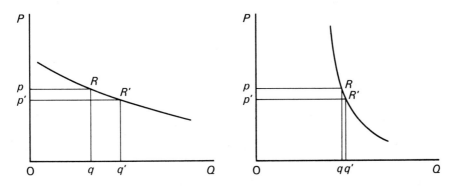

Figures 3.8 and 3.9 Changes in revenue with price.

52 Introduction to microeconomics

The way in which quantity sold responds to changes in price is called the elasticity of the demand. At any given point (q, p) on a demand curve the elasticity coefficient E is given by

$$\text{Elasticity coefficient} = \frac{\% \text{ increase in quantity sold}}{\% \text{ decrease in price}}$$

or

$$E = -\frac{p}{q}\frac{\delta q}{\delta p}$$

$\delta q/\delta p$ is, of course, the reciprocal of the gradient of the demand curve at the point (q, p). A flattish curve, as in Fig. 3.8 therefore has a high elasticity coefficient and a steep one as in Fig. 3.9 has a low one. Figure 3.10 shows four extreme demand curves. If $E > 1$, demand is said to be elastic, if $E < 1$ it is said to be inelastic and $E = 1$ is the case of 'unitary elasticity'.

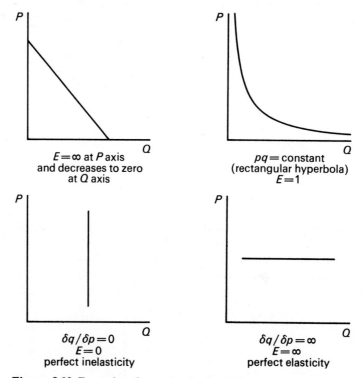

Figure 3.10 Examples of very simple elasticities.

Of course, what concerns a businessman is not the elasticity of demand but the change in his total revenue when he changes the price. A little juggling with calculus shows that the change of revenue δR for a small change of price δp is given by $\delta R/\delta p = q(1 - E)$. If $E > 1$, $\delta R/\delta p$ will be negative and a price increase will not increase revenue. If $E = 1$, $\delta R/\delta p = 0$ and revenue will be unchanged and if $E < 1$, $\delta R/\delta p$ will be positive and an increase in price will bring an increase in revenue.

Knowledge of elasticity of demand is important to a businessman wondering whether or not to change his prices, and also for example to a chancellor of the exchequer contemplating an increase of tax on cigarettes and whisky. One does not need to be a skilled economist to realize that the demand for these two commodities must be highly inelastic.

As a final point, it is important to distinguish between elasticity of demand as it faces the total population and an individual firm. If a firm has many competitors it may find zero elasticity for its own products since a small price increase will result in customers switching to competitors' products. On the other hand, a general price increase throughout the trade may result in all-round increases in revenue.

3.5 Production

We have considered at some length the problems of the consumer and how he relates his purchases to his income, his scale of preferences and the prices of available commodities. We now turn to the problems of production and consider how a producer supplies the market in order to satisfy its demands.

Production consists of using resources to obtain goods and services. This is a broad definition and includes such disparate activities as making steel from iron ore, 'farming' mink, shipping pineapple from Hawaii to London, storing wine till it matures and giving haircuts. The resources used in production are called the factors of production and may be classified into three groups – land, labour and capital.

Capital is property of all kinds – buildings, machines, tools, etc. – used in production. It includes intangible assets such as patents and trade marks. A person who owns capital is a capitalist. In centrally planned economies, there are no capitalists, but capital still exists, its ownership being vested in the state. For the use of his capital, the capitalist is rewarded either by receiving interest on the money he has lent to buy the capital goods, or by a share in the profits on the sale of what is produced.

Labour is the work expended in the act of production. It is provided not only by the shop-floor workers but also by those involved indirectly in production such as foremen or plant maintenance men. In return for their labour, people receive wages, salaries or fees. The distinction is related to types of work. Wages are usually geared to production/incentive schemes, salaries are paid to staff on fixed annual remuneration, and fees are claimed by self-employed professionals.

Land is taken to include the space in which production is carried out, and any natural resources used in production such as oil or iron ore. People who own land usually receive a rent for permitting its use in production. Land on which a factory, shop or other premises stands may have been purchased outright as a freehold in which case unrestricted future development on the site is ensured, or it may be leased from its owners for a time in return for regular payments of rent. A company owning a freehold will sometimes sell it to an institution such as a bank and lease it back again, so as to realize its capital value for further development of production facilities.

Land, labour and capital – but we should not forget the entrepreneur who starts up a business, or the organizing genius who transforms an ailing concern. The drive to succeed in business – what John Maynard Keynes described as 'animal spirits' – is scarcely recognized in economic theory but is basic to a country's prosperity. In the past, it has flourished primarily in Christian Protestant societies whose populations are presumably imbued with the so-called Protestant Ethic. Furthermore, certain groups, in certain milieux, appear particularly successful as entrepreneurs – Jews outside Israel, Armenians, Asians in East Africa, Chinese in Malaya – and are much resented for this reason. The sources of the talent are far from obvious. A large company has less need of entrepreneurial talent than a small one, since it is run on largely bureaucratic lines, but many large companies were built up by such people in the past. Our society benefits more than we like to admit from men whose aim in life, while staying within the law, is to make the odd ten million as quickly as possible. We owe more to Marks and Spencer than to Marx and Engels.

3.5.1 Diminishing returns

If we wish to increase our production of a given commodity, the most obvious way is for us to increase our input of all three factors of production. This is not always possible since, to take an agricultural example, good farming land is always in short supply. We can still increase food production, however, by increasing the amount of machinery (i.e. capital) and labour we employ on a given piece of land.

Suppose a man can produce 300 bushels of corn from a given field in a year. The advent of a second man might increase this to 1 000 bushels since each man could specialize on what he does best. A third man might increase output to 3 600 bushels, but eventually there must come a time when each extra man adds less than 300 bushels/year to the total output, a time when he does not produce enough extra corn to pay his wages and ultimately a time when he gets in the way to such an extent that his advent causes production to fall.

Table 3.1 Production schedule

Men employed	Production (bushels/year)	Increase in production per extra man employed (bushels/year) = marginal product	Average production (bushels/man/year)
1	300	300	300
2	1 100	800	550
3	2 600	1 500	867
4	4 600	2 000	1 150
5	6 900	2 300	1 380
6	9 300	2 400	1 550
7	11 600	2 300	1 657
8	13 600	2 000	1 700
9	15 100	1 500	1 678
10	15 900	800	1 590
11	16 200	300	1 473
12	16 200	0	1 350
13	16 100	− 100	1 238

We can represent what happens by a production schedule (Table 3.1) or a graph (Fig. 3.11) which represents the first two columns of Table 1 plotted against one another. The third column shows the extra production obtained by the employment of each successive man. This quantity is known as the marginal product and is shown in Fig. 3.12 plotted against number of men employed. (Cf. marginal utility section 3.3.)

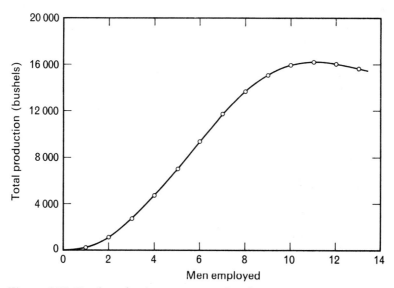

Figure 3.11 Total production *vs.* men employed.

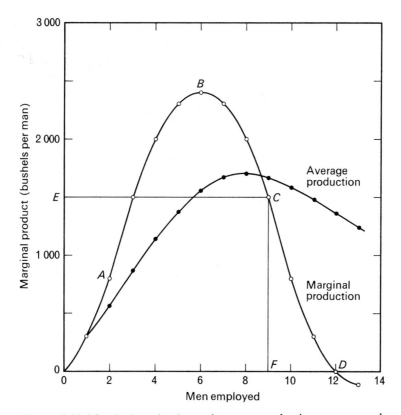

Figure 3.12 Marginal production and average production *vs.* men employed.

* In Figs. 3.11 and 3.12 a smooth curve has been drawn through the points rather than a stepped one, thus implying the possibility of employing a fraction of a man. Readers who are worried by this should multiply all figures by 1 000 and consider millions of bushels of corn produced by thousands of men.

56 *Introduction to microeconomics*

We can draw certain conclusions from Figs. 3.11 and 3.12:

1. Assuming land and capital remain constant, there are no circumstances under which it would be sensible to employ more than twelve workers. After this point, production actually declines and the marginal product becomes negative. This is in spite of the fact that the average production per man (Table 3.1, col. 4) remains high. As can be seen from Fig. 3.12 average production is not a good guide to the optimum level of the labour force.

2. If the standard wage is equivalent to 1 500 bushels corn/year, then the rational employer will go on hiring labour as long as the marginal product – the amount produced by the extra man – is greater than 1 500 bushels/year. In Fig. 3.12 this is the point at which the 1 000 bushels/year ordinate intersects the *downward* sloping section of the marginal product curve, and indicates an optimum labour force of nine. In general an employer goes on hiring labour until the marginal product (at a time when it is decreasing) becomes equal to the wage.

3. Because the curve in Fig. 3.12 is the differential of the curve in Fig. 3.11, the production if N men are employed is the area bounded by the axes, the curve and the vertical line through the point corresponding to N men. Thus if nine men are employed, total production is the area $ABCFO$. On the other hand, the wages which must be paid will correspond to 1 500 bushels $\times$ q, i.e. the area $OECF$. The difference between these two areas is the amount of corn which an employer has left over to cover his other expenses (new seed corn for next year, payments on money borrowed for purchase of equipment, rent on the field, etc.), and to give him his profit.

It should be stressed that the production schedule in Table 3.01 only applies if all factors other than labour remain constant. If the capital employed were increased, perhaps by the purchase of another combine harvester, the production schedule would change and so would the number of workers it would be sensible to employ. Furthermore, it is possible to plot graphs similar to Figs. 3.11 and 3.12 for variations in capital with land and labour held constant, and for variations in land with labour and capital constant. If the availability of the resources is also known, it should then be possible to estimate the optimum balance of resources for the enterprise concerned.

The above agricultural example illustrates the point that if we keep two factors of production constant and increase the third, the marginal product resulting from these increases will eventually go through a maximum and then decline, so that the slope of the marginal product curve becomes negative. That is to say that after a certain point, each additional man (or unit of capital or acre of land) adds less to the total output than did the previous man (unit of capital, acre of land) to be added. This idea is enshrined in economics textbooks as the law of diminishing returns. It is of great importance in situations where one or two of the factors of production is strictly limited. For example, in pastoral communities, both land and capital are in short supply, and diminishing returns soon set in as the labour force increases. The relevance of the law in a modern industrial environment is open to certain doubts some of which will be discussed in the next section.

3.5.2 Economies of scale

The downward sloping section of Fig. 3.12 represents the setting-in of diminishing returns, as the number of employees increases. The early section of the graph, however, slopes upwards and indicates that each successive employee in that region of the curve adds *more* to the total output than did the previous man. This is due to the technological fact that many enterprises can be operated more efficiently as their scale increases. It becomes possible to employ experts who specialize in particular aspects of the enterprise, and to break down complex processes into simple repetitive operations. This is the basis of mass production and of the production line. Increases of production resulting from such activities are called economies of scale, and they come into full play when commodities are being produced on a scale large enough for it to be worthwhile to set up an elaborate organization to carry out production. Economies of scale also result from the fact that cost of plant does not increase linearly with size. For example, the cost of distillation columns increases approximately with their capacity to the power two thirds because costs are related more or less to surface area while capacity is related to volume. This is called the square-cube law. The cost of instrumentation is virtually independent of plant size and so on. Overall, plant costs are approximately proportional to (capacity)$^{0.6}$.

In Fig. 3.12 economies of scale are overtaken by diminishing returns after the employment of a mere six men. This example may seem to imply that economies of scale exist only when relatively few resources are being employed. Such a conclusion is erroneous because in our example two of three factors of production are being held constant. In a

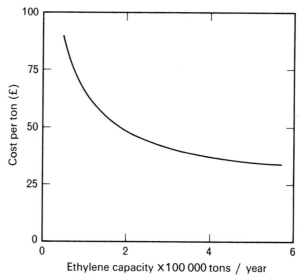

Figure 3.13 Unit production cost of ethylene obtained from the cracking of naphtha plotted against capacity. The numerical value of the unit production depends on the price of naphtha, the severity of cracking and the price at which by-products can be sold in addition to factors such as wage rates, interest rates and the cost of land.

Little significance should be attached to the actual figures above, therefore, but only to their relative values. If by-products were sold as chemicals instead of fuel, as assumed above, then a typical 1969 UK ethylene production cost from a 200 000 tons/year plant would be about £32/ton.

modern industrial society, all three factors of production can be increased simultaneously and to all intents and purposes without limit. The chemical industry requires little land for its activities and employs a small proportion of the national labour force. Almost any desired amount of capital can be raised for a sufficiently attractive project.

Figure 3.13 shows the unit production cost of ethylene made by the cracking of naphtha in plants of various sizes and illustrates the apparently endless economies of scale which are possible. While in the early 1950s a 30 000 tons/year ethylene cracker was considered large, present designs are an order of magnitude larger. Costings of most chemical industrial processes show a similar dependence on scale. The only limit to the scale on which a firm can operate a process would thus appear to be the problem of selling the output. At the end of the 1960s, however, there were signs of other limiting factors. The larger a plant, the more complicated the entire technology of building and operating it. Because it takes longer to build, the capital spent on it is tied up for a longer time before producing any return. If interest rates are high, the extra time taken can be expensive; indeed late commissioning of such plant can wreck the entire economic basis on which it was designed.

The economic consequences of breakdown in very large plants are also more serious, particularly as the majority of chemical plants are built on a 'single-train' philosophy in which the possibility of breakdown is countered by careful design of supposedly reliable components rather than by the building-in of additional equipment which can be phased in in an emergency. Many experts now consider that the economies of scale in ethylene production no longer outweigh the above drawbacks and that no ethylene crackers with a capacity greater than 500 000 tons/year will ever be built. These predictions would carry more weight were it not that similar expert forecasts in years gone by suggested that much lower capacities represented the end of the road for economies of scale in ethylene production.

Whatever the outcome of the above debate, we can conclude that while the law of diminishing returns is conceptually important in understanding economic theory, it is of little practical relevance to the chemical industry where economies of scale are all-important.

3.5.3 Alternative production possibilities

Much of the discussion in the previous section hinged on one or all of the factors of production being in limited supply. The availability of such resources depends on how far the resources of a community are already being fully used and how readily they can be transferred from one production job to another. To transfer an engineer in a polyethylene plant to a polypropylene plant is simple; to transfer him to the electronics industry would be much more difficult.

To discuss the allocation of resources between the countless productive activities in society would be very complicated. Let us suppose therefore that only two goods, bread and wine, are being produced. If we devote all our resources to bread, let us suppose we can produce 10 m. loaves; if we devote them all to wine we can produce 5 m. bottles. There are then intermediate possibilities: we can have *some* bread and *some* wine. Indeed, this has obvious advantages; land more suitable for wheat than for vines can be used for bread production and vice versa. If we were to estimate the amounts of bread and wine

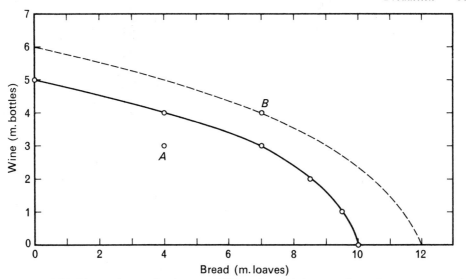

Figure 3.14 Alternative production possibilities – Bread/wine transformation curve.

we could produce using our total resources we would get a production possibility schedule – Table 3.2 – which can also be represented as a graph – Fig. 3.14. The graph is called a production possibility or 'transformation' curve, because we can think of bread being 'transformed' by switching of resources into wine. Any point on the curve represents total employment of resources. On the other hand, if we produce, say, 4 m. loaves and 3 m. bottles, this is represented by point A which lies inside the curve and shows that resources are being under-used; for example there might be some unemployment. A

Table 3.2 Alternative production possibilities

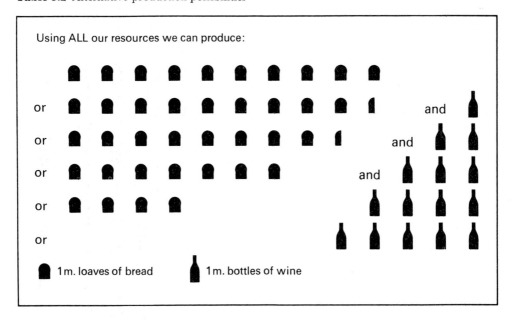

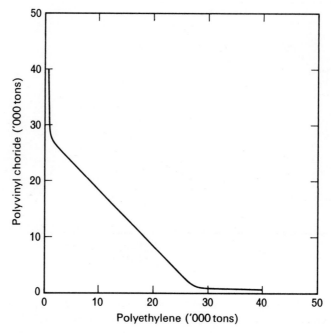

Figure 3.15 Polyethylene/polyvinyl chloride transformation curve.

point such as *B* outside the curve represents production which we cannot attain without increasing our resources in some way. If we do this, the production possibility curve itself will shift outwards to a new position, (represented by the dotted line) appropriate to the new resources.

The convex shape of the bread/wine transformation curve is a consequence of the fact that bakers and wheatfields cannot be transformed into vintners and vineyards with complete efficiency. The same sort of curve would be obtained if we devoted our total resources to the production of ethylene *or* propylene which are inevitably obtained together from the cracking of naphtha. On the other hand, if we were dividing fixed total resources between polyethylene and polyvinyl chloride production in a poor country with a relatively small population, economies of scale would probably make it worthwhile for us to make only one or the other, so that bits of the transformation curve would almost coincide with the axes of the graph. Once both materials are being produced, however, switching of resources from one to the other is fairly easy and the transformation curve will be more or less linear – Fig. 3.15. It is possible that the shape of Fig. 3.15 is more typical of an industrial situation that Fig. 3.14 and this could well explain the diminishing choice available to the consumer in many areas.

3.6 Supply

In the same way that we can draw a demand curve which relates the quantity of goods which people will buy at any given price, we can also construct an exactly analogous curve showing the quantity of goods which people are willing to produce and sell at any

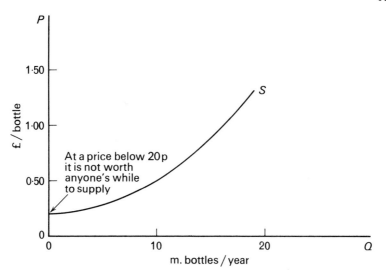

Figure 3.16 Supply curve for wine.

given price. The higher the price of a particular article, the more willing people will be to switch resources to its production, so that in general supply curves will have a positive gradient. They will also have a positive intercept on the price axis since there must be a price below which production will simply not be worthwhile under any conditions (Fig. 3.16).

The axes of supply and demand curves are identical so we can superimpose Figs. 3.16 and 3.7 to give Fig. 3.17. We can see that if the price of wine is £1 per bottle suppliers will set themselves up to produce 16 m. bottles, but consumers will be prepared to buy only 3·5 m. of them. Thus stocks will build up and occupy cellar space which has to be

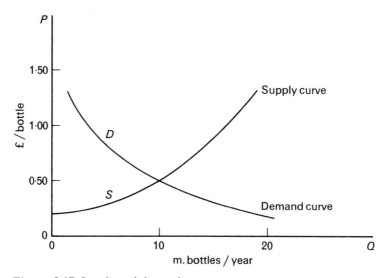

Figure 3.17 Supply and demand.

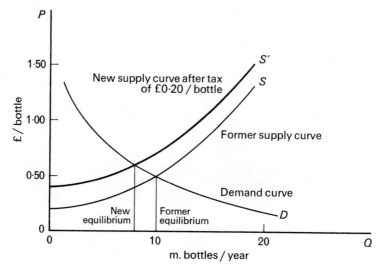

Figure 3.18 Effect of taxation.

paid for. Meanwhile money to pay for production will not materialize and a hard-pressed producer is going to offer his wine at a lower price. Ultimately the other producers will be forced to follow suit; the price will fall and consumption will increase.

Conversely if the price is £0·30 per bottle, customers will be prepared to buy 15 m. bottles but only 6 m. will be available. Wine will go 'under the counter', there will be long waiting lists, and producers will find they can charge higher prices and still sell their total output.

The point of equilibrium is clearly where the two curves cross – in this case the 10 m. bottles at £0·50 point – and, as long as other factors remain constant, the price can re-

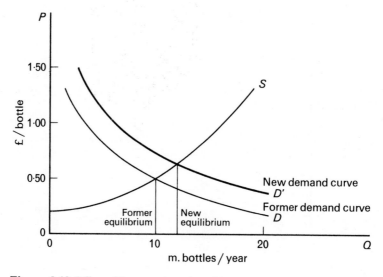

Figure 3.19 Effect of increase in price of beer.

main steady at this level indefinitely. All the laws of supply and demand reduce to this idea – that the equilibrium price of an article is the price at which supply and demand are equal.

The effect of changes in other conditions are shown in Figs. 3.18 and 3.19. In Fig. 3.18 the government has put a tax of 20p per bottle on wine. This moves the supply curve 20p upwards to a parallel curve S' and results in equilibrium being reached at a price of 60p (of which 40p goes to the producer and 20p to the tax collector) and a consumption of 8 m. bottles.

In Fig. 3.19 the price of beer has risen so it competes less effectively with wine and a new demand curve D' is established. Equilibrium is reached at 65p and 12 m. bottles. Occasionally, in an emergency a government decides it must hold prices down and either establishes price controls or rationing. Suppose it wishes to hold the price of wine at 25p per bottle. This will result in a measure of unsatisfied demand and the possible emergence of a black market. This could be countered either by a subsidy to the wine growers to persuade them to make more wine (i.e. the opposite of Fig. 3.18) or by severe penalties for black marketeers. These would aim to take away so much of the extra money which they make by selling wine above the legal price that their return on investment would be lower than that obtained by an honest wine merchant.

3.6.1 Long-term balance of supply and demand

The above discussion assumes naively that an increase in price will be met immediately by an increase in production. In fact it might take several years for producers to adjust themselves to increased demand. Economists distinguish three cases, which merge into one another:

1. Short-run equilibrium when supply is fixed.
2. Intermediate run equilibrium when expedients are used to increase production, i.e. by an increase in the output of existing plant.
3. Long-run equilibrium when firms have had a chance to enter and leave an industry, build new plant, etc.

We can demonstrate this by considering further the supply and demand for wine. Suppose that because of a long-run increase in prosperity and in population the demand for wine rises steadily. Each year a new demand curve will be established to the right of that of the previous year. These are shown in Fig. 3.20. The short run supply curve S_S is vertical as supply is fixed. The intermediate run supply curve for 1970 is S_I and the long term is S_L.

By 1972, demand has risen but the wine industry has not yet realized it, and supply has remained constant, so the price has risen to B and the same quantity of wine has been sold. In 1972 the industry realizes demand is rising. Unfortunately it takes about six years from the planting of a vine for it to yield drinkable produce, so nothing the growers can do in terms of planting new vines and increasing acreage will have any effect till 1978. Meanwhile, they can introduce intermediate-term expedients to increase output, such as diverting grapes from the fruit market, using better fertilizers, and employing more labour. By 1974 these have had their effect and equilibrium is now reached on the intermediate-run supply curve at C. The intermediate-term measures have more or less stabilized the higher price which was reached two years before.

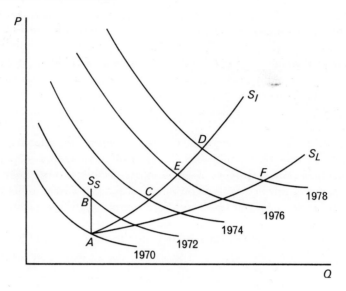

Figure 3.20 Long-term balance of supply and demand.

By 1976, the new vineyards are maturing nicely but excess of demand over supply is pushing up the price and equilibrium is reached at E. By 1978 the new vines are producing wine. The quantity sold increases sharply and the price drops but, assuming the vintners have calculated correctly, the decrease in profit per bottle will be more than outweighed by the increase in the number of bottles sold. This is not necessarily so, and a situation where increased production does not lead to increased profits is said to be one of 'over-production'. This is the sort of situation in which grapes are left to rot on the vine, and cabinet ministers are pelted with rotten surplus vegetables as a protest against farm policy.

It is also a situation which occurs frequently in the chemical industry where economies of scale tempt producers to build giant plants, and trust to luck to sell the products. In 1958, UK low density polyethylene capacity (72 000 tons) matched production (68 200 tons) at a polyethylene price level of £285/ton. By 1960 capacity had virtually doubled to 135 000 tons with production limping behind at 105 400 tons in spite of a £35/ton price cut to £250/ton. It took a swingeing price cut to £172·5/ton in 1962 before the overcapacity was absorbed by increased consumption (1962 capacity 170 000 tons, production 155 300). Even if we allow for a drop in the price of ethylene and economies of scale in polyethylene manufacture, it is difficult to see how producers could have made higher profits in 1962 than in 1958.

3.7 Competition

The demand curve which faces an individual producer of a commodity is not the same as the demand curve for all producers. It might well be that users of polyethylene would still buy half of their present purchases of polyethylene if the general price increased by £20/ton, but if a single firm were to increase its prices by this amount, its sales might drop to zero. Other manufacturers might well have sufficient spare capacity to satisfy the total demand at the lower price.

On the other hand, if the single firm had a complete monopoly of polyethylene production, then by increasing its prices its sales would indeed halve because as a monopoly producer its demand curve would be identical with the national demand curve. Between these two extremes is a number of intermediate cases where a firm can increase its prices and lose some but not all of its sales. The extent to which the national demand curve is also the demand curve of the individual firm is a measure of the degree of competition in the industry in question. We can visualize two extreme cases and one intermediate one. Perfect competition is the situation where a small increase in price by an individual firm causes it to lose all its market; monopolistic competition is where the firm loses a part of its market and complete monopoly is where the firm retains 100% of the market though of course the size of the total market might decrease. Some of the characteristics of these forms of competition are given in Table 3.3.

Table 3.3

	Pure (perfect) competition	*Monopolistic or imperfect competition*	*Complete monopoly*
Examples	Agriculture	Most firms and industries	GPO Atomic Energy Authority Int. Computers Ltd. (but they face competition from abroad)
Control by individual firm over prices	None	Some, depending on similarity of substitute products and on number and location of rivals	Limited only by fear of government intervention and possibility of substitute products
Size of firm	Each firm accounts for only a small % of output and this could easily be produced by others	Each firm has substantial slice of market. Withdrawal of one would reduce available supply	One firm supplies whole market
Selling	Organized marketing system or auction. No advertising	Advertising but little price rivalry	Public relations and informative advertising
Demand curve for individual firm	Infinitely elastic	Rise in price will reduce sales	= national demand curve

The chemical industry does not fit neatly into the above classification. On the one hand the major producers such as ICI, BP Chemicals, Shell Chemicals and Monsanto are very large and one would therefore expect the industry to be close to a state of monopoly. On the other hand, examination of firms' balance sheets and of the profits they are making suggests that the industry is not far from a state of perfect competition. There are two

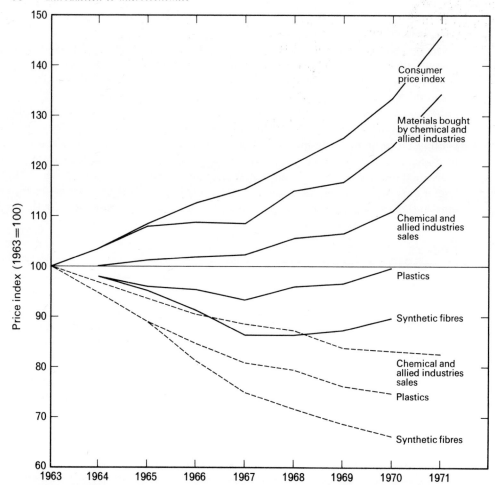

Figure 3.21 Prices in the UK chemical industry 1963–71. The chemical industry being international, indices of prices in other countries show very similar trends except for the spurt in prices in 1967 due to devaluation. Figures for chemical and allied industries are for home sales only. Inclusion of exports raises the index by less than 1%. Full lines show actual prices; dotted lines show prices based on constant value of the £ (=actual price ÷ consumer price index).

related reasons for this anomaly. The first is that the chemical industry operates on a world scale. Many chemical firms spread their investments over a range of countries and all of them look to world markets rather than national markets for their products. Thus a near-monopolist in the UK must still face intense international competition. Second, there appears to be endemic overcapacity throughout the world for certain major chemicals. Economies of scale have become so important that many firms prefer to build plants twice as large as they think they will need in the hope that market expansion over and above what they have banked on and an increase in their own share of the market will eventually take up the slack. At a time of currency inflation, it is very tempting to say, 'Let's build it this year; it will cost us a lot more if we leave it for a year'. In short, producers would rather be faced with over-capacity than under-capacity and over-capacity is consequently what faces them. In addition, many under-developed countries are trying to industrialize and some of these are building their own chemical industries partly in order to reduce the need for chemical imports and partly for reasons of national prestige. This trend adds to the prevailing over-capacity, and when economies of scale and advances in technology are taken into account it is not surprising that the past 20 years have seen an almost uninterrupted drop in the real prices of basic chemicals. Figure 3.21 shows the change in price indices in the chemical industry from 1963.

Returning to the factors mentioned in Table 3.3, we can say that a firm like ICI has some control over prices nationally but none over world prices, and as it exports about 50% of its production it cannot be said to be in a monopolistic situation. Similarly, though it is a giant firm even by world standards, its share of the world market is relatively small and could be made up quite easily by other producers – though not in the UK alone. Advertising is also a guide to the type of competition under which a firm operates, but the chemical industry proper does very little advertising. This is consistent with a state of near-perfect competition, but on the other hand what advertising there is tends to be of the public relations sort which is normally associated with monopolies or near monopolies.

A fair conclusion seems to be that the chemical industry operates internationally under conditions of almost perfect competition, but that the firms involved in this are so large in national terms that they have some of the characteristics, though rarely the profits, of monopolists.

3.8 International trade

In addition to the transactions between individual members of a society, trade also occurs between different societies, that is, across international boundaries. In the same way that individuals specialize in one job and sell their surplus production to fill their other needs, nations can adopt an international division of labour and trade the surpluses of their specialized activities. For example, the UK can make Scotch whisky which it can exchange for French champagne. Though at least one US president has quaffed American rather than French wines on ideological grounds, there can be little doubt that a world drinking British champagne and French whisky would be worse off than our present one.

Trade between nations is thus as logical and desirable as trade between individuals, and there would seem to be no reason for it to be hindered by customs duties and the like. Unfortunately, nations approach it with a selection of bizarre and irrational attitudes. One of these is the desire for a perpetual positive balance of trade. For all governments wish to send out of their country goods of a higher value than they bring in. A moment's reflection shows that this is not only eccentric but impossible. The total positive trade balances of the nations in surplus must be exactly equal to the negative balances of nations in deficit. Short of goods being tipped into the Marianas Trench, there is no way in which all nations can export more than they import. Nonetheless, they like to try, and most governments apply tariffs to deter goods from being brought into a country, and offer incentives to exporters.

Governments also impose tariff barriers in response to political pressures at home. An inefficient industry, rather than putting its house in order, normally tries to get the government to impose tariff barriers so that the home producers can charge higher prices without fear of foreign competition. Governments, worried about possible unemployment if the industry goes to the wall, tend to give in to this pressure and though the industry stays in being, its products are more expensive than they need be and, consequently, efficient firms buying such goods find it more difficult to compete.

Tariffs are also imposed for strategic reasons since nations like to feel that they possess sufficient basic industries to allow them to continue if blockaded. Governments setting up new industries will often give them tariff protection to save them from foreign competition during their infancy. Such tariffs are to some extent defensible but all too often they continue into an industry's middle age and even act as a prop in its declining years.

A further barrier to international trade is the vague feeling that foreigners do not play fair. According to spokesmen for the UK shipbuilding industry, for example, their industry cannot compete with the Japanese because the latter allegedly pay low wages. On the other hand, the Swedes pay very high wages and UK shipbuilders cannot compete with *them* because they have so much modern equipment and are so productive.

3.8.1 Comparative advantage

A feeling related to the above which is prevalent in countries of middling wealth like the UK is that it is no use our trying to sell to poor countries because they cannot afford our prices and no use buying from rich countries such as the USA because all their goods are so expensive.

This is a fallacy. International trade is worthwhile even between countries of disparate standards of living so long as they produce different goods with different *relative* efficiencies. This is known as the law of comparative advantage.

Suppose for example that Spain, with low labour costs, can produce oranges at £100/ton and washing machines at £50 each (a cost ratio of 2:1), while Germany, being more industrialized and colder, can produce oranges at £400 per ton and washing machines at £80 each (a cost ratio of 5:1). Germany is thus *relatively* better at producing washing machines.

At first sight it would appear that Spain would be foolish to trade with Germany as she can produce both commodities at a lower price. Consider, however, Germany investing £2 m. in production: she could produce 5 000 tons of oranges, or 25 000 washing machines or, to take an intermediate case, 3 000 tons of oranges and 10 000 washing machines, this balance approximately satisfying her market.

Similarly, Spain, with a smaller budget of £1 m., could produce 10 000 tons of oranges or 20 000 washing machines, or perhaps 5 000 tons of oranges and 10 000 washing machines.

If Spain and Germany were to trade oranges for washing machines on the basis of 1 ton of oranges for 3 washing machines (a ratio of 3:1), then Germany could specialize and manufacture only washing machines. Of the 25 000 machines produced, 15 000 could be exported in exchange for 5 000 tons of oranges, and 10 000 retained for the home market. Germany would thus be 3 000 tons of oranges better off than if she had tried to be self-sufficient.

Similarly, Spain would exchange 5 000 tons of her total orange crop of 10 000 tons and in exchange receive the 15 000 washing machines thus finishing 5 000 washing machines better off than in the absence of trade.

This example oversimplifies in that it does not allow for transport costs and assumes the possibility of substituting oranges for washing machines and vice versa. It also treats world trade as if it were bilateral rather than multilateral. There is no need for a country to sell to the country from which it buys so long as trade flows smoothly. A classical though deplorable example is the eighteenth century shipping of cotton goods to Africa where they were exchanged for blacks; the shipping of blacks to America for use as slaves, payment being made in raw cotton, and the shipping of the cotton back to Liverpool or Bristol for conversion into cotton goods. It is nonetheless true to say that if countries manufacture two commodities with different relative efficiencies, it will pay each of them to specialize in what it does best and to exchange its surplus for its requirements of the other commodity at a ratio of prices somewhere in between the cost ratios obtaining in the two countries. Wage rates, absolute production costs, etc., are irrelevant; all that matters is relative costs.

The imposition of tariffs distorts this picture by inhibiting trade. Countries divert resources into areas where they lack comparative advantage and, as a consequence, are mutually impoverished. Rational countries would each concentrate on what they did best and would exchange their surpluses freely. In the past Great Britain grew rich on free trade, but few politicians support it nowadays since the votes of workers in declining and inefficient industries are so important. Thus is classical Marxism confounded in that politics is supreme and economics disregarded.

3.8.2 Import duties on chemicals

Tariffs are applied to shipment of chemicals by almost all countries. Trade is nonetheless free inside free trade areas such as the European Free Trade Association and the European Economic Community. The tariff schedules are extremely complicated and often apply to individual chemicals rather than broad groups. The approximate levels of tariffs imposed by the USA, the UK and the EEC which came into effect on 1 January 1972 are shown in Table 3.4. The existence of two sets of figures dependent on whether or not the ASP (American Selling Price) system is abolished calls for further comment.

Since 1967 there has been a worldwide reduction in import duties on chemicals as a result of the Kennedy rounds. The full reductions negotiated, however, were made conditional by the UK and EEC on the abolition of the ASP system and as long as it remains in force only 40% of the agreed reductions will take place. Under the ASP system, tariffs are levied on the price at which American manufacturers sell the material being imported. This figure may be an unrealistic 'list' price or may even be entirely notional and the resulting tariff is rarely under 50%. The system is, not surprisingly, supported by the powerful US chemical lobby and whether or not it will survive (tariffs in column A) or be abolished (column B) is an open question.

Between 1973 and 1977 the UK will have to adopt the common external tariff of the EEC and join in free trade within the community. There will be a 40% move on 1 January 1974 and three subsequent shifts of 20% in the three succeeding years, by which time the schedules will have been harmonized. In general EEC tariff levels are lower than UK levels, and UK accession to the Common Market will make exporting into the UK easier both for countries inside and outside the free trade area.

3.8.3 Chemical exports and imports

In spite of these restrictions, trade in chemicals is truly international. Until the end of 1971 Norway for example made chlorine by electrolysis using cheap hydro-electric power, shipped it to Ireland where it was converted into vinyl chloride, and then re-imported the monomer for the production of PVC. Developed countries as a whole export a sizeable proportion of their chemical production. The figures for 1969 are given in the final column of Table 3.5. The Low Countries and Switzerland export a particularly high proportion of their chemical output. The UK chemical industry exported some 21% of its turnover in 1969 compared to 13·3% for manufacturing industry as a whole. ICI export about half their UK production and are responsible for the vast majority of British chemical exports, being the only British firm large enough to have an effective international selling organization.

Table 3.4 Import duties after Kennedy round (as percentages of value)*

	USA*		UK		EEC	
	A	B	A	B	A	B
Inorganic chemicals	0–(5)–14	0–(5)–14	8–23	5–12½	2½–13½	1½–11
Organic chemicals	9–(20)–27	9–20	0–23	0–12½	2½–18½	1½–12½
Plastics	12	12	10–(16)–23	6½–10	7–(16)–18½	5½–(10)–11½
Dyes	17–49	30	8–(10)–16	5–(10)–16	7–(10)–17	4½–(10)–13½
Paints	4–7½	4–7½	10½–12½	7½	12	7½
Drugs	9–(23)–53	9–(20)–27	7–23	5–12½	8–17½	6–12
Synthetic rubber	3	3	4	4	0–5	0–5
Synthetic fibres	7½–25	7½–25	13	13	9	9
Fertilizers	0	0	0–(12½)–23	0–(7½)–12½	0–8	0–6
Pesticides	15–24	15–20	8	5	6½–9½	4–6
Soaps and detergents	3–15	3–11	10–20	10–12½	12	7½
Photographic chemicals	5–30	5–14–20	8–16	7½	9½	6

A. ASP not abolished.   B. Conditional on abolition of ASP.
Bracketed figures show rates which apply to a very wide range of products.
* In the US the value taken is the f.o.b. price. In other cases it is the c.i.f. In practice figures given in the table should therefore be lowered in the US since freight and insurance add up to 20% of prices.

In order to sell abroad, it is necessary for a company to match the prices prevailing in the foreign market. To a first approximation, production costs are similar in all developed countries so that in general a company selling over a 20% tariff barrier will receive 20% less money than it would if it sold at the same price in the home market. It is thus unlikely to make the same profit from exporting, and it will usually be satisfied if it has something left over after covering its marginal costs. Exporting chemicals is usually a way of keeping one's plant operating at full capacity. A corollary of this is that a company will have to make its main profits in the home market and it will do this by pushing up its home prices knowing that it is protected by tariffs from foreign competition. If the gap between home and foreign selling prices becomes too large however, a company risks being accused of dumping and all countries have strict anti-dumping legislation.

The exports and imports of chemicals in 1969 between the main OECD countries (section 6.2) are shown in Table 3.5 together with Russian data in Table 3.6. The countries with a favourable balance of trade in chemicals are: Belgium/Luxembourg, France, Netherlands, United Kingdom, West Germany, USA and Japan. Net importers are Italy, Spain, Sweden, Switzerland, Canada and USSR. Of the OECD countries not specifically mentioned, all except Norway are net importers of chemicals.

There is a great deal of intra-European trade in chemicals. The USA exports mainly to Canada, Japan and Europe but imports relatively little except from Canada. The coun-

Table 3.5 OECD exports and imports of chemicals 1969 (figures in $ m.)

Exports to	European member countries of OECD	North America	Japan	Total to OECD countries	Other countries	Grand total	Exports as % of turnover
EXPORTERS							
Belgium–Luxemburg	668	20	9	697	112	809	61
France	977	81	28	1 086	503	1 589	23
Italy	447	42	13	502	318	820	15
Netherlands	986	33	15	1 034	263	1 297	56
Spain	57	10	2	69	39	108	6
Sweden	184	10	4	198	37	235	26
Switzerland	556	73	46	675	311	430	82
United Kingdom	724	113	46	883	761	1 644	21
West Germany	2 287	190	110	2 587	1 003	3 590	36
Other European OECD countries§	454	47	6	507	206	713	—
European total	7 340	619	279	8 238	3 553	11 791	30‡
Canada	81	300	17	398	58	456	21†
USA	1 229	511	304	2 044	1 339	3 383	8
Japan	142	137	—	279	737	1 016	10

Imports from	European member countries of OECD	North America	Japan	Total from OECD countries	Other countries	Grand total
IMPORTERS						
Belgium–Luxemburg	604	97	4	705	20	725
France	1 080	187	14	1 281	68	1 349
Italy	793	121	10	924	51	975
Netherlands	707	141	9	857	60	917
Spain	340	82	12	434	26	460
Sweden	428	54	5	487	35	522
Switzerland	451	58	8	517	28	545
United Kingdom	658	288	23	969	142	1 111
West Germany	1 125	301	31	1 457	103	1 560
Other European OECD countries§	1 396	144	12	1 552	159	1 711
European total	7 582	1 473	128	9 183	692	9 875
Canada	140	542	8	690	13	703
USA	535	310	121	966	266	1 232
Japan	327	361	—	688	95	783

† 1967 figure.
‡ Average figure for OECD European members.
§ Austria, Denmark, Finland, Greece, Ireland, Iceland, Norway, Portugal, Turkey.

try with the most impressive balance of trade is West Germany whose exports of chemicals exceed those of the USA.

Classical economic theory pictures countries with developed industries exporting manufactured goods to less developed countries and in return importing raw materials from them. It is apparent from the data in Table 3.5 however, that OECD countries trade mainly among themselves. The European members export $8 238 m. to OECD members and $3 553 m. to the rest of the world. The USA and Japan redress the balance

Table 3.6 USSR international trade in chemicals (includes dyestuffs/paints, explosives, photographic materials, agricultural chemicals and rubber/asbestos). Figures are for 1969 and are in millions of roubles. For purposes of comparison, 1 rouble ≈ $1.

	W. European Countries	North America	Japan	OECD Total	Other	Grand Total
USSR Exports	58·8	0·2	8·5	67·5	272·4*	339·9
USSR Imports	145·7	6·9	19·7	172·3	297·3†	469·6

* Including 169·6 m. to Eastern Europe, 43·2 m. to Cuba and 13·6 m. to communist countries of Asia.
† Including 154·6 m. from Eastern Europe and 107·4 m. imports of natural rubber from S.E. Asia.

slightly, bringing the figures for total OECD exports to $10 959 m. and $5 687 m. respectively. The import statistics show the disparity between the charmed circle of OECD members and the rest of the world even more clearly. OECD imports from other members total $11 527 m. and from non-members a mere $1 066 m. Non-OECD countries, by and large, lack the ability to manufacture and market sophisticated chemicals to the standards necessary to penetrate OECD markets. The comparative advantage undoubtedly lies with OECD countries and non-OECD countries sell other commodities to buy OECD's chemicals. Within OECD, the seven net exporters may be thought of as having a comparative advantage and these are, not unexpectedly, the most highly developed industrial countries in the group.

Notes and references

General There are many excellent introductions to microeconomics. In this and the following chapter, we have drawn heavily on P. A. Samuelson, *Economics*, 8th edn. McGraw-Hill, New York, 1970. Another important basic text is R. G. Lipsey, *An Introduction to Positive Economics*, 3rd edn., Weidenfeld and Nicolson, London, 1971, and there is a useful paper-back, C. F. Carter, *The Science of Wealth*, Arnold, London, 1967. For self-tuition, our students speak highly of K. Lumsden, R. Attiyeh and G. L. Bach, *Microeconomics – a Programmed Book*, Prentice-Hall, New Jersey, 1966.
Section 3.5.2 Figure 3.13 comes from J. P. and E. S. Stern, *Petrochemicals Today*, Arnold, London, 1971. See also A. F. Orlicek, *Large-size Ethylene Plants*, ECMRA Budapest, 1970.
Section 3.7 Figure 3.21 is based on data from the *Annual Abstract* and *Monthly Digests of Statistics*, HMSO, London.
Section 3.8.2 The tariff situation for chemicals was discussed by Dr A. Henfrey in a Stirling *et al.* investment seminar, Paris, February 1972. Table 3.4, a miracle of compression, is taken from his paper.
Section 3.8.3 Tables 3.5 and 3.6 are derived from the OECD *Annual Report 1969–70* and G. Hemy, *The Soviet Chemical Industry*, Leonard Hill, London, 1971.

4 The decision to produce

Having developed these various concepts of supply and demand we are now in a position to consider the factors that face a firm when trying to decide whether or not to enter the market and produce a given commodity. Some of these factors are of a strategic nature and we deal with them here. Others are more technical and are discussed in section 5.7. It would be untrue to say that all the factors are always taken into account or that they can all be calculated with any precision, but any businessman contemplating production must weigh them up to some extent, even if he performs the operation in a largely intuitive fashion.

4.1 The demand schedule

A manufacturer must first of all establish some sort of demand schedule for his product both nationally and in terms of his own prospects. If he faces perfect competition, this is simple. He must be able to sell at the prevailing market price. If he is likely to have a monopoly by reason perhaps of a patent then his demand curve will be the same as the national demand curve which, if his product has not been on the market before, will have to be deduced from the prices and demands for the various alternative products. It may thus be necessary for him to offer pilot plant samples of the material around the trade to try to establish some sort of opinion on its value.

A manufacturer offering any product will, of course, charge what he thinks the market will bear. This is unrelated to his costs except that if he is not going to make a profit he will not enter the market. On the whole, a manufacturer will pitch his initial price fairly high to test the market as he will then gain goodwill by subsequently reducing it. Buyers who are tempted to install new equipment to process a cheap new material do not like to see its price raised after they have committed their capital.

If a state of imperfect competition exists, the potential manufacturer will have some idea of what to charge for his product because he knows the prevailing market price, but this only gives him a maximum price because his entering the market may bring down prices generally by increasing the supply. He may feel that undercutting other manufacturers will enable him to gain a large share of the market but he would probably only do this if he held some manufacturing advantage which would enable him to sell at prices which would not allow his competitors to make a profit. Other things being equal, a long-established manufacturer has an advantage because he is actively selling and has partly amortized his plant so that he can stand a price war better than a new entrant to the market. Indeed he may very well promote one in order to force the 'new boy' out of the market altogether.

An assessment of these factors should lead to some sort of approximate demand schedule. Supposing conditions of imperfect competition, this might look like Table 4.1

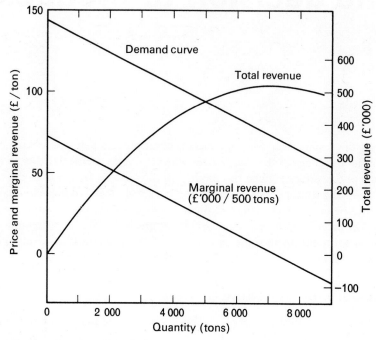

Figure 4.1 Demand, revenue and marginal revenue.

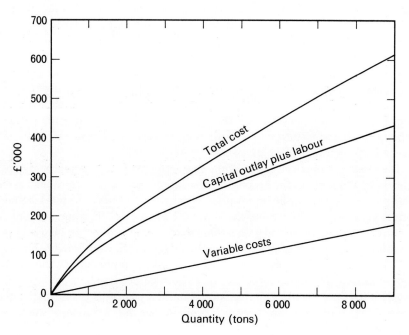

Figure 4.2 Total, capital outlay plus labour, and variable costs.

Profits 75

Table 4.1 Calculation of optimum capacity when plant has not been erected

1 Quantity (tons) q	2 Price (£/ton) p	3 Revenue (£'000) $R = p \times q$	4 Capital + labour costs for plant of designated capacity (£'000) FC	5 Variable cost (£'000) VC	6 Total cost (£'000) $TC = FC + VC$	7 Profit (£'000) Pro	8 Marginal revenue (£'000/500 tons) MR	9 Marginal cost (£'000/500 tons) MC	10 Marginal profit (= marginal revenue − marginal cost) (£'000/500 tons) $MPro = MR - MC$	11 Average cost per ton (£) $AC = \dfrac{TC}{q}$
0	144	0	0	0	0	0				—
500	139	69.5	63	10	73	−3.5	69.5	73	−3.5	146
1 000	134	134	100	20	120	14	64.5	47	17.5	120
1 500	129	193.5	131	30	161	32.5	59.5	41	18.5	107
2 000	124	248	159	40	199	49	54.5	38	16.5	100
2 500	119	297.5	184	50	234	63.5	49.5	35	14.5	94
3 000	114	342	208	60	268	74	44.5	34	10.5	89
3 500	109	381.5	231	70	301	80.5	39.5	33	6.5	86
4 000	104	416	252	80	332	84	34.5	31	3.5	83
4 500	99	445.5	272	90	362	83.5	29.5	30	−0.5	80
5 000	94	470	292	100	392	78	24.5	30	−5.5	78
5 500	89	489.5	312	110	422	67.5	19.5	30	−10.5	77
6 000	84	504	330	120	450	54	14.5	28	−13.5	75
6 500	79	513.5	348	130	478	35.5	9.5	28	−18.5	74
7 000	74	518	366	140	506	12	4.5	28	−23.5	72
7 500	69	517.5	383	150	533	−15.5	−0.5	27	−27.5	71
8 000	64	512	400	160	560	−48	−5.5	27	−32.5	70
8 500	59	501.5	417	170	587	−85.5	−10.5	27	−37.5	69
9 000	54	486	433	180	613	−127	−15.5	26	−41.5	68

76   The decision to produce

columns 1 and 2 for the manufacturer in question. It implies that he could sell 1 000 tons/year of the material at £134/ton but that if the price dropped to £94/ton, 5 000 tons could be sold. Column 3 shows the revenue (price × quantity sold) he would receive at each of the various levels of production. It will be seen that rather than the revenue increasing steadily with sales, it goes through a maximum at 7 000 tons. Even if the manufacturer could produce his product at zero cost, he would be unwise to reduce his revenue by 'swamping' the market with his product. He would, in fact produce about 7 000 tons. Column 8 shows the marginal revenue which is taken as the *extra* revenue acquired by sale of each successive 500 tons of product. These three quantities are shown in Fig. 4.1. It will be seen that marginal revenue = 0 at the point of maximum revenue, as would be expected from calculus considerations.

4.2   Costs and marginal costs

Having investigated his market, the potential manufacturer estimates his costs. These must include all costs of production such as amortization of plant, land and labour costs, and raw material costs. Column 6 of Table 4.1 shows total costs at various outputs, which are made up of capital and labour costs (column 4) and raw material costs (column 5). The latter increase linearly with scale, but the former (typically in the chemical industry) are taken to increase as $(\text{capacity})^{2/3}$. (See Fig. 4.2.)

Column 11 gives the average cost per ton, which can be thought of as the unit cost. It will be seen that the average cost is high at low outputs but decreases as the output increases. This represents economies of scale. Column 9 shows the marginal cost which is taken as the additional cost of each successive 500 tons. The marginal cost drops rapidly at first, then more slowly. If economies of scale were overtaken by 'diminishing returns' it would start to rise again, but as has been noted, this happens rarely in the chemical industry. Columns 9 and 11 are shown graphically in Fig. 4.3.

4.3   Profits

The profit is obtained by subtracting the total costs from the total revenue and is shown in column 7 and graphically in Fig. 4.4. Like the revenue, it goes through a maximum, but at a somewhat lower tonnage because the costs which are subtracted rise all the time with output. Column 10 shows the marginal profit, which is the difference between the marginal revenue and the marginal price. In the same way that revenue reaches a maximum when marginal revenue = 0, profits reach a maximum when marginal profit = 0. This of course is the maximum which the manufacturer is seeking and where he will try to operate. In this case it will occur at an output a fraction above 4 000 tons per year and a price of just under £104/ton.

The various marginal quantities are shown in Fig. 4.3. The position of a maximum profit corresponds on the graphs to the output at which:

1. The profit curve goes through a maximum.
2. The total revenue and total cost curves have identical slopes.
3. The marginal profit = 0.
4. The marginal cost and marginal revenue lines intersect.

The size of the optimum profit can be estimated graphically by drawing a vertical line

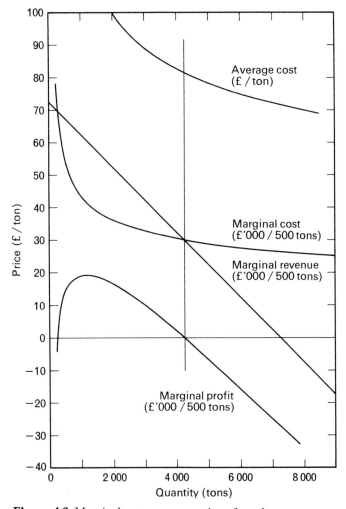

Figure 4.3 Marginal cost, revenue and profit and average cost.

through the optimum output and then drawing horizontal lines through the points where it intersects the average cost and demand curves. The intersection with the demand curve is the optimum price per unit, so the difference between this and the average cost will represent the unit profit. The area of the shaded rectangle formed by the various lines will then represent the optimum total profit which can be made (Fig. 4.5).

The decision to produce would be based on whether or not the optimum profit is satisfactory on the basis on which the industry works. In this case, the profit is about 20% of revenue. It is not possible to calculate the return on capital from the data given, because the proportion of column 4 which corresponds to capital investment is not known. The average return on capital in British industry in 1969 was 15% and in view of the uncertainties involved, a manufacturer would be hardly likely to enter a new field unless he

78 *The decision to produce*

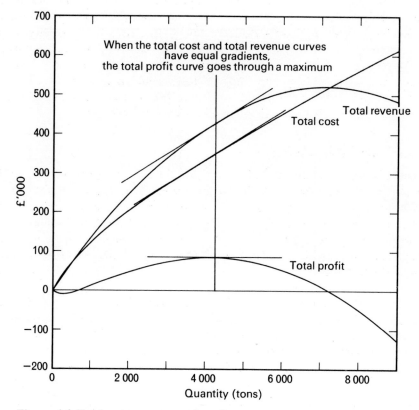

Figure 4.4 Total costs, revenue and profits.

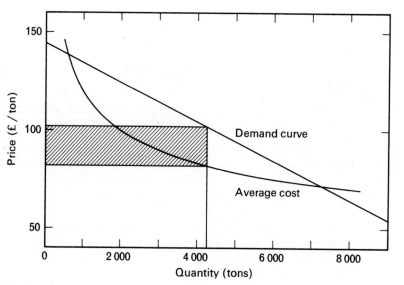

Figure 4.5 Calculation of profit from demand and average cost curves.

could get at least this figure. Hence he would certainly not go ahead with the above project if the capital outlay required was more than about £360 000 in all. The fixed costs of £260 000 per year in Table 4.1 seem very high in comparison with this, but the problem has to be studied in more detail (Chapter 5) before decisions can be taken.

The return on capital on which a manufacturer insists before deciding to produce depends on the prevailing bank rate, i.e. what it would cost him to borrow money. Below a certain level of profitability, it would be both safer and more profitable for him to invest in a building society or lend it to someone else at the bank rate.

4.4 Fixed and variable costs

Once a firm has committed itself and entered the market, its position changes in that its money is invested in plant which costs money for its upkeep and which is depreciating even if it is not operating. It is customary to divide the costs of running plant into two groups. The costs which are independent of output and which are incurred at any output short of a complete shutdown are called fixed costs. They include plant amortization, wages of the labour force and land rental. Other items such as raw material costs and the costs of heat, and electricity and payment for overtime by the operators depend critically on output and are known as variable costs.

Suppose the manufacturer confronted with Table 4.1 has anticipated an expanding market which will take up surplus capacity in the future and built plant designed to produce 6 000 tons/year and which costs £330 000/year in fixed costs. In addition to these, he has the output-based variable costs shown in Table 4.2 and Fig. 4.6. At outputs below 6 000 tons/year, his costs are higher than they would have been had he built plant of

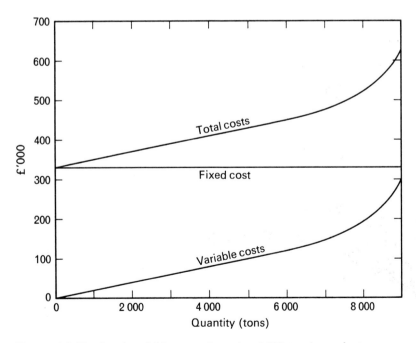

Figure 4.6 Fixed and variable costs of running 6 000 tons/year plant.

80   *The decision to produce*

a more suitable size and similarly at outputs substantially above the optimum, costs rise very steeply as various expedients (overtime working, perhaps restarting obsolete plant, etc.) are used to increase the rated capacity. The new total, average and marginal costs are shown in columns 4, 8 and 5 of Table 4.2 and the total and marginal profits are shown in columns 9 and 7. These data are plotted in Fig. 4.7 from which it will be seen that the extra costs of operating above rated output are such that it is more profitable for 6 100 tons/year to be manufactured and sold at £83/ton to bring in a profit of £55 000. Thus the policy to be followed by a manufacturer whose plant is already built may well be different from one who has not yet taken the decision to produce. This is true of his choice of output – in general, it pays a manufacturer who has already built plant to operate it at full capacity under almost any circumstances – and also of his response to changes of price for his product.

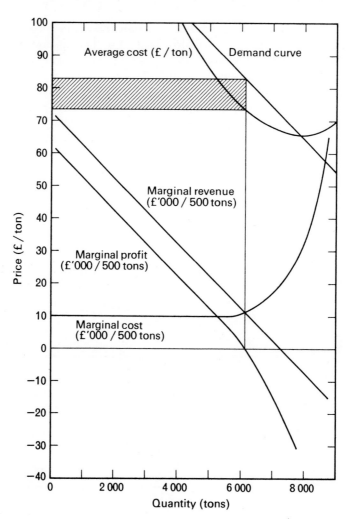

Figure 4.7 Optimum output and profit from 6 000 tons/year plant.

Fixed and variable costs   81

Table 4.2 Calculation of optimum production when plant already exists

1 Quantity (tons) q	2 Fixed cost of 6 000 tons/year plant (£'000) FC	3 Variable cost of 6 000 tons/year plant (£'000) VC	4 Total cost (£'000) $TC = FC + VC$	5 Marginal cost (£'000/500 tons) MC	6 Marginal revenue (£'000/500 tons) (from Fig. 4.7) MR	7 Marginal profit (£'000/500 tons) $MPro = MR - MC$	8 Average cost per unit (£) $AC = TC/q$	9 Total profit (£'000)
0	330	0	330				∞	−330
500	330	10	340	10	69·5	59·5	680	−270·5
1 000	330	20	350	10	64·5	54·5	350	−216
1 500	330	30	360	10	59·5	49·5	240	−166·5
2 000	330	40	370	10	54·5	44·5	185	−122
2 500	330	50	380	10	49·5	39·5	152	−82·5
3 000	330	60	390	10	44·5	34·5	130	−48
3 500	330	70	400	10	39·5	29·5	114	−18·5
4 000	330	80	410	10	34·5	24·5	102	6
4 500	330	90	420	10	29·5	19·5	93	25·5
5 000	330	100	430	10	24·5	14·5	86	40
5 500	330	110	440	10	19·5	9·5	80	49·5
6 000	330	120	450	10	14·5	4·5	75	54
6 500	330	132	462	12	9·5	−2·5	71	51·5
7 000	330	147	477	15	4·5	−10·5	68	41
7 500	330	167	497	20	−0·5	−20·5	66	20·5
8 000	330	195	525	28	−5·5	−33·5	65·6	−13
8 500	330	235	565	40	−10·5	−50·5	66·5	−63·5
9 000	330	300	630	65	−15·5	−80·5	70	−144

4.5 Response of a firm to varying prices

Given the demand schedule in Table 4.1, the manufacturer does best to sell 6 100 tons/year at a price of £83/ton. The demand schedule shows that he is operating under conditions of imperfect competition or monopoly, because it has a finite slope. Suppose, instead, that conditions of perfect competition exist and that prices are decided by external factors and that our manufacturer can sell as much or as little as he likes without the price changing.

Figure 4.8 shows the marginal cost curve taken from Table 4.2 and converted to units of £/ton. If the price were fixed at £83/ton instead of decreasing as output increased, it would clearly pay the manufacturer to produce more than 6 100 tons. How much more? The answer arises from the fact that at a fixed price, the marginal revenue curve becomes the horizontal line BG and this cuts the marginal cost curve at G (i.e. at G marginal revenue = marginal cost) so that optimum output is 8 300 tons leading to a profit of £138 900 (revenue = £8 300 × 83; costs = £550 000). This apparently anomalous result that a manufacturer appears to make higher profits in the presence of greater competition is, of course, an artifact of the assumption that his additional output will have no effect on prices.

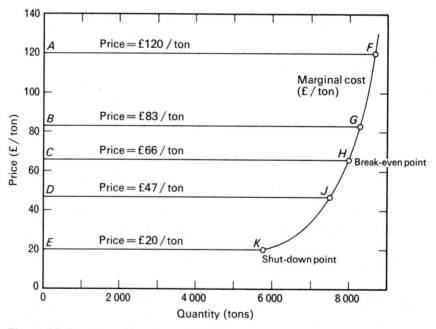

Figure 4.8 Response of firm to varying prices.

It is more than likely, however, that the price will vary because of external factors. Economic recession or the discovery of a new almost identical product will tend to lower prices; economic expansion or the development of new uses for a product will tend to raise them. In either case, the established manufacturer will try to produce an amount such that the marginal cost is equal to the new price. That is, he will produce until the cost of the next unit is more than he receives for it.

Suppose the price rises to £120/ton. The line AF (Fig. 4.8) cuts the marginal cost curve at 8 700 tons which is the new optimum output (revenue = £8 700 × 120 = £1 044 000; costs = £590 000; profit = £454 000). Suppose on the other hand the price drops to £66/ton. Then the optimum output is 7 950 tons and the revenue and cost are both £525 000 so that the profit is nil. How can this be an optimum profit? Because at any other output, a loss would be made. For example at an output of 6 000 tons, revenue = £396 000; costs = £450 000 so there is a loss of £54 000. The point at which zero profits are made, even at optimum output, is called the break-even point.

If the price drops any lower, the firm is bound to lose money. At £47/ton it will manufacture 7 500 tons and lose £144 500/year. Under these circumstances, it might seem more sensible for the firm to close down the plant but this is not so since fixed costs would still have to be paid and the firm would lose £330 000/year which is worse. So long as the revenue is greater than the variable costs, it is worthwhile for the firm to stay in business since, though it is making a loss, the loss is smaller than it would be if the firm closed down. The surplus of revenue over variable costs goes to offset at least some of the fixed costs. It is thus said to 'make a contribution to overheads'.

If the price drops even further to £20/ton, then the revenue is equal to the variable costs. The manufacturer loses £330 000 whether he sells 5 700 tons or does not produce at all. This is called the 'shut-down' point and at this price or below it pays the firm to shut down rather than to continue to produce.

It will be seen, therefore, that in the situation described above a new manufacturer would only start to produce if he could get more than £66/ton for his product, whereas an established manufacturer would remain in business until the price dropped below £20/ton. This underlines the difference between their two positions.

The above considerations apply in the short run. In the long run, all costs are variable costs because employees retire or change jobs or can be declared redundant, and plant wears out and need not be replaced. A firm will still produce until marginal cost is equal to price, but firms will enter and leave the industry until price is equal to average cost. In the long run, for conditions of perfect competition:

$$\text{marginal cost} = \text{price} = \text{average cost}$$

It is hardly necessary to add that it is very rare for economic conditions to remain stable enough for this long-term equilibrium to be achieved.

Finding the resources

Having decided that a certain manufacture is a profitable enterprise, a manufacturer has to set about finding the resources which will be devoted to its production. He will have considered the cost of these as part of his original costings. They will have been based on certain general considerations.

Projects may be classified as labour intensive, capital intensive or land intensive depending on which of these resources is used the most lavishly. The characteristics of these kinds of project are summarized in Table 4.3. There is an important difference, however, in the effects of increasing the use of either land, labour or capital. If graphs are plotted of output against number of men employed (land and capital being held constant) or against land used (labour and capital being held constant) then such graphs will rarely curve upwards. They will probably be linear at the start and then flatten out

Table 4.3 Capital, labour and land intensive projects

Project	Characteristic	Examples
Labour intensive	Many people employed relative to capital and land usage. Characteristic of underdeveloped countries but also found in research laboratories, service industries, etc. Wages are often low but form a high proportion of costs. Productivity (output/man) is also frequently low	Subsistence farming in underdeveloped countries, hairdressing, restaurants, university teaching, craft industries, building the pyramids
Capital intensive	Machines replace men. Mass production and use of labour to perform simple repetitive operations increases productivity. High level of automation. Return on capital relatively low; wages may be high	Large-scale industry especially the chemical industry. Car manufacturing, oil refining, some branches of engineering
Land intensive	Confined to areas of low population density. Usually involves farming of some sort	Maize growing in the USA (yield per acre is lower than in many countries but profitability is higher), nomadic grazing, sheep farming, timber growing

as diminishing returns set in. The graph of output against capital employed, however, usually curves upwards for quite a long way before it bends over. That is to say that there are few economies of scale possible with use of labour and land in most realistic situations, and diminishing returns set in fairly quickly. It is the intensive use of capital which brings the major economies of scale, and in industries such as petroleum refining it is still uncertain where diminishing returns set in if they ever do.

The chemical industry is, of course, an example of a capital intensive industry. Much of it is automated and it pays relatively high wages. Labour costs represent only about 10% of total costs and the industry would rather hire competent labour and be free from labour troubles than save an insignificant proportion of its total costs. The chemical industry therefore has little trouble in hiring labour. To do so, in the UK, would involve going to the local employment exchange, advertising in the press, etc. Of course labour is more readily available in certain parts of the country than in others. In particular, areas of high unemployment such as the north-east of England are also areas to which it is difficult to attract graduates and managers. There are also problems of training of labour. Nonetheless, the obtaining of this resource is a problem which can be solved relatively simply by the chemical industry.

The cost of land is also trivial compared with other chemical industry costs. It is important, however, for an appropriate site to be found. A petrochemical complex should be sited near adequate facilities for transport and shipping. Sea transport is particularly important and the main centres of the chemical industry in the UK (Fig. 6.3) are all accessible to ships. Most chemicals are dispatched by road nowadays so rail facilities are not as important as they were. Other factors affecting choice of site include facilities for effluent disposal, the possibility of obtaining adequate supplies of cooling water, whether the site is in an area qualifying for development grants or other government aid, and

Finding the resources 85

whether there is a lobby opposed to the intrusion of chemical plant into their midst which could prevent the granting of planning permission.

The resource which the chemical industry uses most lavishly is capital, the money to buy capital equipment, and this can be raised in a number of ways:

1. It can come out of a firm's depreciation and undistributed profits; that is some of the profits made by a firm on its past projects can be ploughed back into the new one instead of being distributed to shareholders. So also can money set aside for depreciation. This is obviously an attractive way for a firm to finance new projects and indeed is the major source of fresh capital (section 5.3).

2. The firm can borrow money from the bank, as an overdraft. For the use of this money, it will have to pay interest to the bank at a level about that of the so-called 'bank rate'. This is in many ways a short term expedient since the bank may decide to call in the overdraft. In any case it will probably demand some security for the loan in terms of the firm's capital equipment.

3. The firm can borrow money from the general public by issuing loan stock. This can be dated, undated or convertible. If it sells an undated loan stock, the firm guarantees that it will pay a fixed rate of interest (which has to be high enough to attract investors) on the money. This will be paid as long as the firm exists and if it goes bankrupt the investor will participate in the shareout. There is no way in which he can recover his money from the firm but he can sell his stock to someone else. If interest rates in general have risen since he bought it, he will be unable to recover all that he paid for it because a potential buyer could get a better rate of interest elsewhere. He will thus make a loss. If the opposite is true, he will make a profit.

If a firm issues dated loan stock, this means that it will pay fixed interest on the money it has borrowed until a certain date when it will repay the loan. As repayment is certain unless the company goes bankrupt, the price at which dated loan stock can be sold to other investors will not fluctuate as much as that of undated loan stock.

Convertible loan stock is where the firm guarantees to pay a fixed rate of interest but also offers at various specified dates to convert the loan stock into ordinary shares (see below). Convertible loan stock is a halfway house between loan stock and ordinary shares.

Loan stock is the safest way in which an investor can put money into industry because it places the greatest obligation on the firm which has to pay the fixed interest no matter whether or not it is making any profits. On the other hand, even if the firm makes a very large profit, the investor will only get his agreed interest rate. Interest rates in the UK have risen more or less steadily since the Second World War and people holding undated loan stock from that period have lost heavily. The classic example is 'War Loan' which was an undated $3\frac{1}{2}\%$ loan stock issued by the government during the war when interest rates were very low. An investor holding this stock only receives about 40% of current interest rates and consequently could only sell it for about 40% of the price he paid for it. As inflation has reduced the value of the reduced sum of money he could get back, it is clear that his investment has lost in value by about an order of magnitude.

4. The firm can sell preference shares to the general public. Like loan stock, these carry a fixed rate of interest but the firm is obliged to pay it only if it makes a profit. If a firm consistently loses money, the preference shares will drop in value. The purchase of a preference share involves a greater risk than the purchase of loan stock, so the firm issuing it

will have to offer a higher rate of interest. On the other hand a firm like ICI which is likely to go on making profits for the foreseeable future could, if it wished, issue a preference share at only a very slightly higher rate of interest than a loan stock.*

5. Finally, a firm can raise money by issuing ordinary shares, otherwise known as equity shares. These make the investor truly a part-owner of the firm and at the end of each year he will be paid a dividend out of what profits remain after the preference shareholders have taken their fixed rate of interest. Loan stock holders, of course, were paid before the profit was calculated, so the ordinary shareholder comes at the end of quite a long queue. If the firm goes bankrupt, he is not entitled to a penny; if it makes only a small profit, the preference shareholders will take it. On the other hand, if the firm prospers he might very well be paid dividends of 25 or 50% of his original investment or even more. In addition, the value that other investors are prepared to pay for his share will rise and he may be able to sell his share for many times what he paid for it. Often, a share in a firm with good prospects will command a very high price even if the firm is not currently making high profits or paying a good dividend. The price at which an ordinary share changes hands is only partly related to its financial performance – it is fixed by the supply of and demand for the shares on the stock exchange. This is determined by how investors think the firm will do in the future compared with other firms.

As the equity shareholder takes a greater risk than the holder of loan stock, especially government loan stock, known as 'gilt-edged', it used to be an axiom of economics that the average rate of return on equity shares (i.e. current dividend per share/current price of a share on the stock exchange averaged over all shares) would be higher than the rate of return on gilt-edged. On 27 August 1959 in the UK this situation reversed itself because investors decided that inflation was proceeding so rapidly that they would rather buy a share in a firm, the value of which in terms of money would increase as the real value of the money decreased. This is called the 'reverse yield gap' and has existed in the UK ever since. Similarly, investors have been reluctant to lend money on a fixed interest basis and issuers of loan stock have had to offer higher and higher rates of interest. There is some evidence that this trend is now changing but the way in which investors view future inflation is uncertain.

Which of the above methods a firm should use to raise money in a given situation is a highly technical matter and large firms have merchant bankers and other advisers who try to decide which method will cost the firm least yet provide the money which it requires.

4.7 Limited liability

There is nothing to stop a small businessman from sinking his savings into founding a small business, keeping accounts only for income tax purposes and financing himself out of his own savings and profits. If he gets into debt, however, all his personal property can be seized by his creditors and he will in any case find considerable difficulty in raising

---

* When he buys a stock, an investor lends his money to a firm. When he buys a share, he notionally buys a part of the firm, i.e. he becomes a part-owner. Stocks and shares are traded on the stock exchange and the majority of transactions involve one investor selling his share in a firm to another investor, i.e. the firm itself is in no way involved. New issues, by which firms themselves raise capital, form a relatively small fraction of transactions.

money as he would not be allowed to sell loan stock or shares and would have to rely on a bank overdraft. Most firms, therefore, choose to turn themselves into limited liability companies. These companies exist as an entity in themselves and if they get into debt only the company's property and not that of the directors can be seized. In return for this limited liability, the company has a legal obligation to publish annual accounts. These must be scrutinized by a qualified auditor to ensure that they comply with the requirements of the law.

The shares in a small limited liability company will probably all be owned by the directors and their families and will not be bought and sold. Such companies are said to be private. As it grows, a company may decide to go public and it would do this by offering shares on the open market, and these would be traded on the stock exchange. If a company issued more than 50% of its equity as ordinary shares, outside shareholders would be able to get together and vote the original directors off the board and appoint new ones. If the company were doing badly, this might well happen and there are few large companies left where a particular family or individual still has a majority shareholding. Nonetheless, because of the difficulty faced by dissidents in organizing the vast body of outside shareholders to vote against the *status quo*, directors are rarely sacked.

Notes and references

See Notes to Chapter 3.

Fig. 4.8 In many cases the marginal cost curve shown in Fig. 4.8 rises at low outputs, i.e. it is saucer-shaped. In such a case the horizontal 'price line' cuts the marginal cost curve in two places. Inspection will show that the intersection with the upward sloping section of the marginal cost curve is the significant one as far as these calculations are concerned. The intersections both correspond to situations where marginal cost = marginal revenue, i.e. marginal profit = 0. This condition means, in calculus terms, that a maximum, minimum or point of inflection has been reached. The economic situation rarely produces points of inflection, but it will be found that the intersection between the price line and the downward sloping marginal cost curve indeed corresponds to the point of minimum profit.

5 How do you know that you are making money?

5.1 Accounts and accountants

Companies exist to make money but it is often difficult at first sight to know whether a particular firm is doing so. How should the various incomings and outgoings of money be monitored; how should they be described; how may they be made to yield economically meaningful information? It is the function of accountants to answer these questions and thereby to provide a steady flow of financial information upon which managerial decisions may be based.

The form in which accounts are presented has evolved by trial and error over several hundred years and is now broadly the same all over the world. For some purposes the common practice of accountants has the force of law. In all advanced countries public – and usually private – companies are obliged to publish annual accounts approved by qualified independent auditors according to specified rules. The degree of disclosure required varies considerably; in Britain and the USA the relevant legislation – the Companies Acts of 1948 and 1967 for the former – calls for a considerable degree of frankness, whereas European and Japanese firms are allowed to be much more reticent. The principle of publication is however universally accepted and the trend is towards greater rather than less disclosure.

From the viewpoint of a company's own accountants, the preparation of the annual accounts is of less importance than their internal functions. At a lower level they monitor the multifarious flows of money through the company; at a higher level they analyse the continuous changes in the company's present and future position. They supply financial information at monthly or even weekly intervals to the executives, who are often themselves from an accounting background. A firm can survive without research, it can survive without a personnel department, but it cannot survive without accountants.

Although the principles behind accountancy are simple, their detailed application is of formidable complexity and, like all professions, accountancy enjoys its mysteries. It is nevertheless essential for the technologist to master the terminology and the basic assumptions of the accountant if he is to understand or, even more, if he is to influence the decision-making process. In office warfare knowledge may not be strength but ignorance is weakness.

5.2 The balance sheet

The central principle in accountancy is the separation of capital, the resources employed by the company, from the revenue which their employment generates. The capital accounts are presented in the form of a balance sheet which itemizes both the assets which form the capital employed by the company and the sources from which that capital was obtained. The balance sheet of the ICI group of companies at 31 December 1971 is shown in expanded form in Table 5.1. It is instructive to examine the individual items which it contains.

Assets are of two kinds: long-term assets, which have a long – often an indefinite – lifetime, and current assets, which arise from the day-to-day working of the company. The most important long-term assets are the fixed assets, which comprise land, buildings, plant and equipment. They are valued at their initial cost less accumulated depreciation (cf. sections 5.3, 5.6 below). Interests in subsidiary or other companies are put equal to the value of the shareholders interest involved. Goodwill is a more nebulous concept. In the accounts under examination it is no more than the difference between the price paid for a recently acquired subsidiary (Atlas Chemical Industries Inc.) and ICI's considered estimate of the value of its assets. In more venturesome hands goodwill has been used to cover the value of brand names, lists of customers or even less tangible assets. Unexpired patents may also be assigned a value. It is however common practice to take a conservative view of the value of such items and external auditors may respond unfavourably to attempts to inflate their significance.

Current assets comprise stocks and work in progress, goods that have been made or are being made but have not been sold, together with unused raw materials; debtors, money receivable from those to whom goods have been sold but who have not yet paid; and liquid assets, which comprise cash, bank deposits and instantly realizable investments such as government bonds. The valuation of stocks involves the estimation of future trends in the market; it is therefore customary to value them at cost of manufacture or current market price, whichever is the lower. This apparently conservative procedure may not always be adequate, however: when AEI was absorbed by GEC the estimates of the value of the former's stocks given by the incoming and outgoing managements differed by £14m. ($34m.). Clearly a substantial element of judgement is involved. Similarly, it may be necessary to allow for bad debts in calculating the sums owed by debtors.

The work of a company creates, of course, current liabilities as well as current assets but these are more easily described. Creditors include not merely trade suppliers but dividends declared and taxation incurred which have not yet been paid. Short-term borrowings, of which bank overdrafts are the most important, are also usually included among current liabilities. Unless it feels an urge to enter the chemical industry a bank may well hesitate to demand the immediate repayment of a very large sum but there is no doubt of its legal right to do so.

After current liabilities have been deducted from total assets the net assets which remain form the capital employed by the company. From where did this capital come? In the case of ICI about 9% comes from investment grants and deferred taxation. The former were a premium paid by the central government to companies which invested in areas of high unemployment and have now been partly replaced by tax concessions. The latter represents the employment of funds earmarked for the payment of company taxation in future years and is of interest only to specialists. Far more important as a source of capital are long-term loans which amount to about one-third of the capital employed by ICI. They include both loan stock issued by the company and long-term bank loans. Legally enforceable debts, they do not have to be repaid until a specified date and in fact less than 10% mature within the next five years.

Most of the remaining capital was provided by the shareholders of ICI. About 60% of this shareholders' interest came from the sale of shares while the other 40% – the reserves

Table 5.1 Group balance sheet and profit and loss account for ICI Ltd. for the year 1971

US terminology	Balance sheet UK terminology		£m.		US terminology	Profit and loss account UK terminology		£m.
	Employment of capital							
Land, plant and machinery	Fixed Assets		1 157·1		Sales	Turnover		1 524·4
	Goodwill		15·4		Gross profit	Trading profit		145·4
	Interests in subsidiaries		57·1			– after charging		
	Trade investments		145·2			depreciation	129·4	
	Current Assets		862·7			raw materials and purchased services	879	
Inventories	Stocks	344·8				wages and salaries	354	
Accounts receivable	Debtors	366·4				pension fund, etc.,	30	
	Liquid resources	151·5				*Add* Investment income	12·8	
	Less Current liabilities		463·6			Share of profits of associated companies	22·1	
Accounts payable	Trade Creditors	248·9				*Less* Interest on loans	43·3	
	Dividends	35·8				Employees' bonus	70	
	Current taxation	65·8				Profit before taxation		130·1
	Short-term borrowings	113·5				*Add* Investment grants	18·4	
						Less Taxation	50·7	
	Net Assets		1 773·9		Net profit	Profit after taxation		97·8
						Dividends	65·6	
						Retained profits	19·7	
						Applicable to minorities in subsidiary companies	12·5	
	Capital employed							
Net worth	Shareholders' interest		931·2					
Share capital	– Issued capital	482·6						
Undistributed profits	– Reserves	448·6						
	Minority interests		114·3					
	Long term loans		575·5					
	Deferred taxation		62·2					
	Investment grants		90·7					
			1 773·9					

– represents the current value of assets which arise from the employment of retained profits. Changes in the value of a company's assets produce corresponding changes in the value of the reserves: thus the reserves form a balancing item whose magnitude is such that capital employed is always equal to capital supplied. They are therefore liable to considerable fluctuations: between 31 December 1970 and 31 December 1971 the ICI reserves dropped by £30m. ($72m.) due mainly to a decision that the assets of the wholly owned subsidiary companies Qualitex Ltd. and British Nylon Spinners Ltd. were worth substantially less than had previously been thought. Conversely a decision to increase the value of assets would be reflected by an increase in the value of the reserves.

5.3 The profit and loss account

The revenue and profits earned by a company in a period of time are presented in the profit and loss account. That for the ICI Group for the year ending 31 December 1971 is shown in Table 5.1. With one exception the individual items are self-explanatory: depreciation, however, merits a closer examination.

In essence depreciation charges arise from the separation of capital from revenue. Fixed assets have a limited lifetime: eventually they will wear out or become obsolete and must be replaced. Depreciation charges are an attempt to spread the cost of replacing an asset over its lifetime so that when it falls due for replacement the money is available. If depreciation were not charged, then profits would in part be paid out of capital, whereas the object of investment is to use money to make more money. Two methods of calculating depreciation payments are currently used in Britain. In the straight-line method the value of the asset is reduced by equal amounts in equal periods of time: if the the initial and scrap values are I and S respectively and the lifetime is N years, then the annual depreciation charge is $(I - S)/N$. The reducing balance method reduces the value of the asset by equal proportions of the residual value in equal periods of time. This makes the depreciation charges larger in early years and smaller in later years (cf. Fig. 5.1) which has certain advantages (cf. section 5.7 below). In either method it is essential to make an accurate estimate of the expected life of the asset. Too high a figure will result in profits being paid out of capital as the depreciation charges will not cover the replacement cost; too low a figure will lead to abnormally low profits while the investment is being recovered. In high technology fields change is rapid and life-times are short. Ten years is the period generally assumed in the chemical industry and seven is often used; as Table 5.1 shows the depreciation charged by ICI during 1971 corresponds to an average lifetime of just under nine years for its fixed assets.

Depreciation charges are in practice reinvested in the company as they are made. The total value of the assets remains constant but their nature changes continuously. Together with the undistributed profits they form the cash flow, the stream of uncommitted funds which result from the company's operations and which the directors may employ to maintain, to expand or to diversify the company's activities. In the chemical industry depreciation charges are the larger part of the cash flow and the cash flow is the major source of funds for investment. In the ten years 1962–71 ICI invested £1 937m. in new assets, of which £113m. came from issues of ordinary shares, £511m. from loans, £198m. from retained profits and £849m. from depreciation. This exemplifies the dilemma of a company in a field which is both capital and technology-intensive. To survive requires heavy and continuous investment; depreciation charges must therefore be high even

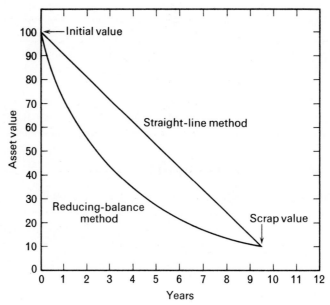

Figure 5.1 The straight-line and reducing balance methods of charging depreciation.

though dividends suffer; today's profits must always be sacrificed if tomorrow's challenges are to be met.

What is the connection between the profit and loss account and the balance sheet? Depreciation charges do no more than maintain the value of the assets. Profits and losses however affect both the assets and the shareholders' interest. Retained profits increase the value of the net assets and therefore of the reserves. Conversely a company which makes a loss will cover it by realizing some of its assets – by reducing its liquid resources or selling some property for example – or by increasing its liabilities by borrowing. In either case net assets and reserves are correspondingly reduced.

5.4 Measures of profitability

Firms exist to make a profit for their owners. How may their efficiency in this respect be measured? From the managerial standpoint the fundamental measure of the efficacy with which assets are used is the return on capital, the ratio of profits to capital employed. This concept requires careful definition. Capital employed has a specific meaning but several measures of profit may be used. British firms favour the use of profit before charging tax and interest for the purpose of international comparison and argue that this eliminates the effect of different interest rates and systems of taxation. It results however in deceptively high rates of return and, more seriously, introduces an appreciable bias in favour of companies which borrow much of their capital. American companies generally use profit after taxation as the yardstick and this is also used in the annual surveys published in *Fortune* and *Chemical and Engineering News*. Either definition is acceptable but it is essential to compare like with like. In the case of ICI after-tax profits varied between 4·5 and 7·4% of capital employed in the period 1962–71 (cf. Table 5.2) which was near the

Table 5.2 Profitability of the ICI group 1962–71

	1962	1963	1964	1965	1966	1967	1968	1969	1970	1971
Pre-tax, preinterest profits as % capital employed	9·1	10·4	12·0	10·3	8·1	8·8	11·9	12·4	10·3	9·4
After-tax profit as % capital employed	4·5	4·8	6·3	7·4	4·7	4·6	6·3	6·9	6·1	5·5
After tax profit as % shareholders interest	5·7	6·0	8·1	9·3	6·4	7·1	10·1	11·0	9·6	9·1
Dividend per £1 ordinary share (p)	9·2	10·0	12·5	12·5	12·5	12·5	12·9	13·8	13·8	13·8

average for large European chemical companies. Comparable American firms made rather more successful use of their capital and averaged 8–13% during this time.

From the shareholders' point of view, however, more significance attaches to the return on the shareholders' interest and, above all, to the dividend per share. These rates of return are substantially higher than the after-tax return on capital employed; this is because a substantial part of ICI's capital is in the form of fixed-interest long term loans. It should be noted that in a company in which loans make up most of the capital a small change in profits will produce a large change in the rate of return on the shareholders' interest. Such a company is said to be highly geared. To managers, however, these latter measures of profitability are chiefly of importance because they indicate both the ease with which money may be raised by the sale of ordinary shares and the propensity of the shareholders to rebellion. The humdrum nature of the British chemical industry in the past decade is perhaps indicated by the absence of both boardroom revolutions and successful large-scale issues of shares.

5.5 Measures of financial stability

The accounts may also be interpreted to yield information about the stability of the company. A firm must be able to meet its current liabilities at all times; failure to do so is the cause of bankruptcy. Further, it must be able to do so from its current assets; its fixed assets are necessary for its continued existence and may in any case be unsaleable at short notice.

The fundamental requirement for stability is therefore that current assets should exceed current liabilities. In practice the current ratio – the ratio of current assets to current liabilities – varies widely from firm to firm: during the past five years it has been 1·8–1·9 for ICI compared to 3·0–3·5 for DuPont. There is no particular significance in the absolute magnitude of the current ratio as this reflects both the nature of the firm's activities and its management practices. Systematic changes in the current ratio are much more important: a downward trend indicates a decrease and an upward trend an increase in the ultimate margin of safety. That current assets should exceed current liabilities is, however, a necessary but not a sufficient condition of financial stability. Assets differ widely in liquidity: cash and bank deposits are completely liquid, debtors may take some time to pay their bills, and stocks may be difficult to sell at their true value in a crisis. The proportions of the current assets composed of liquid assets and of stocks are therefore important measures of financial stability. Provided that a firm maintains a certain minimum liquidity – enough cash to meet payments due in the next seven days is a common yardstick – the absolute values of these parameters are of less significance than their trends. An increase in the ratio of stocks to current assets indicates a decline in liquidity and therefore in the ability to respond instantly to a sudden crisis. Conversely a rise in the ratio of liquid assets to current assets indicates an increase in liquidity and, other things being equal, in stability.

The onset of a liquidity crisis in a large firm is a complex process. The case of Rolls-Royce is instructive. Several years of extremely expensive research and development are necessary before a jet engine is ready for series production: during this period money is spent on the largest scale without any immediate return. In the case of the RB-211 engine Rolls-Royce was quite unable to finance the development stage from its own cash flow and was forced to borrow to the limit. Both the cost and the time required were seriously underestimated and the company ran out of cash before the engine was ready. The management, composed largely of engineers, was gravely misled by the peculiar accounting methods used. Research and development expenditure was capitalized rather than charged against profits: it was treated as the creation of an intangible asset which would be written off when the project proved successful. As a result the company appeared to be profitable even though it was moving rapidly towards total illiquidity. It is doubtful whether this policy deceived the financial world but it certainly blinded the Rolls-Royce directors.

5.6 The impact of inflation

The techniques described so far assume that money has a constant value but, of course, we live in an era of inflation. How does this affect the accounts? In terms of current prices the balance sheet undervalues non-monetary assets; in a capital-intensive industry this is equivalent to understating the capital employed. Moreover, since the replacement value of fixed assets is understated, depreciation charges are too low and profits therefore too high. These points are illustrated by the following highly simplified example. Con-

Table 5.3 The influence of inflation upon accounts

	Historical cost basis	Current purchasing power basis
Balance sheet		
Fixed assets	100	110
Current assets	25	25
Less Current liabilities	10	10
	115	125
Shareholders' interest	65	75
Long-term loan	50	50
	115	125
Profit and loss account		
Trading profit before depreciation	20	20
Less Depreciation	10	11
Profit before taxation	10	9
Less Tax at 40%	4	3·6
Profit after taxation	6	5·4

sider (cf. Table 5.3) a company with fixed assets of £100m. acquired at the beginning of a year in which inflation is 10%. During the year it makes a profit of £20 m. of which £10 m. is charged to depreciation and invested in new plant. It reports an after-tax profit of £6 m. from £115 m. employed. If inflation is taken into account the picture is different. Fixed assets are now £110 m., depreciation is £11 m. and the after-tax profit is £5·4 m. from a capital of £125 m., i.e. a rate of return of 4·3% compared to the 5·2% predicted by orthodox bookkeeping. In real life the differences are much larger because the assets have been acquired over a long period of inflation. It has indeed been argued that many nominally profitable firms would be shown to have lost money continuously if their accounts were expressed in contemporary rather than historical currency.

What can be done to correct for inflation? Some companies, such as the Dutch electric firm Philips N.V., use replacement cost depreciation which generates an increased cash flow for asset replacement but does not convey the effect of inflation upon capital employed. The English Institute of Chartered Accountants favours the more comprehensive approach of expressing all items in terms of current purchasing power and this will probably be the procedure ultimately adopted. Until such a time extreme care in the analysis of accounts is advisable.

5.7 The choice of projects

5.7.1 The discounted cash flow method

Hitherto we have been concerned with the analysis of the past. Accountants have also developed procedures to assist in the selection of projects for the future. The most elegant and powerful of these is the discounted cash flow method.

Money has a time value; we feel intuitively that a £ today is worth more than a £ in the future. How may this be expressed in quantitative terms? Consider a sum of money, A, invested at an annual interest rate r: after one year the value of the investment will be $A(1 + r)$, after two years $A(1 + r)^2$ and after N years $A(1 + r)^N$. Let us now consider the magnitude of the sum which must be invested now at an interest rate r so as to produce a sum A in a year's time. Clearly this sum P_1 is given by:

$$A = P_1(1 + r) \quad \text{whence} \quad P_1 = \frac{A}{1 + r}$$

P_1 is called the present value of a sum A due in a year's time when the interest rate is r, or, more tersely, the present value of A discounted at rate r. Similarly, the present value of a sum A due in two years time is given by $P_2 = A/(1 + r)^2$ and the present value of a sum A due in N years time is $P_N = A/(1 + r)^N$.

Now let us glance at an investment which, as far as we can tell, will yield trading profits before depreciation of $T_1, T_2, \ldots, T_N$ in the first, second, ..., Nth years of its life. What is the present value of these trading profits for an interest rate r? Obviously T, where:

$$T = \frac{T_1}{1 + r} + \frac{T_2}{(1 + r)^2} + \cdots + \frac{T_N}{(1 + r)^N} = \sum_{n=1}^{n=N} \frac{T_n}{(1 + r)^n}$$

The project must also replace its capital cost: if the depreciation payments are $D_1, D_2, \ldots, D_N$ then their present value is similarly D where:

$$D = \sum_{n=1}^{n=N} \frac{D_n}{(1+r)^n}$$

If for the moment we neglect taxation, it is obvious that trading profit and depreciation is a cash flow similar to that defined in section 5.3 above. Thus if the cash flows expected from the project are $C_1, C_2, \ldots, C_N$ their present value C is given by:

$$C = T + D = \sum_{n=1}^{n=N} \frac{T_n}{(1+r)^n} + \sum_{n=1}^{n=N} \frac{D_n}{(1+r)^n} = \sum_{n=1}^{n=N} \frac{(T_n + D_n)}{(1+r)^n} = \sum_{n=1}^{n=N} \frac{C_n}{(1+r)^n}$$

There exists a value of r, the interest rate, such that C, the present value of the cash flows expected from the project is just equal to the initial investment I. This value of r is the internal rate of return, so-called because it is determined only by the magnitudes of I, $C_1, C_2, \ldots, C_N$ and not by anything outside the project. The significance of the internal rate of return is that it is the maximum rate of interest on the initial investment I which the project will stand, for:

$$I = \frac{C_1}{1+r} + \frac{C_2}{(1+r)^2} + \cdots + \frac{C_N}{(1+r)^N}$$

or

$$I(1+r)^N = C_1(1+r)^{N-1} + C_2(1+r)^{N-2} + \cdots + C_N$$

Now, if we borrowed I at this rate r for N years, allowing the interest to accumulate, we would have to pay back $I(1+r)^N$ after N years. But we could equally well invest the successive cash flows from the project at the same rate of interest to give after N years the sum

$$C_1(1+r)^{N-1} + C_2(1+r)^{N-2} + \cdots + C_N = I(1+r)^N$$

i.e. a sum exactly equal to that which we would owe on the initial investment. In a similar way it can be shown that if the prevailing interest rate is greater than the internal rate of return, C will be less than I and the project will lose money, whereas if it is less than the internal rate of return C will be greater than I and the project will make money.

5.7.2 How to use the DCF method

Two general methods may be used which usually give the same answer. In the first and simpler one the current rate of interest r is taken as given, the present value of the cash flows is estimated, and the project which gives the sum which most exceeds the proposed initial investment is chosen. Alternatively the internal rate of return may be determined and compared with the current interest rate. Projects for which the internal rate of return is less than the interest rate are rejected and that with the highest internal rate of return is chosen.

This latter method requires the solution of the equation

$$I = \sum_{n=1}^{n=N} \frac{C_n}{(1+r)^n}.$$

This may be solved exactly as a polynomial in $1/(1 + r)$ but this is excessively tedious in most cases. It is usually more convenient to use Figs 5.2–3. Figure 5.2 shows the present value of £1 discounted at various rates and receivable N years hence. It may be used to calculate C for various values of r; the results may be plotted against r, and the value of r for which $C = I$ read off the curve. A convenient and not uncommon case is that in which the cash flows are the same each year; Fig. 5.3, which shows the present value of £1 receivable for N years at an interest rate r, may be used.

Let us consider an example. A car which costs £1 000 is used as a taxi for three years and then sold for £400. It may be operated in three different ways: to yield cash flows of £300 per year; to yield cash flows of £200, £300 and £400 in successive years; and to yield cash flows of £400, £300 and £200 in successive years. Which is the most profitable mode of operation? In each case we must add the scrap value of the car to the operating cash flow for the last year; the cash flows are then:

	Year 1	Year 2	Year 3
Method 1	300	300	300 + 400 = 700
Method 2	200	300	400 + 400 = 800
Method 3	400	300	200 + 400 = 600

For method 1 the internal rate of return r is given by:

$$1000 = \frac{300}{1 + r} + \frac{300}{(1 + r)^2} + \frac{700}{(1 + r)^3}$$

and solving as a cubic in $1/(1 + r)$ we obtain $r = 12 \cdot 4\%$. Methods 2 and 3 give r as $11 \cdot 5\%$ and $13 \cdot 3\%$ respectively. Clearly method 3 is the most profitable and should be chosen.

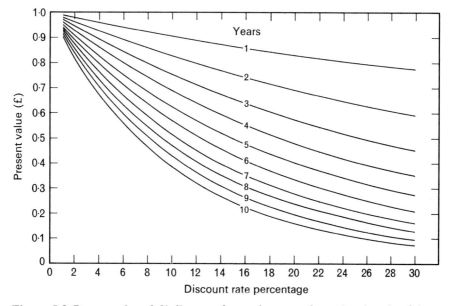

Figure 5.2 Present value of £1 discounted at various rates for various lengths of time.

A second example illustrates the use of Figs. 5.2–3. A chemical plant costs £10 000 and has a lifetime of 10 years. It produces annually 100 tons of an ester which sells for £100 per ton. Direct costs are £75 per ton and indirect costs £500 per year. Assuming maximum production, what is the maximum rate of interest which the project will stand? The annual cash flow is:

$$£ (100 \times 100) - (75 \times 100) - 500 = £2\,000$$

and so if r is the internal rate of return:

$$10{,}000 = \sum_{n=1}^{n=10} \frac{2\,000}{(1+r)^n}$$

or dividing both sides by 2 000:

$$5 = \sum_{n=1}^{n=10} \frac{1}{(1+r)^n}$$

From Fig. 5.3 the present value of £1 received annually for 10 years at a discount rate of 15% is £5·012 and so the internal rate of return is a little over 15%.

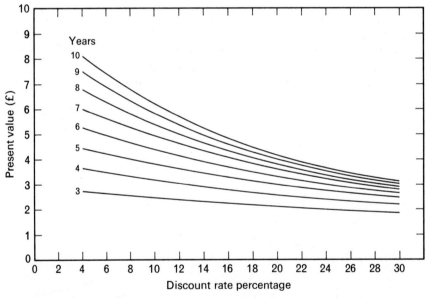

Figure 5.3 Present value of £1 received annually for various lengths of time discounted at various rates.

5.7.3 Implications of the DCF method

The application of DCF techniques is of course normally a much more complex process than these simple examples suggest. The answers obtained depend critically on the magnitude of the cash flows which are inevitably difficult to forecast. Taxation cannot be neglected: companies are primarily interested in after-tax profits and different projects may well incur different tax liabilities. It may well be desirable to allow for inflation. Similarly the life of a project is difficult to evaluate in advance. Yet this is merely to say that mathematical techniques cannot eliminate the manager's responsibility to use his judgement. The DCF method does, however, enable him to compare a wide range of scenarios in a quantitative way.

The technique does, moreover, give results in agreement with common sense. As the first example of section 5.7.2 shows it discriminates in favour of projects which give a rapid return and against projects which yield most of their profits late in life. This is reasonable: the further we look ahead the more uncertain are the prospects. Further, it brings out the penalties attached to delays in completing and starting up large plants. Consider a plant which is immediately available at cost I and will produce cash flows of $C_1, C_2, \ldots, C_N$ for N years. At a discount rate r the present value P_1 of the cash flows is:

$$P_1 = -I + \frac{C_1}{1+r} + \frac{C_2}{(1+r)^2} + \cdots + \frac{C_N}{(1+r)^N}$$

If, however, the plant takes one year to build, I_1 being paid now and I_2 in a year's time, where $I = I_1 + I_2$, then the present value P_2 of the cash flows becomes:

$$P_2 = -I_1 - \frac{I_2}{1+r} + \frac{C_1}{(1+r)^2} + \frac{C_2}{(1+r)^3} + \cdots + \frac{C_N}{(1+r)^{N+1}}$$

and for any positive value of r, P_2 must be less than P_1, i.e. the latter project must be less profitable. This has been abundantly confirmed in many actual cases. Holroyd's study of ammonia plants, summarized in Table 5.4, shows that the difference between the best and worst cases is genuinely dramatic. Like nothing else the DCF method emphasizes the time value of money and the desirability of a rapid return on investment.

Table 5.4 Internal rates of return for a 1 000 ton per day ammonia plant (%)

Annual output % capacity	No delay (2 years construction time)	6 Months delay	12 Months delay
Year 1 onwards 100	26	23	21
Year 1 60 2 80 3 onwards 100	16	14	13
Year 1 30 2 80 3 90 4 onwards 100	12	10	9
Year 1 30 2 70 3 onwards 90	7	6	5

A major criticism of the method is that it does not allow for the possibility of reinvestment. Consider a chemical which could be made by different processes with plant lives of five and ten years respectively. Though the DCF method might show the latter to give a better return, the former might still be preferable because it gives the manufacturer the option of leaving the market after five years and investing elsewhere the money which would have been spent on a second plant. The value of this option is impossible to assess, however, and would vary with country, industry and time.

5.7.4 Alternatives to the DCF method

Some old-fashioned companies use more primitive methods to choose projects. The payback period is often used as a criterion; this is the number of years required for the sum of the cash flows to equal the initial investment. In effect this is no more than the time taken for the project to pay for itself and tells one nothing about its actual profitability. Another measure is the average return on capital invested. The average net profit after depreciation is divided by the average capital used. If the straight-line method of depreciation is used, the latter is given by $\frac{1}{2}$ (initial investment − scrap value). This completely ignores the time element and is therefore highly unsatisfactory. It may be noted that both these methods give the same answer to all three cases in the first problem given in section 5.7.2 above.

Notes and references

Most texts on accounting are written for accountants. Some which are useful for managers are: R. Mathews, *Accounting for Economists*, F. W. Cheshire, Melbourne, 1965. J. Batty, *Management Accountancy*, 3rd edn., Macdonald and Evans, London, 1970. E. B. Jones, *Finance for the Non-Financial Manager*, Pitmans, London, 1972. The BBC Publication, I. Gray and J. Smith, *Management Accountancy*, 1968, London, which accompanied the Hardy Heating Company T.V. series is also very helpful.
For the impact of inflation see *The Economist*, 1971, **238**, 58–59 and 1972, **243**, 67–68.
Information about ICI was taken from the *Report and Accounts* for 1971.
Section 5.7.3
Table 5.4 is from Sir Ronald Holroyd, *Chem. and Ind.*, 1967, 1310.

6 The world chemical industry

There is a chemical industry, however small, in almost every country in the world. In developing countries it tends to be a traditional industry producing, for example, fertilizers, sulphuric acid, soda and soap. Figure 6.1 shows the chemical-producing industries in the world around 1970. In spite of wide distribution of the industry, it is especially concentrated in the USA, Europe and Japan.

A useful guide to the size of the chemical industry in various countries is the level of sulphuric acid production. This is shown in Table 6.1. The size of the modern chemical industry may be gauged from production of plastics and resins and these data are also shown. 'League tables' compiled from these figures should not be taken too literally since the figures may be distorted by local industrial patterns. For example, Japanese sulphuric acid production reflects partly the atypical pattern of Japanese fertilizer production.

This chapter deals with the structure of the chemical industry in the UK and the developed countries of the world, and considers some of the problems facing underdeveloped countries trying to industrialize.

6.1 The UK chemical industry

6.1.1 The national income

What is the National Income? To understand this concept it is first necessary to distinguish between productive activities which contribute to the flow of goods and services, and non-productive activities which simply redistribute these goods and services among the various members of the community. The latter are called transfer payments: they consist of payments for which there is no corresponding provision of goods or services, such as old-age pensions, student grants and taxes.

The national product is the value of the goods and services produced by the nation. The measurement of productive activity is however less straightforward than might first appear. Some goods and services are bought for their own sake: others are merely required to make other goods. It is essential to differentiate between intermediate and final products. To evaluate the national income by adding the value of all productive transactions would be to incorporate an element of double counting. If a plastics fabricator buys £100 of moulding powder from which he makes £200 of ashtrays the economy receives £200 of ashtrays. It does not receive £100 of moulding powder and £200 of ashtrays as the former was sacrificed to make the latter. The national product must therefore be expressed in ways which reflect this principle.

One way of doing so is by the expenditure method: the national product is taken as the value of all final products, i.e. all goods and services that are not consumed in the pro-

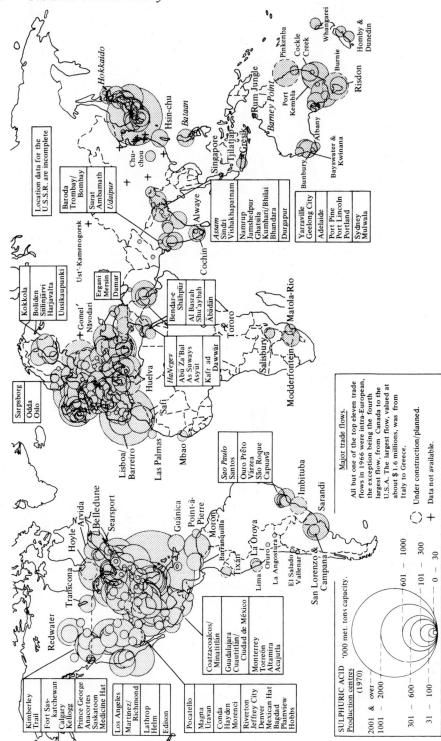

Figure 6.1 Centres of the world chemical industry 1970.

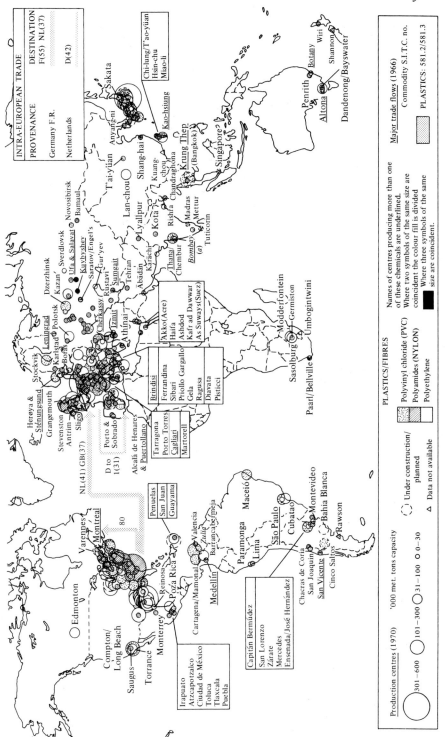

Table 6.1 World production of sulphuric acid and plastics and resins 1970

	Sulphuric acid ('000 tons/month)	Plastics and resins ('000 tons/month)	Population (m.)	Sulphuric acid (lb/head/year)	Plastics and resins (lb/head/year)
Argentine	15·0	n.a.	23·21	17·1	
Australia	148	18·6	12·55	312	39·2
Austria	n.a.	15·6	7·39		55·7
Belgium	150	n.a.	9·68	410	
Brazil	35·0	n.a.	95·31	9·72	
Bulgaria	41·9	n.a.	8·49	131	
Canada	206	23·7	21·41	254	29·3
China (Taiwan)	23·5	15·1	14·04	44·3	28·4
Czechoslovakia	92·5	20·0	14·47	169	36·6
Egypt	24·8	n.a.	33·33	19·7	
Finland	70·2	n.a.	4·70	395	
France	307	81·9	50·78	160	42·7
E. Germany	91·6	30·8	17·25	140	47·2
W. Germany	370	361	60·50	162	158
Hungary	40·4	4·57	10·31	94·9	11·7
India	89·8	6·61	550·38	4·32	0·32
Israel	16·9	2·73	2·91	153	24·8
Italy	271	139	53·67	134	68·5
Japan	577	505	103·39	148	129
S. Korea	3·51	n.a.	31·79	2·92	
Mexico	102·9	n.a.	50·67	53·7	
Netherlands	130	96·0	13·02	264	195
Norway	24·2	n.a.	3·88	165	
Pakistan	3·42	n.a.	126·74	0·71	
Phillipines	16·3	3·27	38·49	11·2	2·25
Poland	158	22·4	32·47	129	18·2
Portugal	31·8	2·92	9·63	87·4	8·02
Romania	82·9	17·2	20·25	108	22·4
Spain	173	26·7	33·65	136	21·0
Sweden	58·6	n.a.	8·04	193	
Tunisia	34·4	n.a.	5·14	177	
Turkey	2·30	n.a.	35·23	1·73	
USSR	10·05	139	242·77	110	15·1
UK	279	124	55·71	133	58·8
USA	22·36	679	204·80	289	87·7
Yugoslavia	48·6	8·09	20·37	63·1	10·5

duction of other goods and services. An equivalent alternative is the production method, in which the national product is evaluated as the sum of the values added by each industry. The value added by an industry is the difference between the value of its output and the value of the goods and services which it buys from other industries. It forms the fund from which salaries and wages, profits, depreciation, taxes and reserves are paid and must therefore be divided between taxation and the factors of production, capital and labour. Clearly the sum total of such payments must be equal to the total value added and so to the national product; this is the income method of evaluating the national product.

These methods should give the same result but due to the limitations of the statistics used they usually do not do so; published estimates are usually based on a conflation of the results obtained by the income and expenditure methods.

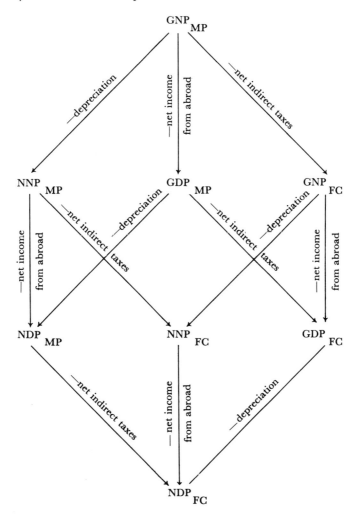

Figure 6.2 Relationship between eight national product concepts (MP = market prices, FC = factor cost).

The national product is by definition spent on consumption and investment. Both may be private or public. Private consumption includes the purchases of households, such as food, clothes, domestic heating and lighting and consumer durables such as private cars which are used for non-productive purposes. Public consumption comprises current public expenditure on defence, health, education, administration and the maintenance of order. Investment represents additions to the national assets. A nation may invest at home by devoting some of its products to the creation of new fixed assets or increased stocks of raw materials or finished products. Alternatively it may import less than it ex-

Table 6.2 The British Chemical Industry in 1968

	Gross output (£m.)	Net output (£m.)	Number employed ('000)	Operatives ('000)	Other employees ('000)	Value added per employee (£)	Exports* (£m.)	Imports* (£m.)	New capital investment (£m.)
All manufacturing industry	39 854	16 012	8 077	5 999	2 041	1 982	5 278	3 773	1 565
Chemical and allied industries	3 246	1 390	411	245	166	3 385	645	435	223
Organic chemicals	} 1 143	} 441	} 112	} 71	} 41	} 3 937	99	118	87
Inorganic chemicals							50	47	16
Other general chemicals							94	59	n.a.
Pharmaceuticals	368	219	57	29	28	3 839	93	22	18
Synthetic resins and rubbers	477	196	55	33	22	2 557	121	96	40
Paint	194	92	33	16	17	2 765	16	4	n.a.
Soap and detergents	190	68	17	9	7	4 071	19	7	n.a.
Toilet preparations	131	68	23	14	9	2 897	17	4	n.a.
Dyes and pigments	142	69	22	13	9	3 140	45	25	7
Fertilizers	210	58	16	10	6	3 532	7	29	20
Other	391	179	76	50	26	2 355	84	24	35

* Manufactured goods only

ports and so create physical or monetary assets abroad. It will also have to make some provision for depreciation so as to maintain the value of its capital stock (cf. section 5.3).

Although simple in concept the measurement of the national product is complex in practice and several different yardsticks are commonly used. Their interrelationships are shown in Fig. 6.2. The gross national product (GNP) is the total gross output of goods and services produced by the nation. The gross domestic product (GDP) is the gross national product less net income from abroad. The net national product (NNP) or national income is the gross national product less depreciation. The net domestic product (NDP) is similarly the national income less income from abroad. These measures may be expressed in market prices (MP) or at factor cost (FC), that is after the elimination of indirect taxes and subsidies. Since the latter are a special form of transfer payment – the firms are simply collecting taxes for the government – it is preferable to use factor cost where this is possible and national income in particular is usually calculated on this basis.

6.1.2 The chemical industry and the UK economy

At the time of the 1968 census of production the gross national product of the United Kingdom at market prices was £42 899 m. Manufacturing industry accounted for £16 012 m. of which the chemical and allied industry contributed £1 390 m. In terms of value added (cf. Table 6.2) general chemicals was the most important sector, followed by pharmaceuticals – because of the large ratio between sales and raw materials (cf. Chapter 13) – synthetic resins and rubbers, paints, dyestuffs and pigments, soaps and detergents, toilet preparations and fertilizers. Defined in this way the chemical industry is the fourth largest manufacturing industry and is exceeded only by mechanical engineering, food, drink and tobacco, and vehicles.

The position of the chemical industry in the British economy is most clearly illustrated by the input–output table of which a portion is shown in Table 6.3. This refers to the year 1963 but it is believed that the structure of industry has not changed so dramatically as to destroy its value. A table for 1968 will be published in 1973. Expenditure by the chemical industry is shown in the column: thus in 1963 it bought £3 m. from agriculture, forestries and fishing; £166 m. from mining and quarrying; and so on. Sales by the chemical industry are shown in the row: in 1963 it sold £95 m. to agriculture, forestries and fishing; £13 m. to mining and quarrying and so on. Space precludes the inclusion of the entire table in which each industry has a row, a column and a set of entries in the final demand and payment sectors. It is clear that although these transactions are expressed in terms of sales and purchases the value added by an industry is the sum of the items listed as payments (cf. section 6.1.1 above).

An examination of this table shows that the chemical industry buys a wide range of goods and services. It consumes coal, oil, gas, water and electricity, transport services, paper and printing, and the products of the engineering trades in large amounts. The apparently huge purchases of coal are an artefact of classification: in the summary table shown, coke production is included as part of the chemical industry. In the main table which divides the economy into 70 sectors rather than 20 it is obvious that the chemical industry proper consumed about £30 m. of coal and £10 m. of other minerals. It is similarly possible to analyse the sales of the industry. Just over half the production is sold to other domestic industries, 26% to private and public consumers and 20% is exported. In descending order of importance the main industrial customers are plastic, rubber and

Table 6.3 Part of the Input–Output table for the UK 1963 (£m.)

	Chemicals, etc.
Chemicals, etc.	—
Agriculture, forestries, fishing	3
Mining and quarrying	166
Food manufacturing	21
Drink and tobacco	4
Mineral oil refining	30
Metal manufacture	29
Mechanical engineering	24
Electrical engineering	7
Other metal goods	37
Vehicles	4
Textiles, leather, clothing	7
Paper and printing	46
Rubber, plastic and other mfrng	45
Construction	22
Gas, electricity, water	62
Transport and communications	100
Distributive trades	41
Miscellaneous services	172
Public administration, etc.	—
Imports	256
Sales by final buyers	4
Taxes on expenditure and subsidies	35
Income from employment	390
Gross profits	354

Sales of Chemicals, etc. to:

Sector	£m.
Agriculture, forestry, fishing	95
Mining and quarrying	13
Food manufacturing	102
Drink and tobacco	8
Mineral oil refining	21
Metal manufacture	125
Mechanical engineering	31
Electrical engineering	53
Other metal goods	26
Vehicles	31
Textiles, leather, clothing	92
Paper and printing	48
Rubber, plastic and other mfrng	150
Construction	53
Gas, electricity, water	8
Transport and communication	22
Distributive trades	5
Miscellaneous services	72
Public administration, etc.	—
Total intermediate output	955
Final demand — Current expenditure: Consumers	348
Final demand — Current expenditure: Public authorities	141
Final demand — Domestic capital formation: Fixed	36
Final demand — Domestic capital formation: Stocks	−1
Final demand — Exports	380
Total final output	904
Total output	1 859

other manufacturing; food manufacturing; agriculture; and textiles and clothing, each of which absorbs more than 5% of the total output. Once more we must neglect sales to the steel industry which consist almost entirely of coke.

The input–output table is more than a striking way to represent the structural interdependence of industries. Suppose, as is true for small changes and short periods of time, that the consumption of inputs is directly proportional to output, i.e. that to increase output by 1% requires an increase of 1% in all inputs. Adepts will recognize that we may then express the output of all industries in a set of simultaneous equations which form a square matrix. By routine mathematical operations much additional information may be obtained. Thus inversion of the matrix yields the direct and indirect requirements for £1 000 of goods delivered by any industry to the final demand sector; this comprises not merely the inputs to a particular industry but also the inputs to all the direct and indirect suppliers of that industry. Such a figure may be used to measure the sensitivity of the chemical industry to changes in the output of other industries. To deliver £1 000 worth of agricultural goods to final demand requires £78 of chemicals; to deliver £1 000 of textiles requires £80 of chemicals; and to deliver £1000 of plastic, rubber and other goods requires £159 of chemicals. On the other hand £1 000 of motor vehicles, coal and aerospace equipment require only £35, £15 and £17 of chemicals respectively. Many other applications of the input–output matrix are possible.

6.1.3 The nature of the UK chemical industry

The chemical industry is strikingly different from most parts of manufacturing industry. It is capital-intensive: fixed assets employed by the industry in 1968 were worth about £2 200 m. and the capital per employee was twice the national average. Investment continues to be high: in 1968 the chemical industry accounted for 8·7% of the value added by manufacturing industry but for 15% of gross fixed capital expenditure. Heavy organic chemicals (39%) and plastics (15%) attracted the majority of this expenditure, as has usually been the case for many years.

As in most other industrial countries, capital expenditure in the chemical industry followed a four-year cycle in the 1960s, with sharp peaks in 1962, 1966 and 1970. The expectation of profitable opportunities led to new investment in large plants; as these plants came on stream capacity temporarily overtook demand; prices were depressed and investment was cut back. Eventually conditions improved and the cycle began again. Whether this pattern of behaviour will continue in the 70's is a moot point; a recent study suggests that cash flow may prove the limiting factor in the future.

The industry is also research-intensive. Since the First World War it has invested heavily and continuously in research and development: in 1967–8 it employed 8 700 graduates and 3 800 technicians on such work, upon which it spent £71 m. Over 90% of this money was provided by the companies themselves, whereas all other research-intensive industries relied on the government for the large majority of such funds. The pharmaceutical sector spent £17 m. and plastics £13 m.

Conversely the chemical industry uses men relatively sparingly. In 1968 employment was 410 500 – that is to say 5·1% of the total for manufacturing industry. Value added per head was 71% higher than the national average. The composition of the labour force, also differed from that of other industries: whereas operatives comprised 75% of those employed by manufacturing industry as a whole, they were only 60% of those employed by the chemical industry. The industry requires fewer manual workers than average and proportionately more managers, salesmen, research workers and clerical staff. The staff is notably better educated than is usual: 12% are qualified scientists or engineers compared to 5·4% for all manufacturing industry.

It is not surprising to find that the chemical industry is an industry dominated by large companies. Although 64% of all UK chemical companies employ less than 100 people and only 3% more than 1 000, the latter account for 39% of total employment and approximately 80% of total output. Only large companies can afford the high capital investment and research expenditure which is the price of survival in many parts of the industry.

The UK chemical industry is geographically very widely spread, some of it even being located in the favoured dormitory areas of south-east England. Land requirements, as previously noted, are not critical, though access to docks is useful. The organic and inorganic chemical industries are inter-related within themselves to such an extent that it is usual to make a large number of products on a single site. This ensures use of steam, by-product hydrogen, etc. Also, the product from one process is frequently the raw material for another and it is more convenient simply to be able to pump it over a fence. Such large complexes of chemical plant have been built for example at Baglan Bay in Wales, Wilton/Billingham on Tees-side, Grangemouth in Scotland, Saltend near Hull. The overall distribution is shown in Fig. 6.3.

Unlike the car industry, most of the British chemical industry is British owned. The only substantial foreign interests are in the pharmaceutical industry and in Shell and Unilever both of which are partly Dutch. The UK subsidiaries of Esso and Monsanto do not yet threaten the British giants like Imperial Chemical Industries and BP Chemicals. The US giant Procter and Gamble shares the soap and detergent market with Unilever, but this is a relatively small sector of the chemical industry.

There has been a tendency in recent years towards vertical integration in the chemical industry. Companies making chemicals have been anxious to ensure their supplies of petrochemicals by merging with oil companies; oil companies have tried to ensure outlets for their petrochemicals by taking over chemical companies. Further down the chain of processes, firms making plastics have tried to ensure their markets by going in for plastics fabrication and firms making paint raw materials have decided to enter the paint market itself.

Examples of the oil companies entering the chemicals field include the setting up of Esso Chemicals and Shell Chemicals by Esso and Shell, and the purchase by British Petroleum in 1967 of the chemicals and plastics interests of Distillers Company which then became BP Chemicals. Examples of chemical companies processing their own raw materials include the entry of ICI into the domestic paint market after the Second World War, and the takeover by Distillers Company of British Xylonite, a plastics fabricating firm, in 1962.

There are arguments for and against vertical integration. It is currently very fashionable but one of its chief opponents is Marks and Spencer whose efficiency is a by-word. There is some evidence that most vertical integration is taking place not because it leads to greater efficiency by reduction of selling and administrative costs, but in order to restrict competition. It is noticeable that the West German and United States chemical industries, which are legally prohibited from participating in this sort of arrangement, have easily the best profit records of any such industries.

The UK chemical industry

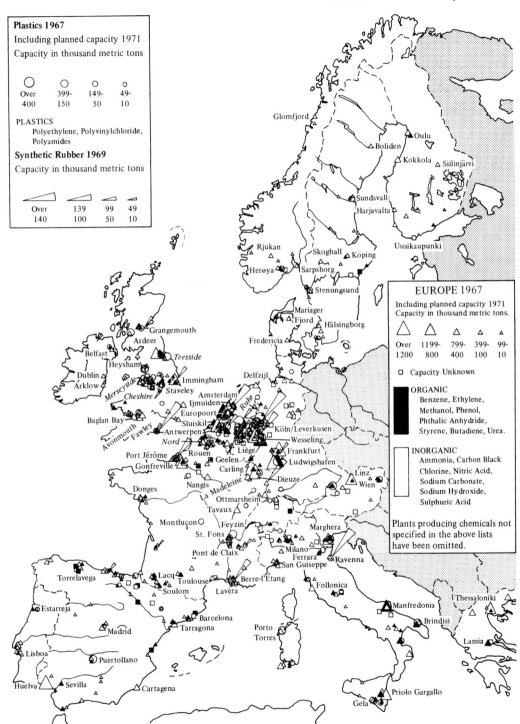

Figure 6.3 Location of the European chemical industry.

6.1.4 The performance of the UK chemical industry

In many ways the history of the British chemical industry is a success story. Its contribution to the Gross National Product has risen from 1·1% in 1907 to 2·1% in 1935 and 3·3% in 1968. Since the Second World War it has grown much more rapidly than industrial production in general: thus between 1958 and 1967 it grew at more than twice the rate of British manufacturing industry as a whole (Table 6.4). Since 1970, however,

Table 6.4 Growth of chemical and industrial production 1958–67

	Industrial production (excl. construction)	Chemical production	Ratio
UK	3·1*	6·6	2·1
W. Germany	5·1	11·0	2·2
France	5·1	9·9	1·9
Italy	8·9	13·4	1·5
USA	6·0	8·3	1·4

* Figures are % per year growth rate

this growth rate has been sharply reduced and it is now only marginally greater than the average for the national economy. A capital intensive industry like chemicals is far more sensitive to slumps than a labour intensive industry which can lay off men. As Fig. 6.4 shows, growth has been most rapid in synthetic resins, in organic chemicals – especially those used to make polymers – and in pharmaceuticals. Older products, such as fertilizers, inorganic chemicals, and soaps and detergents have grown much more slowly.

The chemical industry is the nation's third largest exporter, exceeded only by 'machinery and transport equipment' and 'articles classified by material', and in 1968 accounted for about 10% of the total. The most important categories (cf. Table 6.2) were general chemicals, synthetic resins and rubber, and pharmaceuticals. Between 1965 and 1970 exports of chemicals increased considerably more rapidly than did UK sales, and exports became of even greater importance than previously. In 1968 the main markets were the Commonwealth (24%), the Common Market (18%), EFTA (14%), other West European countries (8%) and North America (8%); in recent years exports to Western Europe have been growing much more rapidly than those to the Commonwealth. The UK remains the third largest exporter of chemicals in the world after the USA and West Germany.

Imports of chemicals are substantially smaller than exports. In 1968 exports were £645 m. and imports were £435 m., so yielding a favourable balance of £210 m. It is interesting to note that general chemicals and synthetic resins and rubber are both imported and exported on a large scale whereas most other sectors of the industry export much more than they import. A more detailed examination of the trade statistics suggests that Britain tends to import intermediates and export finished products. The main suppliers were the Common Market (41%), North America (29%) and EFTA (14%); once again the trend is toward greater trade with Western Europe at the expense of other areas.

Rapid growth has not led to great prosperity. Between 1957 and 1966 the pre-tax

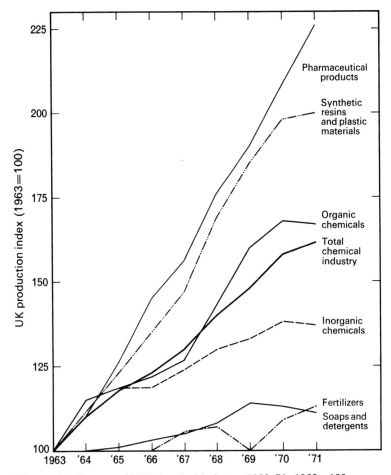

Figure 6.4 Output of UK chemical industry 1963–71. 1963 = 100.

profit margins of companies in the chemical and allied industries varied between 8·3 and 12·6% of turnover, figures appreciably but not dramatically higher than those for manufacturing industry as a whole. Moreover, these modest margins must be seen in the context of a capital-intensive industry. In the period 1960–70 the pre-tax profits of quoted chemical companies were about 12% of capital employed. Once more this figure is close to that for manufacturing industry as a whole. There are obvious reasons for this profitless prosperity. The world chemical industry is intensely competitive and price wars have been frequent, especially in fibres, plastics and fertilizers. Chemical prices have therefore risen less rapidly than those of other products. At the same time capital investment has to be high, depreciation charges are heavy, and profits must necessarily suffer.

How does the UK chemical industry compare to that of other European countries? Not brilliantly. During the 1960s almost all European nations recorded a faster growth in chemical output than did Britain. Value added has also grown more slowly. Much of this comparative failure is due to the slow growth of the UK economy: indeed the ratio between the rates of growth in chemical production and in all industrial production in the period 1958–67 was higher in Britain than in any other country except West Germany (cf. Table 6.4). Given its opportunities the UK industry has done quite well. There are, however, some disturbing features about its performance. Both the productivity of labour and the productivity of capital appear to be lower than in other West European countries. Value added per employee was above the West European average in 1958 but about 10% below it by 1969; measured in this way British productivity was lower than that of any other major chemical producer except Italy. As a percentage of output, UK investment in the chemical industry has been only slightly below the European average, but the productivity of investments has been substantially less. Between 1958 and 1966 the annual increase in value added was between 16 and 55% of new investment compared to 31 to 73% for West European countries as a whole. It is difficult to avoid the conclusion that the British industry uses its resources less efficiently than do its continental rivals.

6.1.5 Individual companies

The history of the British chemical industry and the various firms which constitute it has been dealt with in detail elsewhere and a number of references are given in the bibliography. To provide some idea of the relative importance of the different companies, Table 6.5 gives financial data about the largest of them. In addition, details are given at the end of the section of the two Anglo-Dutch giants, Royal Dutch Shell and Unilever. Paint and textile companies (e.g. Courtaulds) are excluded by the *Chemical Age* definition as are foreign owned subsidiaries such as Esso Chemicals or Monsanto. These, and subsequent, financial data come from a feature on the 'top two hundred' world chemical companies which appears annually in *Chemical Age*.

6.1.5.1 Imperial Chemical Industries The origins of ICI were covered in Chapters 1 and 2. Since its acquisition of the US firm Atlas and the devaluation of the dollar, it has become the largest chemical company in the world, and it is larger than any other British chemical company by an order of magnitude. It has never been as profitable, however, as its sheer size would suggest. Between 1962 and 1972 its return on capital (before tax and interest) oscillated between 8·1 and 12·4%, figures which compare unfavourably with Du Pont or its German rivals, and are not much better than those which could be obtained with less risk from investment in a building society.

As ICI participates in almost every sector of the chemical industry, it is instructive to examine the profits currently being made in each sector and these data are given in Table 6.6. The problems of world overcapacity for fibres and plastics are clearly illustrated as is the present high profitability of pharmaceuticals.

The reasons for ICI's comparative failure are complex. As the only British company which makes all kinds of chemicals, ICI has inevitably been hindered by the slow growth rate of the economy. There is evidence, in addition, that it makes too many chemicals; it offers perhaps twice as many products as Du Pont. The reason for this is related to ICI's semi-monopolistic position in the UK and its consequent fear of government intervention should it fail to serve the 'national interest'. It also reflects the reduced opportunities for specialization offered by the smaller size of the UK market.

ICI's problems were compounded from 1950 onwards by repeated failures of management. From 1930 to 1950 the Company Chairman was Lord McGowan, a tough veteran

Table 6.5 The top ten British chemical companies in 1970

World rank in 'Chemical Age 200'		Group sales ($ m.)	Assets ($ m.)	Net profits after tax ($ m.)	Capital spending ($ m.)	R & D spending ($ m.)	Employees	$\frac{Profits}{assets}$ (%)	$\frac{Profits}{sales}$ (%)	$\frac{Profits}{employees}$ ($/head)	$\frac{Turnover}{employees}$ ($ '000 head)
2	Imperial Chemical Industries	3 509.8	3 979.2	237.6	412.8	122.6	194 000	6.0	6.8	1 224	18.1
48	British Oxygen Co.	498.9	574.1	18.2	62.7	6.2	39 400	3.2	3.7	460	12.7
69	British Petroleum Co.*	*390.0*	n.a.	n.a.	199.2	n.a.	n.a.	n.a.	n.a.	n.a.	n.a.
73	Glaxo group	378.0	233.0	29.8	48.9	9.6	29 214	12.8	7.9	1 020	12.9
89	Albright and Wilson	293.5	259.2	2.7	18.3	n.a.	15 000	1.4	0.9	180	19.6
98	Fisons	264.2	162.1	8.3	10.3	7.9	8 419	5.1	3.1	988	31.5
112	Wellcome Foundation	205.9	80.8	11.2	n.a.	13.9	13 242	13.9	5.5	845	15.6
123	Rio Tinto Zinc Corp.*	182.2	187.4	19.4†	n.a.	n.a.	n.a.	10.3	10.6	n.a.	n.a.
147	Beecham Group*	132.2	n.a.	38.4†	n.a.	n.a.	n.a.	n.a.	29.0	n.a.	n.a.
163	Laporte Industries	111.1	146.9	4.6	15.4	n.a.	5 573	3.1	4.1	827	19.9
ANGLO-DUTCH COMPANIES											
14	Royal Dutch/Shell*	1 318.1	n.a.	n.a.	173.0	n.a.	n.a.	n.a.	n.a.	n.a.	n.a.
50	Unilever*	477.6	n.a.	40.8†	65.5	n.a.	n.a.	n.a.	8.5	n.a.	n.a.

* Figures are for chemicals only   † Operating profits
Italic figures are estimates   n.a. not available

The UK chemical industry 115

The world chemical industry

Table 6.6 ICI 1971 – Analysis of sales and profits by sector

	Sales (£ m.)	Profit* (£ m.)	Profit as % of sales
United Kingdom Divisions and Subsidiaries			
Agricultural chemicals	158	18	11·4
Fibres	149	5	3·4
General chemicals	282	34	12·0
Paints	65	8	12·3
Petrochemicals and plastics	274	9	3·3
Pharmaceuticals	33	11	33·3
Imperial Metal Industries	180	15	8·3
Miscellaneous	—	—	
Overseas subsidiaries	649	45	6·9
Total	1 810		
Less inter-class sales	286		
Total as in profit and loss account	1 524	145	9·5

* Profit after depreciation but before tax

from the Glasgow slums, who knew nothing about chemistry and everything about market-sharing agreements. After his departure, a demoralized management found itself facing a totally new situation in which the cartel situation had broken down and technologies were changing rapidly. The emphasis had to change from inorganic chemicals to petrochemicals, but it was not till Sir Paul Chambers (Chairman 1960–68) that a major effort was made to drag the company into the modern era. The aim was achieved but only at the expense of some costly blunders. For example, a new acrylonitrile plant based on acetylene and hydrocyanic acid was mothballed on the day it was completed because the process was obsolete (section 10.7.4). A vinyl acetate monomer plant based on the Wacker process (section 10.5.1) corroded away and had to be shut down because it was built in mild steel and stainless steel instead of the recommended stainless steel and titanium, and a gross overexpansion of nylon capacity to 250 000 tons when the UK market was only 40 000 tons resulted in the subsequent write-off of huge amounts of equipment. Overlicensing of its polyester process added to the problems with fibres.

Cash flow problems exacerbated the technical ones. A company of the size of ICI would normally finance expansion out of retained profits and depreciation but ICI had to issue £50 million $7\frac{1}{4}$% loan stock in 1965 and another £60 m. 8% stock in 1966. There has been a question for some years as to whether investors would take up an equity issue if one were made. At first sight, too, the £66 m. ICI had to pay for Atlas (1970 turnover £65 m.) is a high price, especially as the US Federal Trade Commission insisted that ICI divest themselves of Atlas's explosives and aerospace sectors, the first of which accounted for 19% of turnover. The purchase had to be financed by yet more loan stock, and the servicing of this long-term debt must be a constant drag on profitability.

By comparison with other giants, though not with other UK companies, ICI suffers from low capital and low labour productivity. It remains to be seen whether the planned 10% reduction in staff (about 20 000 people!) improves the latter, but it may simply be that ICI has become too large for its own good. It is too big to be taken over and is effectively free from government or stock market control. Yet its middle management and research and development are excellent and it is an oversimplification to see ICI as a blind dinosaur crashing through the industrial jungle. All companies make mistakes and large companies are always beset by organizational problems. The organic chemical industry has changed so fast in the past 30 years that it is perhaps inevitable that some plants should become obsolete between the decision to build and start-up. It may be significant that ICI has weathered the current recession better than many and its 1971 figures compare quite well with those of its main rivals.

6.1.5.2 British Oxygen Company BOC has been a manufacturer of industrial gases since its incorporation as Brins Oxygen Co. in 1886. In addition to the main industrial gases, oxygen, nitrogen, hydrogen, argon and acetylene, it makes synthetic resin emulsions and melamine, refines tall oil and has extensive engineering interests involving welding equipment and plant for producing gases and using them. It was harmed in the 1950s by its conviction that acetylene was going to be the key intermediate in industrial organic synthesis, but profited in the 1960s from developments in welding technology and from the changeover of the steel industry to the Linz–Donawitz (LD) process of steelmaking in which oxygen is blown through steel. The average consumption of oxygen per ton of steel rose from 100 ft^3 in 1958 to 1 600 ft^3 in 1971.

BOC makes calcium carbide and dicyandiamide in Norway where electricity is cheap. It also set up the Wulff process for acetylene (section 7.7) in Northern Ireland and while this will never reach rated capacity, £0·5 m. of modifications have given hope that it will break even.

The management were in trouble with the Monopolies Commission in the 1960s for selling their gases at a loss in order to put a competitor out of business and then raising prices sharply. They were compelled to allow a competitor – Air Products, which by 1970 held about one-third of the industrial gases market – into their field, and the resulting trauma led to a largely effective overhaul of management. Meanwhile, in 1970, the chemicals sector of their activities (8% of sales) made zero profit while the gases sector (50% of sales) made 65% of total profits and the returns on capital or per employee can scarcely be regarded as satisfactory.

6.1.5.3 BP (Chemicals) International The Anglo–Persian Oil Co. was set up in 1909 and changed to the Anglo–Iranian Oil Co. in 1935. After expropriation by the late Dr Mossadeq, it changed its name to the British Petroleum Co. in 1954. It edged cautiously into chemical manufacture in the late fifties and early sixties forming joint companies (British Hydrocarbon Chemicals, Grange Chemicals, Forth Chemicals, Border Chemicals etc.) with established chemical manufacturers, especially the Distillers Company Ltd. (DCL). It came as no surprise when it took over almost the whole of DCL's chemical interests in 1967, and BP Chemicals International consists largely of the old DCL Chemicals and Plastics divisions.

Distillers themselves were the result of the amalgamation in 1877 of six grain distillers with a view to production of standardized blended whiskies of the type which are now internationally consumed. They got into industrial organic chemicals in 1928–29 when their Hull factory was built alongside an established distillery producing industrial alcohol and vinegar. Their early chemical activities, which included operation of the Weizman Process, were fermentation based.

DCL's experience of deep-fermentation techniques led to its being asked to build the Government deep-culture penicillin factory at Speke, Liverpool, towards the end of the war. This later became the nucleus of DCL's biochemicals division, but the company was unable to provide sufficient funds to finance the research activities required by a modern pharmaceutical company and it sold out to Eli Lilly and Co. in 1962. The biochemical division was involved in the thalidomide tragedy though it was only importing and not manufacturing the product. The sale of the division left DCL active principally in heavy organic chemicals and plastics.

Distillers were fortunate in having their research directed by H. M. Stanley, the greatest British industrial chemist since W. H. Perkin. He can claim the major share of the credit for the cumene-phenol process (section 9.5.1.1), the PFD route to acetic acid (section 9.4), one version of the ammoxidation route to acrylonitrile (section 10.7.4), butylene based chloroprene (section 10.6.7) and one of the propylene-based routes to acrylates. He pushed Distillers into petro-chemicals before any other British company, but it seemed that the whisky and gin magnates on the board of directors would never back his processes without at least one other company sharing the risk, and between 1945 and 1967 they spawned joint ventures with a wide variety of British, French and American companies. With most of its capital tied up in stocks of maturing whisky, DCL was continually faced by funding problems and never managed to capitalize on its brilliant research and development record. The takeover by BP has certainly led to better organizational structure and financial control but, as chemicals profits are not divulged, it is impossible to assess if these changes have even stemmed the tide which is eroding chemical profits.

In a management newsletter circulated at the end of 1971, the Managing Director of BP Chemicals International, D. G. L. Bean, said that in view of the world situation 'it is not altogether surprising that I have to report a poor profit performance, . . . but it is a matter for concern that by most standards we have to admit that we are less successful than our main competitors . . .'.

BPCI are active in what are currently the least profitable areas of the chemical industry. Could it be that they are actually losing money?

6.1.5.4 Glaxo Group Glaxo Group is an association of companies of which Glaxo Laboratories is the largest unit. The group makes pharmaceuticals, including veterinary medicines and agricultural products, and also baby and health foods, surgical equipment and related products. The group has strong management and an excellent growth and profit record, but a reputation for being weak on the selling side. This led to a takeover bid by Beechams at the beginning of 1972, contested by Boots Pure Drug Co. who proposed a merger. Both courses of action were vetoed by the Monopolies Commission.

The Company's main pharmaceutical is Betnovate (betamethasone), a steroid used for skin conditions, some allergies and general steroid therapy. In 1970/71 it accounted for 28·5% of Glaxo's UK sales of ethical drugs. Major contributions are expected in the future from Ceporex, a cephalosporin antibiotic related to the penicillins, and Ventolin, a bronchodilator for use in asthma and bronchitis.

6.1.5.5 Albright and Wilson Arthur Albright started to manufacture white phosphorus for matches in 1844 and perfected the non-toxic red variety in 1852 prior to going into partnership with his fellow Quaker, J. E. Wilson, in 1856. The company remained primarily in the phosphorus industry till the formation of Midland Silicones Ltd. in 1950 and the takeover of Marchon products in 1955, and Boake Roberts in 1960 which took them into detergents, plasticizers, and flavours and essences.

In the late 1960s, a management consultant recommended the closure of the company's scattered phosphorus plant in the UK and its rebuilding in Newfoundland. The advice was taken but proved a disaster, for the new plant was beset by formidable technical and pollution problems which were exacerbated by its remoteness. It is claimed that many of these problems had been solved by 1972, but they are clearly reflected in the 1970 and 1971 profit figures. The company had to be rescued from its dire financial straits by the US company Tenneco (*Chemical Age* ranking 104 on chemical sales only) which paid £17·5 m. in return for convertible loan stock and which now has effective financial control of the group.

6.1.5.6 Fisons Fisons had its roots in the manufacture of fertilizers (mainly superphosphates) in the middle of the nineteenth century. It now manufactures a wide range of fertilizers, agrochemicals, pharmaceuticals, industrial chemicals and scientific apparatus.

Fisons return on capital has been unimpressive. Fertilizer margins have been squeezed for many years and Fison's attempts to diversify have been successful in pharmaceuticals but very much less successful in heavy chemicals.

For their present profits, Fisons rely heavily on pharmaceuticals (1969/70 14·7% of sales but 44·4% of profits) particularly their anti-asthmatic and anti-hay fever drug Lutal (mainly disodium cromoglycate). This accounted in 1969 for £2·3 m. out of £3·35 m. of their UK ethical pharmaceutical sales, but may suffer from competition in the future with Glaxo's Ventolin.

6.1.5.7 Wellcome Foundation When he died in 1936, Sir Henry Wellcome, sole owner of the Wellcome Foundation, bequeathed the company's shares to the Wellcome Trust, a recognized public charity whose aims include the advancement of research in human and veterinary medicine and the financing of research museums and libraries. The foundation trades as Burroughs Wellcome and Co. and makes pharmaceuticals, agricultural chemicals, etc. Its profits record is excellent though the return on capital is probably exaggerated by undervaluation of assets. Nonetheless it has the world market for a number of important vaccines.

6.1.5.8 Rio-Tinto Zinc Corporation The Rio-Tinto Zinc Corporation is a large, somewhat chaotic conglomerate comprising many virtually independent subsidiaries in many countries. Its chemical interests include production of inorganic compounds of metals such as titanium, zinc, aluminium, thorium, barium and the rare earths. A takeover of Borax Consolidated in 1968 took RTZ into boron chemistry (borax, sodium perborate, etc.) and into inorganic acids, pigments and related chemicals.

RTZ has large mining interests. In 1970 the earnings came 33% from copper production, 20% from iron ore, 18% from aluminium, 12% from boron chemicals and 9% from lead and zinc. It is currently diversifying into exploration for gas and oil in the North Sea. Its profitability is fairly good – probably because it is not competing in the oversupplied commodity markets. On the other hand, its Bristol lead and zinc smelting plant was closed for ten weeks early in 1972 because of toxic hazards. Also, pollution problems in Japan have reduced the level of Japanese copper smelting and as a consequence the production of RTZ's copper mines in New Guinea has been cut. Thus the danger to RTZ's profitability comes less from overproduction than from environmental controls.

6.1.5.9 Beecham Group The Beecham Group has developed in the period from the late nineteenth century as manufacturers of proprietary medicines (the famed Beecham Pills), toilet products (Brylcreem), drinks (Idris, Ribena, Horlicks), and canned foods. On this somewhat unlikely base, H. G. Lazell, chairman of Beecham Products, decided on the

£3·5 m. research budget which in 1963 led to the semi-synthetic penicillins (section 13.6.1). By 1967/68 pharmaceuticals accounted for 22% of total sales of the Beecham Group. In 1971 they made a takeover bid for Glaxo which was contested by Boots Pure Drug Company and was eventually vetoed by the Monopolies Commission. Beecham's has a dynamic management and a splendid growth and profit record. They have great hopes for their new semi-synthetic penicillin, Amoxil, but badly need some new pharmaceuticals to back up the penicillins.

6.1.5.10 Laporte Industries   Laporte were importers of chemicals for textiles in the 1880s, and in 1888 started to produce hydrogen peroxide in Shipley, Yorkshire, as a bleach for straw used in straw hat manufacture. They expanded by a variety of takeovers but remained in the field of hydrogen peroxide and related materials. They also produce acids, fluorine chemicals, phthalic anhydride, fuller's earth and titanium dioxide. In the late 1960s they decided to build plant to manufacture whiter-than-white titanium dioxide by the chloride route (section 10.5.6). The investment was large for a company the size of Laporte and the project was dogged by technical and cash-flow problems. The plant should have been commissioned early in 1970 but had still not operated near full capacity early in 1972.

The company was partly rescued by Solvay who arranged for joint marketing of perborate, but the problems of the titania plant are likely to depress the company's financial position for many years to come.

6.1.5.11 Royal Dutch/Shell   The Shell Company was originally an importer of sea shells which were much in demand for decorative purposes in the Victorian era. Importing led to trading; trading led to dealing in petroleum products. In 1885 Shell produced the first sea-going tanker in which kerosene was shipped in bulk rather than in drums – with a consequent saving in transport costs. From transporting petrol, Shell diversified slowly into production and refining. The original Shell was a one-man company under the autocratic and unpredictable control of Marcus Samuel. During the 1890s he became increasingly interested in political life and the company ran into difficulties and was forced to amalgamate with the Royal Dutch Petroleum Co. in 1904.

Since that time, Royal Dutch/Shell has been one of the seven major oil companies. Shell Chemical Co. was formed (as Technical Products Ltd.) in 1928 to market petroleum based products and before the Second World War they marketed and then manufactured a secondary alkyl sodium sulphate detergent called 'Teepol' after the company. Like most oil companies, it diversified into petrochemicals, the impetus coming mainly from the US subsidiary. Petrochemicals and plastics are now manufactured together with a range of agricultural chemicals. Financial data for the chemical activities are not available, but the company has elaborate management training schemes and a reputation for sound management. They have responded to recent overcapacity in the chemical industry by closing certain plants and cancelling many of their proposed expansion plans. It seems probable that in 1971 they lost money on their chemical activities and they are obviously prepared to be ruthless about cutting such losses.

6.1.5.12 Unilever   Unilever was formed in 1929 by the amalgamation of the Dutch Margarine Union with Lever Bros. of England. William Lever, an embodiment of the Protestant ethic, was a Liverpool grocer who started to sell Sunlight Soap in 1883 and afterwards expanded into production and marketing on a national scale. By the First World War he controlled half of the UK soap capacity, and a post-war buying spree gave him control of Wall's Sausages, Mac Fisheries and the United Africa Company among others. The acquisitions led to severe problems and professional management moved in in 1922.

Unilever and its foreign associates are one of the biggest companies in the world selling all kinds of cleaning and toilet products, many kinds of foods (Bird's Eye frozen foods, margarine, ice-cream, sausages) wood and paper products and a range of chemical products apart from detergents, which are excluded from the figures given. The United Africa Co. remains the biggest trading company in Africa.

Unilever has suffered from many of the managerial problems which afflict ICI and has never proved as profitable as its small, more single-minded competitor, Procter and Gamble (section 6.2.1.1).

6.2 The chemical industry in the developed world

In the world chemical industry it is the big, highly industrialized countries which count. Countries containing 30% of the world's population produce 90% of the chemicals; six countries (USA, West Germany, UK, France, Italy, Japan) account for 75%. Three-quarters of the 250 largest chemical companies in the world in 1970 (excluding countries with centrally planned economies) were in only five countries (Fig. 6.5). Furthermore, trade in chemicals occurs largely between the countries of the developed world.

The developed countries outside the Iron Curtain belong to a club, the Organization for Economic Co-operation and Development (OECD), which was set up in 1960. It has suitably lofty ideals – high employment and economic growth, rising standards of living – and aims also at expansion of world trade on a multilateral non-discriminatory basis so as to help both non-members and its poorer members in the process of economic expansion. The members of this charmed circle are Austria, Belgium, Canada, Denmark, Fin-

Table 6.7 Turnover, valued added, investment and number of employees in the chemical industry in some OECD countries in 1969. (Financial data in $m.)

	Turnover	Value added	Investment	Employees, '000
USA	48 760	27 000*	3 100	1 049
Japan	11 245	4 625	1 892	428
W. Germany	10 600	6 095	1 100	503
France	7 475	3 130	560	270
UK	7 470	3 575	616	418
Italy	5 680	1 920	480	245
Netherlands	2 485	910	359	91
Canada	2 274	1 270*	206	71
Spain	2 175	685	98	107
Belgium	1 520	560	226	60
Switzerland	1 205	n.a.	n.a.	50
Sweden	930	485	102	36

* estimated

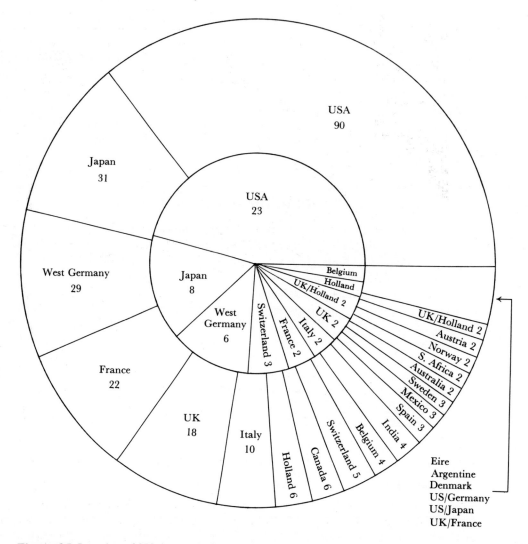

Figure 6.5 Location of 250 (outer circle) and 50 (inner circle) largest chemical companies in the world in 1970.

land, France, Federal Republic of Germany (West Germany), Greece, Iceland, Ireland, Italy, Japan, Luxembourg, the Netherlands, Norway, Portugal, Spain, Sweden, Switzerland, Turkey, the United Kingdom and the United States of America.

Figure 6.6 shows the consumption of chemicals per head in some of the OECD countries and Table 6.7 shows turnover, value added, investment and labour forces.

The outstanding changes in the OECD chemical industry over the past twenty years have been the rapid industrialization of Japan and Spain, the increasing prosperity of the Netherlands and Belgium, the slowing down of Italy's growth and the tendency of the UK to lose ground on almost every economic indicator.

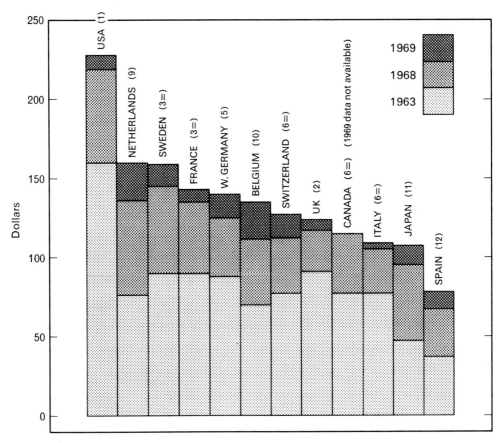

Figure 6.6 Estimated consumption of chemicals per head of population in 1963, 1968 and 1969. Numbers in parentheses show 'positions' in 1963.

In such a welter of statistics, it is reasonable to ask if there is a single economic indicator which measures the efficiency of an industry. There are in fact a number of indicators each useful for different purposes. For example, an investor considering buying shares would want to know the distributed profits per unit of equity capital, or the share price to company earnings ratio. For the purpose of a global look at the chemical industry, the two most important indicators are value added per employee (=productivity) and value added per unit of capital employed. The latter figure, however, is almost impossible to obtain for individual companies because of differences in accounting procedures, and the former figure will be used here. Columns 3 and 5, Table 6.7, show value added and number of employees for the major OECD countries and Fig. 6.7 shows how their ratio has varied over the past 12 years. The USA is spectacularly more productive than any other country. The UK and Italy are losing ground and Japan is gaining it.

The location of chemical industries in Western Europe and in the USA is illustrated in Figs. 6.3 and 6.8. There are obvious advantages in building refineries and petrochemical plant adjacent to oil wells and this situation obtains on the US Gulf Coast. Crude oil is cheap to ship compared with manufactured products, however, and in

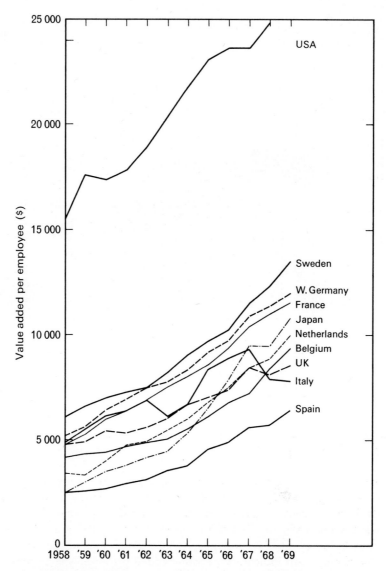

Figure 6.7 Value added per employee for the chemical industries of some OECD countries.

Europe refineries and petrochemical plant tend to be adjacent to one another and to markets rather than to oil wells, a fact not unrelated to political instability in the Middle East. Location of chemical plant, however, is also influenced by historical factors (where a company built its first units), availability of labour, government pressures, access to deep-water ports, cost of electricity and many other factors, and hard and fast rules cannot be laid down.

The growth of the chemical industry in Europe, the USA and Japan is illustrated in

The chemical industry in the developed world 125

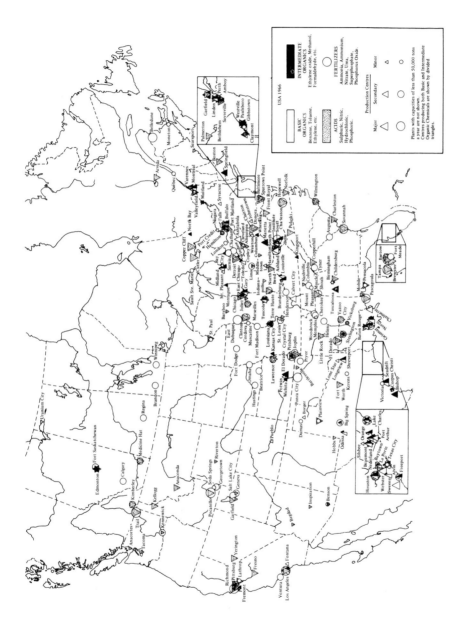

Figure 6.8 Location of chemical industry in North America in 1966.

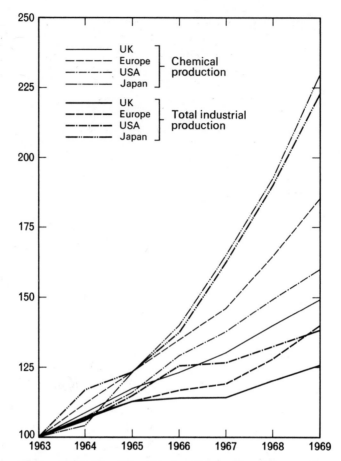

Figure 6.9 Indices of production for the European member countries of OECD, the United States and Japan. 1963 = 100.

Fig. 6.9, and the development of various sectors in Fig. 6.10. These emphasize points made previously about the rapid expansion of Japan, and the boom in organic chemicals and plastics compared with the relatively dull performance of dyes, paints and heavy inorganic chemicals.

6.2.1 Individual countries

The chemical industry in all West European developed countries is broadly similar to that in the UK. It is capital intensive; it grows at about twice the rate of the overall economy. Its fastest growing sectors have been organic chemicals and plastics but these are now suffering from overcapacity. The USA differs in terms of both scale and availability of raw materials. In the following section, individual countries and their major chemical companies are briefly discussed.

Table 6.8 shows the world's fifty largest chemical companies taken from the *Chemical Age* 'top two hundred'. Table 6.9 shows some of these companies listed by country to-

The chemical industry in the developed world 127

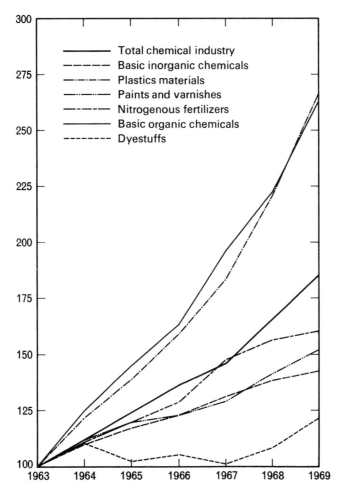

Figure 6.10 Development of the chemical industry and of certain branches in Western Europe. 1963 = 100.

gether with various ratios which enable their profitabilities to be compared. Most of them emphasize the dominance of the American chemical industry in world markets.

On a time basis, most of these ratios are falling. The ratio of profits to sales for the major US companies was 8·8% in 1961, rose to 10·1% in 1965 and dropped to 6·4% by 1970. German and UK figures show the same peak and subsequent fall. That the patterns follow so closely underlines the point that the chemical industry is truly international.

Data exclude the so-called 'centrally planned economies' – Russia, China and their satellites. These countries have no chemical companies in the capitalist sense and are discussed in sections 6.2.1.8, 6.2.1.9 and 6.3.

6.2.1.1 United States of America The chemical industry of the United States is the largest in the world. Its size is impressive: in 1969 it accounted for 48% of chemical sales and

55% of the value added by the chemical industry in the OECD countries. To a considerable extent this reflects the size of the American economy. The American chemical industry, like that of Japan, but unlike those of Western Europe, depends almost entirely on domestic demand. Foreign trade is of minor importance in the American chemical economy: exports account for only 7% of sales, and imports – largely excluded by high tariffs – to a mere 2·5%. This is, however, a limitation as well as an opportunity, as the fortunes of the chemical industry are even more than normally at the mercy of trends in the national economy. During the past ten years, economic growth in the USA has been

Table 6.8 Fifty largest chemical companies in the world

Ranking '70	'69	Company	Country	Group sales ($m.)	Assets ($m.)	Net profits ($m.)	Capital spending ($m.)	R & D spending ($m.)	Employees
1	(1)	Du Pont, E. I. de Nemours and Co.	US	3 618·4	3 566·6	328·7	471·0	250·0	110 685
2	(2)	Imperial Chemical Industries Ltd.	Britain	3 509·8	3 979·2	237·6	412·8	122·6	194 000
3	(3)	Union Carbide Corp.	US	3 026·3	3 563·8	157·3	393·7	78·0	102 144
4	(5)	Farbwerke Hoechst AG	Germany	2 882·2[1]	3 584·7	88·1[2]	393·0†	99·0†	139 460
5	(6)	Badische Anilin und Soda-Fabrik AG	Germany	2 874·3	1 969·1[2]	73·2	444·5	116·1	106 776
6	(4)	Montecatini-Edison SpA	Italy	2 841·6	3 332·8[2]	0[2]	n.a.	79·2	162 791
7	(7)	Farbenfabriken Bayer AG	Germany	2 602·2	2 176·7[2]	80·8[2]	511·4	127·6	110 200
8	(12)	Akzo NV	Holland	2 014·0	2 341·0	67·0	278·0	56·0	100 800
9	(9)	Rhône-Poulenc SA	France	1 984·9	2 439·5	82·7	335·9	85·0	120 514
10	(8)	Monsanto Co.	US	1 971·6	2 144·7	66·5	300·8	98·1	62 940
11	(11)	Grace, W R & Co.	US	1 917·6	1 575·2	30·2	110·4	20·8	61 700
12	(10)	Dow Chemical Co.	US	1 911·1	2 779·8	103·4	348·1	91·5	47 400
13	(—)	Ciba-Geigy AG[3]	Switzerland	1 612·5	1 507·9	19·5	179·3	104·4†	68 455
14	(14)	Royal Dutch/Shell*	UK/Holland	1 318·1	n.a.	n.a.	173·0	n.a.	n.a.
15	(—)	Warner-Lambert Co.[1]	US	1 256·6	1 006·8	98·3	58·2	54·1	55 500
16	(13)	Allied Chemical Co.	US	1 248·5	1 582·0	68·3	139·2	28·5	32 766
17	(17)	American Cyanamid Co.	US	1 158·4	1 065·9	91·5	93·0	46·0	38 020
18	(19)	Mitsui & Co.*[5]	Japan	1 156·0	n.a.	n.a.	n.a.	n.a.	n.a.
19	(21)	Ugine-Kuhlmann	France	1 152·6	753·6	39·4	138·3	n.a.	46 500
20	(18)	Hoffmann La Roche, F. & Cie AG	Switzerland	1 130·0†	n.a.	n.a.	n.a.	n.a.	n.a.
21	(16)	Esso Chemical Co.[6]	US	1 111·0	n.a.	n.a.	n.a.	n.a.	n.a.
22	(15)	Celanese Corp.	US	1 037·0	1 610·7	48·5	132·0	53·4	n.a.
23	(22)	Henkel & Cie	Germany	943·4	n.a.	n.a.	n.a.	n.a.	32 000
24	(23)	Toray Industries[7]	Japan	872·9	1 073·1	44·2	168·8	20·1	20 477
25	(24)	Pfizer Inc.[8]	US	870·4	558·8	81·1	54·9	30·6	23 100
26	(—)	Mitsubishi Corp.*[5]	Japan	861·8	n.a.	n.a.	n.a.	n.a.	n.a.
27	(25)	Solvay & Cie	Belgium	840·5	1 147·0	41·5	164·2	22·0	42 914
28	(26)	Hercules Co.	US	798·6	1 080·0	49·6	91·0	22·0	22 112
29	(34)	Mitsubishi Chemical Industries Ltd.	Japan	780·1	1 368·3	18·9	n.a.	n.a.	14 574
30	(30)	Merck & Co.	US	747·6	634·4	113·5	67·8	69·0	22 300
31	(31)	Asahi Chemical Industry Co. Ltd.	Japan	679·5†	n.a.	n.a.	n.a.	n.a.	n.a.
32	(32)	Occidental Petroleum Corp.*	US	658·0	n.a.	35·0	48·0	n.a.	n.a.
33	(38)	Sumitomo Chemical Co. Ltd.	Japan	646·0	1 072·0	16·0	157·0	16·0	18 069
34	(34)	Sterling Drug Co.	US	643·9	403·2	57·3	18·9	n.a.	23 600
35	(36)	Sandoz AG	Switzerland	638·3	753·8	12·3[2]	57·3	57·3	30 399

Ranking '70 '69	Company	Country	Group sales ($m.)	Assets ($m.)	Net profits ($m.)	Capital spending ($m.)	R & D spending ($m.)	Employees
36 (29)	Degussa[5]	Germany	623·9	131·4[2]	13·0	26·1	n.a.	14 248
37 (33)	FMC Corp.*	US	602·2	n.a.	n.a.	n.a.	n.a.	n.a.
38 (40)	Lilly, Eli & Co.	US	592·3	675·7	94·4	78·0	61·0	19 300
38 (41)	Eastman Kodak*	US	588·0	n.a.	n.a.	n.a.	n.a.	n.a.
40 (—)	Teijin Ltd.[5]	Japan	586·0	678·2	32·1	87·3	10·2	14 000
41 (43)	Ethyl Corp.	US	556·9	621·0	35·6	64·2	18·0	13 700
42 (39)	Diamond Shamrock	US	555·8	662·5	30·0	56·6	1·0	10 617
43 (37)	SNIA Viscosa	Italy	552·5	943·2	n.a.	6·9	n.a.	42 753
44 (46)	Sherwin Williams[9]	US	525·8	388·1	15·1	n.a.	n.a.	20 465
45 (52)	Veba-Chemie AG[5]	Germany	509·8	235·6	23·2	62·5	n.a.	8 492
46 (—)	Mitsubishi Rayon[7]	Japan	509·3	470·4	8·9	27·9	7·0	10 675
47 (44)	International Minerals & Chemical Corp.[10]	US	505·9	529·2	4·5	13·8	n.a.	7 800
48 (58)	British Oxygen Co. Ltd.[5]	Britain	498·9	574·1	18·2	62·7	6·2	39 400
49 (45)	Stauffer Chemical Co.	US	482·5	476·7	26·0	n.a.	12·7	10 245
50 (54)	Unilever*	UK/Holland	477·6	n.a.	40·8	65·5	n.a.	n.a.

* Figures are for chemicals only
† Estimate
[1] Group sales estimated on previous basis, less VAT
[2] Parent company only
[3] Formed by merger of Geigy (No. 27 a year ago) and CIBA (No. 28)
[4] Includes Parke-Davis & Co. (No. 88 a year ago)
[5] Financial year ends September 30
[6] Formerly listed as Standard Oil Co. (New Jersey)
[7] Financial year ends March 31
[8] Name changed from Chas Pfizer & Co. Inc.
[9] Financial year ends August 31
[10] Financial year ends June 30

relatively slow and the American chemical industry has grown less than those of most other advanced countries.

Since the Second World War, the American chemical industry has been consistently more efficient than those of Western Europe or Japan. The structure of the US industry is very similar to that of other nations; the prices of chemical products and raw materials are much the same, although fuel costs are lower, but value added by the industry averaged 53–56% of turnover between 1958 and 1969 compared to 43–46% for Western Europe. Moreover, the productivity of both capital and labour is higher in the USA. A study of the incremental effect of investment showed that between 1958 and 1967 each dollar invested increased the annual value added by an average of 67¢ in the USA and 43¢ in Europe. The difference in labour productivity between America and Western Europe is even more remarkable: the value added per employee was four times as great in 1958 and still over twice as great in 1970. Even allowing for the high rates of pay which prevail in the USA, American firms spend proportionately less on wages and salaries than their European competitors.

It is clear that resources are employed more effectively in the USA than elsewhere. This is mainly due to economies of scale. Until recently, American plants were much larger than those in Europe – they still enjoy some advantage in this respect – and large installations are notably more economical in their use of capital, labour and raw materials. Good management is also important. American companies use their resources in an intelligent and flexible way which is still far from common in Europe. These advantages are, however, a declining asset: as European and American plants become similar in size and American management techniques become more widely diffused, so the European chemical industry – and still more, that of Japan – has approached American standards of efficiency. It is also clear that neither high tariffs nor high efficiency have saved the American chemical industry from the overcapacity, narrowing margins and low profits which have made the early 1970s so unpleasant in other countries. When the world economy slows down all are at risk, though some are more vulnerable than others.

In practice the major US companies have been more profitable than their European counterparts. During the 1960s the four large general chemical firms averaged 8–13% after tax on capital employed and 10–18% on the shareholders' interest – respectable if not spectacular results. Preeminent among them is E. I. Du Pont de Nemours Inc., the oldest and until very recently the largest chemical company in the world. Founded by a French refugee in 1802 to make gunpowder, it dominated the US explosives trade by 1900. From cordite it expanded into nitrocellulose plastics and lacquers in the early years of the present century, into dyes and phenolic resins during the First World War, and into ammonia and fibres in the 1920s. Heavy expenditure on research and development in the interwar era led to the discovery of nylon and chloroprene and has been continued ever since. Du Pont is perhaps unique in combining continuous innovation with high profits. It has always marketed a limited range of constantly improved products; it has concentrated its attention on rapidly developing fields; and its management, which still contains a powerful family element, has maintained a firm grasp on the company's multifarious operations. There are no better formulae for success.

The other general chemical companies have shorter histories and were of only minor importance before 1930. The Dow Chemical Company was founded in 1888 by H. H. Dow, a university student who had discovered a route to bromine, and for many years its production was limited to the halogens and their derivatives, and to magnesium compounds. During and after the First World War it diversified into the manufacture of dyes and phenol – by the well-known Dow process – and assumed major status with the commercial introduction of polystyrene in 1937. Monsanto is more recent in origin. Founded in 1902 to produce saccharin and other fine chemicals, its fortune was made by the allied blockade of Germany between 1914 and 1919 (section 1.9.3), and a vigorous policy of acquisition in the interwar period made it of general importance in the industry. As its name suggests Union Carbide was formed in 1898 to produce calcium carbide by the electrothermal process. Greatly enlarged by amalgamation in 1917 with four other companies in related fields, it subsequently developed the manufacture of organic solvents and vinyl polymers on a large scale, and has continued to diversify its interests since the Second World War. In contrast, the Allied Chemical Company has declined in relative importance since its formation in 1920. An ICI-like fusion of alkali and dyestuff manufacturers, it was second only to Du Pont between the wars but now occupies no better than a highly unprofitable seventh place in the table of American chemical firms.

The specialists are more profitable than the generalists. The drug companies are of particular significance in that they have dominated not merely the US but the world pharmaceutical industry since the Second World War. Despite heavy and continuous research and promotional expenditure, profits remain high and often amount to 20% or

Table 6.9 The biggest chemical companies in various countries

	Rank in 'Chemical Age 200'		1970 sales ($ m.)	Profit/ sales (%)	Profit/ assets (%)	Profits/ employee ($ '000)	Sales/ employee ($ '000)
USA							
	1	Du Pont	3 618	9·1	9·2	2·97	32·69
	3	Union Carbide	3 026	5·2	4·4	1·54	29·63
	10	Monsanto	1 972	3·4	3·1	1·06	31·32
	11	W. R. Grace	1 918	1·6	1·9	0·49	31·08
	12	Dow Chemical	1 911	5·4	3·7	2·18	40·32
	15	Warner-Lambert	1 257	7·8	9·8	1·77	22·64
	16	Allied Chemical	1 248	5·5	1·5	2·08	38·10
	17	Cyanamid	1 158	7·9	8·6	2·45	30·47
	21	Esso Chemical	1 111	n.a.	n.a.	n.a.	n.a.
	22	Celanese	1 037	4·7	3·0	n.a.	n.a.
		USA Average	—	6·4	5·0	1·82	32·03
WEST GERMANY							
	4	Hoechst	2 882	5·0	2·4	0·63	20·67
	5	BASF	2 874	6·3	3·7	0·68	26·92
	7	Bayer	2 602	4·8	3·7	0·73	23·61
	23	Henkel	943	n.a.	n.a.	n.a.	29·48
	36	Degussa	624	2·1	9·9	0·91	43·79
FRANCE							
	9	Rhône-Poulenc	1 985	4·2	3·4	0·69	16·47
	19	Ugine-Kuhlmann	1 153	3·3	5·2	0·85	24·79
JAPAN							
	18	Mitsui & Co.*	1 156	n.a.	n.a.	n.a.	n.a.
	24	Toray Industries	873	5·1	4·1	2·16	42·63
	26	Mitsubishi Corp.	862	n.a.	n.a.	n.a.	n.a.
	29	Mitsubishi Chem. Ind.	780	2·4	1·4	1·30	53·53
	31	Asahi Chem. Ind.	679	n.a.	n.a.	n.a.	n.a.
	33	Sumitomo Chem. Co.	646	2·5	1·5	0·89	35·75
SWITZERLAND							
	13	Ciba-Geigy	1 612	12·1	1·3	0·28	23·56
	20	Hoffmann la Roche	*1 130*	n.a.	n.a.	n.a.	n.a.
	35	Sandoz	638	1·9	1·6	0·40	21·00
OTHERS							
	2	ICI	3 510	6·8	6·0	1·22	18·09
	6	Montecatini-Edison	2 842	0	0	0	17·47
	8	Akzo	2 014	3·3	2·9	0·66	19·98
	14	Royal Dutch Shell*	1 318	n.a.	n.a.	n.a.	n.a.
	27	Solvay	841	4·9	3·6	0·97	19·59

* Chemicals only   *Italic* = estimated figure

more of the shareholders' interest. The two largest companies are American Cyanamid and Pfizer Inc. The former was organized in 1907 to manufacture the now defunct fertilizer calcium cyanamide and later diversified by the acquisition of other companies among which was the Lederle Antitoxin Laboratories. Its main growth post-war has been in pharmaceuticals but it retains substantial interests in other areas. Pfizer is more single-minded and more profitable. Active in fine and fermentation chemicals since the nineteenth century, it pioneered the large-scale production of penicillin and streptomycin and, with American Cyanamid, played a major part in the introduction of the tetracyclines. It continues to specialize in ethical and non-ethical drugs and related products such as diet foods.

Two other firms on the border of the chemical industry proper merit attention. Eastman–Kodak is the world's largest manufacturer of photographic film and equipment and a major producer of dyes, isotopic tracers and fine chemicals. As soap and detergent makers, Procter and Gamble Inc. rank behind Unilever Ltd. in size, but are far more profitable. It was founded in 1837 and by a process of organic growth had become the largest US soap firm by 1914. It expanded into the UK by buying the Newcastle soap-makers, Thomas Hedley and Co., in 1930 and has since set up subsidiaries in many other countries. Procter and Gamble's policy has always been to market a strictly limited range of products selected for their profitability. The quality of management has been exceedingly high since the early 1920s – R. R. Duepree was almost as important a figure in the development of management techniques as Alfred P. Sloan – and the company's record of profits, particularly in the UK, is an enviable one.

One of the major growth areas in the chemical industry in recent years has been US investment abroad, and often the US-owned firms in other countries are the only ones able to compete with home companies back in the US. One corollary is the growth of anti-American feeling directed against so-called economic imperialism. Another is the outflow of dollars from the US which led to the balance-of-payments crisis and devaluation of 1971.

6.2.1.2 Germany The top five German firms are shown in Table 6.9. The German chemical industry is the oldest in the world apart from the UK and its history was outlined in section 1.7. The three top companies, Farbwerke Hoechst, Badische Anilin und Soda Fabrik and Farbenfabriken Bayer, were, as their full names imply, originally dyestuffs manufacturers. For example BASF, founded in 1865, started large-scale production of alizarin in the 1870s and, in 1897, first produced synthetic indigo, followed by the indanthrene range of dyes.

During the First World War the German firms had to switch to processes of military significance and afterwards found it difficult to switch back in the face of world competition. In 1925 they merged to form IG Farben. In case this book should give the impression that money and profits should be the sole end of industrial endeavour, it is proposed at this point to outline the history of IG Farben.

Between 1932 and 1943, the sales of IG rose from 900 m. reichmarks (RM) (profit 71 m. RM) to 3 000 m. RM (profit 571 m. RM). In 1932 Hitler met the leading industrialists in Germany and declared his intention of seizing power by violence and not by votes. Goering appealed for money and IG contributed 400 000 RM, the largest single contribution.

From 1932 onwards, Hitler prepared for war and IG collaborated closely with him. It developed BUNA rubber (sections 10.6.3, 10.6.6; in 1942 IG made 91·1% of the world's synthetic rubber) and synthetic gasoline from CO/H_2 mixtures, by the reverse of steam reforming, an uneconomic process which would have otherwise been abandoned. In addition it increased explosives capacity, displacing Chile as the world's major nitrate producer, and boosted production of methanol, poison gas, lead tetraethyl and light metals. Between 1930 and 1942 magnesium production increased forty times and aluminium production thirteen times. IG became 'a state within a state'.

In 1941 it began to employ slave labour and numbers rose in successive years from 18 000 in 1941 to 100 000 by 1945. Labour turnover was as high as 300% per year, and as many as 100 people per day died at work from exhaustion.

Experiments on human beings were carried out with poison gases, vaccines, etc. Food and rest for the slaves were grossly inadequate and if, upon cursory examination, it appeared that an inmate of the Monowitz camp at Auschwitz (the labour camp for IG) would not be restored to productive capacity in a few days, he was sent to the nearby Birkenau camp for extermination. Hitler remarked of conditions in IG Farben factories 'What does it matter to us? Look away if it makes you sick'.

In 1945, IG was liquidated and its leaders were arraigned as war criminals. The three main companies from which it had originally been formed were re-established in 1953. The new management of the German chemical industry have proved more efficient and very much more humane than their predecessors, and it is now one of the best in the world both in terms of economic efficiency and employee relations. Meanwhile, at a time when businessmen are fond of saying that businessmen should be allowed to run the country on sound commercial principles, it is salutary to remember that it was the German industrialists who brought Hitler to power and their sound commercial principles that led them to employ slave labour and experiment on human beings. The importance of profits is a theme of this book, but they are only worth pursuing in combination with humane social and political institutions.

The record of the three major German companies for many years has been one of unparalleled expansion and profitability by almost all economic measures. 1970, however, saw their profits slump from a total of $332·4 m. in 1969 to $242·1 m. and clearly even they are not immune from the general malaise afflicting the chemical industry. BASF alone reported a fall in pre-tax profit of 45%, while sales growth slowed down considerably from previous years and there have been drastic cut-backs in future investment.

6.2.1.3 France France boasts two giant chemical companies – Rhône-Poulenc and Ugine-Kuhlmann with the others a long way behind.

Rhône-Poulenc was formed in 1928 by the fusion of two fine-chemical companies, Poulenc Frères of Paris and Usines de Rhône of Lyon. The new company, strong in pharmaceuticals, photographic chemicals and cellulose acetate, expanded rapidly into plastics and synthetic fibres after the Second World War and bought a controlling interest in Péchiney-St future.' the leading French fertilizer company, in 1969 to become undoubtedly the most important French chemical company. Rhône-Poulenc, in spite of its size, has low profits and sales per employee and appears to be lavish in its use of labour. Ugine-Kuhlmann appear much better in this respect.

Ugine-Kuhlmann is the second French chemical giant. Kuhlmann started in 1824 with a small sulphuric acid plant at Lille while Ugine, founded in 1889, was concerned with electrochemicals and extraction of metals. The plant was therefore concentrated in the Savoy Alps where cheap hydroelectric power was available. These two companies plus the Société des Produits Azotes merged at the end of 1966 to form Ugine-Kuhlmann. In 1967 it derived 53·3% of its turnover from chemicals, the rest coming from metals.

6.2.1.4 Japan Japan has been a large producer and consumer of fertilizers for many years because her high population density makes intensive farming essential. She was also a large textiles producer and hence had large rayon interests.

In 1955, Sumitomo Chemical and Mitsui and Co. (a newly formed 'Zaibatsu' of seven Mitsui companies and Kao Oil Co.) entered the petrochemical field, Mitsui first making Ziegler polyethylene and then cumene–phenol. Mitsubishi Petrochemical Co., which makes mainly plastics and organic chemicals, was formed in the same year.

Meanwhile, the major rayon producers, Asahi, Mitsubishi Chemical Industries (founded 1934 as the Nippon Tar Co. but buying rayon producers Shinko Jinken in 1944) and Toray (founded in 1926 as Toyo Rayon) were moving into synthetic fibres then diversifying further. They licensed ICI's polyester process in 1958. Asahi were the first licensees of Dow Chemical's polyvinylidene chloride route.

Aided by US capital and know-how, the Japanese chemical industry boomed. There was a swift rise in market demand, extensive technical innovation and intense domestic competition. The annual growth rate of the industry from 1962 to 1968 was 16% per year and the wages of the once impoverished Japanese worker rose at 12% per year. Being relatively young, the Japanese chemical industry is still fragmented; there are many small and medium-sized firms and no giants (Table 6.9).

The Western world is currently gasping with admiration for the Japanese economic miracle as it gasped in the 1950s at the West German miracle. While the achievement of Japan in raising herself from the status of an underdeveloped country is an impressive one, three caveats should be uttered before anyone tries to imitate her.

First, the industry has been able to capitalize on the traditional stability and cohesion of Japanese society. The violence of Japanese student protest suggests that industrialization may destroy this sooner than anyone imagines. Second, the profitability of the Japanese chemical industry is low even by the standards of the beleaguered West European giants, and third, the Japanese have paid a high price for their chemical industry in terms of environmental deterioration. The dynamic young managers of the late 1950s, now fifteen years older, are going to have to do something about this before expansion can go much further.

It would be no surprise therefore if Japan were soon to find herself with all the political, social and economic problems which afflict other developed countries. Indeed, without them, she hardly has a right to membership of the club.

6.2.1.5 Switzerland Switzerland is the world's fine-chemical manufacturer. The Swiss boast three giant companies all operating in this area (Table 6.9). On the other hand their largest heavy chemical firm, Lonza, is 176th in the *Chemical Age* rating with a turnover of a mere $97·2 m.

Hoffman la Roche are something of a legend. They are probably the largest manufacturers of pharmaceuticals in the world, and their shares change hands at a price in the region of £20 000 each. Swiss laws on disclosure of company information are lax and Hoffman la Roche publish no accounts, investment figures or costs. In 1967 they announced a net profit of 39 700 000 francs (£3·3 m., $9·3 m.) and declared laconically:

> 'The result for the year 1966 again shows improvement over the year before. Sales and earnings have increased in approximately equal proportions. The global development of the concern shows a somewhat slower rate of expansion than in previous years but remains at a high level ... the volume of investment was again large and will hardly diminish in the foreseeable future.'

Ciba and Geigy, who were previously 27th and 28th in the *Chemical Age* ratings, merged in 1970 to form a new giant company. Although pharmaceuticals are the brightest sector of the chemical industry at present, the Swiss firms have not shared in this prosperity. Sandoz profits have dropped sharply from $43·8 m. in 1969 and in general performances have been disappointing.

6.2.1.6 Italy The Italian chemical industry grew at great speed in the 1950s and 1960s. In part this was a result of the remarkable growth of the Italian economy as a whole. The industry also profited from the energetic regional policies of the Italian government, which were intended to bring work to the impoverished south of the country. As happened with investment grants in the UK, the financial inducements offered encouraged the capital-intensive petrochemical industry to move south while the labour-intensive engineering trades stayed in the north. The Italian chemical industry has been strongly export-oriented during the past decade and its recent severe losses are due in part to the general demoralization of international markets.

Confusion surrounds the affairs of Montecatini-Edison, the largest Italian chemical firm. The origins of Montecatini are curious: a small pyrites-mining company, it became the vehicle for a complex series of takeovers during and after the First World War which left it, by 1931, in control of the Italian fertilizer and dyestuff industries. Further rapid expansion in plastics and petrochemicals after the Second World War has made it a giant by world standards but, because of the lack of effective central direction, a highly unprofitable giant. In 1966 the controlling interest was bought by the Edison group whose generating stations had been nationalized, but this seems to have deepened rather than reduced the chaos which has characterized the company for many years.

The next largest Italian company in 1970 was SNIA Viscosa which was less than a fifth of the size of Montedison and was originally based on rayon. Early in 1972 it was taken over by Montedison, but there seems no reason to suppose that this will improve the performance of either party.

6.2.1.7 Netherlands and Belgium During the 1960s the Netherlands turned itself into the entrance to Europe for transatlantic cargoes. The port facilities in Rotterdam are among the best in the world and huge tonnages of chemicals destined for other European countries are unloaded there. In addition, the Netherlands has a chemical industry of her own in the shape of a Dutch share in two joint ventures with Britain – Royal Dutch/Shell and Unilever (sections 6.1.5.11, 6.1.5.12) – and her own giant company Akzo formed at the end of 1969 by a merger between two other acronymous conglomerates Aku and Kzo. It is Europe's largest synthetic fibre producer and second only to Du Pont in the world. Though it has other chemical interests, tumbling fibre prices and chronic over-capacity have reduced its profits, and its fibre capacity was pruned early in 1972 by a ruthless 25%.

The Belgian chemical industry is slightly smaller than the Dutch even allowing for her smaller population. Her major company is Solvay, eponymous owner of the important turn of the century soda process. Solvay was a private company from 1863 and only went public in 1967. Soda ash is still made for glass manufacture by the traditional route, and in 1970 Solvay sold 3 m. tons which accounted for 25% of turnover. The rest was made up of plastics, fertilizers, halogen derivatives and peroxy compounds, the last being marketed jointly with Laporte (section 6.1.5.10).

6.2.1.8 Union of Soviet Socialist Republics Until 1958, the Soviet chemical industry was small, old-fashioned and based on coal. J. V. Stalin, intent on building up heavy industry after the devastation of the war and seeing coal miners as pillars of the proletariat, was either unable or unwilling to experiment with a petrochemical industry. After his death, the policy slowly changed, but early Russian attempts to design and build petrochemical plant were only partly successful and various processes, know-how and construction skills were imported from abroad. For example, the UK contractors Humphreys and Glasgow have recently built crackers for ethylene and propylene at Kazan and Polotsk.

Meanwhile, growth of the Soviet chemical industry since 1963 has been at the remarkable rate of about 15% per year, a figure equalled only by Japan among the developed countries. Comparison of the turnover of the Soviet industry with industries in the West is difficult because of widely different accounting methods, but in 1970 the USSR produced 45% as much sulphuric acid and 20% as much resins and plastics as the USA. These figures show that the Russians have a vast traditional chemical industry and a rapidly developing modern one.

6.2.1.9 Eastern Europe Reliable statistics about the other countries of the Communist bloc are difficult to find, and unreliable when found. Values assigned to products are notional and published exchange rates largely fictitious. A study of United Nations figures for 1966, for example, reveals that while the USA (population 200 m.) invested $2 970 m. in the chemical industry, Hungary (population 10 m.) invested $3 190 m. While the former figure came from OECD, the latter came from the Hungarian government! Meanwhile, among the centrally planned economies, European Russia must be regarded as a major producer of chemicals and Hungary, Poland, Rumania, Czechoslovakia, Jugoslavia and East Germany can be regarded as developed countries.

All these countries have built up modern chemical industries fairly recently, and many still use soft coal, of which there are extensive deposits in Eastern Europe, as a raw material. Bulgaria, Czechoslovakia, East Germany, Hungary, Poland, Rumania and Jugoslavia together have a population about half that of the USSR and in 1970 produced 55% as much sulphuric acid and 74% as much plastics and resins. East Germany in particular can be considered to have a more developed chemical industry than the USSR, although comparisons based on population are somewhat misleading because only European Russia can truly be considered a developed country.

6.3   The chemical industry in developing countries

Most of the chemical industry and most of the readers of this book are located in the prosperous countries of the so-called Western world, that is Western Europe and the North American continent and, somewhat anomalously, Japan. The populations of these countries, if not static, are growing at about $\frac{1}{2}$% per annum and by and large they are free to worry about such things as pollution, police brutality and the dangers of overeating. Table 6.10 shows 1964 turnover in the chemical industry by region together with data on petrochemical plant and population. The 30% of the world's population in the developed countries produces 90% of the world's chemicals. This leaves about 70% of the world's population with only a rudimentary chemical industry producing about 10% of the world's chemicals. Indeed they have only rudimentary industry of any sort, and such countries are designated as 'underprivileged', 'under-developed' or, more optimistically, as 'developing'.

Table 6.10 Turnover in the chemical industry by regions in 1964

	Turnover ($ m.)	Turnover (%)	Population (m.)	Population (%)	Petro-chemical plants, 1963
1. OECD European member countries	28 200	29·20	312·8	9·60	226
2. USA and Canada	33 600	34·80	211·4	6·50	573
3. Japan	5 200	5·37	97	3	52
4. USSR	11 400	11·80	228	7·03	n.a.
5. Other East European countries	(6 000)	6·22	121·5	3·76	n.a.
6. Latin America	3 100	3·25	234	7·20	48
7. Asia excluding Japan and Communist China	(3 000)*	3·10	895·3	27·80	10
8. Oceania	(2 000)	2·07	17·6	0·50	n.a.
9. Communist China	(2 300)†	2·59	775	24	n.a.
10. Africa excluding South Africa	(600)	0·62	295	9·10 ⎫	1
11. South Africa	(700)	0·72	18	0·50 ⎭	
12. Middle East	(250)	0·26	32·7	1·01	2
Total	96 550		3 238·3		

* Later re-calculated as 2 000. † Later re-calculated as 1 800. Figures in parenthesis are estimates.

The year 1964 for which Table 6.10 was compiled was significant because it marked the first meeting of the United Nations Conference on Trade and Development (UNCTAD). At these meetings, held every four years, the rich and poor nations get together and discuss the economic problems that divide them. The poor nations' equivalent of OECD is 'The Group of 77' which in 1964 embraced 77 countries but has since expanded to 96 without change of name. Has UNCTAD succeeded in closing the gap, illustrated in Table 6.10, between rich and poor? Figure 6.11 provides the answer. Between 1960 and 1970, the developing countries' share of world trade actually dropped.

The reasons for this are complex, and some of them will emerge later in this chapter. 'The Group of 77' has asked that developed countries should give 1% of their GNPs as development aid and think this might help. Meanwhile few of the developed countries seem inclined to donate more than about half this. Communist China, taking part in her first UNCTAD conference in Santiago in 1972, denounced all aid as 'exploitation' anyway, and ecologists mutter darkly about there being insufficient world reserves of minerals for everyone to have a 'western' standard of living. The Santiago conference was generally agreed to be a fiasco with the developing countries squabbling among themselves for so long that there was no time for dialogue with their richer brethren.

The problem of whether it is likely or even possible in terms of world resources that developing countries should succeed in western terms in industrializing themselves and raising their standards of living is outside the scope of this book. The point is that most of them wish to do so and that their rulers are interested in setting up a chemical industry, which in this context usually means a petrochemical industry, or expanding one which already exists. In the second case, the existing chemical industry is usually a

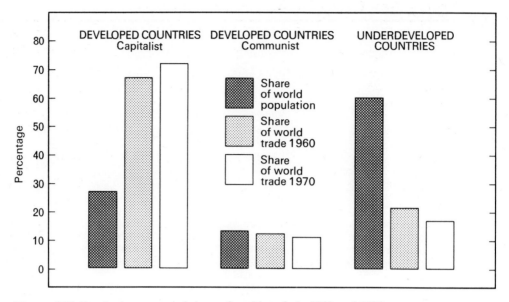

Figure 6.11 Developing countries' share of world trade in 1960 and 1970.

traditional one which produces salt, soda, sulphuric acid and sometimes nitric acid, nitrates, ammonia, ammonium nitrate, phosphates etc. The production of fertilizers is obviously an important function in primarily agricultural countries but even here the differences between the 'haves' and 'have-nots' are apparent. In 1963, Japan applied 200 lb fertilizer per acre to her rice crops, compared with India's 1 lb per acre, and as a consequence obtained four times the yield of rice.

In this section the problems facing an underdeveloped country considering the construction of a chemical industry will be considered.

6.3.1 Differences between under-developed countries

On general demographic grounds, the countries of the world may be classified into four groups according to their density and rate of growth of population (Table 6.11).

That the developed countries are all in the low-population-growth sectors is no surprise. If a country whose population increases at $\frac{1}{2}\%$ per year can raise its gross national product by 3% per year, it can double its standard of living in thirty years; a country whose population increases at $3\frac{1}{2}\%$ per year will find its standard of living dropping at the same rate of economic growth. Furthermore, if the supply of arable land is limited, more and more people will have to live off a given area. It has been said that the economic benefits which Egypt expected from the Aswan dam were entirely swallowed up by the population increase in the years it took to build. Thus the possibility of under-developed countries attaining a 'western' standard of living hinges critically on their achieving a lower rate of population growth and developing a suitable political and economic infra-structure.

In the discussion of whether or not a chemical industry is an appropriate vehicle for

Table 6.11 Division of countries by demographic characteristics

Density of population	Rate of growth of population	Examples (Population/sq. mile in 1960)	State of development
High	Low, $\sim\frac{1}{2}\%$ per year	Japan (650), Europe (excluding USSR) (229)	Developed or, like Spain, developing rapidly
Low	Low, $\sim\frac{1}{2}\%$ per year	USA and Canada (26), Australia and New Zealand (4)	Developed
		Southern South America (Argentine, Chile, Uruguay, Paraguay) (20)	Under-developed
		USSR	European Russia is developed, Asian Russia is not
High	High, $\sim 3\%$ per year	Caribbean (208), Indian Sub-continent (284), South-East Asia (132), China and other countries of East Asia (193)	Under-developed
Low	High, $\sim 3\%$ per year	Africa (21), Middle East (38), Central America (47)	Many are under-developed; some like Israel, South Africa and Turkey are industrializing rapidly
		Northern South America (20)	

raising a country's standard of living, it is convenient to divide countries into four categories:

1. Countries with large petroleum resources, limited domestic markets and a generally favourable balance-of-payments position enabling them to invest in capital-intensive petrochemical plant the products of which must be exported. Libya, Kuwait and the Sheikhdoms of the Persian Gulf are in this position, but until now have contented themselves with selling oil rather than processing it.

2. Countries with potentially large domestic markets and a reasonably good supply of raw materials. Such countries include Nigeria, Iran and Venezuela. Their problem is to keep the increase in demand for chemical products as the standard of living rises, in step with the expansion of industry which causes the standard of living to rise in the first place. It requires considerable skill for a government to balance its desire for rapid economic expansion against the impossibility of achieving more than a certain rate of the social change which will consume the products of the expansion and provide the skilled manpower to run it. Given an incompetent government, the beneficent spiral may not start at all. For example, Iraq, site of the garden of Eden, has been blessed with fertile land and large deposits of oil. The land has been rendered infertile by bad farming methods and neglect of the traditional irrigation system, and the oil revenues have been frittered away.

3. Countries with potentially large domestic markets but not well supplied with raw materials. India falls into this group. If a market can be developed, it may well be worthwhile to import raw materials (e.g. crude oil) and process them. This is the pattern

followed in many developed countries and is related to the ease with which, say, crude oil may be transported compared with, say, fabricated plastic goods.

4. Countries with limited domestic markets and few raw materials. The number of such countries is depressingly large and it is tempting to say that there is no hope for a viable chemical industry in any of them. Some small countries such as Switzerland and Israel have built up chemical industries, however, and if a country is in a favourable social and economic position there is no reason why it should not successfully manufacture selected chemical products particularly those with a high value, such as pharmaceuticals (Switzerland) and isotopically labelled compounds (Israel).

The government of an under-developed country or even an individual firm intent on setting up a chemical industry is faced with the classic economic problems but in a somewhat special form. It starts with certain advantages and certain drawbacks compared with a producer in a developed country. The following does not apply to all under-developed countries but is typical of many of them.

The advantages include a simple demand structure (the customer is relatively unsophisticated and is prepared to accept goods which elsewhere might be thought inadequate) and cheap raw materials if the country comes in groups 1 or 2. Also there is the fact that the technology of all the processes will have been fully worked out so that development can be planned in detail and there is no need for every stage of the industrial revolution to be re-enacted. The major drawbacks include the small size of local markets, a shortage of foreign exchange for buying equipment if the country comes in groups 3 or 4, a lack of skilled personnel to operate plant and the fact that 'transformation' industries which turn chemical products into consumer goods grow only slowly as the demand structure of the population alters. Some of these factors require consideration in more detail.

6.3.2 Small size of local markets

The small size of local markets, resulting more often from the low *per capita* income of the population than from its lack of numbers, means that only small capacity plant can be justified and this prevents the producer from achieving economies of scale. Some comparative costings for unspecified under-developed countries are given in Table 6.12. A

Table 6.12 Economies of scale (United Nations data for unspecified countries in 1964)

Process	Ethylene from naphtha	Ammonia from natural gas	Carbon black	PVC from ethylene
Annual output, '000 tons	10 60	36 180	10 50	6 20
Capital investment, $/ton	570 250	179 89	300 160	285 170

producer might consider building large plant and exporting his surplus but exporting requires selling skills and a degree of quality control which are not readily available. Furthermore, many exporters from developed countries are prepared to sell at a price which covers only their marginal costs plus a small contribution to overheads and thus keep their plants working at full capacity. In order for a producer to compete in this sort of market, a solidly-based home market is essential.

On the other hand, economies of scale occur most spectacularly in the transition from small to medium-sized plant and a home producer, particularly if he has tariff protection, can hope to stay in business even if he operates on a scale an order of magnitude less than that of the North American and European giants.

The fact that the giants tend to have surplus capacity which they are prepared to export at absurdly low prices, however, raises the question as to whether it would not be better for a government to buy chemicals from abroad for its 'transformation' industries, rather than trying to produce them at home. The economic arguments in favour of this course are strong. Its drawbacks are that it does not provide employment, that a rise in demand in the exporting country might cause interruption of supplies, and most important of all, that a large chemical industry even working at a horrifying loss is good for national prestige. In countries where changes of government are normally accomplished by violence, no leader can afford to neglect national prestige, and developed countries engaged in building supersonic airliners or sending rockets to the moon can scarcely sneer at this preoccupation. There is, however, a very strong argument for the building-up of 'transformation' industries using imported chemicals before plant for the manufacture of the basic chemicals is even considered.

It is unfortunate that basic petrochemicals and plastics on the one hand and heavy inorganic chemicals on the other are best made on huge integrated plant. The difficulties of developing such a complex and being able to market all of its products are formidable and can really only be undertaken if a substantial home market exists. In this connection, economic unions like an African common market could be of tremendous value but it is difficult to see the political obstacles being overcome.

6.3.3 Labour

Many under-developed countries have huge reserves of unskilled labour who can be paid relatively low wages, and few skilled workers. This is not an invariable pattern, but a very common one. Exceptions include Israel and India both of which have larger numbers of graduates than can readily be employed, and Kuwait which has been desperately short of any kind of labour and has imported Palestinian refugees to operate its oil installations.

Unfortunately the modern chemical industry has no great need for unskilled labour. Indeed, being capital-intensive, it does not employ very many people of any kind. United Nations figures suggest that a $72 m. complex for synthetic rubber and related products requires only 700 workers and that in an ammonia fertilizer plant, of the 263 employees required, 22 are unskilled labourers while 145 are technicians, engineers and scientists. Carbon black and polyethylene manufacture employ about 45% unskilled labour but the proportion rarely exceeds this for any chemical manufacturing process.

Furthermore, the cheap labour has a low productivity and requires expensive training. Qualified technicians and skilled workmen often have to be imported from abroad at very high cost, so that the real cost of labour in a developing country is usually 30–50% higher than in an industrialized country even though wages are much lower. The temptation to use under-trained local labour both for political and economic reasons has ensnared many constructors and was to some extent responsible for Albright and Wilson's debacle in Newfoundland in 1970. US firms, building plant abroad, in general operate on the principle that no new technology should ever be tried out for the first time in a developing country and that everything down to the paper clips should be brought in from outside. This means that their plant actually works but that there is virtually no educational spin-off for the under-developed country. A corollary of this is that the USA and the big oil companies are usually reluctant to undertake joint ventures involving an under-developed country while British, West German, Italian and French companies are more inclined to do so.

All this means that construction and running costs of sophisticated chemical plant are likely to be higher than in an industrialized country and quality control worse. Table 6.13 gives approximate comparative construction costs for petrochemical plant in different countries.

Table 6.13 Comparative construction costs for petrochemical plant in different countries

France	100	Algeria	111	
USA	105	Turkey	115	(at coast)
United Kingdom	95	Iran	116	(at coast)
Italy	95	India	121	(at coast)
Germany	100		125	(inland)
Belgium	100			

The above comments do not apply to the 'transformation' industries which convert, say, raw plastic to consumer goods. These industries are frequently labour intensive and might even be able to carry their 'parent' industry – although again the question of import of cheap raw materials arises.

6.3.4 Choice of product

The structure of demand for chemicals in an under-developed country may be quite different from that in a developed one both because of its location (refrigerators to Eskimos?) and its way of life. From a distance it may seem that mud huts should be replaced by some other material, but mud is cheap and in a dry country it is an excellent construction material.

The choice of product hinges both on availability of raw materials and the possibility of selling the product. A country with spare agricultural capacity and no oil would probably do better to make Nylon 11 from castor oil than Nylon 6 from oil. It would certainly do better to manufacture fertilizers and pesticides than sophisticated plastic consumer goods.

Similarly an oil-rich sheikhdom in the Persian Gulf would apparently do well to manufacture petrochemicals. That such countries have not done so is partly a measure of their backwardness and also reflects the smallness of possible home markets, the difficulty of transporting finished goods compared with crude oil, and the shortage of skilled manpower. There are signs, however, that this situation may be coming to an end. One Sheikh has recently started to convert imported bauxite to aluminium using natural gas which he had previously allowed to go to waste.

The difference in the pattern of demand between developed and under-developed countries is well exemplified by the chlorine–soda problem. Electrolysis of brine is bound, by the stoichiometry of the process, to give 80 tons of sodium hydroxide for every 71 tons of chlorine. In an under-developed economy, the soda is in great demand for glass and soap manufacture. Chlorine, however, has few uses apart from bleaching and as a raw material for bleaching powder. The shortage of caustic soda must be made up either by imports (but it is awkward to ship as a solid and only a little pleasanter as a 50 or 73% aqueous solution) or by construction of Solvay plants.

Developed countries, on the other hand, use massive tonnages of chlorine in organic chemicals and plastics (PVC, chloro-fluoro organics per- and trichloroethylenes, ethyl and methyl chlorides, DDT, etc.), and have an excess of soda. Far from building Solvay plants, they are forced to manufacture chlorine by methods which do not produce sodium hydroxide as a by-product, or to use oxychlorination processes.

There would obviously be an interest in under-developed countries selling chlorine to developed ones, or soda being transported in the opposite direction, but chlorine is even nastier to ship than caustic soda so this desirable solution is not generally applicable. The result in Peru, to take a specific example, is that W. R. Grace, who need sodium hydroxide for paper-making, make it by electrolysis and find themselves with surplus chlorine. This they react with ethylene to make PVC (section 10.3.1), but as Peru has no ethylene cracker, they are forced to make the latter by dehydration of ethyl alcohol which they in turn make by fermentation of molasses. As Peru is a sugar producer, this is readily available.

6.3.5 Choice of process

In general, a process which is good for a rich country is also good for its poor relation, especially as it is easier to buy the technology for a widely used process. In a choice between more or less equivalent processes, one of which is more economic at high outputs and the other at lower outputs, an under-developed country will tend to choose the latter, and similarly it will tend to choose processes which make fewest demands on highly skilled labour, greatest demands on cheap labour and which have a fair margin of error in case anything goes wrong.

The problem of educational spin-off is hard to assess. In crude terms: does it pay to have a man sitting on top of a distillation tower reading the temperature when an additional lump of capital expenditure could buy a remote-control thermometer? A firm of constructors from a developed country would favour automation but the training which an additional man would get simply by being around the plant might be of social benefit to the under-developed country. Thus, for example, one can make out a good case for building hydroelectric rather than nuclear power stations on the grounds that the educational spin-off from the former is much greater.

It is sad that governments of under-developed countries also tend to choose highly automated plant. When Ché Guevara, myth figure of the New Left, was Minister of Industry in Cuba, he commissioned two highly automated factories for making screws and pencils. These provided little employment but were so productive that one week's output satisfied the Cuban screw and pencil market for two years, and the factories had to cease production.

144 The world chemical industry

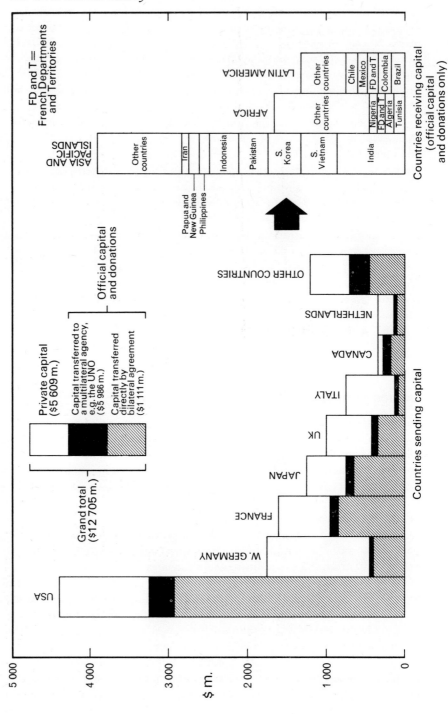

The chemical industry in developing countries 145

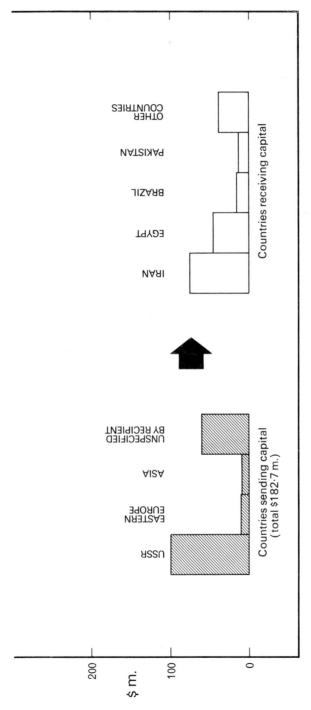

Figure 6.12 Long-term capital flow to developing countries from the western world, Japan and the centrally planned economies (1969). Note that the vertical scale of the lower diagram is ten times that of the upper, and that aid from the centrally planned economies undeclared by the recipient is not listed.

6.3.6 Raising the capital

It is very difficult for a private investor in an under-developed country to raise the capital to build modern chemical plant. The scale of even a relatively small enterprise means that it must be undertaken by a large corporation, and such organizations are rare in under-developed countries. This is less true for the traditional chemical industry, and sectors such as soap, soda and fertilizers already exist in many under-developed countries and may well be privately owned.

Nonetheless, the scale of finance required, combined with the vaguely Socialist leanings of the governments of many under-developed countries, means that governments usually play an important if not a dominant role in building a national chemical industry. They could think of raising capital by taxation but in a poor country, the tax base is small and the necessary amount of money is simply not available. Citizens might be invited to invest their savings in the enterprise but they will demand a high rate of return and the number of people rich enough to invest will probably be small. Life insurance syndicates, building societies, banks and similar non-government financing organizations might be persuaded to put up money, again supposing that the country in question is developed enough to have such organizations with sufficient money to invest.

Money from outside the country can come as foreign aid (technical and/or financial assistance, long term loans or outright gifts) from individual countries or through the United Nations and such bodies as the World Bank. Alternatively it can come as foreign investment.

To be able to attract funds from almost any of the above sources, but especially the last, a government must be able to offer an attractive rate of return combined with a low risk. No one is going to invest in a country where he feels there is a strong possibility that his investment will be expropriated. A degree of political stability is vital if a country is to attract foreign investment at a reasonable rate of return.

One country which has succeeded in industrializing since the Second World War is Japan and Table 6.14 shows its main source of finance.

Table 6.14 Sources of capital for Japanese petrochemical industry 1955–63

	$ m.	%	
Japanese-owned capital	180	20	
Financed by Japan Development Bank	50	6	
Non-governmental co-operative financing	590	65	
Life insurance syndicates	30	3	
Foreign investment	50	6	[know-how 20, cash 10, loans 20]
	900		

Since 1955, the flow of private capital to under-developed countries has been of the order of $5 000 m. per year. This is similar in magnitude to the flow of public capital (i.e. government aid, world bank loans, etc.) to such countries but the direction has been different. The private investor prefers Latin America to Asia and Africa presumably because it has been politically more stable in recent years and because of the relatively high per capita income. For example, in 1957, Latin America received $1 500 m. private investment while ten major Asian countries with $3\frac{1}{2}$ times the population of Latin America between them received only $100 m. Conversely, public capital tends to flow to Asia and Africa where the political pay-off is seen as greater.

In terms of foreign aid, the United States is far and away the greatest benefactor though there are signs that she is becoming inclined to withhold money from countries which are too vocally critical of her policies. The Communist bloc has provided very little aid apart from Russia's financing the Aswan High Dam. Figure 6.12 shows the flow of capital to developing countries.

6.3.7 Conclusions

The previous pages have been pessimistic in tone and have implied that the difficulties of building up a chemical industry or indeed any sophisticated industry in an under-developed country are well-nigh insuperable. This is regrettably true; if it were not true the under-developed countries would no longer be under-developed. Indeed all the evidence suggests that the technological gap between rich and poor is widening. Industry gets more sophisticated all the time. In 1900 an under-developed country could make chemicals by destructive distillation of the wood which most of them have in abundance. Now they have to undertake elaborate high pressure cracking technology. However, in spite of the difficulties, some progress is being made and two possible approaches have become apparent.

In the first approach, which is gradualistic and involves the so-called 'intermediate technology', the country re-enacts certain stages of the industrial revolution. Any 'traditional' chemical industry is encouraged to expand and a sort of nineteenth century chemical industry producing bleach, fertilizer, explosives, etc., is encouraged to evolve. Its products can be used by whatever rudimentary industries already exist. Then fabricating industries are set up which use imported plastic, next the plastic is made locally and finally petrochemicals are made from petroleum. 'Intermediate' skills are required; technicians are of more use than highly trained graduates in academic subjects. The absence of prestige projects makes this approach unattractive to many politicians but it tends to be recommended by cautious Western experts.

In the second approach, which is revolutionary, a large prestige project is established with government money and people are then more or less forced to use its products. These are sold very cheaply so that the main unit bears the loss and the user industries make a profit or break even. The Aswan High Dam is an example of this approach and to some extent a test of it. Definite conclusions are not possible yet but in some ways the dam has been the greatest Egyptian ecological disaster since the Ten Plagues. It appears to have wiped out the fishing industry in the Nile Delta and beyond it. The reduced flow of the Nile is allowing salt water to creep up the delta. The debilitating disease bilharzia has increased dramatically and much of the prized water for irrigation is being lost by evaporation. The extent to which these unexpected ill-effects outweigh the benefits is not yet clear. On general grounds, however, we tend to favour the gradualist approach with particular emphasis on the prior building up of transformation industries using imported raw materials.

Is the construction of a chemical industry in an under-developed country not particularly blessed with natural resources a profitable undertaking anyway? In most cases it probably is not, at any rate in the early stages. In situations of this sort, however, while economics are important they are not all-important. If they were, no country would ever get off the ground at all. Instead one has to try and evaluate the social benefits which accrue to a developing country and set these against possible financial losses. This technique, known as social costing, was used in London to justify the building of the Victoria Line on the London Underground and takes into account not only the cost of building and running against the receipts but also assesses, for example, the value of the time saved by passengers brought more quickly to their destinations. Developments of the technique should provide under-developed countries with a method of distinguishing between expensive and useless prestige projects, and other projects in which financial losses are balanced by social and educational benefits.

6.4 Industries of special importance to developing countries

Certain branches of the chemical industry are of great importance to developing countries. Fertilizers (section 14.2) and pesticides (section 18.1.2) are mentioned briefly in subsequent chapters but desalination and manufacture of synthetic protein are outside the mainstream of the chemical industry and will be discussed here.

6.4.1 Desalination

Water and agriculture go together like drought and famine. Throughout history, civilization has grown up round sources of fresh water and even nomads wander from water hole to water hole. The Babylonians, the Romans, the Nabateans all built up elaborate irrigation systems to conserve what water they had. Nowadays, the many dams around the world testify to the need for and practice of water conservation. The leader in this field is probably Israel where it is estimated that 90% of the water potentially available is now being used. In many countries, demands on natural water resources have now reached the point at which the marginal cost of water is increasing with scale. Each successive unit of water is more difficult to obtain than the last.

The need of industry for water is less widely realized than that of agriculture. The manufacture of 1 ton of paper requires 100 tons of water, the extraction of a ton of iron from natural ore requires 300 tons. Many countries, even if they used their total available water resources, would be unable to support an optimum level of agriculture let alone a modern industry. This fact has led to an international effort through the United Nations to develop a reasonably economic system for desalinating salt and brackish waters. (Brackish water is any water containing less dissolved solid than sea water.) Sea-water typically contains 3·5% (35 000 ppm) dissolved solids; inland seas contain much more: the Dead Sea contains some 25%. The requirements for human consumption are less than 500 ppm dissolved solids. Farm animals can survive on up to 5 000 ppm (cows) or 8 000 ppm (sheep) while agriculture is impossible at more than 2 500 ppm and most crops fail at less than half that salinity especially if the water is alkaline. Industrial requirements vary. Power stations require water of much lower salinity than is required for human consumption while other industries have successfully used sea water for cooling purposes by employing equipment that can withstand its corrosive effect.

Many methods of desalination have been evaluated in detail but the following are the main ones of interest.

6.4.1.1 Solar evaporation Solar radiation is used to heat saline water and the resultant water vapour is allowed to condense on a cool surface whence it runs into gutters and thence to storage. This is an imitation of nature's evaporation/condensation cycle and the largest unit to be built was in a mountain mining camp in Chile 80 years ago. It had about 5 000 m^2 of surface and produced about 23 tons of water per day. Though the fuel (solar energy) is free, and labour costs are low, land requirements and capital costs are huge in comparison with other methods.

6.4.1.2 Multiple effect distillation If saline water is boiled and the vapour condensed, the distillate will be free from salt. This principle was known to Aristotle and has been used by sailors at sea since the time of Sir Richard Hawkins (1593). A simple distillation plant was installed by the British Navy at Aden in 1869.

In the simplest distillation apparatus, such as that used in a laboratory, the condenser is cooled by the circulation of cold water and the heat which it absorbs goes to waste. If the saline liquid in the evaporator boils at a sufficiently high temperature and pressure, however, and if no water is circulated through the cooling jacket, the steam will heat the cooling water until it boils. The vapour from this second boiling can then be passed on to another similar condenser to give a two-effect evaporation sequence. Several more stages can be added if desired, and in each of them the heat requirement can be largely satisfied by the latent heat given up on condensation of vapour from the previous stage. The process is known as multiple-effect distillation (Fig. 6.13).

The thermal efficiency increases with the number of stages but so does the capital cost. Above twelve stages, improvements in thermal efficiency become insignificant. On an average, the number of tons of water per ton of heating steam in 1, 2, 3, 4 and 5 effect evaporators is 0·9, 1·75, 2·5, 3·2 and 4·0 respectively.

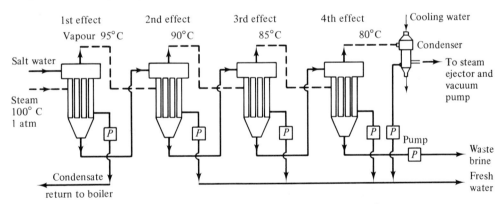

Figure 6.13 Four-stage, multiple effect evaporator. Pressure and boiling temperature decrease from left to right.

6.4.1.3 Flash distillation In conventional distillation processes, the salt water is heated and boils in the same vessel and this can lead to formation of solid deposits (scale) on surfaces whose thermal conductivity is thus much reduced. In the flash evaporation process, sea-water is first heated in tubes and then made to evaporate in chambers where there is a lower pressure than in the heating tubes. As the vapour flashes off the surface of the warm liquid, any solids are formed in the body of the liquid rather than on heated surfaces.

A four-stage multiflash evaporator is shown in Fig. 6.14. Essentially it is an elaborate counter-current heat exchanger (section 15.4.3.4). Brine is heated to its boiling point (say 70°C) at a given pressure in vessel A. It flows to vessel B which is at a slightly lower pressure where some of it flash-evaporates instantly to give vapours in equilibrium with the liquid at the new pressure and temperature (say 60°C). The vapours condense on the tubes C which bring fresh brine into the installation and the condensate, which is distilled water, is withdrawn. Meanwhile the remaining saline flows to further chambers D, E and F at lower pressures where the process is repeated. Each condensation process helps to warm up the incoming brine.

A mechanical breakthrough in the design of the multistage system and flash chambers was made in England in the late 1950s and from 6 to 10 tons of fresh water can now be produced per ton of steam used. As a consequence, multiflash distillation is currently the preferred desalination technique for large-scale operation and much of the plant is designed in England.

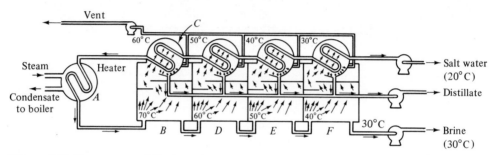

Figure 6.14 Four-stage multiflash evaporator.

6.4.1.4 Freezing processes To vaporize 1 g of water starting at ambient temperature requires about 620 cal (2 600 J). To cool and freeze 1 g of water requires only 100 cal (420 J). Furthermore, when salt water is frozen, pure water separates out as the solid phase. Consequently there has been much interest in the past in desalination by freezing. Pilot plants have been built but they met many problems of which corrosion proved the most serious. The UK pilot plant has been abandoned and no full-scale plant is thought to be operating anywhere in the world.

6.4.1.5 Electrodialysis Membranes of synthetic resins have been developed which are highly selective to the passage of negative ions and others have been developed which are equally selective to positive ions. A stack is formed of hundreds of these membranes arranged alternately and salt water flows parallel to the membranes (Fig. 6.15). A voltage is applied across the stack. In compartments such as A, the positive ions can flow into the compartment above and negative ions into the one below; in compartments such as B such movement cannot occur. Thus in half the compartments, fresh water is produced while in the other half the saline becomes more concentrated.

The cost of electricity for electrodialysis plant depends on the number of ions to be removed, and their electrochemical equivalents. If the cost of equipment and of replacement of the rather fragile membranes is added in, it is clear that electrodialysis is only economic on a large scale for purification of brackish waters with a low mineral content. Electrodialysis plants of small and intermediate sizes (5–10 000 tons/day) have been operated successfully in South Africa and the USA with brackish water feeds.

6.4.1.6 Ion exchange Ion exchange has long been used to soften 'hard' water, i.e. to replace calcium ions by sodium. It can also be used for desalination. The salt water is first passed through a bed containing a cationic ion-exchange resin in its acid form. The sodium ions interchange with the hydrogen to give the sodium form of the resin while the effluent contains hydrochloric acid. It flows to another bed containing an anionic resin in its hydroxyl form which replaces the chloride ions by hydroxyl so that the water has been de-ionized. When all the hydrogen and hydroxyl ions have been removed from the resins, they must be regenerated. The cationic resin is regenerated by passage of dilute sulphuric acid, giving sodium sulphate as a waste material, and the anionic resin by passage of sodium hydroxide giving sodium chloride waste.

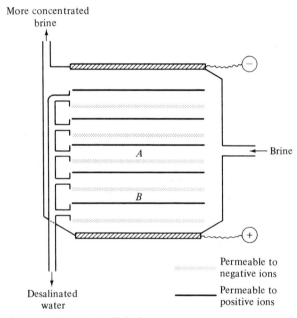

Figure 6.15 Electrodialysis.

The process thus requires acid and alkali in amounts chemically equivalent to the amount of salt removed. The cost of large-scale desalination of sea-water by this method is therefore prohibitive but it can be used for brackish water. For example, purification of water containing 1 750 ppm of salt would require only about 5% of the chemicals that sea-water would require. On the other hand, the energy costs of distillation would be similar for the two salinities.

Ion exchange is more expensive than electrodialysis but removes a higher proportion of ions. It is thus used to purify water for use in boilers in power stations or the manufacture of television tubes where the impurity content of typical drinking water would be far too high. Packets of ion-exchange resins are also included in emergency kits and are designed to have a single use.

6.4.1.7 Reverse osmosis If a solution and the pure solvent are separated by a semipermeable membrane, the solvent diffuses through the membrane to dilute the solution and will do so unless a pressure equal to the osmotic pressure is exerted on the solution side of the membrane. If a pressure greater than this is applied, however, solvent diffuses through the membrane in the reverse direction leaving a more concentrated solution.

In theory this process, called reverse osmosis, offers an ingenious and thermodynamically very efficient method of desalination. All one needs to do is to sink a tube with a semipermeable membrane at its end 12–15 m. into the sea. Hydrostatic pressure would force pure water through the membrane and it could be recovered merely by expenditure of the work required to pump it to the top of the tube.

The two intractable problems which have so far prevented any success with the technique are first the difficulty of constructing membranes with the mechanical strength necessary to withstand the pressures required but which will still allow reasonably fast passage of water, and second the difficulty of preventing concentration gradients near the membrane and indeed the actual crystallization of salt on the membrane itself.

Until a solution of these problems can be found, reverse osmosis will remain a scientific curiosity, at any rate as far as desalination of sea-water is concerned.

6.4.1.8 Comparison of methods Relative costs of desalination methods are difficult to establish because they vary widely in different parts of the world. Furthermore, desalination techniques have to compete with the possibility of conveying fresh water from the nearest available source. Nonetheless, it is possible to say that the capital cost of solar evaporation plant is impracticably high for large-scale operation. Ion exchange is useful only for brackish water, and electrodialysis is more economic unless an exceptionally pure product is required. Electrodialysis is most useful for brackish water and relatively small-scale operation, for example a few hundred litres/day. The world's most economic electrodialysis plant was built in 1962 in Buckeye, Arizona and produces 2 500 tons/day at a cost of 0·18 $/ton.

For large-scale desalination of sea-water, only the distillation processes are being operated on other than a pilot plant basis, and in 1969, multiflash distillation plants accounted for more than two-thirds of the total production capacity of plants of all sizes built since 1957 throughout the world. Many distillation plants are associated with electricity generators and the coupling of these with nuclear power seems to offer the best hope for economic operation in the future.

The most economic plant in the world in 1970 was a 4 500 ton/day unit built in Kuwait in 1960 producing water at $0·174/ton ($0·66/1 000 US gallons; £0·33/1 000 imperial gallons). The cost analysis is shown in Table 6.15. No other desalination unit purifies sea-water for less than $0·3/ton and many have costs up to four times this.

Desalinated water, therefore, although of tremendous value as drinking water, is unlikely to be of industrial or agricultural use in the foreseeable future except as a supplement to natural rainfall. In the complete absence of rain, 1 ton of irrigation water produces only about 3 kg (about ten) oranges or enough grain for seven slices of bread. One ton of steel would require at least $50 of desalinated water.

Certain high-cost crops could prove an exception to this, and of course spectacular breakthroughs in agricultural or desalination technology could alter the economic argument. Cheap nuclear power might cut energy costs, and the thermal efficiency of distillation units is only about 5%. A practicable reverse osmosis process would probably be much more efficient than this.

On the other hand, desalination is vitally important for domestic water supplies in areas such as Kuwait or Elath where military, commercial or political considerations dictate the need for a settlement. Tourists and oil wells, after all, provide a financial return per cubic metre of water consumed very much larger than do agriculture or industry.

Table 6.15 Cost analysis of desalination unit mew 'E' Shuwaik, Kuwait

	$/ton/day
Investment	217
Annual load factor (%)	85
Fixed charges at a rate of 5% (interest is not charged on the capital. A more conventional rate would be 10% which would double the figure given here.)	0·034
Labour plus overheads	0·045
Maintenance and materials	0·029
Chemicals	0·032
Fuel (cheap in Kuwait) and electricity	0·034
Total water cost	0·174

6.4.2 Semi-synthetic protein

The main dietary deficiency in developing countries is lack of protein. High protein foods such as meat and fish are scarce and vegetable protein is dilute and of a poor nutritional quality. Approximately 100 m. people subsist mainly on cassava while another 1 500 m. live mainly on cereal grains. The protein deficiency disease, kwashiorkor, is common particularly among the former group. Also, protein deficiency in the first three years of life is believed to inhibit brain development and 80% of total brain development occurs in that period.

Proteins cannot yet be synthesized completely, but there is much interest currently in the production of semi-synthetic proteins. These must not only be cheap but also culturally acceptable to the communities for whom they are intended. It is ironic that their acceptance so far has been much greater in developed countries, where the need is least. Work has so far been concentrated in three areas.

6.4.2.1 Yeast protein from petroleum   Various processes have been developed in which yeast, a so-called single-cell protein, is grown on a petroleum fraction. The burning of another petroleum fraction provides the warmth required for the propagation. BP are operating one of the processes on a pilot plant scale but the product is useable only as an animal foodstuff because of traces of possibly harmful impurities. At an estimated cost of 35¢/lb protein it is still much more expensive than vegetable protein (12¢/lb) and only slightly cheaper than skim-milk powder (41¢/lb protein).

6.4.2.2 Textured vegetable protein   Soya beans, groundnuts, cottonseed and similar crops all contain a moderate amount of vegetable protein but are not palatable in quantity, and sometimes not at all. Meanwhile, an acre of land can produce some 800 000 calories in terms of plant protein but only 200 000 calories of animal protein. Work has therefore been done on preparing a palatable form of vegetable protein, and the most hopeful product is known as textured vegetable protein (TVP).

Oil is first removed from soya beans, the preferred raw material, by solvent extraction. This is saleable. The residue is dispersed in aqueous sodium hydroxide, allowed to 'age' and then extruded through spinnarets of the kind used for nylon spinning into an acid bath. The protein is precipitated as fibres and these are wound into a hank, mixed with binders, ingredients and flavours, and cut into slices across the grain. Depending on the texture and the flavouring, the product can simulate beef, fish, chicken or even stringy mutton.

TVP is commercially the most advanced semi-synthetic protein and is already being used as a constituent of a number of convenience foods in the UK. It is still not acceptable on its own.

6.4.2.3 Fungal protein   Moulds are very efficient converters of starch to protein and grow with a fibrous structure. The mould *mycelia* is said to give rise to a product with a texture resembling that of meat. It is claimed that such a mould, rich in protein, could be made for as little as 6¢/lb. Meanwhile, the very word 'mould' gives rise to misgiving and extensive acceptability and toxicity tests would have to be carried out before a product of this kind could be put on the market.

Notes and references

General   Much of the historical data about individual companies in this chapter comes from sources listed in Chapters 1 and 2, which we will not cite again, except to mention that the company histories are the best section of D. W. F. Hardie and J. Davidson Pratt: *A History of the Modern British Chemical Industry*, Pergamon, Oxford, 1966. We have also drawn freely on company annual reports which are not separately listed.

Section 6   Figure 6.1 is taken from the *Oxford Economic Atlas of the World*, 4th edition, O.U.P. Oxford, 1972. Table 6.1 is derived from data in the *UN Monthly Bulletin of Statistics*.

Section 6.1.1   W. Beckerman: *An Introduction to National Income Analysis*, Weidenfeld and Nicolson, London, 1968, from which Fig. 6.2 is taken, is an admirably clear introduction.

Section 6.1.2   Table 6.2 is mainly compiled from the preliminary report on the 1968 Census of Production, *Board of Trade Journal*, **197**, 1758, 1969. Capital invested in 1968 is from *ibid*, **198**, 154, 1970; the breakdown among particular sectors is an estimate based on the *Chemical Age* new project surveys. Imports and exports are from the *Annual Statement of Trade* for 1968.

A good book on the theory of input–output analysis is W. H. Miernyk: *The Elements of Input–Output Analysis*, New York, Random House, 1965. The best illustrations of what this approach can do are in *Input–Output Tables for the United Kingdom*, Studies in Official Statistics No. 16, HMSO, London, 1970, from which Table 6.3 and other data are taken.

Section 6.1.3   Data on total fixed assets employed by the chemical industry are taken from National Economic Development Office (NEDO): *Investment in the Chemical Industry*, HMSO, London, 1972, upon which the discussion of the investment cycle is based. For the research and development effort of the chemical and other industries see *Statistics of Research and Development, 1970*, HMSO, London, 1971.

Section 6.1.4   Table 6.4 is from NEDO: *Economic Assessment to 1972 – An Industrial Report by the Chemical Industry EDC*, HMSO, London, 1970. Figure 6.4 is based on *Trade and Industry*, **7**, 488, 1972. Profits as a fraction of turnover were given in the *Annual Reports of the Commissioners for Inland Revenue* until 1968. Pre-tax returns on capital for the chemical industry as a whole are given in NEDO: *Investment in the Chemical Industry*, cited above.

Notes and References 155

The discussion of productivity is based on the *Supplement on Investment in the Chemical Industry* to the OECD *Annual Report* 1968–69, and on Duncan Burn: *Chemicals under Free Trade*, Trade Policy Research Centre, London, 1971.

Section 6.1.5 Table 6.5 is based on the annual *Chemical Age* feature on 'the top two hundred chemical companies in the world'. (CA–200)

Section 6.1.5.1 Table 6.6 is taken from the ICI annual report for 1971. ICI's troubles have been mulled over repeatedly in the national press. See, for example, *The Times* 4, 5 and 6 November 1968; The *Sunday Times* 5 March, 1972. For a discussion of the expensive takeover of the US firm Atlas, see the *Financial Times*, 21 March 1972.

Section 6.1.5.2 Stirling and Co. have produced an excellent report on the British Oxygen Company up to 1971.

Section 6.1.5.5 Albright and Wilson's problems with new plant in Newfoundland were discussed in *Chemical Age*, 1 January 1971.

Section 6.1.5.6 Fisons were also the subject of a 1971 Stirling and Co. report.

Section 6.1.5.10 The Stirling and Co. report on Laporte is a bit dated and missed the titanium dioxide debacle. Otherwise it covers the activities well.

Section 6.1.5.11 For the early days of Shell see R. D. Q. Henriques: *Marcus Samuel*, Barrie and Rockcliffe, London, 1960.

Section 6.1.5.12 The best source is C. Wilson: *The History of Unilever*, 3 vols., Cassell, London, 1954, 1968. The third volume, covering 1945–65, is inevitably inferior to the other two.

Section 6.2 Figure 6.5 is based on CA–200.

Figures 6.5, 6.7, 6.9 and 6.10 and Table 6.7 are based on the OECD *Annual Report*.

Figures 6.3 and 6.8 are taken from the *Oxford Regional Economic Atlas of Western Europe*, 1971 op. cit. and of the *USA and Canada*, O.U.P., Oxford, 1967. Financial data on the larger firms are to be found in *Jane's Major Companies of Europe*, published annually by Sampson Low, London.

Section 6.2.1 Table 6.8 is taken from CA–200 and Table 6.9 based on it.

Section 6.2.1.1 Differences in productivity between the US, UK and European countries are discussed in detail in *Manpower in the Chemical Industry*, NEDO, HMSO, London, 1967, and in the *Supplement on Investment in the Chemical Industry* to the OECD *Annual Report* 1968–69.

Section 6.2.1.2 Details of I.G. Farben come mainly from *Trials of War Criminals before the Nuremberg Military Tribunals*, Vols. VII and VIII.

Section 6.2.1.4 The best review in English of the Japanese chemical industry appeared as a supplement to *European Chemical News*, 29 November 1968.

Section 6.2.1.5 We are not sufficiently privileged to have seen the Hoffmann la Roche annual report, and we are quoting at second hand from the Fortune magazine annual feature on the world's top companies, published in 1967.

Section 6.2.1.6 The SNIA Viscosa takeover was discussed in *Chemical Insight*, No. 2, 1972.

Section 6.2.1.8 The only book on Russia is E. I. Hemy: *The Soviet Chemical Industry*, Leonard Hill, London, 1971. Some quoted statistics were given in *Pravda*, 4 February 1971.

Section 6.3 There is a desperate shortage of reliable statistics on under-developed countries and little incentive for anyone except the local governments to assemble them. Consequently most available data come via the United Nations and the following publications have been drawn on in this chapter (UN Sales Numbers are given, the first digits of which refer to the year; UN publications may be obtained from HMSO or direct from United Nations Sales Sections, New York or Geneva).

Report of the first United Nations Inter-regional Conference on the development of petrochemical industries in Developing Countries (66.II.B.14).

Petrochemical Industries in Developing Countries (E.70.II.B.23).

(These two volumes are reports of conferences held in Tehran, Iran in 1964, and Baku, USSR in 1969. The former was republished by L'Institut Français du Pétrole in 1966 under the title *Petro-chemical Industry and the Possibility of its Establishment in the Developing Countries* edited by C. Mercier and produced by Editions Technip, Paris 15.)

Proceedings of the Seminar on the Development of Basic Chemical and Allied Industries in Asia and the Far East (64.II.F.9).

La Industria Quimica en America Latina.

Advisory Committee on the application of Science and Technology to Development – 3rd report.

Market trends and prospects of Chemical Products (A report by the Economic Commission for Europe) (E.69.IIE/Mim 6). This report is in three volumes, the first two dealing with Europe and the third with under-developed countries. UNIDO (the United Nations Industrial Development Organisation) publishes many other studies of the chemical industry in developing countries and its publications list should be consulted. Table 6.10 is taken from *Market Trends and Prospects for Chemical Producers*, op. cit. and Fig. 6.10 from the *UNCTAD Review of International Trade and Development 1971* reproduced in the *Sunday Times*, 16 April 1972.

Section 6.3.1 Table 6.11 was compiled with the aid of the *Great World Atlas*, Reader's Digest, New York, 1962.

Section 6.3.5 Table 6.13 comes from *Petrochemical Industries in Developing Countries*, op. cit. A slightly different version of the story of Ché Guevara and the pencil factory says that the factory closed because no graphite was available in Cuba. This may be true and reflects equally on Guevara's competence, but raises the question as to why graphite could not be imported.

Section 6.3.6 Figure 6.11 is based on the *UN Statistical Year Book*, 1970.

Section 6.4.1 Technical aspects of desalination are handled excellently in K. S. Spiegler's classic *Salt Water Purification*, Wiley, New York, 1962, though his cost estimates now appear over-pessimistic. Important UN publications include *Water Desalination in Developing Countries* (64.II.B.5); *Water Desalination: Proposals for a costing procedure and Related Technical and Economic Considerations* (65.II.B.5); *Proceedings of the inter-regional seminar on the Economic Application of Water Desalination* (66.II.B.30); *The Design of Water Supply Systems based on Desalination* (E.68.II.B.20); *First United Nations Desalination Plant Operation Survey* (E.69.II.B.17), from which Table 6.15 is taken and *Solar Distillation* (E.10.II.B.1).

Section 6.4.2 Magnus Pyke's excellent book *Synthetic Food*, John Murray, London, 1970, deals briefly with semi-synthetic protein and the Arkady Review, **47**, 21 (1970) discusses textured vegetable protein from the point of view of the manufacturer. The general problems of protein deficiency are under continuous study by the United Nations Protein Advisory Group (P.A.G.) who produced a comprehensive documents list early in 1972 which is an excellent guide to source material.

Part III The products of the chemical industry

29ru The products of the chemical industry

7 Raw materials: petroleum

7.1 The petroleum industry

Petroleum is the lifeblood of the modern chemical industry, the indispensable raw material from which organic chemicals and hydrogen are now made. Yet the chemical industry is only a minor consumer of petroleum: in no country is as much as 10% of refinery output composed of chemicals or chemical feedstock. Petroleum is primarily a source of energy — it is indeed the major source of energy in most industrial countries — and over 90% is eventually burned in furnaces or in internal combustion engines. The major products of the petroleum industry are fuels — gasoline, the middle distillates (kerosene and diesel fuel) and fuel oil — and it is the demand for these products which has determined the technical and economic structure of the industry. The manufacture of organic chemicals from petroleum is by contrast both a recent and a secondary development. Much of the technology employed was originally used to make gasoline; even more important, the economics of petrochemical processes depend crucially on those of the petroleum industry as a whole. Thus to understand how and why petroleum is used as a source of chemicals it is first necessary to study the petroleum industry.

7.2 Supply and demand

The production of petroleum involves many activities: exploring for new petroleum deposits, bringing the crude oil to the surface, moving it to refineries where it is fractionated, purified and chemically modified, and distributing the final products. The industry is dominated by seven large vertically integrated companies — Standard Oil of New Jersey (Exxon, formerly Esso), Standard Oil of New York (Mobil), Standard Oil of California, Gulf Oil, The Texas Company (Texaco), Royal Dutch Shell and British Petroleum (BP) — generally referred to as the majors, which are engaged in all stages from exploration to distribution. These companies operate on a world-wide basis but the nature of their activities varies markedly from country to country. As far as petroleum is concerned the world may be divided into four parts.

7.2.1 The exporters

The world production of petroleum in 1970 was 2 353 m. tons (cf. Table 7.1), of which the equivalent of 1 055 m. tons (45%) was consumed in the country of origin and 1 298 m. tons (55%) was exported as crude oil or as refined products. Two areas, the Middle East – North Africa and the Caribbean, account for 82% of the petroleum entering international trade. The former is much the more important: it produces 39% of world output, is responsible for 67% of world exports, and possesses 56% of the world's proven reserves. In the sedimentary areas around the Persian Gulf and in Libya oil is easy to find and cheap to produce. It was estimated in the early 1960s that production costs in the Persian Gulf area were no more than $1·5–2·2 a ton as opposed to $15–16 a ton in the USA. Moreover the Middle Eastern countries are poor, backward and thinly populated. They consume less than 10% of their current petroleum output. Their main interest is to obtain the maximum return on the oil which they export. In the past their position was weak. The very speed with which the majors discovered and exploited new Middle Eastern fields in the 1950s and early 1960s tended to depress oil prices. With potential supply outrunning demand the bargaining power of the Middle Eastern governments was limited: faced with demands for higher royalties the majors

Table 7.1 Petroleum Production, Consumption, Trade and Reserves in 1970 (m. tons)

	Production	Exports*	Imports*	Consumption†	Reserves
UK	—	—	103	103	n.a.
Other Western Europe	23	—	502	525	600
USA	538	13	170	697	6 100
Caribbean	212	172	20	50	2 300
Middle East	689	631	—	51	47 000
North Africa	236	232	5	30	8 100
South-East Asia	50	39	46	58	1 900
Japan	—	—	2	211	200
USSR, E. Europe, China	380	47	5	346	13 700
Other	225	96	163	228	4 400
Total	2 353	1 263	1 263	2 288	84 100

* Includes refined products.
† Differences between production and consumption represent stock changes and unknown miliary liftings.

could – and did – switch production to more cooperative countries. These days are now past. The demand for oil is so large that the production of the Middle East – North Africa area is indispensable. The host governments, more sophisticated and more aggressive than before, have been quick to exploit their advantage. The latest round of negotiations has increased the posted prices of crude oil – the notional prices at which oil is exported – from $11–14 to $15–17 a ton at the Persian Gulf and from $16–19 to $23–26 a ton on the Mediterranean coast. These prices are scheduled to rise to $17–20 and $24–28 a ton by 1975. For the foreseeable future Middle Eastern crude oil seems certain to be an increasingly expensive commodity.

7.2.2 The importers

The situation of Western Europe and Japan is therefore delicate. Economic growth in these countries since the Second World War has taken their primary energy consumption from 450 m. tons of oil equivalent in 1950 to 1 300 m. tons in 1970. Nearly two-thirds is now derived from crude oil (Table 7.2), almost all of which is imported, 82% coming from the Middle East or North Africa. The predominance of oil is a comparatively

Table 7.2 Primary energy consumption in four major industrial areas in 1970 (m. tons oil equivalent)

	Oil	Natural gas	Solid fuels	Other	Total
UK	103	11	91	2	207
Other Western Europe	525	52	210	33	820
USA	697	513	330	24	1 563
Japan	200	3	60	7	270
USSR*	347	151	480	30	1 008

* 1965.

recent development since until the 1950s coal remained the main source of energy. The policies of national self-sufficiency pursued by most countries in the interwar era favoured the use of indigenous fuels, while the pricing system then imposed by the majors made oil relatively expensive. This situation changed rapidly after 1945. The collapse of the international oil cartel, the advent of Middle Eastern oil and the development of the supertanker combined to reduce the price of oil at a time when the price of coal was rising rapidly because of its high labour content and the exhaustion of the most productive seams. Oil replaced coal as the most important single source of energy in both Western Europe and Japan in the early 1960s and oil consumption has continued to grow while that of coal declines. In tonnage terms the main demand in both areas is for fuel oil for heating and power generation, and for the middle distillates (cf. Table 7.3). Gasoline is less important. Although by world standards the *per capita* motor vehicle population is high – 197 per 1 000 in Western Europe and 122 per 1 000 in Japan – the heavy taxes imposed on gasoline by European governments strongly favour small private cars and diesel-powered freight and public service vehicles.

Table 7.3 Pattern of demand for petroleum products in major industrial areas in 1970 (% consumption by weight)

	Gasoline	Middle distillates	Fuel oil	Other*
USA	40	26	16	18
Western Europe	18	33	40	9
Japan	17	16	55	12
USSR†	23	36	34	7

* Includes refinery fuel and loss.
† 1965.

7.2.3 The USA

The petroleum industry of the USA is largely isolated from that of the rest of the world. Always a very large producer of oil, America was once a great exporter, but since 1948 has been a net importer. For strategic and political reasons the Eisenhower administration of 1953–61 was anxious to maintain US oil production at a high level and in 1959 it imposed a quota system which for most of the 1960s limited oil imports to about 12% of domestic production. The results of this decision have been profound. American production has indeed remained at a high level – the USA is still the largest single producer in the world – but American oil has become relatively expensive. The USA is a high-cost area where oil is by world standards difficult to find and expensive to produce: in the mid-1960s the average well-head price of American crude oil was about $22 a ton at a time when Middle Eastern oil could have been imported for about half as much. Even today imported oil is $4–7 per ton cheaper than the native product. In consequence natural gas and even coal remain strongly competitive sources of energy and US industry is far less dependent on oil than is that of Western Europe and Japan. The market for petroleum products is in fact primarily a market for gasoline. With a *per capita* motor vehicle population of 495 per 1 000 and negligible rail services the USA is uniquely dependent on motor transport. Morover, low sales taxes make gasoline cheap to the consumer and favour its use for freight as well as for passenger vehicles. In contrast, the competition from other energy sources is so severe that fuel oil often has to be sold at prices actually below that of the crude oil from which it is made (cf. Table 7.4).

Table 7.4 Cost of fuel to major industrial users in the UK and USA 1969–70 (US $ per ton of fuel-oil equivalent)

	Fuel oil	Natural gas	Coal
UK	22·4*	18·1	26·5
USA	13·4	10·3	10·5

* Includes hydrocarbon fuel tax of £2·33 per ton.

7.2.4 The Soviet bloc

Like that of the USA the Soviet petroleum industry is largely self-contained. Production is large, imports are negligible and exports are small. The USSR itself, which accounts for 90% of the output, appears to be a medium-cost area: it has been estimated that in the early 1960s average production costs were about $6 per ton of crude oil. The pattern of consumption, however, is more like that of Western Europe, with fuel oil and middle distillates predominating.

7.3 The composition of petroleum

Crude petroleum is a glutinous green, brown or black liquid with a rank smell. All crudes contain the following constituents although the proportions present may vary considerably.

7.3.1 Hydrocarbons

Hydrocarbons make up 50–95% of the total and are by far the most valuable constituents. The hydrocarbon fraction is composed of alkanes, *cyclo*alkanes and aromatics, alkenes being absent. As Table 7.5 shows, the large majority consists of alkanes and *cyclo*alkanes, with aromatics making up only 8–15% of the total; the higher-boiling fractions contain proportionately more aromatics and *cyclo*alkanes. It is also obvious that the ratio of alkanes to *cyclo*alkanes depends markedly upon the origin of the oil; thus whereas the 45–200°C fraction of Iranian and Kuwait crudes contains more than 70% of alkanes, that from the East Texas oil contains only 50% and that from Baku-Surachny only 27%. A very detailed investigation has been made of the individual compounds present in the crude oil from a single well at Ponca City, Oklahoma. The alkane fraction is composed of *n*-alkanes and, to a lesser extent, methyl-*n*-alkanes; only very small amounts of the more highly branched isomers are present. The *cyclo*alkane fraction consists of *cyclo*pentane and *cyclo*hexane derivatives which may contain up to four fused rings. The aromatic fraction is mainly composed of alkylbenzenes and naphthalenes. Other petroleums are similar.

Table 7.5 Composition of some typical crude oils (wt % of hydrocarbon fraction)

Origin	Boiling range °C	Alkanes	Cycloalkanes	Aromatics
Oklahoma City	40–180	62	29	9
	200–300	30	50	20
	350–500	20	50	30
East Texas	45–200	50	41	9
	200–300	30	55	15
	350–500	20	55	25
Kuwait	15–95	87	11	2
	95–175	68	19	13
	149–232	62	20	18
Baku-Surachny	60–200	27	64	9

7.3.2 Oxygen, nitrogen and sulphur compounds

Oxygen compounds are present to the extent of 0·2–3·0%. The most important are the naphthenic acids – alkyl*cyclo*alkyl carboxylic acids – although alkylphenols also occur in certain crude oils. Nitrogen compounds make up 0·1% or less of the total; they are mainly alkylpyridines and quinolines. Sulphur compounds are more important. Present to the extent of 0·1–5·0% as mercaptans, sulphides and disulphides they are catalyst poisons, give rise to sulphur dioxide when burned, and may impart an unpleasant smell to petroleum products. They are therefore removed during refining.

7.3.3 Resins and asphalts

These rather ill-defined materials make up 5–15% of the total. Resins are hydrocarbon-soluble, probably linear polymers, and asphalts are broadly similar hydrocarbon-insoluble, probably cross-linked compounds. Both are believed to contain fused aromatic and hydro-aromatic rings connected by methylene, ether and thioether groups.

The composition of petroleum indicates a probable biological origin. Almost all petroleum was formed in marine sedimentary deposits and many of the compounds which it

164 Raw materials: petroleum

contains are closely related to those found in living organisms. All specimens contain substantial amounts of n-alkanes, and in many, although not all, there is a predominance of those which contain odd numbers of carbon atoms and these obviously may have arisen from the decarboxylation of the naturally occurring even-number fatty acids present in fats and oils. Other indications of a biological origin are the presence in certain crude oils of isoprenoid and steroidal hydrocarbons and of porphyrins related to the blood pigment haemin and to chlorophyll.

7.4 Refining

The first stage in petroleum refining is 'topping' or 'stripping'. The crude oil is distilled at atmospheric pressure to give a series of fractions boiling at temperatures up to 350°C. The main fractions collected are conventionally termed liquid petroleum gas (LPG), b.p. below 40°C; light naphtha (straight-run gasoline), b.p. 40–180°C; heavy naphtha, b.p. 130–220°C; kerosene, b.p. 160–250°C; and gas oil, b.p. 220–350°C (cf. Fig. 7.1). The

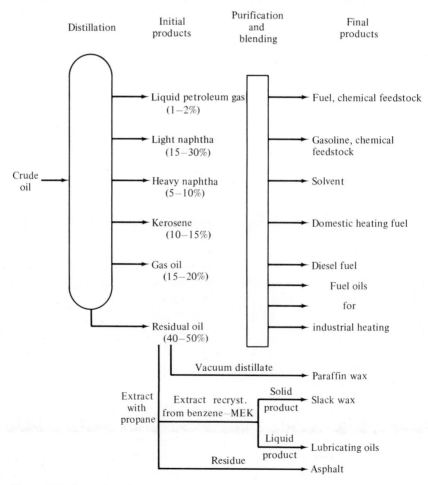

Figure 7.1 Operations in a simple 'hydrostripping' oil refinery.

operation is carried out continuously in a single column with several side-draws or in a series of columns which remove successively higher-boiling products. The various fractions are then redistilled or stripped with steam to remove excessively volatile constituents.

The residual or 'topped' oil boiling above 350°C contains high-boiling hydrocarbons together with asphalts and other relatively insoluble and involatile constituents. Most of this material goes into fuel oil but a proportion is used to make waxes and lubricating oils. In the latter case the hydrocarbons are separated from the asphalts by distillation *in vacuo*, which normally gives both waxy and liquid products; alternatively the residual oil is extracted with liquid propane in which asphalts are insoluble. The extracts and distillates are further purified by counter-current extraction with furfural, in which heavy aromatics and other undesirable constituents are preferentially soluble. The saturated residues are then mixed with benzene and methyl ethyl ketone (MEK) and cooled to $-5°C$ to precipitate slack wax, which is purified by distillation. The dewaxed oils are blended to form lubricants.

Sulphur may be removed from any of the liquid hydrocarbon products by catalytic dehydrosulphurization with hydrogen at 300–400°C and 15–50 atm over a cobalt–molybdenum catalyst. Under such conditions carbon–sulphur bonds are hydrogenolysed with the formation of hydrogen sulphide which is extracted with monoethanolamine (MEA) or diethanolamine (DEA) solutions and converted to sulphur.

7.5 Cracking and reforming

Many European and Asian refineries are of this simple 'hydrostripping' type. Crude petroleum is split into broad fractions which are purified, desulphurized and finally blended to give the desired products. Such a refinery is, however, rarely able to match its output to the demands of the market. Its operational flexibility is limited by the composition of the crude oil used: typically it produces too much light naphtha for the West European gasoline market but not enough for that of the United States. Moreover straight-run gasolines, rich in *n*-alkanes, have relatively poor ignition characteristics; if economical high compression-ratio engines are to be used it is necessary to increase the proportions of *iso*alkanes and aromatics present. The demand for gasoline in the USA led to the development of cracking and reforming processes. Cracking is used to convert high-boiling fractions to gasoline. The principal reactions involved are:

n-alkane ⟶	*iso*alkane	Isomerization
alkane ⟶	smaller alkane + alkene	Fragmentation
alkane ⟶	alkene + hydrogen	Dehydrogenation
alkane ⟶	*cyclo*alkane + hydrogen	Cyclodehydrogenation
*cyclo*alkane ⟶	aromatic + hydrogen	Aromatization

In thermal cracking, introduced as early as 1912, heavy oils are subjected to temperatures of 600°C and upwards in the absence of a catalyst. For gasoline production this process has been largely superseded by catalytic cracking which is carried out at 450–550°C over an acidic silica–alumina catalyst. As Table 7.6 shows, catalytic cracking favours the formation of *iso*alkanes; moreover, the use of zeolite catalysts, widely adopted since 1965, largely suppresses the reactions leading to the excessively volatile C_3–C_4 products. A useful variant of catalytic cracking is hydrocracking, in which the cracking reactions proceed

Raw materials: petroleum

Table 7.6 Products from the thermal and catalytic cracking of n-hexadecane at 500°C and one atmosphere pressure

	Moles product formed per 100 moles cracked during 40% conversion				
	Thermal	Catalytic		Thermal	Catalytic
Hydrogen	16	14	sec-pentene	12	10
Methane	51	4	t-pentene	3	14
Ethylene	84	9	n-pentane	1	6
Ethane	57	7	i-pentane	1	13
Propylene	59	85	C_6 products	27	38
Propane	17	27	C_7 products	15	7
Butadiene	2	1	C_8 products	13	8
n-Butene	19	40	C_9-C_{14} products	40	17
i-Butene	2	28			
n-Butane	3	15	Total	423	375
i-Butene	1	32			

over a silica–alumina–platinum catalyst in the presence of hydrogen at 60–100 atm pressure. Under such conditions the alkenes formed by fragmentation are immediately reduced and the products consist exclusively of alkanes, *cyclo*alkanes and aromatics. Reforming processes, on the other hand, are used to upgrade gasolines by increasing their aromatic content. In catalytic reforming techniques such as 'platforming' the feedstock is reacted at 400–500°C with platinum–alumina or platinum–rhenium–alumina catalysts which promote cyclodehydrogenation and aromatization reactions.

7.6 Alkenes from petroleum

7.6.1 General considerations

Cracking and reforming processes are of interest to the chemical industry because they are a cheap and effective way to convert petroleum to alkenes, acetylene and aromatics. Cracking reactions may be conducted so as to maximize alkene rather than gasoline formation. Thermal cracking is particularly suitable for this purpose as, unlike catalytic cracking, it leads to the formation of substantial amounts of ethylene, the most versatile and the most desirable of alkenes. The thermal cracking of hydrocarbons to ethylene and other low molecular weight alkenes has therefore become the central process of the petrochemical industry.

There are, however, large regional differences in the feedstock used. In Western Europe and Japan crude oil is cheap and the demand for gasoline is limited (section 7.2.2 above). Far more light naphtha is produced than is required for gasoline production; its price is therefore low – currently about £10 ($24) per ton in Britain – and it has become the universal raw material for the manufacture of organic chemicals. The situation in the USA is quite different. Crude oil is more expensive and the demand for gasoline is much larger than the supply of light naphtha (section 7.2.3 above). Most refineries have large cracking sections to convert heavy fractions into gasoline. In these circumstances the price of light naphtha is relatively high – about $35 per ton at the present time – and it is cheaper to make alkenes from the C_2–C_4 off-gases of the catalytic crackers, or, more usually, from the ethane and propane of natural gas (section 8.2).

These differences have to some extent influenced the location of petrochemical installations. American plants are of necessity linked to oil and natural gas processing and are therefore concentrated in the Gulf Coast area of Texas and Louisiana. European plants are generally self-contained units which operate on bought-in naphtha and often have no connection with particular refineries.

7.6.2 Chemistry

The chemistry of cracking is well understood. The overall reactions involved in both thermal and catalytic cracking are, with the exception of isomerization, markedly endothermic but are accompanied by favourable entropy changes (cf. Table 7.7). They are therefore favoured by the use of high temperatures and low partial pressures of reactant.

Table 7.7 Thermodynamic quantities at 25°C of cracking processes

	$\Delta G°$		$\Delta H°$		$\Delta S°$	
	kcal/mole	kJ/mole	kcal/mole	kJ/mole	cal/°C/mole	J/°C/mole
Isomerization	0–1	0–4	−1--−3	−4--−12	2–4	8–16
Fragmentation	6–12	25–50	17–23	71–96	32–34	134–142
Dehydrogenation	21–24	88–100	29–33	121–138	29–31	121–130
Cyclodehydrogenation	6–10	25–42	10–12	42–50	10–12	42–50
Aromatization	20–25	84–105	50–60	209–300	80–90	334–376

The mechanisms of the two processes are, however, different: thermal cracking is a free radical chain reaction whereas catalytic cracking involves carbonium ion intermediates. The reaction sequences displayed in Table 7.8 nevertheless show a strong similarity. In both cases the radicals or ions rearrange, fragment to alkenes and smaller radicals or ions, and are eventually destroyed by chain transfer or termination processes. Why then are the products formed substantially different? The answer lies in the fact that under cracking conditions free radicals rearrange and fragment at roughly comparable rates while carbonium ions rearrange much more rapidly than they fragment. Hence the alkenes formed in thermal cracking derive from the fragmentation of primary and secondary radicals whereas those formed in catalytic cracking derive from secondary and especially tertiary ions. What underlies this behaviour? The directions and velocities of the rearrangements of both radicals and ions reflect the relative stabilities of the reactive intermediates involved. Primary alkyl radicals are less stable than secondary radicals and secondary radicals are less stable than tertiary radicals but the differences in stability are quite small (cf. Table 7.9). In contrast the differences in the stability of primary, secondary and tertiary carbonium ions are very large. Carbonium ions have therefore a much greater tendency to rearrange than have free radicals and, since suitable reaction paths exist, they will rearrange much more rapidly.

It is possible to analyse the later stages of cracking in a similar way. The heats of fragmentation of primary and secondary free radicals are in the range 18–25 kcal/mole (75–105 kJ/mole); fragmentation and chain transfer remain competitive processes for all free radicals in which fragmentation is possible; and so thermal cracking yields appreciable amounts of all alkanes from methane upwards. The heats of fragmentation of large secondary and tertiary carbonium ions are in the range 15–45 kcal/mole (63–188 kJ/mole) but those of ions containing less than six carbon atoms are much larger

168 *Raw materials: petroleum*

Table 7.8 Reaction sequences for thermal and catalytic cracking

Thermal cracking	*Catalytic cracking*

Initiation:

$RCH_2CH_2R' \rightarrow RCH_2^- + R'CH_2$ $X^+ + RCH_2CH_2R' \rightarrow HX + RCH_2\overset{+}{C}HR'$
$X^- + RCH_2CH_2R' \rightarrow HX + RCH_2\overset{|}{C}HR'$

Rearrangement:

$RCH_2\overset{|}{C}H_2^- \rightarrow R\overset{|}{C}HCH_3$ (fast) $RCH_2CH_2^+ \rightarrow R\overset{+}{C}HCH_3$

$RCH_2\overset{|}{C}HR' \rightarrow R\overset{|}{C}HCH_2CH_3$ (fast) $RCH_2\overset{+}{C}HR' \rightarrow R\overset{+}{C}HCH_2R'$ ⎫

$RCH_2\overset{|}{C}HR' \rightarrow R-\overset{|}{C}-R'$ (slow) $RCH_2\overset{+}{C}HR' \rightarrow R-\overset{+}{C}-R'$ ⎬ (very fast)
$\quad\quad\quad\quad\quad\quad\;\; |$ $\quad\quad\quad\quad\quad\; |$
$\quad\quad\quad\quad\quad\quad\; CH_3$ $\quad\quad\quad\quad\; CH_3$ ⎭

Fragmentation:

$RC\overset{|}{H}_2CH_2^- \rightarrow R^- + CH_2{=}CH_2$ $RC\overset{+}{H}_2CH_2^+ \rightarrow R^+ + CH_2{=}CH_2$
$RCH_2\overset{|}{C}HR' \rightarrow R^- + CH_2{=}CHR'$ $RCH_2\overset{+}{C}HR' \rightarrow R^+ + CH_2{=}CHR'$
$RCH_2\overset{|}{C}R'R'' \rightarrow R^- + CH_2{=}CR'R''$ $RCH_2\overset{+}{C}R'R'' \rightarrow R^+ + CH_2{=}CR'R$

Chain transfer:

$R^- + R'H \rightarrow RH + R'^-$ $R^+ + R'H \rightarrow RH + R'^+$

Chain termination:

$R^- + R'^- \rightarrow R{-}R'$ $R^+ + A^- \rightarrow RA$
$R^- + R'^- \rightarrow$ alkane + alkene $R^+ + A^- \rightarrow HA +$ alkene

because of the extreme instability of the small ions ejected. Chain transfer to give alkanes is energetically a much more favourable reaction than fragmentation for small ions, and catalytic cracking therefore gives only those alkanes which contain at least three carbon atoms.

Table 7.9 Heats of formation at 25°C (kcal/mole) of radicals and cations

$\Delta H_F°(R-)$		R	$\Delta H_F°(\overset{+}{R})$	
kcal/mole	kJ/mole		kcal/mole	kJ/mole
52	217	H	364	1 521
32	134	CH_3	255	1 066
25	105	C_2H_5	227	949
22	92	$n\text{-}C_3H_7$	222	928
17	71	$i\text{-}C_3H_7$	193	807
19	79	$n\text{-}C_4H_9$	218	911
13	54	$s\text{-}C_4H_9$	196	819
6	25	$t\text{-}C_4H_9$	171	715

Alkenes from petroleum 169

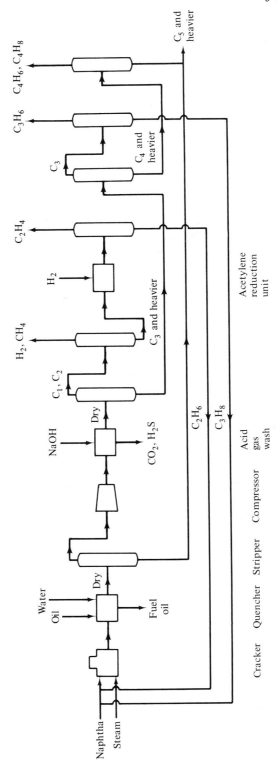

Figure 7.2 Flowsheet for a steam-naphtha cracker.

170 *Raw materials: petroleum*

7.6.3 Technology

In practice the thermal cracking of hydrocarbon feedstocks is carried out at 800–900°C and atmospheric pressure. Ethylene yields increase with increasing temperature but those of the coproducts, especially butadiene, decline; the temperature used will therefore depend on the relative values of the compounds formed. Current practice favours severe conditions so as to maximize ethylene output. Steam is used as a diluent so that a low partial pressure of reactant may be combined with a high flow rate: the residence time in the reactor must be no more than 0·5 s if the decomposition of the products to carbon and hydrogen is to be minimized. A typical steam cracking installation is shown in Fig. 7.2. A hydrocarbon-steam mixture in the ratio 2:1 is vaporized and passed through a vertical tubular reactor. The hot product stream is rapidly cooled and quenched with water and oil to stop secondary reactions. Distillation of the volatile products at atmospheric pressure separates cracked gasoline from the gaseous compounds which are then compressed, scrubbed with sodium hydroxide solution to remove acid gases, dried by passage through molecular sieves, and fractionated at high pressure and low temperature. The ethylene so produced is contaminated with acetylene which is removed by selective catalytic hydrogenation. The C_2–C_4 alkanes formed have little value in themselves and are recycled to extinction or burned as fuel. Butadiene and isoprene are recovered from the C_4 and C_5 streams by extractive distillation with acetonitrile.

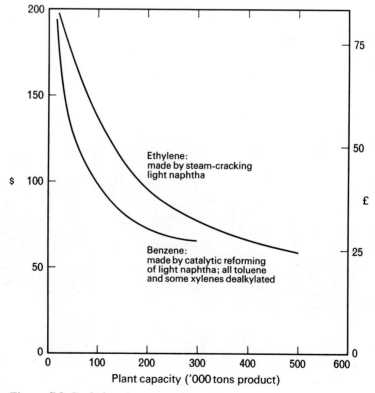

Figure 7.3 Capital cost per annual ton of ethylene or benzene capacity in the UK.

7.6.4 Economics

Steam cracking is a typical capital-intensive process. As Fig. 7.3 shows, a 450 000 ton cracker costs £25–30 ($60–72) per annual ton of ethylene capacity; accordingly capital charges comprise about 25% of the selling price of ethylene. Feedstock and utilities are the major variable costs while labour and supervision is by comparison a trivial expense. (cf. Fig. 7.4).

The capital cost of a cracking unit is divided approximately equally between the cracking and quenching, the compression and the separation sections. The cost per ton of capacity falls as the size of the plant increases, rapidly up to 100 000 annual tons of ethylene and more slowly thereafter. Not only does the square–cube law hold but large economies are possible in crackers of over 50 000 tons capacity by the use of centrifugal compressors. Most recent installations have been in the 300 000–500 000 ton range. Companies have, however, been reluctant to move to yet larger plants. Further economies of scale are indeed possible but the penalties imposed by delays in construction and in

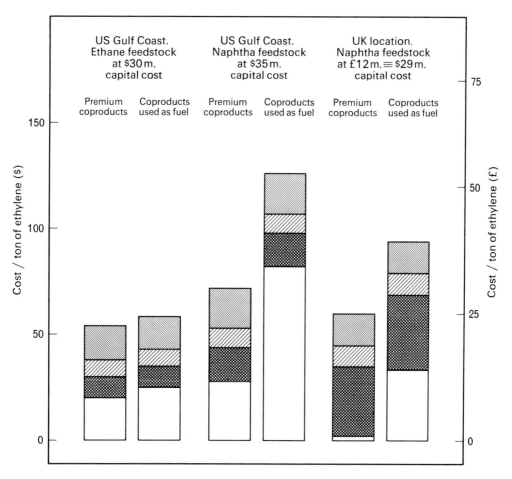

Figure 7.4 Economics of making ethylene in a 450 000 ton steam cracker from ethane and light naphtha under various conditions.

172   Raw materials: petroleum

working up to full output are severe. A consequence of the DCF method of estimating profitability (cf. section 5.7) has been to make executives acutely aware of the drawbacks as well as the advantages of the very large single-stream plant.

Table 7.10 Products from the steam cracking of various hydrocarbon feedstocks

	Tons per ton of ethylene produced				Value in US $ per ton			
					UK		USA	
	Ethane feedstock	Propane feedstock	Naphtha feedstock	Gas oil feedstock	Premium	Fuel	Premium	Fuel
Feedstock	1·312	2·380	2·90	4·286	—	—	—	—
Off-gases	0·212	0·714	0·51	0·429	29	24	17	14
Propylene	0·038	0·385	0·51	0·614	44	19	55	9
Butadiene	0·017	0·076	0·125	0·173	104	19	110	9
Butenes	0·007	0·033	0·15	0·188	22	19	33	9
Cracked gasoline	0·036	0·143	0·55	0·591	29	18	31	8
Fuel Oil	—	0·029	0·06	1·292	13	13	13	8
Feedstock cost (US $ per ton)								
— UK	—	—	25	17				
— USA	22	22	35	17				
Feedstock cost per ton ethylene								
— UK	—	—	72·5	72·9				
— USA	28·9	52·4	101·5	72·9				
Net feedstock cost per ton ethylene								
— UK:								
premium coproducts	—	—	2·2	−22·6				
fuel coproducts	—	—	33·7	16·6				
— USA:								
premium coproducts	20·0	4·5	28·3	−28·5				
fuel coproducts	25·1	36·6	82·4	43·1				

The effect of feedstock costs is complex. As Table 7.10 shows ethylene is only one of the products formed in steam cracking. When naphtha is the feedstock these coproducts form the majority of the output: high-severity cracking gives 35% by weight of ethylene, 27% of propylene, butenes and butadiene, 18% of residual gases – mainly hydrogen and methane – 18% of cracked gasoline and 2% of fuel oil. For accounting purposes it is customary to treat ethylene as the sole product and to deduct the value of the coproducts from the cost of the feedstock. The net cost of the feedstock and therefore of ethylene depends critically on what use is made of the coproducts. If they can be used as chemicals or, where appropriate, as gasoline constituents the net feedstock cost of naphtha under UK conditions is approximately *nil* and the selling price of ethylene would be about £25 per ton. If on the other hand the coproducts are merely burned as fuel then the net feedstock cost is about £15 per ton and the selling price of ethylene around £40 per ton.

If ethylene is to be cheap, premium outlets must be found for the other products. As a major constituent of synthetic rubbers (section 10.6) butadiene is much in demand and commands a relatively high price, but chemical markets for propylene have only been found at prices substantially lower than that of ethylene. The current nominal selling price of ethylene – most production is captive – is about £32 per ton, which suggests that the actual value of the coproducts is somewhere between the extreme cases considered above. It should be noted that the coproducts are only worth separating when they are produced on a large scale; this is a further reason to favour the use of very large cracking plants.

The situation in the USA is quite different. Even on the most optimistic assumptions naphtha is too expensive a feedstock (cf. Table 7.10) and most ethylene is in practice made from ethane which gives relatively small amounts of coproducts. Propylene, the butenes and butadiene are normally obtained by cracking or dehydrogenating propane and butane or as the byproducts of catalytic cracking. The production costs of the alkenes depend mainly on the price of natural gas and are less intimately connected than is the case when naphtha is the feedstock.

7.7 Acetylene from petroleum

Thermal cracking may also be used to produce acetylene. At temperatures above 1 200°C the formation of acetylene from other hydrocarbons is thermodynamically favoured; however, under such conditions acetylene itself has a marked tendency to decompose to carbon and hydrogen. The residence time in the reactor must therefore be very short – preferably no more than 0·01–0·1 s – if substantial yields are to be obtained.

Several techniques have been developed for this purpose. In the BASF one-stage process the combustion of part of the methane feedstock in oxygen heats the remainder very rapidly to 1 400–1 500°C; alternatively, as in the Hoechst two-stage process, the hydrocarbons may be separately injected into a hot gas stream. Much interest has been shown in the cyclic Wulff process. A brick chequerwork surface is heated to 1 500–2 000°C by the combustion of a hydrocarbon–air mixture; in the separate product stage the feedstock is drawn rapidly through the hot reactor. Much of the heat is supplied by the combustion of the process off-gases (cf. Table 7.11) and of the carbon inevitably formed by thermal decomposition. Any hydrocarbon may be used in the Wulff process and by suitable variation of the reaction temperature ethylene-rich or acetylene-rich streams

Table 7.11 Products from the Wulff and BASF acetylene processes

	Tons per ton of acetylene produced	
	Wulff process	*BASF process*
Feedstock	4 tons naphtha	3·9 tons methane +4·2 tons oxygen
Hydrogen	0·42	0·47
Methane	0·92	0·34
Ethylene	0·43	—
Other hydrocarbons	0·20	0·22
Carbon monoxide	0·75	2·99
Carbon dioxide	0·33	0·94
Nitrogen	0·54	—
Oxygen	0·12	—
Feedstock cost	$100	$107
—byproduct credits (as fuel)	61	51
Net feedstock cost	39	56

174 *Raw materials: petroleum*

may be produced at will. On the other hand the residence times are longer than in competitive processes and 15–20% of the feedstock is degraded to carbon. Moreover the system of valves required by a cyclic process is mechanically complex and does not scale up well. Other less widely used methods for acetylene production include the Huls–Dupont electric arc process and the BASF submerged flame technique. In all cases the output stream, rich in acetylene, carbon monoxide and hydrogen, is rapidly quenched with oil and water and the acetylene recovered by extraction with acetone, N-methylpyrrolidone or dimethylformamide.

Regardless of the method used, acetylene is relatively expensive (cf. Fig. 7.5). The processes just described are inherently best suited to outputs of no more than 30 000–50 000 tons per year and most plants are in fact of this size. Capital costs are therefore as

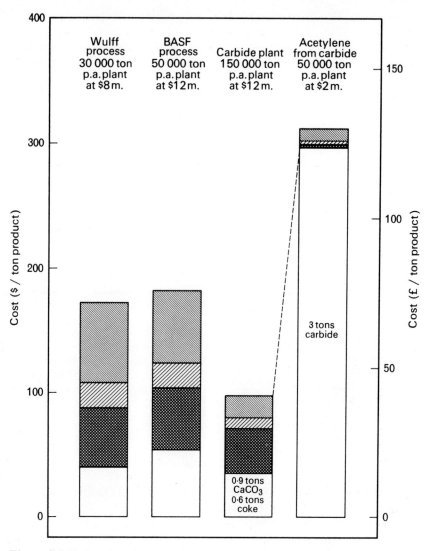

Figure 7.5 Economics of the three routes to acetylene in the UK.

high as £100–120 ($240–300) per annual ton of acetylene capacity. The absence of valuable coproducts leads to a high net feedstock cost and the severe conditions used cause heavy expenditure on utilities. Many users have had serious start-up problems: on the rare occasions that the ICI plant at Runcorn could be induced to work it was both visible and audible from a distance of ten miles. In practice petroleum-derived acetylene is at least twice as expensive as ethylene and it is hardly surprising that it is unable to compete with the latter.

7.8 Aromatics from petroleum

Just as cracking processes have been adapted to the production of alkanes, so reforming reactions now provide the large majority of the aromatic hydrocarbons required by the chemical industry. The catalytic reforming of the C_6–C_8 fraction of light naphtha at 400–500°C and 25–35 atm pressure over a platinum–alumina catalyst gives a 45–55% yield of aromatics rich in benzene, toluene and the xylenes (cf. Table 7.12). Middle

Table 7.12 Products obtained from the catalytic reforming of light naphtha

	Tons per ton of benzene produced			
	C_6–C_7 naphtha feedstock	C_6–C_8 feedstock; hydrogen evolved used to dealkylate all toluene, as much xylene as possible	Toluene dealkylation	Value in US $ per ton
Feedstock	7·1	3·7	1·12 + 0·03 tons H_2	
Toluene	0·5	—	—	40*
Xylenes	—	0·25	—	40*
C_9† aromatics	—	0·20	—	40*
C_5† raffinate	5·0	1·18	—	25†
C_4 and lighter	0·6	1·07	0·15	25†
Feedstock cost	$177·5	92·5	48·8	
—byproduct credits	160·0	74·25	3·7	
Net feedstock cost	17·50	18·25	45·1	

* Valued as gasoline constituent. † Valued as fuel.

distillates may be similarly reformed to napthalenes. Platinum–alumina catalysts are unstable at the low pressures which favour cyclodehydrogenation and aromatization but the recently introduced platinum–rhenium–alumina systems function satisfactorily at 10–20 atm pressure and increase the yield of aromatics to 75%. Catalytic reforming is a carbonium ion reaction closely related to catalytic cracking but in which reactions leading to the formation of aromatics are favoured at the expense of other processes. Platinum functions mainly as a dehydrogenation catalyst but it is possible that Pt^{2+} compounds may facilitate the cyclization of alkanes to *cyclo*alkanes. Some physical constants of aromatics are given in Table 7.13.

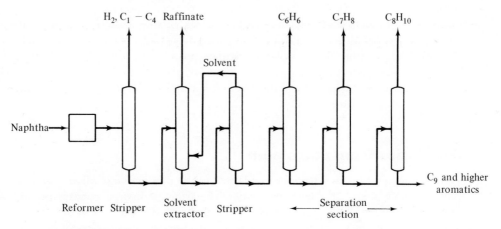

Figure 7.6 Flowsheet for a BTX plant.

A benzene–toluene–xylene (BTX) plant is shown schematically in Fig. 7.6. A desulphurized naphtha feedstock is passed through a fixed-bed reformer; the product mixture is condensed, stripped to remove C_4 and lighter gases, and the aromatic compounds extracted with a diethylene glycol–water mixture (Udex process), N-methylpyrrolidone–ethylene glycol (Aerosolvan process), or sulpholane (Shell process); and the extract fractionated to yield benzene, toluene and mixed xylenes. The separation of the isomeric xylenes is more difficult. Distillation gives pure o-xylene and an apparently inseparable

Table 7.13 Physical constants of aromatics

		Melting point (°C)	Boiling point (°C)
C_8	o-Xylene	−25·2	144·4
	m-Xylene	−47·9	139·1
	p-Xylene	13·2	138·3
	Ethylbenzene	−95·0	136·2
C_7	Toluene	−95·0	110·6
C_6	Benzene	5·5	80·1

mixture of the m- and p-isomers; when this material is cooled to −60°C however, pure p-xylene (m.p. 13·2°C) separates as a crystalline solid, leaving a mother-liquor rich in m-xylene. The o- and p-xylenes are convenient starting points for the manufacture of phthalic and terephthalic acids respectively but m-xylene is of little importance; it is therefore isomerized to an equilibrium mixture of the isomers by the action of a silica – alumina catalyst at 450°C and recycled through the xylene separation plant.

As Fig. 7.7 shows the cost structure of BTX production is similar to that of alkene production. Taking benzene as the main product, capital costs are 35% of the selling price and the main variable cost is the feedstock, which depends on the value assigned to toluene, xylene and the other coproducts. The main outlets for toluene and xylene are as solvents and as gasoline additives; as such their price is considerably less than that of benzene, for which the chemical market is large. Substantial amounts of benzene are indeed made in the USA by the dealkylation of toluene. In a typical process the reaction of toluene and hydrogen at 550–650°C over a silica–alumina catalyst quantitatively converts toluene to benzene and methane in what appears to be a reverse Friedel-Crafts reaction. Such routes are viable whenever the price of benzene exceeds that of toluene by £7 ($17) or more. Since the reforming of C_6–C_8 naphthas gives much more toluene and xylene than benzene, most recent aromatics plants have included a dealkylation unit in which the hydrogen evolved in the reforming stage is used to convert all the toluene and as much of the xylenes as possible into benzene.

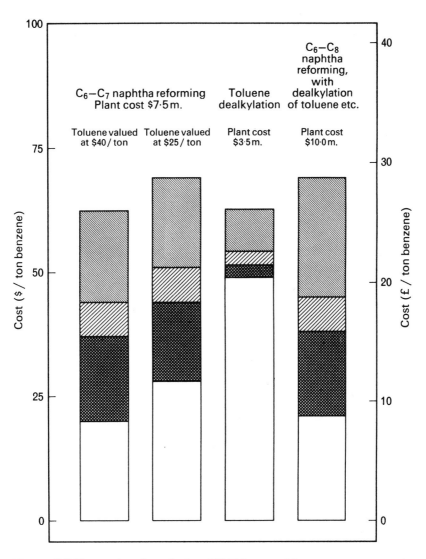

Figure 7.7 Economics of producing 100 000 tons of benzene per annum by various routes in the UK.

7.9 Hydrogen from petroleum

The reaction of hydrocarbons with steam over a nickel catalyst is a cheap and convenient source of hydrogen. The manufacture of hydrogen by the steam reforming of methane was introduced in the USA in 1930 and was adapted by ICI in the 1950s to the use of a light naphtha feedstock. The technology of the process is straightforward. The desulphurized feedstock is reacted with steam over a fixed-bed nickel–alumina–silica catalyst to give a mixture of methane, carbon monoxide and hydrogen. The composition of the product mixture depends on the conditions used: high temperatures (800–1 000°C), low pressures and high (5:1) steam:hydrocarbon ratios favour the formation of hydrogen and carbon monoxide, while lower temperatures (600–700°C), high pressures and low (2:1) steam:hydrocarbon ratios produce a methane-rich mixture. The mechanism of the process is uncertain but appears to involve the cracking of hydrocarbons to methane and alkenes which then react with steam on the catalyst surface. Carbon monoxide may be removed by the shift reaction with steam at 250°C over ferric–chromic oxide or, better, chromic–cupric–zinc oxide catalysts to give carbon dioxide which is then dissolved in water or extracted with DEA.

Steam reforming
$$CH_4 + H_2O \longrightarrow CO + 3H_2, \triangle H = 49 \text{ kcal/mole } (205 \text{ kJ/mole})$$
$$CH_{2.1} + H_2O \longrightarrow CO + 2.05H_2, \triangle H = 59 \text{ kcal/mole } (247 \text{ kJ/mole})$$

Shift reaction
$$CO + H_2O \longrightarrow CO_2 + H_2, \triangle H = -10 \text{ kcal/mole } (-42 \text{ kJ/mole})$$

A major attraction of steam reforming is its versatility. The high-temperature reaction is used in the chemical industry for the production of hydrogen and, suitably modified, for the direct production of mixed gases for the Haber process, for methanol synthesis and for the OXO reaction. In the first case air is introduced into the output stream from the reformer and the oxygen removed by combustion until the hydrogen:nitrogen ratio is 3:1. For methanol synthesis, which requires a 2:1 hydrogen:carbon monoxide mixture, and for the OXO reaction, in which the desired ratio is 1:1, the gas stream is fractionated at low temperature to give pure hydrogen and an appropriate carbon monoxide–hydrogen mixture. Alternatively carbon dioxide is added and the mixture subjected to a reverse shift reaction. The low-temperature reaction on the other hand gives a methane–hydrogen mixture eminently suitable for town-gas production and has been widely used for this purpose.

A further advantage of steam reforming is that methane, natural gas and light naphtha are equally acceptable feedstocks. Methane or natural gas have always been preferred in the USA for economic reasons but until recently light naphtha has been used in Europe. The recent discovery of large deposits of natural gas in the North Sea (cf. section 8.2) has altered the picture completely and most British steam reforming plants are now being converted to use this cheaper raw material.

An alternative route from petroleum to hydrogen is the incomplete combustion of hydrocarbons in an atmosphere of pure oxygen. No catalyst is required and any feedstock from methane to fuel oil may be used. The cost of making or buying pure oxygen is however substantial and the economics of the process are doubtful (cf. Fig. 7.8).

$$CH_{2.1} + 0.5O_2 \longrightarrow CO + 1.05H_2, \triangle H = -21 \text{ kcal/mole } (-88 \text{ kJ/mole})$$

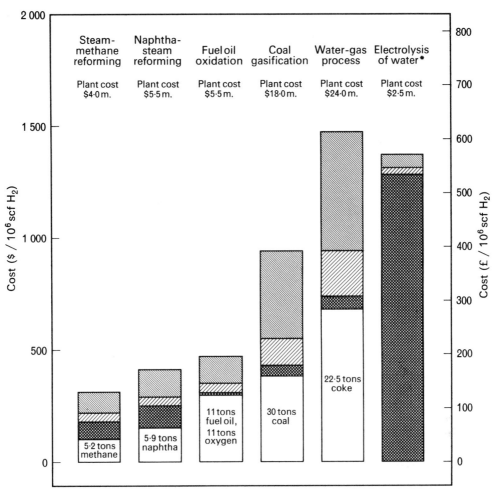

Figure 7.8 Economics of producing 30×10^6 scf. hydrogen per day by different routes.

Notes and references

Units Crude oil and refined products are usually traded in *barrels* (bbl.) of 42 US gallons (=35 imperial gallons). Since the barrel is a unit of volume its weight depends on the material in question; approximate conversion factors are one metric ton = 7·33 bbl. crude oil, 8·45 bbl. gasoline, 7·80 bbl. kerosene, 7·50 bbl. gas oil and 6·70 bbl. fuel oil. Prices are expressed as US dollars per barrel.

General The best introduction to the whole field is the BP publication, *Our Industry – Petroleum*, London, 1970. Intended as a primer for recruits to management it is both highly informative and superbly produced. Earlier editions of this book, called simply

180 Raw materials: petroleum

Our Industry, and the rival Shell *Petroleum Handbook*, 4th edn., London, 1966, are less satisfactory.

Petrochemicals are covered very thoroughly in R. B. Stobaugh, *Petrochemical Manufacturing and Marketing Guide*, 2 vols., Gulf Publishing Co., Houston, 1966, 1968 and in A. V. Hahn, *The Petrochemicals Industry – Production and Markets*, McGraw-Hill, New York, 1970. An excellent short introduction is E. S. and J. P. Stern, *Petrochemicals Today*, Edward Arnold, London, 1971.

The weekly *Oil and Gas Journal* is the international trade paper of the industry. The fortnightly *Petroleum Times* gives more emphasis to the European scene. By far the most useful magazine in the petrochemical field is the monthly *Hydrocarbon Processing* (cited as *HP*): it includes a quarterly *Boxscore* of current construction projects and roughly biennial *Refinery Process* and *Petrochemical Process Handbooks*.

Statistics Outline statistics of the production and consumption of crude oil are given in the *UN Statistical Yearbook*. A much more detailed survey of the production of crude oil and refined products, with details of refining capacity, types of installation and construction projects is given each year in the final issue of the *Oil and Gas Journal*. British consumption figures are given in the Department of Trade and Industry publication *Digest of Energy Statistics*, HMSO, London. American figures are in the US Department of the Interior: Bureau of Mines, *Minerals Handbook 1969*, Washington D.C., 1971.

Two extremely useful – and free – collections of data are the *BP Statistical Review of the World Oil Industry*, published annually, and the Institute of Petroleum Information Service booklets, published intermittently.

The output of basic organic chemicals derived from petroleum and other sources is given in the OECD *Annual Report*, 1969/70.

Sections 7.1–7.2 A handy survey is P. R. Odell, *Oil and World Power*, Penguin, Harmondsworth, 1970; the same author's *An Economic Geography of Oil*, Bell, London, 1963, is also useful but contains many factual errors. C. Tugendhat, *Oil – The Biggest Industry*, Eyre & Spottiswoode, 1968, London, is bland and chatty. The estimates of production costs are those of M. M. Edelman, 'Oil Production Costs in Four Areas', *Proceedings of the Council of Economics*, Washington DC, 1963. On US policy see E. H. Schaeffer, *The Oil Import Programme of the U.S.A.*, F. A. Praeger, New York, 1968. The Russian industry is best studied in R. W. Campbell, *The Economics of Soviet Oil and Gas*, Resources for the Future, Baltimore, 1968.

Section 7.3 A relatively old but very thorough book on the chemical aspects of petroleum is *The Chemistry of the Petroleum Hydrocarbons*, eds. B. T. Brooks, C. E. Boord, S. S. Kurtz and L. Schmerling, 3 vols., Reinhold, New York, 1955. A. N. Sachenen, vol. 1, p. 5 deals with the composition of crude oils and Table 7.5 is based on his data. The Ponca City study is described in F. D. Rossini, B. J. Mair and A. Streiff, *Hydrocarbons from Petroleum*, Reinhold, New York, 1953.

Sections 7.4–7.5 For detailed descriptions of refinery processes see W. L. Nelson, *Petroleum Refinery Engineering*, 4th edn., McGraw-Hill, New York, 1958. The same author has produced a *Guide to Refinery Operating Costs*, Petroleum Publishing Co., Tulsa, 1970. P. H. Frankel and W. Newton, *J. Inst. Petroleum*, 1968, **54**, 23, discuss world trends in refining.

Section 7.6 The chemistry of thermal and catalytic cracking is described by E. W. R. Steacie and S. Bywater and by B. S. Greensfelder respectively, both in Brooks *et al.*, *op. cit.*, vol. 2, pp. 1, 137. Tables 7.6 and 7.7 are based on these reviews. The economics of particular feedstocks have been repeatedly considered; particularly useful are J. G. Freiling, B. L. Huson and R. N. Summerville, *HP*, 1968, **47** (11), 145, from which Table 7.10 is taken, and J. L. De Blieck, H. J. Cijfer and R. R. J. Jungerhans, ibid, 1971, **50** (9), 177. The layout of a large ethylene cracker is described by R. L. Duckworth, *CPE*, 1968, **49** (2), 67, who gives extensive cost data.

Section 7.7 The various processes are described in S. A. Miller, *Acetylene – Its Preparation, Properties and Uses*, vol. 1, Benn, London, 1965. J. M. Reid and H. R. Linden, *Chem. Eng. Progr.*, 1960, **56** (1), 47, discuss gasoline pyrolysis in detail. The economics of acetylene manufacture are considered by R. B. Stobaugh *op. cit.* and, in the UK context, by Anon, *CPE*, 1966, **47** (2), 71.

Section 7.8 The chemistry of catalytic reforming is discussed rather inconclusively by E. L. Pollitzer, J. C. Hayes and V. Haensel in *Refining Petroleum for Chemicals*, eds. L. J. Spillane and H. P. Leftin, *Advances in Chemistry*, No. 97, American Chemical Society, Washington D.C., 1970. For recent technical developments see R. L. Jacobson and C. S. McCoy, *HP*, 1970, **49** (5), 109, and S. Field, ibid., 113. Rather sketchy accounts of the economics of aromatics production are given by A. V. Hahn *op. cit.*, by J. W. Andrews and R. E. Conser, *HP*, 1965, **44** (3), 127, and by Anon, *ECN*, 1965, No. 165, 30.

Section 7.9 Hydrogen is usually measured in standard cubic feet (scf) at atmospheric pressure and 60°F. One scf equals 0·0268 m^3 at NTP. One million scf of hydrogen weighs 2·42 tons.

For the chemistry and technology of naphtha reforming see S. P. S. Andrew, *Chem. and Ind.*, 1965, 826 and G. W. Bryder and W. Wyras, *CPE*, 1967, **48** (9), 101. Economic aspects are discussed by J. Voogd and J. Tielrooy, *HP*, 1967, **46** (9), 115, and by K. K. Bhattacharyya and S. Patil, *Chem. Age of India*, 1968, **17** (3), 169. There is a good comparative account of steam reforming and partial combustion by G. J. Van den Berg and I. N. Van Lookeren-Campagne, *Advancement of Science*, 1971, **27**, 364.

8 Raw materials: natural gas, coal, fermentation

8.1 Petroleum's competitors

Petroleum is not the only source of organic compounds. Hydrocarbons occur in other forms. Natural gas is already of major importance in the USA (cf. Chapter 7) and of increasing significance in Europe. Alternatively, the vast reserves of hydrocarbons contained in shales and tar sands might be exploited. Other sources of fixed carbon are also available. In the recent past chemicals were made from coal, from wood, or by fermentation; the routes used were practicable but expensive and they have been very largely displaced by petroleum-based processes. Yet methods change with circumstances. A desire to exploit indigenous resources for strategic reasons, to conserve foreign exchange, or to increase local employment may favour the use of coal or agricultural products by the chemical industry. This was the case in Western Europe in the interwar era and is true of South Africa today. More generally, although petroleum is currently much the cheapest source of organic compounds, this situation may not last forever. Crude oil prices are rising steadily and seem likely to continue to do so. The world's reserves of petroleum are limited and it may be that the twentieth century will see both the rise and the decline of the international oil industry. A whole range of alternative technologies already exists which could become economically attractive if the relative price of petroleum rose sufficiently. The chemical industry could survive an enforced change in raw materials even though its continued growth might be jeopardized.

8.2 Natural gas

At present the main rival to petroleum as a chemical feedstock is natural gas. Deposits of hydrocarbon gases occur frequently in sedimentary areas (see Table 8.1), sometimes as part of an oil reservoir (associated gas) but more usually in isolation (non-associated gas). All natural gases are largely composed of methane together with smaller amounts of higher alkanes, carbon dioxide, nitrogen and water vapour (see Table 8.2). Certain deposits – notably that at Lacq in the foothills of the Pyrenees – are also rich in hydrogen sulphide. The proportions of these constituents vary considerably. Associated gas is inevitably 'wet': it contains comparatively large amounts of readily liquefied hydrocarbons. Non-associated gases are usually 'dry' although they may nevertheless contain appreciable quantities of the C_2 and higher alkanes.

The technology of gas treatment is broadly similar to that employed in petroleum refining. At the well-head the gas is normally at ambient temperature but at a pressure of 30–100 atm. If present in substantial amounts, propane and higher alkanes are removed by absorption in a high-boiling oil from which they are subsequently recovered and purified by low-temperature fractionation. Ethane may be similarly extracted by absorption at $-50°C$. Acid gases such as hydrogen sulphide and carbon dioxide are removed by treatment with aqueous monoethanolamine (MEA) or diethanolamine (DEA). The residual gas is then dried by extraction with the hygroscopic liquid diethylene glycol in order to inhibit the formation of solid methane–water clathrates which have been known to choke valves and lines at low temperatures.

Table 8.1 World production and reserves of natural gas 1969

	Production	Reserves
UK	188*	28 548
Netherlands	814	90 098
Other W. Europe	1 071	25 668
Middle East	294	207 204
North Africa	113	153 301
USSR	6 695	339 227
USA	22 803	289 788
Other	5 438	247 380
Total	37 416	1 381 105

* Figures in units of thousand million standard cubic feet.

Table 8.2 Composition (volume %) of some representative natural gases

	North Sea	Slochteren, Holland	Lacq, France	Po Valley, Italy	Texas Panhandle	Hassi R'Mel, Algeria	Kuwait Associated gas
CH_4	94.4	81.8	69.6	95.9	80.9	83.5	76.7
C_2H_6	3.1	2.7	3.1	2.8	6.8	7.0	13.2
C_3H_8	0.5	0.4	1.1		2.7	2.0	5.3
$\geqslant C_4H_8$	0.4	0.2	1.3	0.9	1.6	1.2	2.5
N_2	1.1	14.1	—	0.4	7.9	6.1	—
CO_2	0.5	0.8	9.6	—	0.1	0.2	2.2
H_2S	—	—	15.3	—	—	—	0.1
Calorific value (Btu/scf)	1 024	880	830	1 015	1 064	1 130	1 258

The utilization of natural gases by the chemical industry depends on their composition and their price. Most American gases contain recoverable amounts of ethane and higher alkanes: in 1970, 80% of US gas production was processed to yield 67 m. tons of such products. Moreover, the price of US gas is controlled by the Federal Power Commission, whose policy it is to keep gas prices low. On the Gulf Coast natural gas is available at 30¢ (£0·125) per thousand standard cubic feet (scf) and ethane, propane and butane at $20–25 (£8–10) per ton; since light naphtha costs $35 (£15) per ton, it is not surprising that natural gas alkanes are the preferred feedstock for both thermal cracking

184   *Raw materials: natural gas, coal, fermentation*

and steam reforming (cf. Chapter 7). Gas from the North Sea or from Holland, on the other hand, is a dry gas which does not contain worthwhile amounts of ethane. Available to large users at 35–45¢ (£0·145–0·19) per thousand scf, it is steam reformed to hydrogen or synthesis gas or burned as fuel. As it is nearly pure methane it is unsuitable for the manufacture of ethylene; it might be a cheaper feedstock than naphtha for acetylene production but it is doubtful if it could make the latter fully competitive.

8.3 Mineral hydrocarbons

Certain solid minerals contain hydrocarbons. The most commercially attractive type are the tar sands, which are mixtures of quartz, water and a heavy bituminous petroleum. The world's reserves of tar sand bitumen are believed to be about 150 000 m. tons – nearly twice the proven reserves of petroleum – of which approximately 70% are in the Athabasca region of North-Eastern Alberta. The Athabasca tar sands contain 83% quartz and 17% bitumen and water; the bitumen is separated by distillation, by solvent extraction, or, most usually, by hot water flotation. The crude material has a boiling range of 260–550°C and is composed of roughly equal proportions of hydrocarbons and of resins and asphalts, with an overall sulphur content of 4–5%. It may be cracked and dehydrosulphurized to give a range of synthetic products comparable with those obtained from a hydrostripping oil refinery. Processes based on tar sand appear to be about economically feasible under North American conditions since the Great Canadian Oil Sands Co. has recently opened a 2 m. ton per year plant.

Even larger reserves of hydrocarbon exist in the form of oil shales, laminated sedimentary deposits from which the organic material is released by destructive distillation. The world's oil shale contains perhaps 450 000 m. tons of oil, about half being in Colorado and Wyoming. Most oil shales contain only 15–20% of organic matter which is mainly in the form of polymers rich in aromatic, alicyclic, and nitrogen and sulphur containing ring systems. On pyrolysis at 500°C this material is cracked and depolymerized to yield 20–30 gallons of oil per ton of shale. The crude oil distils in the range 200–600°C and contains about 40% of hydrocarbons and 60% of nitrogen, oxygen and sulphur compounds; in contrast to petroleum the hydrocarbon fraction contains a high proportion of alkenes and of aromatics. Further high-temperature thermal cracking and catalytic reduction are required if useful products are to be obtained. Shale oil has been produced intermittently in many countries – notably Scotland – but has everywhere been displaced by petroleum. American interest has revived in recent years but informed opinion favours the tar sands on economic grounds.

8.4 Coal

8.4.1 Structure

Coal is a black combustible solid, rich in carbon, which occurs in layers (seams) in sedimentary areas. It was formed by the anaerobic decomposition of plant debris in swampy areas to give peat, which was subsequently transformed by heat and by the pressure of the accumulated overburden of younger sediments into materials of higher carbon content (higher rank; see Table 8.3).

Table 8.3 Composition of the main types of coal

	Carbon* %	Hydrogen* %	Oxygen* %	Nitrogen* %	Sulphur* %	Moisture† %	Volatiles*	Calorific value (Btu/lb)	
Peat	45–60	3·5–6·8	20–45	0·7–3·0	2–7	70–90	45–75	7 500–9 600	↑ Increasing rank. ↓
Brown coals and lignites	60–75	5·5–4·5	17–35	0·7–2·1	2–3	30–50	45–60	12 000–13 000	
Bituminous coals	75–92	4·0–5·6	3–20	0·7–2·0	1–2	1–20	11–50	12 600–16 000	
Anthracites	92–95	2·9–4·0	2–3	0·7–2·0	0·5–0·6	1·5–3·5	3·5–10	15 400–16 000	

* Dry, mineral-free basis.   † As found.

Bituminous coals are the most important variety and the most studied. The microscopic examination of bituminous coals shows that they contain three types of organic mineral: vitrinites (anthraxylons), black lustrous substances derived from woody tissues; exinites, formed from spores, pollens and cuticle; and inertinites, dull black largely fibrous materials with much higher carbon contents than the other components. The recent work of Vahrman has made clear the nature of these substances. Vitrinites contain 10–20% of low molecular weight materials of two different classes: O-compounds, which have hydroxylated and alkylated aromatic and hydroaromatic skeletons, and H-compounds, which are n-, methyl and polymethylalkanes, alkylbenzenes and naphthalenes, and higher aromatics and neutral oxygen-containing compounds. These small molecules are trapped within a porous, highly condensed and largely aromatic structure from which they may be liberated by pyrolysis or by exhaustive extraction. Exinites are similar in character, while the inertinites have the same polymeric framework but do not contain either H- or O-compounds.

Lignites (brown coals) represent a stage of coal formation intermediate between peat and bituminous coals. They are both richer in oxygen and appreciably less aromatic than the latter, and contain substantial amounts of humic acids – complex brown decomposition products of lignin – together with resins and waxy carboxylic esters.

8.4.2 Production

Coal is produced by underground or by open-cast (strip) mining. Underground mines require large capital investments and much of the coal has to be sacrificed if the workings are not to collapse. When the seams are near the surface, open-cast mining permits the cheap and rapid extraction of the entire deposit but inflicts severe environmental damage. Despite extensive mechanization, both methods are labour intensive by the standards of the oil industry. Even in the USA where the output is as high as 20 tons per man-day, labour is 35% of total costs, and in the less efficient West European industry the corresponding figure is 50% or even higher. This fact is of more than local significance. Large in numbers, concentrated in limited areas, and with strong traditions of discipline and solidarity, miners are a powerful political force in the coal-producing countries and exert a considerable influence over national industrial policies.

186 Raw materials: natural gas, coal, fermentation

As Table 8.4 shows, the tonnage of coal and lignite produced annually is still greater than that of oil, while available reserves vastly exceed those of any other fossil fuel. Production has, however, risen only slowly in recent years and overall world figures conceal large variations between individual nations.

In the USA coal seams are thick, the industry is highly mechanized, and open-cast mining accounts for 40% of total production. Available in most industrial areas at about $11 (£4·5) per ton, coal is a competitive source of heat: it has been estimated that in 1968 the costs of obtaining 1 m. Btu. from coal, natural gas and fuel oil were respectively 25·2, 24·7 and 32·2 cents (10·5, 10·3 and 13·4p). US coal production has risen by 20% in the past ten years and it seems probable that this trend will continue. Russian production is buoyant for similar reasons. The situation in Western Europe and in Japan is less favourable. Seams are thin and deep and productivity is no more than $2\frac{1}{2}$–4 tons per

Table 8.4 World production and reserves of coal and lignite 1969

	Coal			Lignite		
	Production	Measured reserves	Inferred reserves	Production	Measured reserves	Inferred reserves
UK	153*	12 227	3 273	—	—	—
W. Germany	112	70 000	n.a.	107	62 000	n.a.
Other W. Europe	72	6 305	1 345	9	1 080	389
E. Germany	1	n.a.	n.a.	254	n.a.	n.a.
Poland	135	32 425	13 316	31	6 500	8 413
USSR	426	145 123	3 976 480	138	104 000	1 302 026
USA	513	72 000	1 028 000	5	9 400	396 600
Japan	45	5 723	13 525	—	240	1 495
S. Africa	53	36 873	35 592	—	n.a.	n.a.
Other	554	75 723	1 113 530	119	95 780	53 482
Total	2 064	456 399	6 185 061	765	279 000	1 762 405

* Figures in units of m. tons.

man-day. Although wages are low by US standards, coal is more expensive for most purposes than oil or gas. In Britain the industry survives in its present form because of governmental restrictions on coal imports and taxes on oil; it is probable that under free-trade conditions UK coal production would fall from 140 to about 60 m. tons per year. West of the Curzon line, only the Polish coal industry, entirely reconstructed since 1945 under the direction of Totleben, can be said to be fully competitive. For political and strategic reasons, most European governments are anxious to retain their indigenous coal industries in some form, but with the recent discovery of large gas and oil deposits in the North Sea and elsewhere the future of coal appears highly uncertain.

Coal is in any case a less versatile material than petroleum. As a solid fuel it is less easily handled than fuel oil and is best suited to very large scale applications such as electric power generation. Rail transport, town gas manufacture and domestic heating, once major markets, are of rapidly dwindling importance. Coal also suffers from the fact that unlike oil it cannot be desulphurized before use. When burned, its sulphur content therefore appears as sulphur dioxide and so adds to the pollution of the atmosphere.

In the USA this has recently emerged as a major political issue and may well restrict the growth of the coal industry.

Other than as a source of heat, the main outlet for coal is in the manufacture of coke for the iron and steel industry, a process which is also the major source of coal-based organic chemicals.

8.4.3 Carbonization

When heated above 400°C in the absence of air, coal softens and becomes plastic; volatile materials are evolved; and finally the involatile residue coalesces, swells and solidifies to form coke. The nature and yield of the products formed under various conditions are shown in Table 8.5. Low-temperature carbonization leads to large yields of a partly saturated tar and to a reactive coke or 'char' suitable for smokeless fuels such as Coalite. The now obsolete medium-temperature process favours gas formation, while

Table 8.5 Products of the low- and high-temperature carbonization of bituminous coal (yield per ton of coal carbonized)

Gas	Light oil	Tar	Coke	Ammonia
Low-temperature reaction (450–700°C)				
4 000 scf = 30 therms	2·5–3·5 gal	17–90 gal.	0·75 tons	2·5 lb
H_2 10 vol. %	Alkanes 46 vol. %	BTX* <0·5 vol. %	8–20% volatile matter	
CH_4 + higher alkanes 65	Alkenes 16	Cresols 3·5		
	Cycloalkanes 8	Xylenols 6·5		
CO 5	Cycloalkenes 9	Other phenols 13·0		
CO_2 9	Aromatics 16	Other aromatics 3·0		
Other 11	Other 5	Naphtha 36		
		Pitch 26·0		
High-temperature reaction (900–1 100°C)				
10 000 scf = 50 therms	3 gal	6–7 gal	0·7–0·8 tons	3–4 lb
H_2 50	Benzene 72	BTX* 0·6	1–2% volatile matter	
CH_4 + higher alkanes 34	Toluene 13	Cresols 1·0		
	Xylenes 4	Xylenols 0·5		
CO 8	Alicyclics 5	Other phenols 1·5		
CO_2 3	Aliphatics 6	Naphthalene 8·9		
Other 5		Anthracene 1·0		
		Other aromatics 9·0		
		Tar bases 1·8		
		Pitch 60·0		

* Benzene, toluene, xylenes.

188   Raw materials: natural gas, coal, fermentation

high-temperature carbonization gives a hard metallurgical coke and a highly aromatic tar. The volatile products obtained at low temperatures resemble closely the *O*- and *H*-compounds present in the vitrinite and exinite constituents of coal; in all probability carbonization does little more than release them from the molecular sieve structure in which they are trapped. The gaseous and aromatic products formed at higher temperatures presumably arise from the thermal cracking of the compounds initially liberated in this way.

The technology of carbonization is straightforward. Coal is pyrolysed in vertical ovens maintained at the desired temperature. The volatile products are cooled rapidly with dilute aqueous ammonia to precipitate tar; the gas stream is then washed successively with sulphuric acid and light petroleum to remove ammonia and light oil respectively; and the residual gas is treated with iron oxide to remove hydrogen sulphide, dried, and either sold or burned in the ovens. The liquid products are purified by appropriate methods. Tar is distilled to give a range of volatile fractions from which pyridines, phenols, naphthalene and other products are obtained by selective extraction and recrystallization (see Fig. 8.1), together with a residue of pitch. Light oil is fractionated to give benzene, toluene, mixed xylenes and naphtha; the benzene fraction, which is the most valuable product, is contaminated with thiophene, which is best removed by preferential sulphonation with concentrated sulphuric acid.

From an economic point of view the main product of carbonization is coke, and the chemical co-products serve merely to reduce the net cost of the coal from which it is

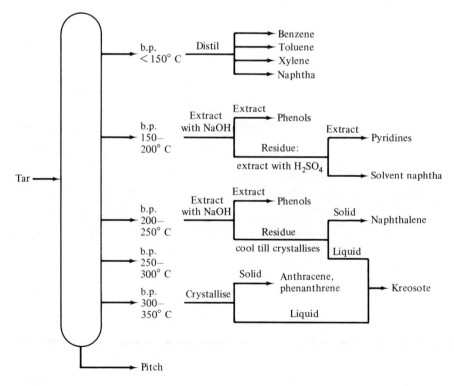

Figure 8.1 Separation of constituents of coal tar.

made. The level of coke production is determined almost entirely by the demands of the steel industry; coal tar and the other co-products are inevitably formed in corresponding amounts and must be made to yield the maximum return. Carbonization remains the major source of certain aromatics, notably the cresols and xylenols, naphthalene and the other polycyclics, and the pyridines. The production of these compounds by tar distillation is reasonably profitable. The other co-products must be sold for what they can fetch in markets dominated by the petrochemical industry, often at prices which cover little more than the direct costs of their extraction.

8.4.4 Acetylene via carbide

Carbonization is only of significance as a source of aromatic compounds. The most practicable route from coal to the reactive aliphatics required by, for example, the polymer industry, is through calcium carbide and acetylene. When calcium oxide and metallurgical coke are heated in an electric furnace at 2 000°C solid calcium carbide of 80–85% purity is formed. As the salt of a very weak acid ($pK_a \sim 20$) it is rapidly and completely hydrolysed when added to water; the acetylene liberated is contaminated only by traces of phosphine and hydrogen sulphide which are readily removed by selective oxidation with sodium hypochlorite:

$$CaO + 3C \longrightarrow CaC_2 + CO, \quad \Delta H = 111 \cdot 2 \text{ kcal/mole } (465 \text{ kJ/mole})$$

$$CaC_2 + 2H_2O \longrightarrow Ca(OH)_2 + C_2H_2, \quad \Delta H = -29 \cdot 9 \text{ kcal/mole } (-125 \text{ kJ/mole})$$

This is a simple but not a particularly cheap way to make acetylene. The production of calcium carbide is expensive for so elementary a process. As Fig. 7.5 shows, it has high capital, raw material and, above all, electric power costs: the extreme temperatures required necessitate the use of electric furnaces which consume 3 000–3 300 kWh per ton of carbide formed. In both Britain and the USA 30–40% of the cost of carbide is that of the electric power used; acetylene is correspondingly expensive and has rapidly lost ground to ethylene as a chemical intermediate. Conversely, in those countries such as Norway, Switzerland and Japan which have very cheap hydroelectric power, carbide-based acetylene has retained much of its former importance (see Table 8.6).

Table 8.6 Production of calcium carbide and acetylene 1969

	Calcium carbide				Acetylene		Ethylene
	Production	Imports	Exports	Consumption	Production	% from carbide	Production
UK	164*	172	<1	335	n.a.	~90?	889
W. Germany	867	24	1	890	353	81	1 717
France	487	2	30	459	161	91	761
Italy	263	9	—	272	243	36	673
Norway	170	—	94	76	25	100	n.a.
USA	756	16	n.a.	?750	444	54	7 453
Japan	1 515	n.a.	n.a.	?1 500	496	97	2 400

* Figures in units of '000 tons.

8.4.5 Water-gas and the Fischer–Tropsch process

Another use of coal is in the production of hydrogen and carbon monoxide. When steam is passed through white-hot coke it is reduced to water-gas (blue gas), a roughly equimolar mixture of carbon monoxide and hydrogen. The reaction is markedly endothermic and a cyclic process has usually been employed. Coke is heated to 1 200°C by partial combustion to carbon monoxide and carbon dioxide (producer gas) in a stream of air; steam is then drawn through the hot coke to give water-gas. Alternatively, the two steps may be combined as in the Lurgi process in which low-rank coals are continuously gasified by a steam–oxygen mixture.

$C + H_2O \rightarrow CO + H_2$, $\Delta H = 31$ kcal/mole (130 kJ/mole)   water-gas reaction

$C + O_2 \rightarrow CO_2$, $\Delta H = -97$ kcal/mole (407 kJ/mole) $\Big\}$ producer gas reaction
$C + \frac{1}{2}O_2 \rightarrow CO$, $\Delta H = -26$ kcal/mole (-100 kJ/mole)

$CO + H_2O \rightarrow CO + H_2$, $\Delta H = -10$ kcal/mole (-42 kJ/mole) shift reaction

These reactions are highly versatile and were widely employed in the chemical industry between 1910 and 1960. By the 'shift' reaction (cf. section 7.9) followed by removal of carbon dioxide, water-gas may be converted to pure hydrogen, and producer gas to a nitrogen–hydrogen mixture. By blending the various streams, mixtures suitable for the Haber process ($N_2:3H_2$) and for methanol synthesis ($CO:2H_2$) are readily obtained. A particularly interesting application is the Fischer–Tropsch process in which hydrocarbons are formed from carbon monoxide and hydrogen in the presence of iron or cobalt catalysts. The reaction appears to involve the reduction of carbon monoxide to methylene radicals which then polymerize on the catalyst surface to give a mixture of predominantly straight-chain hydrocarbons rich in terminal alkenes together with small amounts of alcohols, ketones and acids. In a typical installation a 1:2 carbon monoxide–hydrogen mixture at 20 atm pressure is passed over an alkaline iron catalyst maintained at 250–300°C in both fixed and fluid-bed reactors. At 180–200°C the yield of oxygenated products is increased to 40–50% of the total.

$nCO + 2nH_2 \rightarrow (CH_2)_n + nH_2O$, $\Delta H = -39.4$ kcal/mole (-165 kJ/mole)
$nCO + 2nH_2 \rightarrow H-(CH_2)_n-OH + (n-1)H_2O$, $\Delta H = -58.5$ kcal/mole
$(-245$ kJ/mole)

Despite intensive development, these processes are grossly uneconomic. The weak point is gasification. Since coke is a solid, water-gas reactors are bulky, mechanically complex and therefore expensive (cf. Fig. 7.8); raw material, utility and maintenance costs are also very high. Continuous gasification by the Lurgi process is somewhat more attractive, especially where coal is cheap, but also suffers from excessive capital costs: it is reported that even if coal were free it would be cheaper to make hydrogen from oil. The only Fischer–Tropsch plant currently in operation is that at Sasolburg, South Africa, which is said to gasify 4 m. tons of coal annually to give 1 m. tons of gasoline and other products. Although the coal used costs only $1.00 (40p) per ton this installation is political rather than chemical in origin: South Africa is rich in coal but has no oil, and the Nationalist government lives in constant fear that its racial policies will provoke an oil embargo.

8.4.6 Hydrogenation

Coals and coal tars may be catalytically reduced to liquid hydrocarbons. Extreme conditions are required: in a representative plant a paste of lignite, heavy oil and iron catalyst is treated with hydrogen at 450°C and 700 atm pressure to give an almost quantitative yield of liquid and gaseous products. The main fraction, boiling in the range 180–325°C, is rich in aromatics and may be converted to gasoline by further reduction in the vapour phase. The economics of such processes are distinctly unfavourable, especially when the necessary hydrogen is produced from coal. They were nevertheless widely employed in Germany before and during the Second World War, and annual synthetic gasoline production reached 3·5–4 m. tons in 1943–44. More recently, the rising price of oil has revived American interest in coal hydrogenation and the possibilities offered by reduction under milder conditions or by the use of hydrogen-donating solvents are being explored.

8.5 Fermentation
8.5.1 Nature and scope

A completely different route to organic compounds is provided by fermentation processes. When supplied with suitable nutrients, unicellular organisms such as yeasts, moulds and bacteria grow and multiply; as they do so particular compounds are produced as end-products of the metabolic pathways involved and accumulate in the culture medium. By cultivating particular micro-organisms under appropriate conditions, many compounds may be obtained in commercially significant quantities (see Table 8.7).

Fermentation methods have many advantages. They are remarkably versatile and may be used to make both simple products, such as ethanol and glycerol, and compounds of such extreme complexity as the enzymes and antibiotics. Since the metabolic reactions are specific, high yields are obtained, and since the micro-organisms are living things, the reaction conditions employed are necessarily very mild. On the other hand it is often both expensive and difficult to maintain stable cultures of the desired species. Even more serious, fermentation reactions are slow and, because micro-organisms have only a limited ability to tolerate their own wastes, give very dilute solutions of the desired products. Another major disadvantage is that nutrient prices are often high and always unstable, as is usually true of agricultural materials, while the co-products are rarely saleable and, indeed, often a substantial effluent problem.

For these reasons microbiological methods are now largely confined to the following situations: Where there is no alternative, for example antibiotics; where the product is particularly valuable, for example modified steroids; where the nutrients are cheap, for example in underdeveloped countries; or where a dilute product is acceptable, for example in sewage treatment.

8.5.2 Technology

The technology of fermentation processes is naturally quite different from that of the petrochemical industry and is best illustrated by an example. The antibiotic Penicillin G is obtained by growing a selected strain of the mould *Penicillium chrysogenum* in a carbohydrate medium. The reaction is aerobic and oxygen must be continuously supplied to the

Table 8.7 Typical fermentation processes for the production of organic compounds

Product	Nutrient	Micro-organism	Type of reaction	Temperature (°C)	pH	Reaction time (h)	Yield (%)	Concentration of product (%)
Ethanol	Starch or sucrose	Yeasts	Anaerobic	25–35	4–5	30–72	90	8–10
Butanol (65%) + acetone (35%)	Starch	*Clostridium aceto-butylicum*	Anaerobic	30–35		36–48	30	1·5–2·5
Citric acid	Sucrose	*Aspergillus niger*	Aerobic	28–32	2–3·5	200	70	10–20
Lactic acid	Sucrose	*Lactobacillus delbruckii*	Anaerobic	40–50	5–5·8	140–180	85	10
Penicillin	Sucrose	*Penicillium chrysogenum*	Aerobic	25	5·5	140–180	2–3	0·1–0·3
Proteolytic enzyme	Sucrose and soybean flour	*Bacillus subtilis*	Aerobic	25–32	6·5–8·5	72–120	n.a.	n.a.
Single-cell protein	Gas oil and ammonia	Yeasts	Aerobic	30–40	6·0	10–40	50–60	1–2

organisms. A 10 000 gallon stainless steel reactor is charged with an aqueous solution containing 4% corn-steep liquor solids (a crude form of sucrose), 4% lactose and smaller amounts of other nutrients, at a pH of 5·5. The mixture is sterilized with steam, cooled to 24°C, and inoculated with the mould. Growth continues for 5–6 days during which time the medium is vigorously agitated and oxygenated by a current of sterile air. The fermented liquor is then cooled to $-2°C$ and filtered to give a solution containing 0·2–0·3% penicillin, which is concentrated and purified by countercurrent extraction followed by freeze-drying. Anaerobic processes, in which oxygen is unnecessary or actually harmful, are carried out in a similar way in static media. Continuous fermentation processes have been developed but as yet seem to offer only marginal advantages.

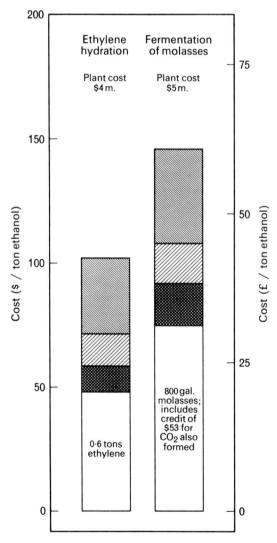

Figure 8.2 Economics of making 30 000 tons of ethanol per annum by different routes in the UK.

194   *Raw materials: natural gas, coal, fermentation*

8.5.3 Applications

In the interwar period, fermentation was the most popular route to aliphatic chemicals and solvent ethanol, butanol, glycerol, acetone and many other compounds were manufactured on a large scale by microbiological methods. With the advent of cheap hydrocarbon feedstocks, these processes lost most of their former importance. As Fig. 8.2 shows, it is economically more attractive to make solvent ethanol from ethylene rather than from carbohydrates because ethylene is substantially cheaper than corn (maize) or molasses. It is interesting to note that the capital cost of a fermentation ethanol plant is quite high: although the conditions employed are mild, the reaction is so slow and takes place in such dilute solutions that an installation for a given output must necessarily be comparatively large. Fermentation is still of immense importance in the manufacture of alcoholic drinks and in sewage treatment (cf. section 17.2) but within the organic chemical industry fermentation methods are now confined to the production of a few medium-tonnage compounds – notably citric and lactic acids – for which no petroleum-based route has yet been found.

However, as the importance of fermentation as a route to bulk chemicals has declined, its importance in the fine chemical sector has increased. Fermentation processes are the only way to obtain antibiotics and enzymes in substantial quantities; they are often the most convenient way to effect such desirable reactions as the 11-hydroxylation of the steroid skeleton in the synthesis of cortisone. Costs may be high, but prices are higher and, where no viable alternative exists, fermentation cannot but flourish. Further, the ability of micro-organisms to synthesize proteins is of interest to others besides biochemists and detergent-powder manufacturers. When supplied with fixed carbon in the form of residual oil, and fixed nitrogen in the form of ammonium nitrate, certain strains of yeast grow rapidly and form a uniquely cheap source of edible protein. In the protein-deficient countries of the third world, better living may come not so much from chemistry as from industrial microbiology (section 6.4.2.1).

Notes and references

Section 8.2
Units Natural gas is usually measured in standard cubic feet (scf) at one atmosphere pressure and 60°F (15·56°C) or in cubic metres at one atmosphere pressure and 0°C. One cubic metre = 37·2 scf; one scf = 0·026795 m^3. Thermal units are often used in Britain: clearly the amount of heat liberated when a volume of gas is burned depends on the composition of the gas but is usually in the range 900–1 000 Btu per scf.

Accordingly one therm = 100 000 Btu = 100–110 scf; one scf = 0·009–0·01 therms.
Statistics Figures for world production and resources of natural gas are given in the *U.N. Statistical Yearbook* from 1970 onwards, from which Table 8.1 is taken. See also the references for section 8.4.2 below.
General News and trends are covered in the *Oil and Gas Journal* and the *Petroleum Times*. A good book on natural gas is badly needed; the best currently available is E. N. Tiratsoo, *Natural Gas*, Scientific Press, London, 1967, which, however, is now appreciably out of date. Table 8.1 is taken mainly from British Petroleum, *Our Industry – Petroleum*, cited in section 7.2, Table 13.1, p. 312.
Section 8.3 There are excellent articles by F. W. Camp, *Tar Sands* and R. E. Gustafson, *Shale Oil* in *Kirk–Othmer*, 2nd edn., vol. 19, p. 682, 1969 and vol. 18, p. 1, 1969, respectively.

R. A. Cardello and F. B. Sprow, *Chem. Eng. Progr.*, 1969, **65** (2), 63, discuss the commercial possibilities of these materials.

Section 8.4.1 This discussion is based mainly on M. Vahrman, *Fuel*, 1970, **49**, 5, which clearly supersedes all previous work. Table 8.3 is from *Kirk–Othmer*, 2nd edn., vol. 5, 1964, p. 627.

Section 8.4.2 Statistics of world production and reserves of coal and lignite are given annually in the *U.N. Statistical Yearbook*, from which Table 8.4 has been compiled. All information about the US Coal Industry is taken from US Department of the Interior – Bureau of Mines, *Minerals Year Book*, 1969 edition, vols. 1–2, Washington D.C., 1971, and that about the British industry from National Coal Board, *Annual Report and Accounts 1970–71*, HMSO, 1971. For Britain see also *Digest of Energy Statistics* (cf. Chapter 7). For the European coal industry see *The Coal Situation in Europe in 1970 and its Prospects*, United Nations, New York, 1971.

Section 8.4.3 The chemistry and technology of coal as a chemical raw material is covered in enormous and uncritical detail in *The Chemistry of Coal Utilisation*, ed. H. H. Lowry, 2 vols, 1945, supplementary volume, Wiley, New York, 1963. Table 8.5 is based on information drawn from the chapters on carbonization in the supplementary volume. *Benzoles: Production and Uses*, ed. G. Claxton, National Benzole and Allied Products Association, London 1961, is also useful.

Section 8.4.4 Carbide-derived acetylene is covered in S. A. Miller: *Acetylene*, vol. 1 cited in Section 7.7 q.v. Statistics of carbide and acetylene production given in Table 8.6 were calculated from the OECD *Annual Report*, 1969–70. The costs of making acetylene via carbide are discussed by W. E. Lobo, *Chem. Eng. Prog.*, **57**, (11), 35, 1961.

Section 8.4.5 Coal gasification processes are described in detail by C. G. Fredersdorff and M. A. Eliott in H. H. Lowry *op. cit.*, supplementary volume, p. 892. The economics of producing hydrogen from coal are discussed by K. K. Bhattacharyya and S. Patil, cited in Section 7.9 above. The Fischer–Tropsch process is covered in H. Pichler and A. Hector: *Carbon Monoxide–Hydrogen Reactions* in *Kirk–Othmer*, 2nd ed., vol. 4, 448, 1964 and in H. H. Storch, N. Golombic and R. B. Anderson, *The Fischer–Tropsch and Related Syntheses*, Wiley, New York, 1951. The mechanism of these reactions is discussed in G. C. Bond: *Catalysis by Metals*, Academic Press, London, 1962.

Section 8.4.6 The best account is by E. E. Donath in H. H. Lowry *op. cit.*, supplementary volume, p. 1041.

Section 8.7.1–8.7.3 There are two good short introductions to fermentation processes: A. H. Rose, *Industrial Microbiology*, Butterworths, London, 1961 and A. Rhodes and D. L. Fletcher, *Principles of Industrial Microbiology*, Pergamon, Oxford, 1966. A much more detailed treatment is given by S. C. Prescott and C. G. Dunn, *Industrial Microbiology*, 3rd ed., McGraw-Hill, New York, 1959, which, however, devotes too much space to the production of simple aliphatic chemicals. None of these books pays much attention to economic factors. Figure 8.2 uses data drawn from Faith, Keyes and Clark *q.v.*; it was drawn by Miss L. C. Burstall.

9 Heavy organic chemicals

The heavy organic chemical industry is nowadays virtually synonymous with the petrochemical industry and the overwhelming bulk of its feedstocks are derived from petroleum, as described in Chapter 7. Figures 9.1 and 9.2 show the production of the most important intermediates in the USA and UK.

9.1 Ethylene

Ethylene is the economic indicator for the modern organic chemical industry in the same way that sulphuric acid is an indicator for the economy as a whole. US and UK production both doubled between 1963 and 1969 (Figs. 9.1 and 9.2) and are measured at a

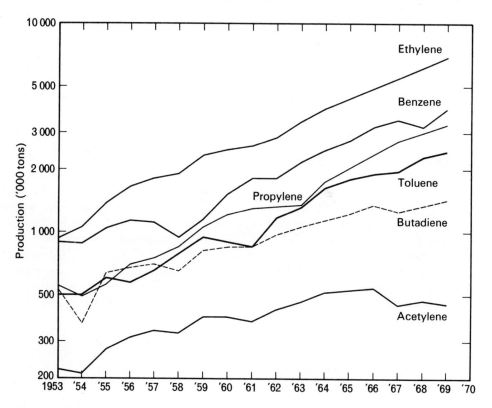

Figure 9.1 US production of some hydrocarbon feedstocks 1953–70.

level of millions of tons per year. Like propylene and the butenes, ethylene is a gas and is inconvenient to transport. It is normally used near its point of production. BP, for example, makes ethylene at Grangemouth in Scotland and hydrates it to ethanol before shipping it to its plant at Saltend, near Hull.

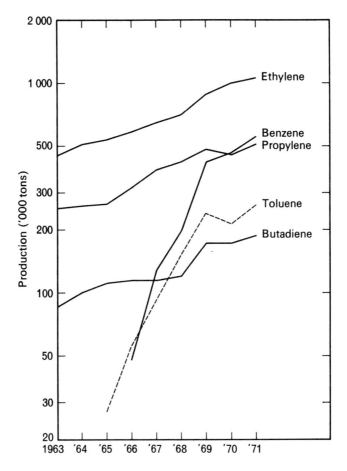

Figure 9.2 UK production of some hydrocarbon feedstocks 1963–71.

Figures for ethylene and propylene are for material for use in chemical manufacture. Benzene and toluene figures exclude coal based material. Benzene production includes material made by dealkylation of toluene and toluene figures include the material dealkylated.

One unexpected consequence of ethylene being used where it is manufactured is that very little is bought or sold except at secret contract prices and it is difficult to know at what price it changes hands in large quantities. A 1969 figure of £30 per ton (3·2 ¢/lb) is probably a good approximation for the USA with the 1971 UK figure a little higher, perhaps £32·5/ton (3·5 ¢/lb).

198 Heavy organic chemicals

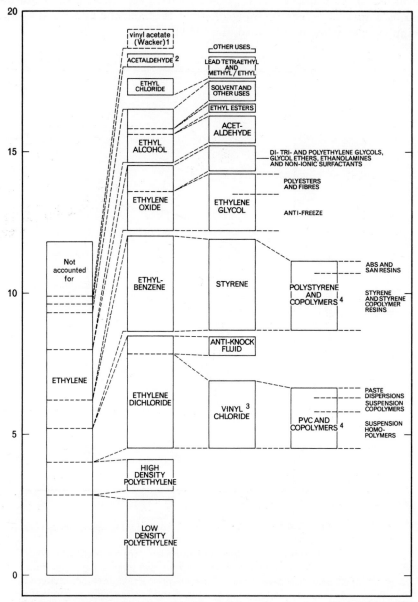

Blocks represent approximate quantities of materials produced and used in the USA in 1967
1 Route not used in USA at the time
2 Acetaldehyde made from ethyl alcohol excluded
3 Some of the vinyl chloride is made by other routes but included in the tonnage here
4 The relatively small amounts of polymer said to be produced from a given amount of monomer suggest that monomer has been exported or has avoided being counted in statistics in some other way

Figure 9.3 Ethylene and its derivatives.

A few of the products made from ethylene are shown in Fig. 9.3 together with an estimated pattern of use. Obsolete or obsolescent processes such as the oxidation of acetaldehyde to acetic acid have been omitted. Market data are given in Table 9.1.

Table 9.1 Ethylene and its derivatives

Material	United States US tariff commission 1969				United Kingdom	
	Production m. lbs	Sales m. lbs	Unit value ¢/lb.	US price ¢/lb. 4/72	Production 1971 '000 tons	Price £/ton. 4/72
Ethylene	16 436	3 877	3·3	3¼	1 040	32·5*
L.D. polyethylene	3 880	3 503	12	14	—	141
H.D. polyethylene	1 610	1 242	15	—	—	197
Ethylene dichloride	6 037	1 227	3	9	—	64·7
Ethyl benzene	4 907	458	3	—	—	—
Ethylene oxide	3 408	4 375	8	—	192	115·0
Ethyl alcohol†	2 361	1 215	6	7·8	136	60·7
Ethyl chloride	6 788	2 675	6	8	—	—
Acetaldehyde	1 652	—	—	9	—	105
Vinyl chloride	3 736	2 358	4	4·7	—	—
Polyvinyl chloride	3 032	2 748	14	23	—	157
Styrene	4 648	1 897	6	8	—	77
Polystyrene (+ copolymers)	2 759	2 504	17	13	170·8	130
Acrylonitrile/Butadiene/ Styrene resins (ABS)	584	332	28	—	28	340
Styrene/Acrylonitrile resins (SAN)				23·5	1·5	230

* 1971 unit value.
† Synthetic alcohol; prices per ton of 100% alcohol content.

Several major outlets for ethylene are in the polymer field and these will be discussed in Chapter 10. They include polyethylene (section 10.2.1), polyvinyl chloride via ethylene dichloride (section 10.3.1), polystyrene (section 10.2.3), ethylene–propylene copolymers (section 10.6.8) and polyvinyl acetate (section 10.5.1).

9.1.1 Ethylene oxide

The largest outlet for ethylene, ignoring polymer outlets, is in the production of ethylene oxide. This was manufactured via the chlorohydrin by a method identical with that used for propylene oxide (section 9.2.2) but a one-stage process is now preferred. Pure ethylene and air or oxygen are passed at atmospheric pressure and about 280°C over a finely divided silver catalyst. Ethylene oxide is formed by the slightly exothermic reaction:

$$CH_2{=}CH_2 + \tfrac{1}{2}O_2 \longrightarrow H_2C{-}CH_2 \underset{O}{\diagdown\diagup}, \quad \Delta H = -24.7 \text{ k cal/mole} \ (-104 \text{ kJ/mole})$$

and in addition carbon dioxide and water are formed by the markedly exothermic reaction:

$$CH_2=CH_2 + 3O_2 \longrightarrow 2CO_2 + 2H_2O, \quad \Delta H = -331 \cdot 6 \text{ kcal/mole } (-1\,390 \text{ kJ})$$

Careful temperature control is necessary to prevent the reaction getting out of hand, and ethylene dichloride is often used as an inhibitor. Even so, yields of less than 70% are normal. The obsolescent chlorhydrin process gave higher yields and could be operated with ethylene of 80–85% purity, but involved two-stages, and used 1·6 tons of chlorine for each ton of ethylene oxide produced. It was thus less economic, as the chlorine accounted for over 40% of production costs. The ethylene oxide end-used pattern is shown in Table 9.2. Its main use is in the production of ethylene glycol (section 9.1.1.1) but other outlets include the manufacture of ethanolamines and of ethylene glycol ethers:

$$H_2C\underset{O}{-}CH_2 + NH_3 \longrightarrow \underset{\text{Monoethanolamine}}{H_2N.CH_2.CH_2OH} \xrightarrow{H_2C\underset{O}{-}CH_2}$$

$$\underset{\text{Diethanolamine}}{HN(CH_2CH_2OH)_2} \xrightarrow{H_2C\underset{O}{-}CH_2} \underset{\text{Triethanolamine}}{N(CH_2CH_2OH)_3}$$

$$ROH + H_2C\underset{O}{-}CH_2 \longrightarrow \underset{\substack{\text{Ethylene glycol} \\ \text{ether}}}{ROCH_2.CH_2OH}$$

Ethanolamines are variously used for gas scrubbing, textile chemicals and non-ionic detergents. Ethylene glycol ethers and their acetates are widely used as solvents, for

Table 9.2 Ethylene oxide derivatives

Material	United States				United Kingdom
	US tariff commission 1969				
	Production m. lbs	Sales m. lbs	Unit value ¢/lb.	US price ¢/lb 4/72	Price £/ton 4/72
Ethylene glycol	2 571	1 936	6	7	99
Diethylene glycol	289	205	8	9	118
Triethylene glycol	91	69	14	21	194
Polyethylene glycol	43	41	23	31	258
Ethylene glycol monobutyl ether	104	91	17	17	182
Methylene glycol monoethyl ether	29	23	16	16½	—
Ethylene glycol monomethyl ether	115	80	12	15	—
Ethylene glycol monoethyl ether	137	65	10	16½	143
Diethylene glycol monomethyl ether	13	9	12	16	—
Monoethanolamine	87	69	12	13	208
Diethanolamine	91	65	11	12	208
Triethanolamine	77	76	13	16	189

example ethylene glycol ethyl ether (Cellosolve) and diethylene glycol monomethyl ether (Carbitol) (section 16.6)

Ethylene oxide can itself be polymerized to give a polyether structure

$$-CH_2-CH_2-O-CH_2-CH_2-O-CH_2-CH_2-O-$$

known variously as polyethylene oxide, polyoxyethylene and polyethylene glycol. If this is added at the 30 ppm level to the water used in fire hoses, friction between the water and the hose walls is reduced to such an extent that the range of the hose is doubled. Similarly, if polyethylene oxide is allowed to trickle into the water in front of a racing yacht, drag is appreciably reduced. This is not a significant end-use, however, as it has been banned by most yacht-racing societies.

Polyethylene oxides have a range of pharmaceutical, cosmetic and other uses, primarily as lubricants, and are also used as the impregnant in archaeological recovery of wooden timbers such as those of the *Vassa* in Stockholm.

Polyethylene oxide chains can be formed on the hydroxyl groups of polyhydroxycompounds, e.g. sorbitol. Such compounds are non-ionic surfactants and are used, for example, to stabilize synthetic whipped cream.

Ethylene oxide and, indeed, the ethylene and oxygen mixture from which it is made are both unstable and liable to explode. Their use is subject to elaborate precautions and most ethylene oxide is used near where it is manufactured.

Ethylene oxide itself is used as a sterilant and fumigant.

9.1.1.1 Ethylene Glycol Ethylene glycol is made by hydration of ethylene oxide with 1% sulphuric acid at 60°C:

$$H_2C\underset{O}{-}CH_2 + H_2O \longrightarrow \underset{OH\ \ OH}{CH_2-CH_2}$$

A six-fold excess of water is used to discourage the formation of di- and triethylene glycols, which are nonetheless formed in quantities larger than can readily be sold:

$$\underset{OH\ \ OH}{CH_2-CH_2} + H_2C\underset{O}{-}CH_2 \longrightarrow \underset{OH\qquad\qquad OH}{CH_2-CH_2-O-CH_2-CH_2} \xrightarrow{H_2C\underset{O}{-}CH_2}$$

$$\underset{OH\qquad\qquad\qquad\qquad\qquad OH}{CH_2-CH_2-O-CH_2-CH_2-O-CH_2-CH_2}$$
Triethylene glycol

Ethylene glycol is separated by vacuum distillation.

80–90% of ethylene glycol production is used as an antifreeze in cars and as a working fluid in cooling systems. It is also used in polyesters, especially of the 'Terylene' type (section 10.7.3).

The dibenzoate of diethylene glycol finds a small outlet as a high-quality non-staining plasticizer in PVC floor tiles (section 10.3.4.2).

9.1.2 Ethyl chloride

Ethyl chloride is made by reaction of hydrogen chloride with ethylene:

$$CH_2{=}CH_2 + HCl \longrightarrow CH_3-CH_2Cl$$

202   Heavy organic chemicals

The bulk of ethyl chloride production is used in the manufacture of tetraethyl lead, which is added to petrol to reduce 'knocking' and improve the octane number (section 17.4.3). If lead pollution is proved to be a problem in cities, demand may drop sharply.

9.1.3   Ethyl alcohol

Up to 1950, ethyl alcohol made by fermentation was an important feedstock for aliphatic organic chemicals. Indeed, the Distillers Company, which was the second largest chemical company in Britain till its chemical interests were bought by BP in 1967, entered the field of chemicals in the nineteen twenties because they had capacity for more fermentation than they could use for industrial purposes or in gin. When required, ethylene was made by dehydration of ethyl alcohol:

$$C_2H_5OH \longrightarrow C_2H_4 + H_2O$$

and acetaldehyde and acetic acid were made by its oxidation:

$$C_2H_5OH \xrightarrow{O_2} CH_3.CHO \xrightarrow{O_2} CH_3COOH$$

With the advent of cheap ethylene from naphtha, it became more economic to make ethanol by the hydration of ethylene over a catalyst of phosphoric acid on celite at 330°C and 60 atm.

$$C_2H_4 + H_2O \xrightarrow[\text{on celite}]{H_3PO_4} C_2H_5OH$$

This process has displaced the fermentation method as a source of industrial ethyl alcohol and the latter route is used in the UK only for the manufacture of potable alcohol (for gin and vodka). The people who flavour and bottle these beverages appear, for cultural reasons, to prefer a feedstock derived from grain rather than oil and, indeed, this is a legal requirement in some countries. Nonetheless, the production of alcohol from oil has set free a substantial amount of grain for food and in a small way has helped the world's food problem.

Between 1950 and 1964, ethyl alcohol based on oil replaced fermentation alcohol in the traditional aliphatic organic chemical industry and most of it was oxidized to acetaldehyde and acetic acid. Acetaldehyde, however, is now more cheaply made by the Wacker process (section 9.1.4) and acetic acid is obtained directly from a petroleum fraction (section 9.4). Though a number of plants based on ethyl alcohol are likely to continue in use until they require replacement, ethyl alcohol has nonetheless been displaced as the keystone of the aliphatic organic chemical industry and consumption is likely to fall rather than rise. Currently it consumes some 10% of ethylene output. The output of some of its derivatives is shown in Table 9.3.

Apart from being a source of acetaldehyde, ethyl alcohol is widely used as a solvent. Mixed with small percentages of methyl alcohol, colouring matter and other denaturants, it is sold as methylated spirits. It also has a number of outlets as a chemical intermediate, e.g. for ethyl acetate, ethyl acrylate, diethyl phthalate and diethyl ether. In the UK it is used as a base for after-shave lotions, though in Germany isopropanol is preferred.

Table 9.3 Derivatives of ethyl alcohol and acetaldehyde

Material	United States				United Kingdom
	US tariff commission 1969			US price	UK price
	Production m. lbs	Sales m. lbs	Unit value ¢/lb.	¢/lb. 4/72	£/ton 4/72
Ethyl acetate	167	139	8	12	96·5
Ethyl acrylate	200	68	18	20	—
Diethyl ether	103	—	—	11·5	193·2
Diethyl phthalate	22	19	19	21	180·6
Chloral	62	—	—	23·5	—
Pentaerythritol	92	74	23	24	200
α-Picoline	0·8	0·9	36	46	—
Other picolines (mixed)	1·8	1·0	40	32	—

9.1.4 Acetaldehyde

Like ethanol, the importance of acetaldehyde has decreased in recent years with the development of a direct route to acetic acid. Some is still made by ethanol oxidation, but the most economic route is via the Wacker process. Ethylene and oxygen are bubbled at about atmospheric pressure and 60°C through a catalyst solution containing acidified cupric and palladium chlorides. The palladium chloride oxidizes the ethylene to acetaldehyde via a π-bonded complex:

$$C_2H_4 + PdCl_2 + 2HCl \longrightarrow \left[Cl_3Pd \leftarrow \begin{matrix} H \\ | \\ C \\ \| \\ C \\ | \\ H \end{matrix} \begin{matrix} H \\ \\ \\ \\ \\ H \end{matrix} \right]^- + 2H^+ + Cl^-$$

$$\downarrow H_2O$$

$$CH_3.CHO + Pd + 4HCl$$

The palladium metal is oxidized back to its chloride by the cupric chloride which is itself reduced to cuprous chloride and the latter is oxidized back to cupric chloride by the oxygen:

$$Pd + 2CuCl_2 \longrightarrow PdCl_2 + 2CuCl$$

$$4CuCl + O_2 + 4HCl \longrightarrow 4CuCl_2 + 2H_2O$$

This is one of the few catalyst systems whose mode of action is reasonably well understood.

Acetaldehyde can be used as a source of acetic anhydride but this route is declining. Air oxidation of acetaldehyde in the presence of certain metallic acetates gives peracetic acid which reacts with further acetaldehyde to give acetic anhydride:

$$CH_3CHO + O_2 \longrightarrow CH_3COOOH$$

$$CH_3CHO + CH_3COOOH \longrightarrow \begin{matrix} CH_3C \overset{\nearrow O}{\underset{\searrow O}{}} \\ CH_3C \overset{}{\underset{\searrow O}{}} \end{matrix}$$

204   *Heavy organic chemicals*

Another obsolescent process based on acetaldehyde is the production of butanol via the aldol condensation. Two molecules of acetaldehyde condense to give aldol which is dehydrated to crotonaldehyde. The latter is hydrogenated to *n*-butanol:

$$CH_3.CHO + CH_3.CHO \longrightarrow \underset{\text{Aldol}}{CH_3.CHOH.CH_2.CHO} \xrightarrow{-H_2O}$$

$$\underset{\text{Crotonaldehyde}}{CH_3CH=CH_2CHO} \xrightarrow{H_2} \underset{n\text{-Butanol}}{CH_3CH_2CH_2CHOH}$$

This route is being ousted by the OXO process starting with propylene (section 9.2.3).

The declining demand for acetaldehyde from what were previously its main outlets, leaves only a few small-tonnage uses. Acetaldehyde is a possible starting material for the hypnotic and DDT-intermediate chloral (section 16.10.2). It reacts with formaldehyde to give pentaerythritol, a raw material for alkyd resins (section 10.5.3).

$$4HCHO + CH_3CHO + H_2O \xrightarrow{NaOH} HOCH_2-\underset{\underset{CH_2OH}{|}}{\overset{\overset{CH_2OH}{|}}{C}}-CH_2OH + HCOOH$$

It also reacts with ammonia to give a mixture of α- and γ-picolines:

$$3CH_3CHO + NH_3 \longrightarrow \underset{\gamma\text{-Picoline}}{\underset{}{}} + \underset{\alpha\text{-Picoline}}{\underset{}{}}$$

The proportions of the isomers are more or less constant, and in the early days of manufacture, no outlet could be found for the α-picoline so the γ-compound had to bear the total cost of manufacture. It sold at over £1 000/ton and was used as a starting material for the anti-tuberculosis drug, isoniazid. A use was then found for α-picoline as a coccidiostat (i.e. it counters the disease coccidiosis in chickens) and with the development of factory farming of chickens, its consumption increased rapidly. Tuberculosis meanwhile declined, and now the α-picoline is the expensive compound instead of being the unsaleable by-product.

9.2 Propylene

Propylene is produced with ethylene in petroleum cracking processes. It has always found a less ready market than ethylene, and in spite of the propylene/ethylene ratio having dropped steadily with improvements in cracking technology, propylene is still something of a drug on the market. Though its price has been creeping up in recent years it is still sold at as little as £17/ton.

Some of the chemicals derived from propylene are shown in Fig. 9.4 and Table 9.4. The production of polypropylene (section 10.2.2), acrylonitrile (section 10.7.4), acrolein, acrylic acid and acrylates (section 10.5.2), epichlorohydrin via allyl chloride (section 10.4.6), and phenol via cumene (section 9.5.1.1) are described elsewhere.

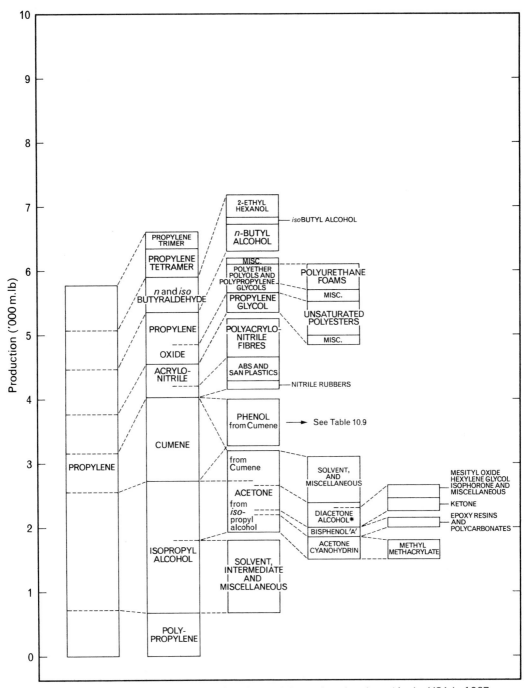

Blocks represent approximate quantities of materials produced and used in the USA in 1967
* Diacetone alcohol figure in statistics appears not to allow for material used to make other chemicals

Figure 9.4 Uses of propylene.

Table 9.4 Propylene and its derivatives

	United States				United Kingdom	
		US tariff commission 1969				
Material	Production m. lbs	Sales m. lbs	Unit value ¢/lb.	US price ¢/lb. 4/72	Production 1971 '000 tons	Price £/ton 4/72
Propylene	7 235	3 288	2·5	3·3	506	17·3†
Polypropylene	1 090	971	22	25	—	204
Isopropyl alcohol	2 013	851	6	7·35	143*	72
Cumene	1 687	909	4	—	—	86·4
Acrylonitrile	1 157	562	12	14·5	—	133
Acetonitrile	—	—	—	24	—	—
Propylene oxide	1 177	151	8	15	—	168
Propylene tetramer	459	269	4·0	—	—	—
Propylene trimer (nonenes)	304	240	3·3	—	—	—
Allyl chloride	—	—	—	16	—	—

* Includes small amounts of *n*-propyl alcohol.
† Unit value 1971

9.2.1 Isopropyl alcohol

Isopropyl alcohol is made by the hydration of propylene from cracking gases. These gases contain a proportion of ethylene which would be expensive to remove. They are desulphurized and passed up an absorption tower at about 25°C down which 85% sulphuric acid trickles. About half the propylene and scarcely any of the ethylene is absorbed, the major products being propylene hydrogen sulphate and di-isopropyl sulphate. The unreacted gases are recycled until the ethylene concentration gets too high, when they are used to make ethyl alcohol. The reaction is exothermic and cooling is necessary.

Dilution of the sulphates generates much heat and with the aid of additional steam, the sulphates are hydrolysed and the crude alcohol boiled off. Initial purification gives a 91% isopropyl alcohol – water azeotrope which is broken with the aid of di-isopropyl ether:

$$CH_3CH=CH_2 \xrightarrow{H_2SO_4} \underset{\substack{\text{Propylene hydrogen} \\ \text{sulphate}}}{CH_3\underset{\substack{|\\O.SO_3H}}{CH}-CH_3} + \underset{\text{Di-isopropyl sulphate}}{CH_3-\underset{\substack{|\\O\\|\\SO_2\\|\\O\\|\\CH_3-CH-CH_3}}{CH}-CH_3} \xrightarrow{H_2O} CH_3-\underset{\substack{|\\OH}}{CH}-CH_3 + H_2SO_4$$

The sulphate moiety adds onto the secondary carbon atom in accordance with the Markownikov rule. Thus no *n*-propyl alcohol is obtained by this route. If required, it can be obtained as a by-product of the air oxidation of hydrocarbons, or by the OXO process (section 16.3) using ethylene, carbon monoxide and hydrogen as a feedstock.

Isopropyl alcohol is used as a solvent, chiefly for oils, gums, shellac and synthetic resins.

It is also used as an intermediate for propyl esters and in the pharmaceutical industry, and can form the basis of after-shave lotions. About half of isopropyl alcohol production, however, is oxidized or dehydrogenated to acetone.

9.2.1.1 Acetone In the oxidation process isopropyl alcohol vapour is mixed with air and passed over a copper catalyst at about 500°C and 3 atm:

$$CH_3.CHOH.CH_3 + \tfrac{1}{2}O_2 \longrightarrow CH_3.CO.CH_3 + H_2O$$

In the dehydrogenation process, a zinc oxide catalyst and a temperature of 380°C are used. Air is not added:

$$CH_3.CHOH.CH_3 \longrightarrow CH_3COCH_3 + H_2$$

The catalyst life in the dehydrogenation process is much shorter but the by-product hydrogen has a certain value if an on-site outlet is available. It therefore appears to be more economic than the oxidation route.

Acetone is also produced in the cumene–phenol process (section 9.5.1.1). The present economic pattern is that this by-product acetone is sold first and the substantial deficit in production is made up by isopropyl alcohol oxidation. It is said that acetone can also be produced directly from propylene via the Wacker process (section 9.1.4) but it is not thought that this route is being used at present.

The major use of acetone until the 1950s was as a solvent. Since then, solvent uses have, if anything, declined and the booming market for acetone reflects its uses as a chemical intermediate. The main reactions are as follows:

1. An aldol condensation of two molecules of acetone with an alkali catalyst gives diacetone alcohol, which may then be dehydrated to give mesityl oxide:

$$2CH_3.CO.CH_3 \xrightarrow[10°C]{NaOH} \underset{\text{Diacetone alcohol}}{\overset{H_3C}{\underset{H_3C}{>}}\underset{OH}{\overset{|}{C}} - CH_2.CO.CH_3} \xrightarrow[110°C]{-H_2O \text{ (weak } H_2SO_4\text{)}} \underset{\text{Mesityl oxide}}{\overset{H_3C}{\underset{H_3C}{>}}C=CHCOCH_3}$$

About 40% of diacetone alcohol production goes into hydraulic fluids and printing inks. It is also a solvent for cellulose-type and epoxy resins. Mild hydrogenation gives methyl isobutyl ketone which is used in large quantities as a solvent for nitrocellulose and vinyl resins:

$$\overset{H_3C}{\underset{H_3C}{>}}C=CH.CO.CH_3 \xrightarrow{H_2} \underset{\text{Methyl isobutyl ketone}}{\overset{H_3C}{\underset{H_3C}{>}}CH-CH_2-CO-CH_3}$$

Further hydrogenation gives methyl amyl alcohol (methyl isobutyl carbinol):

$$\overset{H_3C}{\underset{H_3C}{>}}CH-CH_2-CO-CH_3 \xrightarrow{H_2} \overset{H_3C}{\underset{H_3C}{>}}CH-CH_2-CHOH-CH_3$$

Hydrogenation of diacetone alcohol over Raney nickel at 7 atm, however, gives hexylene glycol:

$$\begin{array}{c} H_3C \\ \diagdown \\ H_3C \end{array} \!\!\! C\text{--}CH_2\text{--}CO\text{--}CH_3 \xrightarrow{H_2} \begin{array}{c} H_3C \\ \diagdown \\ H_3C \end{array} \!\!\! C\text{--}CH_2\text{--}CH\text{--}CH_3$$
with OH groups on the appropriate carbons.

It is used as a coalescing agent in paints, a viscosity reducer in PVC plastisols and other dispersions and a component of hydraulic brake-fluids.

2. When treated with 5% caustic soda at 230°C and 50 atm, three molecules of acetone condense to give isophorone. Some mesityl oxide is also produced:

$$3\ \begin{array}{c}H_3C\\ \diagdown \\ H_3C\end{array}\!\!C\!=\!O \longrightarrow \text{Isophorone}$$

Isophorone is a high boiling ketonic solvent, used in roller coating lacquers based on phenol–formaldehyde resins, epoxy resins and vinyl acetate–vinyl chloride copolymers. It is also used in printing inks and PVC processing.

Table 9.5 Derivatives of Isopropyl Alcohol and of Butyl Alcohols

Material	United States US tariff commission 1969			US price ¢/lb. 4/72	United Kingdom Price £/ton 4/72
	Production m. lbs	Sales m. lbs	Unit value ¢/lb.		
Acetone	1 518	1 078	5	5·85	60*
From Isopropyl alcohol	871	553	5	—	—
Isopropyl acetate	—	45	11	11·5	—
Diacetone alcohol	77	34	13	13·5	134
Methyl isobutyl ketone	200	165	12	13·5	122
Isophorone	—	—	—	18	186
Hexylene glycol	42	—	—	16·5	167
Acetone cyanohydrin	540	—	—	—	—
Methyl methacrylate	433	—	—	18½	—
Biphenylol propane (Bisphenol 'A')	182	64	19	30½	—
Epoxy resins	166	164	51	—	—
n-Butyl alcohol	405	256	9	12	108
Iso-butyl alcohol	101	90	6	8	—
n-Butyl acetate	59	59	12	13·5	123
Iso-butyl acetate	—	—	—	13	—
n-Butyl acrylate	65	34	19	20	—
Di-n-butyl phthalate	34	30	18	22	—
2-ethyl hexanol	405	201	10	11	142

* 1971 production 140 000 tons, unit value £47.

3. Acetone is the starting material for the production of methyl methacrylate, the monomer for 'Perspex' production (section 10.2.4).
4. Acetone and phenol condense in the presence of an acid catalyst to give bisphenol 'A' (biphenylolpropane) which is used in the production of epoxy and polycarbonate resins, two of the fastest growing sectors of the chemical industry (section 10.4.6, 10.2.6).

Total UK acetone production in 1969 was 140 000 tons. Phenol production is not disclosed so it is uncertain now much of this acetone arose as a by-product from phenol manufacture. A rough guess suggests a figure of 60 000 tons leaving 80 000 tons to be derived from isopropyl alcohol. On this basis just under half of the UK production of isopropyl alcohol was used in acetone manufacture. The comparable figure for the US in 1969 was 57%.

9.2.2 Propylene oxide

In 1970 all propylene oxide was made by the chlorohydrin process. Propylene is treated with hypochlorous acid (aqueous chlorine solution) followed by dehydrochlorination of the chlorohydrin with calcium hydroxide:

$$CH_3-CH=CH_2 + HOCl \longrightarrow \underset{\text{Propylene chlorohydrin}}{CH_3-\underset{OH}{\underset{|}{CH}}-\underset{Cl}{\underset{|}{CH}}} \xrightarrow{Ca(OH)_2} \underset{\text{Propylene oxide}}{CH_3-HC\underset{O}{\overset{}{\diagdown\diagup}}CH_2}$$

Ethylene oxide used to be made by this route and many of the present propylene oxide plants are converted ethylene oxide units.

The development of a direct oxidation route to ethylene oxide started a search for a similar route to propylene oxide but this has proved elusive. One new route, which may become commercial, depends on the oxidation of acetaldehyde to peracetic acid. The latter reacts with propylene *in situ* to give propylene oxide, and acetic acid as a by-product:

$$CH_3CHO \xrightarrow{\text{Oxidation}} CH_3C\underset{OOH}{\overset{O}{\diagup\diagdown}} \xrightarrow{CH_3CH=CH_2} CH_3COOH + CH_3-HC\underset{O}{\overset{}{\diagdown\diagup}}CH_2$$

A similar route which is not yet commercial uses ethylbenzene and propylene as feedstock and gives styrene as a by-product:

$$\underset{\text{Ethylbenzene}}{C_6H_5-CH_2-CH_3} \xrightarrow{\text{Air}} \underset{\underset{\text{Ethylbenzene hydroperoxide}}{OOH}}{C_6H_5-\underset{|}{CH}-CH_3} \qquad \text{(cf. oxidation of cumene, section 9.5.1.1)}$$

$$\downarrow CH_3CH=CH_2$$

$$\underset{\text{Styrene}}{C_6H_5CH=CH_2} \xleftarrow{\text{dehydration}} \underset{\underset{\text{Methyl benzyl alcohol}}{OH}}{C_6H_5-\underset{|}{CH}-CH_3} + \underset{\text{Propylene oxide}}{CH_3HC\underset{O}{\overset{}{\diagdown\diagup}}CH_2}$$

Table 9.6 Propylene oxide derivatives

	United States				United Kingdom
	US tariff commission 1969				
Material	Production m. lbs	Sales m. lbs	Unit value ¢/lb.	US price ¢/lb. 4/72	Price £/ton 4/72
Propylene glycol	460	411	9	13	141
Dipropylene glycol	48	38	11	15·5	161
Polypropylene glycols	226	—	—	—	—
Polypropoxy ethers	369	313	16	—	—

The major outlets for propylene oxide are in polypropylene glycols and related polyether polyols, and in propylene glycol itself (Table 9.6).

9.2.2.1 Propylene glycol Propylene glycol is made by hydration of propylene oxide, in a process similar to ethylene glycol production:

$$CH_3-HC\overset{O}{-}CH_2 \xrightarrow{H_2O} CH_3-\underset{OH}{CH}-\underset{OH}{CH_2}$$
$$\text{Propylene glycol}$$

Propylene glycol's greatest market is as an intermediate for the production of polyester resins (section 10.4.4). It is also used for the softening of cellophane, as a humectant for tobacco, in polymeric plasticizers (section 10.3.4.4) and in brake fluids.

9.2.2.2 Polypropylene glycol and other polyether polyols Polypropylene glycol is made by base-catalysed addition of propylene oxide to propylene glycol:

$$\begin{array}{l} CH_2OH \\ | \\ CHOH \\ | \\ CH_3 \end{array} + (n+m)CH_3-HC\overset{O}{-}CH_2 \longrightarrow \begin{array}{l} CH_2-O(CH(CH_3)-CH_2)_nOH \\ | \\ CH-O(CH(CH_3)-CH_2)_mOH \\ | \\ CH_3 \end{array}$$
$$\text{Polypropylene glycol}$$

Other polyether polyols are made by starting with glycerine or sorbitol rather than propylene glycol.

It is possible to produce so-called 'block' copolymers by reacting propylene glycol with propylene oxide and reacting the subsequent polymer with ethylene oxide to give polyether chains of the type

$$\begin{array}{l} CH(OCH(CH_3)CH_2)_m O(CH_2CH_2)_n OH \\ | \\ CH_2(OCH(CH_3)CH_2)_x O(CH_2CH_2)_y OH \end{array}$$

These are used mainly as non-ionic surface-active agents.

Propylene 211

In 1968, over 85% of all flexible and 95% of all rigid polyurethane foams were based on polyether polyols. The latter are also used as non-ionic surfactants, and lubricants.

9.2.3 n-Butyl and isobutyl alcohols

n-Butyl and isobutyl alcohols may be made from propylene by the OXO process (section 16.3). Propylene, carbon monoxide and hydrogen are passed at several hundred atmospheres and 170°C over a cobalt naphthenate catalyst. A mixture of *n*-butyraldehyde and isobutyraldehyde in a ratio of about 2:1 results. Hydrogenation over nickel or copper chromite gives the corresponding alcohols.

$$CH_3CH=CH_2 + CO + H_2 \longrightarrow \underset{\underset{\text{Isobutyraldehyde}}{\text{CHO}}}{CH_3\overset{|}{C}HCH_3} + \underset{n\text{-Butyraldehyde}}{CH_3CH_2CH_2CHO}$$

$$\downarrow H_2 \qquad\qquad \downarrow H_2$$

$$\underset{\underset{\text{Isobutyl alcohol}}{CH_2OH}}{CH_3\overset{|}{C}HCH_3} \quad \underset{n\text{-Butyl alcohol}}{CH_3CH_2CH_2CH_2OH}$$

A little *n*-butyl alcohol is still made from acetaldehyde via crotonaldehyde (section 9.1.4). *n*-Butyl and isobutyl alcohols and their acetates are widely used as lacquer solvents, the acetate being used especially with cellulose nitrate. Butyl esters are used as plasticizers (dibutyl phthalates), and monomers (butyl acrylates), etc. In general, the straight chain primary alcohol and its derivatives have the preferred properties and this is reflected in their prices (Table 9.5).

9.2.4 2-Ethylhexyl alcohol

If *n*-butyraldehyde is treated with alkali, it undergoes condensation to give an aldol which loses water to give an unsaturated aldehyde. This can readily be reduced to give 2-ethylhexanol, an important plasticizer alcohol, which is used as its phthalate and adipate (in PVC, section 10.3.4) and its acrylate (in PVA, section 10.5.2):

$$2CH_3CH_2CH_2CHO \longrightarrow CH_3CH_2CH_2\underset{\underset{\underset{CH_3}{|}}{\overset{|}{CH_2}}}{\overset{\overset{|}{OH}}{C}H}-\overset{}{C}H_2-CHO \xrightarrow{-H_2O}$$

$$CH_3CH_2CH_2CH=\underset{\underset{\underset{CH_3}{|}}{\overset{|}{CH_2}}}{C}H-CHO \xrightarrow{H_2} \underset{\text{2-Ethylhexanol}}{CH_3CH_2CH_2CH_2CH_2CH_2OH}\!\!\begin{array}{l}\\[-0.3em]\underset{\underset{CH_3}{|}}{\overset{|}{CH_2}}\end{array}$$

9.2.5 Propylene tetramer

The catalytic polymerization of propylene over a phosphoric acid catalyst at 70 atm and 200°C gives propylene tetramer, a mixture of dodecenes, which used to be reacted with benzene to give detergent alkylate (section 11.4.2). Problems of pollution by such detergents have resulted in the use of straight chain dodecenes, made from other feedstocks, as a basis for detergents in developed countries, and propylene tetramer is now used only in areas where the size of the waterways and the sparseness of the population or the infrequency with which they wash make pollution problems insignificant.

9.3 Butenes and butadiene

Steam cracking processes yield a C_4 stream which contains butadiene and the three isomeric butenes (cf. Tables 7.3, 7.5):

$$\text{butene-1} \quad CH_3CH_2CH{:}CH_2, \quad \text{butene-2} \quad CH_3CH{:}CHCH_3 \text{ and}$$
$$iso\text{butene} \quad (CH_3)_2C{:}CH_2.$$

Butadiene may be recovered by extractive distillation with acetonitrile; acid treatment then removes *iso*butene and leaves a mixture of butenes-1 and 2. Naphtha cracking gives a 4–6% yield of butadiene; accordingly this is the main route to this compound in Europe. In the USA on the other hand, the main cracking feedstock is ethane, which yields little C_4 material, and most butadiene is made by the vapour-phase dehydrogenation of 1- and 2-butenes at 600°C over a promoted iron catalyst.

The C_4 hydrocarbons are used on a smaller scale than ethylene and propylene. Production and market data are given in Table 9.7. Much the most important is butadiene. A high proportion of this compound, probably 80%, goes into styrene-butadiene rubbers, while the remainder finds its way into polybutadiene and nitrile rubbers (cf. section 10.6). The *n*-butenes have few uses. Two significant ones are the production of chloroprene (section 10.6.7) and of methyl ethyl ketone (MEK), a solvent for nitrocellulose lacquers and vinyl and acrylic resins. The process for the manufacture of this latter is similar to that used to make acetone from propylene. The alkene is hydrated to *sec*-butyl alcohol by sulphuric acid, and the alcohol is then dehydrogenated over a zinc oxide or copper–zinc catalyst:

$$CH_3{-}CH{=}CH{-}CH_3 \xrightarrow{H_2SO} CH_3{-}\underset{\underset{SO_3H}{|}}{\overset{\overset{O}{|}}{CH}}{-}CH_2{-}CH_3 \xrightarrow{H_2O}$$
$$\text{Butene-2}$$

$$CH_3{-}CHOH{-}CH_2{-}CH_3 \xrightarrow{-H_2} CH_3{-}CH_2{-}CO{-}CH_3$$
$$sec\text{-Butyl alcohol} \qquad\qquad \text{Methyl ethyl ketone}$$

Methyl ethyl ketone is an important solvent and competes directly with *n*-butyl acetate in many fields. *sec*-Butyl alcohol has a few minor uses as a solvent in addition to being a source of methyl ethyl ketone. Tertiary butyl alcohol is a by-product of secondary butyl alcohol manufacture. It is a solid (m.p. 25·5°C) and has few industrial applications.

Table 9.7 C_4 and higher aliphatic hydrocarbons and their derivatives

Material	United States US tariff commission 1969			United Kingdom	
	Production m. lbs	Sales m. lbs	Unit value ¢/lb.	US price ¢/lb. 4/72	UK price £/ton 4/72
Total C_4 fraction	9 800	6 499	4·2	—	—
n-Butane	2 159	1 188	1·1	—	—
1 and 2 Butene mixtures (mainly butadiene feedstock)	1 629	1 515	2·8	6·5(1)* 5·7(2)	—
Butadiene/butene fractions	1 264	280	3·0	—	—
1:3 Butadiene	3 113	1 982	8·4	9·5	59†
Butadiene derivatives:					
Styrene-Butadiene rubbers	2530	2 130	23	23	175
Stereoregular polybutadiene rubber	597	553	18	25	179
Nitrile rubber	—	—	—	50	396
ABS and SAN resins	584	332	28	—	—
sec-Butyl alcohol	101	90	6	13·5	—
Methyl ethyl ketone	484	424	10	10	96
Total C_5 hydrocarbons	751	444	2·8	—	—
Isoprene	260	—	—	—	—
Total Higher aliphatics	3 948	2 197	4·6	—	—
Mixed heptenes	244	135	4·2	—	—
Higher alcohols:					
2-Ethyl hexanol	405	201	10	11	142
Iso-octanol (from heptenes)	109	83	10	25	—
Decyl alcohol (from nonenes)	162	58	10	24	—

* Prices of individual isomers.   † 1971 production 188 000 tons, unit value given.

The total n-butene production in the UK in 1969 was approximately 200 000 tons. Approximately 50 000 tons of isobutene were also produced and of this about 80% went into butyl rubber and 20% into polyisobutenes which are viscous liquids used in the manufacture of additives for lubricating oils.

9.4 Acetic acid

During the 1950s, the production of acetic acid from a petroleum feedstock was a multi-stage process:

Petroleum fraction ⟶ Ethylene ⟶ Ethyl alcohol ⟶ Acetaldehyde ⟶ Acetic acid

This was replaced in the UK early in the 1960s by the one-step 'primary flash distillate' process pioneered by the Distillers Company in which a light hydrocarbon fraction from

214   *Heavy organic chemicals*

petroleum is oxidized in the liquid phase with air under pressure over an antimony oxide–tin oxide catalyst. Acetic acid is the major product; by-products include formic, propionic and succinic acids.

The PFD process cut the cost of acetic acid sharply and also displaced an older method for the manufacture of formic acid, since this by-product had to be sold at whatever price could be obtained for it. Outlets for propionic acid were not obvious but it was found that calcium propionate inhibited 'rope' formation in bread and subsequent government legislation permitted its use. Formic acid is used in textile dyeing and finishing and to inhibit fermentation to butyric acid in silage. Another major market is in the preservation of moist grain, use of formic acid being cheaper than drying. Adequate outlets for succinic acid have still not been found and most of UK production is burned. It is interesting to note that almost any organic chemical can be used as a fuel and has a value of about £8–£10/ton as such.

Acetic acid can also be made from butane by controlled air oxidation. This process is operated in the USA at Pampa, Texas, where large quantities of cheap butane are available from natural gas.

Acetic acid is a major organic chemical. Production and market data are given in Table 9.8. Table 9.9 shows the pattern of end-users, the largest one being the production of acetic anhydride for use in cellulose acetate manufacture (section 10.7.1).

Table 9.8 Acetic acid and its derivatives

Material	United States US tariff commission 1969			US price ¢/lb. 4/72	United Kingdom £/ton UK price 4/72
	Production m. lbs	Sales m. lbs	Unit price ¢/lb.		
Acetic acid	1 770	426	6	{ 9* { 9·41	{ 85·5* { 73·5
Acetic anhydride	1 675	145	9	14	110
Formic acid	—	—	—	12·5	100·5
Propionic acid	41	17	9	14	—
Calcium Propionate	13	13	22	27	—
Ethyl acetate	167	139	8	12	96·5
Isopropyl acetate	—	45	11	11·5	99
n-Butyl acetate	59	59	12	13·5	123
Isobutyl acetate	55†	49†	10†	13	105

* Upper figure glacial acetic acid, lower figure 80% tech. acid.
† 1967 data; includes sec-butyl acetate.

9.4.1 Ketene and diketene

Glacial acetic acid when cracked at a high temperature and reduced pressure gives ketene:

$$CH_3COOH \longrightarrow CH_2CO + H_2O$$

Ketene is a poisonous, powerfully lachrymatory gas with a choking smell. It is invariably used *in situ*. It can be dimerized over trimethyl phosphate to diketene:

$$2CH_2CO \longrightarrow \begin{array}{c} H_2C-C=O \\ | \quad\quad | \\ H_2C=C-O \end{array}$$

Ammonia is added inhibit the back reaction. Diketene can be transported in bulk under refrigeration though it is said to be explosive above 10°C. It has uses of its own and can also be decomposed back to ketene, if the latter is required.

Almost the only industrial use for ketene is in the manufacture of acetic anhydride (section 9.4.2). Diketene, however, is a useful intermediate for the synthesis of several fine chemicals. With methyl and ethyl alcohols it gives methyl and ethyl acetoacetates, respectively:

$$\begin{array}{c} H_2C-C=O \\ | \quad\quad | \\ H_2C=C-O \end{array} + HOCH_3 \longrightarrow \begin{array}{c} H_2C\quad\quad\quad C=O \\ | \quad\quad\quad\quad | \\ H_3C-C=O \quad OCH_3 \end{array}$$

Methyl acetoacetate

Diketene will also react with aromatic amines to give acetoacetarylamides: with aniline, acetoacetanilide is produced:

$$\text{PhNH}_2 + \begin{array}{c} H_2C-C=O \\ | \quad\quad | \\ H_2C=C-O \end{array} \longrightarrow \text{Ph-NH.C(O).CH}_2\text{.C(O).CH}_3$$

o-Anisidine (2-methoxyaniline) gives acetoacet-*o*-aniside,

o-Toluidine (2-methylaniline) gives acetoacet-*o*-toluidide,

2,4-Xylidene (2,4-dimethylaniline) gives acetoacet-2:4-xylidide, and

o-Chloraniline (2-chloroaniline) gives acetoacet-*o*-chloranilide.

These are all used in the production of azo-dyes (chapter 12).

9.4.2 Acetic anhydride

Acetic anhydride was traditionally made by passing ketene into glacial acetic acid:

$$CH_3COOH + CH_2\!:\!CO \longrightarrow (CH_3CO)_2O$$

216 Heavy organic chemicals

This route was replaced, when acetaldehyde became cheap, by the liquid phase oxidation by oxygen of acetaldehyde in the presence of copper, cobalt and nickel acetates:

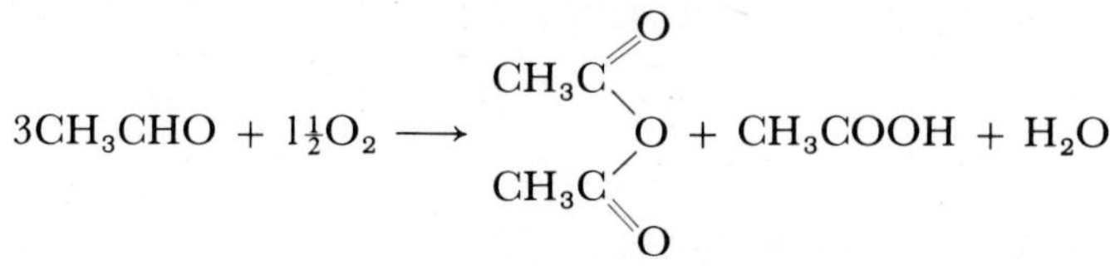

$$3CH_3CHO + 1\tfrac{1}{2}O_2 \longrightarrow \begin{matrix} CH_3C(=O) \\ CH_3C(=O) \end{matrix}\!\!>\!O\; + CH_3COOH + H_2O$$

A drop in the price of acetic acid subsequent to the commercialization of the 'primary flash distillate' process, however, resulted in the reinstatement of the earlier route at least in the UK.

The uses of acetic anhydride are shown in Table 9.9. The major one is the acetylation of cellulose to give cellulose acetate (section 10.7.1) though a certain amount is used in the manufacture of aspirin (section 13.3). The share of cellulosics in the plastics and fibres markets is declining, but the market itself is so large that usage of acetic anhydride is bound to remain significant for the foreseeable future.

Table 9.9 UK acetic acid and acetic anhydride end-use pattern 1971

					37% of acetic acid goes into acetic anhydride	
Pharmaceuticals 6%	8% Vinyl acetate†	7% 2-Chloro-acetic acid	13% Synthetic fibres	15% Acetate esters Ethyl n-butyl Isobutyl Isopropyl and glycol ether*	82% of Acetic anhydride → cellulose acetate	3% ← Miscellaneous
Dyes and dyeing 6%						3% ← Esters
Foodstuffs 4%						3% ← Pharmaceuticals
Miscellaneous 4%						9% ← Analgesics

* In descending order of importance.
† This figure is abnormally low because of shortage of plant in UK. A typical figure would be about double.

9.5 Aromatics

Production and prices of aromatics are shown in Table 9.10. Figure 9.5 shows the general pattern of use of aromatics in Western Europe.

9.5.1 Benzene

Benzene is the most important of the aromatics. Its outlets are shown in Table 9.11 and Fig. 9.6. Three of them involve alkylation of benzene by an alkene – the Friedel–Crafts reaction. Alkylation with ethylene gives ethylbenzene which can be dehydrogenated to styrene (section 10.2.3), alkylation with propylene gives cumene which in turn gives phenol and acetone (section 9.5.1.1) and alkylation with dodocene gives dodecyl benzene – known as detergent alkylate (section 11.4.2). Benzene can be hydrogenated to cyclohexane, the raw material for nylon (section 10.7.2). Cyclohexane is also obtained by fractionation of petroleum streams but this source is inadequate. In 1965 it accounted for only 30% of US cyclohexane and by 1970 this figure had probably dropped to about 16%.

Aromatics 225

because it poisons the catalyst in afterburners or presents a health hazard (section 17.4.3).

Much of the remaining toluene is dealkylated to benzene in the presence of hydrogen at 600–750°C and 30–75 atm depending on whether catalytic or thermal dealkylation is desired (section 7.8).

These uses for toluene together with its uses as a solvent and in phenol manufacture are ways of getting rid of a material which no one really wants. It is genuinely valuable as a raw material for the manufacture of TNT explosives (section 16.8) and tolylene di-isocyanate (section 10.4.5) which goes into the rapidly growing polyurethane manufacturing industry. Market data are given in Table 9.12.

9.5.3 Xylenes

The mixed xylenes obtained from catalytic reforming processes can be used for octane improvement of petrol or as a solvent. Chemical uses require separation of the *o*-, *p*- and *m*-isomers as described in section 7.8. Market data are given in Table 9.13.

There was in 1971 a world shortage of *o*- and *p*-xylenes but few uses had been found for *m*-xylene which is normally returned to be blended with petrol, though some is oxidized to isophthalic acid and used in speciality resins.

Table 9.13 Derivatives of xylenes

Materials	United States US tariff commission 1969			US price ¢/lb. 4/72	United Kingdom	
	Production m. lbs	Sales m. lbs	Unit value ¢/lb.		Production 1971 '000 tons	Price £/ton 4/72
o-Xylene	850	732	4	—	—	} 27.5
p-Xylene	1 628	1 207	7	6·25	—	
Phthalic anhydride	760	446	11	7·5	58·87*	68
Terephthalic acid	1 045	—	—	—	—	—
Dimethyl terephthalate	1 537	752	15	—	—	—
Total organic phthalates	884	821	15	—	93·13	—
Total polyester resins	688	607	27	—	—	—

* 1969 data; unit value 1971, £69.

p-Xylene is the raw material for terephthalic acid and dimethyl terephthalate from which polyester fibres such as 'Terylene' and 'Dacron' are made (section 10.7.3). Air oxidation to the acid is carried out in the liquid phase in the presence of an oil-soluble cobalt salt:

$$\underset{p\text{-Xylene}}{\underset{CH_3}{\underset{|}{\bigcirc}}\!\!-CH_3} \xrightarrow[O_2]{130-145°C} \underset{\text{Terephthalic acid}}{\underset{COOH}{\underset{|}{\bigcirc}}\!\!-COOH}$$

226 *Heavy organic chemicals*

o-Xylene is oxidized to phthalic anhydride in a gas-phase process in which air and o-xylene are passed over a vanadium pentoxide catalyst. Very short contact times, of the order of tenths of a second, and high temperatures are employed:

$$o\text{-Xylene} \xrightarrow[O_2]{700°C} \text{phthalic anhydride}$$

A number of competitive routes are available but their overall chemistry is similar. Phthalic anhydride, being an anhydride, is more easily esterified than iso- and terephthalic acids and is widely used in phthalate plasticizers, polyesters and alkyd resins. Being a solid (m.p. 130·8°C) it presents transport difficulties and large quantities are shipped molten in tank-cars heated to 140°C.

The xylene-based route to phthalic anhydride is in competition with an old route based on naphthalene and thus, indirectly, on coal (section 9.5.4). The petrochemical route has superseded the earlier method rather slowly – partly because of development troubles with plant. At the end of 1968, Laporte and BP/ICI in the UK were still operating naphthalene-based plant, but either operating or planning xylene-based units. Phthalic Anhydride Chemicals, a Company owned by the National Coal Board, had, of course, no intention of switching from the coal-based process.

9.5.4 Naphthalene

Most naphthalene is produced by distillation of coal tar, of which it forms approximately 10% and in which it is the most abundant single chemical compound. It can also be obtained by dealkylation – in the presence of hydrogen – of petroleum fractions high in methyl-naphthalenes. The major use of naphthalene is as a feedstock for phthalic anhydride production.

Naphthalene is oxidized with air in the gas phase over a vanadium pentoxide catalyst, the air being added in two stages:

$$\text{naphthalene} + 4\tfrac{1}{2}O_2 \longrightarrow \text{phthalic anhydride} + 2CO_2 + 2H_2O$$

A by-product of the reaction is maleic anhydride which is itself manufactured from benzene by an analogous reaction.

The extent to which the naphthalene route to phthalic anhydride can compete with the newer xylene-based route is open to question. One drawback is inherent in the stoichiometry of the reactions. At 100% yields, 106 tons of xylene are required to give 148 tons of phthalic anhydride whereas two of the carbon atoms in naphthalene are lost as carbon dioxide and 128 tons are required to give a similar amount of product. Meanwhile, in 1968 Western Europe and the USA both consumed about 250 000 tons of naphthalene in phthalic anhydride production.

9.6 Acetylene

Consumption of acetylene in the USA and UK rose rapidly during the 1950s and early 1960s but is now declining. The shift from the carbide process to petrochemical acetylene arrested the decline in the USA to some extent, but in the UK attempts to operate the same process with naphtha rather than natural gas were largely unsuccessful.

The reasons for the decline of acetylene are not hard to find. In 1965 US acetylene consumption was 530 000 tons. 33% went into vinyl chloride, a use which has been largely displaced by oxychlorination (section 10.3.1). The last UK plant is due to close in 1972. 17% went into neoprene rubber and although the plant still operates, butene-based routes are safer and more economic (section 10.6.7). 19% went into acrylonitrile, a process which has been entirely superseded by ammoxidation (section 10.7.4). 12% went into vinyl acetate and troubles with the Wacker process have left this a current but declining use and the solitary UK plant operates only intermittently (section 10.5.1). Other chemical uses (13% of 1965 consumption) include acrylates via acetylene, currently threatened by a propylene-based route (section 10.5.2), and trichloroethylene (section 16.10.2).

Even when made successfully by the Wulff process (section 7.7) or by the carbide route in a country with cheap hydroelectric power (section 8.4.4), acetylene appears about two to three times as expensive as ethylene and four to five times as expensive as propylene. While this situation continues, it is hard to see any new acetylene-based plant being built. Acetylene chemistry may revive if the price of petroleum rises so high that production from carbide based on nuclear electricity becomes attractive.

Acetylene is still important as a fuel gas for welding (12% of US 1965 consumption) and this outlet continues to flourish.

Notes and references

General The heavy organic chemical industry is described in many books on the petrochemical industry. Hahn has already been mentioned. There are two important books by A. L. Waddams, the doyen of European chemical market researchers: R. F. Goldstein and A. L. Waddams, *The Petroleum Chemical Industry* 3rd edn., Spon, London, 1967 (new edition due in 1972) and A. L. Waddams, *Chemicals from Petroleum*, 2nd edn., John Murray, London, 1968. R. B. Stobaugh, *Petrochemical Manufacturing and Marketing Guide* (vol. 1, *Aromatics and Derivatives*, vol. 2, *Olefines, Diolefines and Acetylene*), Gulf Publishing Co., Houston, Texas, 1966, 1968, apart from being a little dated, is the best book on the subject. It is unfortunately almost unobtainable in the UK; the only copy which we have been able to trace was in the Institute of Petroleum library.

We have also found useful from a chemical viewpoint E. Kilner and D. M. Samuel, *Applied Organic Chemistry*, London, 1960; A. J. Gait, *Heavy Organic Chemicals*, Pergamon, London, 1967, and D. M. Samuel, *Industrial Chemistry – Organic: Advanced Level*; R.I.C. teachers' monographs no. 11. R.I.C. London, 1966. 'Best buy' is a slim paper-back, J. P. and E. S. Stern, *Petrochemicals Today*, Arnold, London, 1971.

Data on chemicals production are assembled mainly from the US Tariff Commission report on synthetic organic chemicals for 1969, and the UK *Business Monitor*. 1972 prices are taken mainly from *European Chemical News*, April 14th, 1972 and *Chemical Marketing*

Reporter, March 27, 1972. Prices of cheapest grades in bulk quantities have been taken in all cases. US Tariff Commission data give production, sales and unit value of certain chemicals. Production figures include intracompany transfers but exclude intermediate compounds which are not isolated. Sales represent amounts of production actually sold and exclude intracompany transfers. The difference between these figures, therefore, represents the extent of captive use. Unit value is the net selling price averaged over all sales. As it includes sales on bulk, long-term contracts, it is generally lower than the list price but higher than the intracompany transfer price which is, of course, not divulged.

Section 9.1 Figure 9.1 is from US Tariff Commission amended for double counting as suggested by Stobaugh *op. cit.* Figure 9.2 is from *Business Monitor* and *Annual Abstract of Statistics*.

S. A. Miller (ed.), *Ethylene and its Industrial Derivatives*, Ernest Benn, London, 1969 is a tedious but comprehensive 1321 pp. anthology, but the first 58 redeeming pages are by H. M. Stanley himself.

Section 9.1.1 We have drawn on R. Landau and A. C. Deprez's paper on Ethylene and Propylene Oxides and their Derivatives. *ECMRA Budapest, 1970*.

9.2 and 9.5.

Section 9.1.3 Ethyl alcohol, in the Brussels Nomenclature used by most countries in the world, is classified as an agricultural rather than a chemical product. In all countries fermentation is still the main method of manufacture and potation the main end-use. In consequence published statistics are well-nigh incomprehensible, but a major clarification (with 63 tables) has been brought about by T. M. D. Ball, *ECMRA London, 1969*.

UK units are proof gallons and continental units are hectolitres of concentration expressed in degrees Gay–Lussac. 95° Gay–Lussac represents 95% by volume at 15°C. One imperial gallon of 100% ethyl alcohol contains the same amount as 1·75 proof gallons, and 1 hectolitre equals 22 imperial gallons. The specific gravity of ethyl alcohol is 0·79 hence 1 proof gallon contains 4·5 lb (2·04 kg) ethyl alcohol and 1 hectolitre of 100% alcohol weighs 174 lb (79 kg).

Section 9.2.2 See R. Landau and A. C. Deprez, *op. cit.*

Section 9.3 European Market Data are given by M. Niel, *ECMRA London, 1969*.

Section 9.5 Figure 9.5 comes from F. Gindici and B. Arcelli's paper, 'Aromatics and their Processing in Europe', *ECMRA Budapest, 1970*.

Section 9.5.1.1 For Fig. 9.7 and the associated discussion, we have drawn especially on A. S. Biancu, *Chem. and Proc. Eng.*, January, 1967. See also R. Cociancich, Phenol in the EEC and UK, *ECMRA London, 1969*.

Section 9.5.3 An excellent survey of the phthalic anhydride market is given by J. P. Allan, *ECMRA London, 1969*. See also R. A. Duckworth, *Chem. and Proc. Eng.*, 1969, **50** (1).

10 Industrial polymers

About 80% of the output of the world organic chemical industry is used in the production of synthetic polymers. These materials – plastics, elastomers and fibres – are the products of the chemical industry with which the man in the street is most familiar and which have had the greatest impact on his life. It is now possible to make, at a price, a polymer with almost any desired combination of properties.

Before the manufacture and properties of commercial polymers are discussed, the principles underlying polymer chemistry will be outlined. Polymers may be divided into two groups: natural, such as proteins and cellulose; and synthetic, such as polyethylene and nylon. Prior to 1925 many chemists believed that polymers were agglomerations of small molecules held together by forces similar to those operating in colloids. They are now known to be high molecular weight materials, consisting of long chains of repeating single molecules, or slight modifications of them, called monomers. Most technically useful polymers have molecular weights between a few thousand and a few hundred thousand (say 10 000–200 000) although others outside this range have been made.

If we consider a polymer of structure —A—A—A—A—A—A—···, then the repeating unit —A— is called a mesomer and the monomer could, for example, be H—A—OH. Molecules of the monomer could combine together with elimination of water to give the polymer.

The above polymer is said to be a linear polymer and the mesomer —A— has two free, or potentially free, valencies, i.e. the monomer contains two functional groups and is said to be bifunctional.

Some polymers consist of more than one mesomer and have the type of structure:

—A—B—A—B—B—A—A—A—B—B—A—A—B—B—B—A— B—A—A—B—

They are said to be copolymers of the mesomers A and B. Sometimes the As and Bs are arranged regularly (e.g. the ethylene glycol and terephthalic acid units in 'Terylene', section 10.7.3) and sometimes randomly (e.g. the vinyl acetate and ethylhexyl acrylate units in copolymer emulsion paints, section 10.5.1).

Certain polymers have branched rather than linear chains:

```
                                    A—A—
                                   /
                            A—A—Y
                           /        \
         —A—A— A—A—Y               A—A—
                          \
                           A—A—
```

and these can build up into three-dimensional structures known as cross-linked polymers:

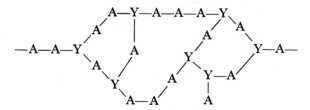

All the units in linear polymers are bifunctional, that is they have two free valencies. In cross-linked polymers, some must be at least trifunctional as exemplified by the mesomer $-\overset{|}{Y}-$ above.

Polymers are classified as elastomers, plastics and fibres depending on their initial moduli of elasticity. A value of 15–150 psi is typical of an elastomer, 1 500–15 000 of a plastic and 150 000–1 500 000 of a fibre. They are also classified according to their behaviour on heating. Thermoplastic polymers soften on heating and are unaltered chemically. Thermosetting polymers degrade. Broadly speaking, linear polymers are thermoplastic and cross-linked polymers are thermosetting.

An addition polymer is said to be formed when the empirical formula of the polymer is the same as that of the monomer, e.g.

$$n\text{CH}_2\!\!=\!\!\text{CH}_2 \longrightarrow (\!-\!\text{CH}_2\!-\!\text{CH}_2\!-\!)_n$$
$$\text{Ethylene} \qquad\qquad \text{Polyethylene}$$

Condensation polymerization involves the elimination of a small molecule during the polymerization process, e.g.

$$n\text{H}_2\text{N}\!-\!\text{R}\!-\!\text{COOH} \longrightarrow (\!-\!\text{HN}\!-\!\text{R}\!-\!\text{CO}\!-\!)_n + n\text{H}_2\text{O}$$
$$\text{Amino-acid} \qquad\qquad \text{Polyamide}$$

But see section 16.11 for a more rigorous definition.

Condensation polymers are usually more readily hydrolysed than addition polymers.

10.1 Properties and molecular structure

The classification of linear polymers on the basis of their moduli of elasticity has already been mentioned. Table 10.1 shows the other mechanical properties associated with elastomers, plastics, and fibres. All the polymers mentioned in this table have a long-chain structure. The question arises as to what causes the differences between them, and what endows them with their unique properties of high viscosity, tensile strength and toughness.

The most obvious factor is molecular weight, and indeed the impact and tensile strengths of polymers depend markedly on chain length. Polymers start to develop mechanical strength when their chain length reaches about 50 mesomer units and this increases up to a chain length of about 500 units, after which, further increases in chain length make little difference to mechanical strength.

The differences between various polymers may be attributed to the extent to which

Table 10.1 Mechanical properties

	Elastomers	Plastics	Fibres
Initial modulus of elasticity, p.s.i.	15–150	1 500–15 000	150 000–1 500 000
Upper limit of extensibility, %	100–1 000	20–100	less than 10
Crystallization tendency	low when unstressed	moderate to high	very high
Molecular cohesion, cal/mole (J/mole)	1 000–2 000 (4 200–8 400)	2 000–5 000 (8 400–21 000)	5 000–10 000 (21 000–42 000)
Examples	Natural rubber, polychloroprene, polybutadiene, polyisobutene	Polyvinyl chloride, polyvinyl acetate, polystyrene	Polyamides, silk, polyvinylidene chloride, Terylene, cellulose

they are crystalline and this in turn depends on two factors – the magnitude of the attractive forces between the molecules of the polymer, and the ease with which the polymer chains will fit into a crystal lattice. A polymer with small intermolecular forces and with a bulky molecule which cannot easily pack into a crystal lattice will show little tendency to crystallize and will usually be an elastomer. Conversely, a polymer with strong intermolecular forces and a compact molecular structure will usually show typical fibre properties.

The properties of a polymer are thus related to its tendency to crystallize. The crystalline regions found in polymers are, however, different from the three-dimensional order found in simple crystalline substances or in metals. Polymers show a microcrystalline structure in which long segments of different chains align themselves in bundles with high lateral order (Fig. 10.1(a)). These ordered regions are small and are scattered throughout the polymer which is otherwise amorphous. Electron microscopy has shown that polymers crystallize in plate-like crystals with surprisingly uniform thickness of about 10 nm. As polymer molecules are typically of the order of 200 nm long, it is likely that many of the chains fold back on themselves at the platelet surface (Fig. 10.1(b)). Parts of the chains still remain unincorporated in the crystal and these form the amorphous regions (see also Fig. 10.2).

In thermodynamic terms, the tendency to crystallize may be identified with the free energy change accompanying it. This is made up of two terms:

$$\Delta G_{\text{crystallization}} = \Delta H_{\text{crystallization}} - T \Delta S_{\text{crystallization}}$$

The enthalpy term is the strength of the intermolecular bonds (hydrogen bonds, van der Waals' forces, etc.) which are formed as the polymer molecules come together, and the entropy term represents the influence of the geometrical bulkiness of the chains and other obstacles to the neat packing of the molecules into a crystalline structure.

If ΔH is large (say, an interaction of about 5 kcal/mole (20 kJ/mole) over a 0·5 nm length of chain, typically the length of one mesomer) then the polymer is likely to behave as a fibre. Between 2 and 5 kcal/mole (8 and 20 kJ/mole) plastic properties would be expected and below 2 kcal/mole (8 kJ/mole) an elastomer should result.

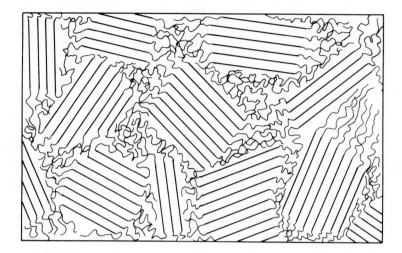

Figure 10.1(a) Illustration of the microcrystalline structure of polymers.

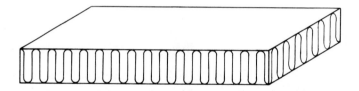

Figure 10.1(b) Polyethylene crystal showing chains folding back on themselves. Such regular crystals are obtained only from dilute solution. Polymers obtained from a melt probably have structures intermediate between this and the microcrystalline structure of Fig. 10.1(a).

The ΔS term modifies this. Chains which fit easily into a crystal lattice will do so under the influence of relatively weak forces and the properties of the polymer will be more fibre-like than might otherwise have been predicted. Similarly, bulky side chains will cause a polymer to be more elastomeric than expected on the basis of the enthalpy term.

Temperature also has an effect. Raw rubber, for example, crystallizes almost completely in a few hours at $-25°C$. The rubber chains also tend to align themselves during stretching, and crystallization has been known to occur suddenly during extension. Similarly, polyethylene becomes more elastomeric as the temperature is raised.

The above principles can be applied qualitatively with good results but there is a certain amount of overlapping between the classes of polymers. Polyester fibres such as

Properties and molecular structure 233

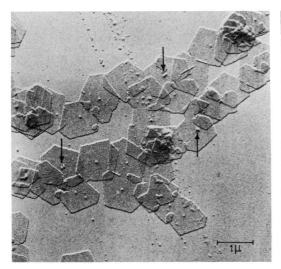

Figure 10.2(a) Electron micrograph of uniform monolayer crystals of polyethylene.

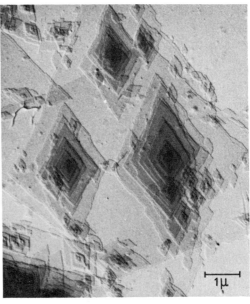

Figure 10.2(b) Electron micrograph of typical crystals of polyethylene grown from solution.

Figure 10.2(c) Polypropylene spherulites, crystallized from the melt, showing deformation of crystal structure. ($\times$ 1000).

'Terylene' or 'Dacron' for example can exist in an almost completely amorphous form and, indeed, when the polymer is first manufactured it has low crystallinity. This is increased by mechanical means by the 'drawing' of fibres of the material. The polymer chains orient themselves parallel to one another and hence the drawn fibres have high crystallinity and characteristic mechanical properties.

The effect of the enthalpy term can be seen by considering the series of polymers polyethylene – polyvinyl chloride – polyvinylidene chloride:

Polyethylene

$$-\underset{\underset{H}{|}}{\overset{\overset{H}{|}}{C}}-CH_2-\underset{\underset{H}{|}}{\overset{\overset{H}{|}}{C}}-CH_2-\underset{\underset{H}{|}}{\overset{\overset{H}{|}}{C}}-CH_2-$$

Polyvinyl chloride

$$-\underset{\underset{H}{|}}{\overset{\overset{Cl}{|}}{C}}-CH_2-\underset{\underset{H}{|}}{\overset{\overset{Cl}{|}}{C}}-CH_2-\underset{\underset{H}{|}}{\overset{\overset{Cl}{|}}{C}}-$$

Polyvinylidene chloride

$$-\underset{\underset{Cl}{|}}{\overset{\overset{Cl}{|}}{C}}-CH_2-\underset{\underset{Cl}{|}}{\overset{\overset{Cl}{|}}{C}}-CH_2-\underset{\underset{Cl}{|}}{\overset{\overset{Cl}{|}}{C}}-CH_2-$$

Polyethylene has low intermolecular forces and consequently a low melting point. It looks as if it should be an elastomer but the ease with which the chains will pack moves it into the class of plastics. Polyvinyl chloride has fairly high dipole–dipole bonds via the C—Cl groups and is therefore a plastic. Polyvinylidene chloride has two C—Cl groups per molecule and is therefore a fibre.

Polyisobutene has a non-polar side chain and packs less easily into a crystal lattice. It is therefore an elastomer. Ziegler polyethylene and polypropylene (q.v.) are stereoregular and pack very readily into lattices. They are thus less elastomeric than polyethylene.

The opacity of polymers is also related to their degree of crystallinity since light is scattered at crystal boundaries. Polymethyl methacrylate ('Perspex' or 'Lucite') is amorphous and transparent. Polystyrene and polycarbonates are similar. Low density polyethylene gives a slightly milky sheet which is widely used for packaging but highly crystalline high density polyethylene is opaque and is less used for this purpose. Similarly nylon and polypropylene are crystalline and opaque. Polyester film ('Melinex') is transparent in spite of being made from a crystalline polymer. This is because the film has been axially oriented by drawing: hence it is effectively one large crystal and there are no crystal boundaries.

Free energies of crystallization are not sufficiently well known for the properties of new polymers to be predicted with certainty. It is nonetheless possible to guess approximately what properties a new material will have and how these properties might be modified by the addition of other materials. The polymer chemist can thus match or improve on the properties of most natural materials and it seems likely that in the future polymers will take over from them to an increasing extent. Furthermore, the tendency for polymers to be 'tailor-made' suggests that there will be a trend towards copolymers where the properties of one polymer are modified by copolymerization with another.

The range of possible copolymers is almost infinite and the number commercially available is growing. On the other hand, the polymers available on a really large scale were nearly all known before the Second World War. The following section will therefore deal only with homopolymers and with the more important and long-established copolymers.

10.2 Thermoplastics (excluding PVC)

Table 10.2 shows the main synthetic linear polymers available commercially. There is, however, an even greater selection which can be obtained, ranging from products manufactured in the UK on a quarter of a million tons per year scale to others which are produced in small laboratory batches of a few kilograms.

Table 10.2 Thermoplastics

Section	United States 1971			United Kingdom 1971		
	Production	Sales[1]	Price 4/72	Production	Con-sumption[1]	Price 4/72
	(m. lb)	(m. lb)	(¢/lb)	('000 tons)	('000 tons)	(£/ton)
Polyvinyl chloride	3 320	3 286[3]	12	335	355	125
Polyethylene (high density)	1 900	1 885	18	63	58	197
(low density)	4 500	4 480	13·8	305	290	141
Polypropylene	1 260	1 230	25	78	98	204
Polystyrene		2 845	13	144	130	130
Expendable polystyrene (for foams)[5]	3 840	264		26·8	15·8	
Styrene–acrylonitrile co-polymers		47	23·5	*1·5*	*1·5*	230
ABS copolymers		660	25	28	23	340
Polyvinyl acetate	426[10]	332[10]	26[10]	40	n.a.	155[6]
Polyvinyl alcohol	57[10]	51[10]	34[10]	n.a.	n.a.	—
Polymethyl methacrylate	—	408	18·5[7]	15	12·5	—
Polycarbonates	—	45	—	n.a.	*1·0*	730[8]
Polytetrafluoroethylene	—	15·4[4]	345	n.a.	*1·0*	2 900
Polyacetals	—	68	61[8]	nil	5	1 600
Nylon[2]	—	111	—	n.a.	*13·7*	580
Polyethylene glycol tere-phthalate[2]	—	—	—	n.a.	8	—
Cellulosics[2]	165	—	36[9]	10	10·2	319[9]

[1] In the US, sales figures include exports but exclude material used internally by the producer. In the UK, differences between production and consumption figures are due to imports and exports.
[2] Non-textile applications.
[3] Includes copolymers.
[4] 1969.
[5] Excludes foamed sheet made from crystal.
[6] Dispersion, 55% solids.
[7] Monomer.
[8] Estimated.
[9] Cellulose acetate
[10] 1969.
[11] 1970.
Italicized figures are approximate.

236   Industrial polymers

The following sections deal with the manufacture of the monomers for making the more important polymers, and mention is also made of polymerization techniques where these differ substantially from the techniques discussed in section 16.11.

Polyvinyl chloride, because it is the most versatile plastic and is frequently used with plasticizers, has been omitted from this section and is discussed in section 10.3.

10.2.1 Polyethylene

Polyethylene was discovered accidentally by ICI in 1933 and developed by them in the late 1930s. They used the trade name 'Polythene' for it, but this has since passed into the language to the extent that all brands of polyethylene and frequently many quite different plastics are referred to as 'polythene'.

Low density polyethylene (LDPE) was the first to appear on the market; it is made by a high pressure polymerization:

$$n\text{CH}_2\!\!=\!\!\text{CH}_2 \longrightarrow (\!-\!\text{CH}_2\!-\!\text{CH}_2\!-\!)_n, \quad \Delta H \text{ per ethylene mole} = -22 \text{ kcal } (-92 \text{ kJ})$$

Le Chatelier's principle suggests the polymerization will be favoured by high pressure, and a temperature as low as is consistent with a reasonable rate of reaction. In fact, ethylene gas of more than 99·8% purity is drawn from the cracking and purification plant at a pressure of about 35 atm and is further compressed to about 1 200 atm. In the presence of organic peroxides or 0·07% gaseous oxygen and at a temperature of 200°C some ethylene polymerizes to give liquid polyethylene which is removed continuously from the reactor and separated from unchanged ethylene which is recycled. The polyethylene is allowed to solidify and is cut into cubes. Yield per pass is about 25%; overall yield about 95%.

The product is a tough waxy mainly linear polymer of molecular weight 10 000–40 000 and density 0·92 g/cm³. There are occasional oxygen bridges between the chains:

$$\begin{array}{c} -\text{CH}_2-\text{CH}_2-\text{CH}-\text{CH}_2- \\ | \\ \text{O} \\ | \\ -\text{CH}_2-\text{CH}-\text{CH}_2-\text{CH}_2- \end{array}$$

and about 2% chain branching arising from intramolecular chain-transfer processes. The reaction is quite exothermic so heat removal is vital and must be carefully controlled.

High density polyethylene (HDPE) is made by a low temperature, low pressure method discovered by Karl Ziegler in 1952–53 and named after him. Ethylene is passed at about atmospheric pressure and 60°C into a suspension of aluminium triethyl and titanium tetrachloride in an aliphatic oil. High purity ethylene is not required, though methane, ethane, sulphur compounds, oxygen, water and carbon dioxide are normally removed. Conversion is almost quantitative. The effluent from the reactor is fed to a series of flash drums where solvent distils off. Water is added to destroy residual catalyst and the polyethylene slurry is centrifuged and dried. The water and the solvent are recycled but the catalyst is lost and its replacement is a major item in the operating costs.

The product is unbranched and crystalline with a density of 0·97 g/cm³. It has a higher tensile strength, greater rigidity and higher softening temperature than low density polyethylene. It will withstand boiling water and is not readily cracked by solvents and these are its two main advantages over the low density material. The latter, with its greater transparency, is more suitable for packaging. As polyethylene has no side chains, the difference between the high and low density materials cannot be one of stereoregularity. It is rather that high density polyethylene lacks chain branching and oxygen bridges and hence can crystallize more easily.

The costs of the low and high pressure processes are compared in Fig. 10.3. The former requires greater capital investment, more expensive catalyst and more power and gives lower yields but of a premium product.

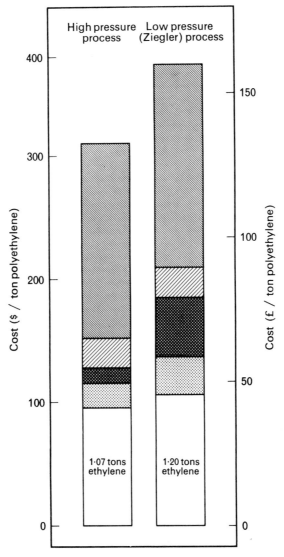

Figure 10.3 Comparative costings for high pressure and low pressure (Ziegler) polyethylene plant (capacity 45 000 tons/year, ethylene at $88/ton) in USA 1965.

A medium pressure process is used in the USA in which ethylene is polymerized in a cyclohexane solvent in the presence of a Cr_2O_3–Al_2O_3 catalyst.

The growth of polyethylene production has been one of the most startling in the history of the chemical industry, and is described in section 2.9.2. In the USA between 1950 and 1963 production rose from 25 000 tons to 1 000 000 tons a year, an annual growth rate of 37%.

Table 10.3 End uses of polyolefines 1971

	LDPE		HDPE		Polypropylene	
	USA (m. lb)	UK ('000 tons)	USA (m. lb)	UK ('000 tons)	USA (m. lb)	UK ('000 tons)
Blow moulding	60	23	825	31 }	<672‖	50
Injection moulding	620	42	425	17·1 }		
Filament and fibres	—	—	10	2·2	345	30
Pipe, conduit, etc.	75	6	115	2·0	12	13·5*
Wire and cable	450	23	35	0·8	12	—
Film and sheet	2 380	167	90	4·5	100‡	14·5†
Extrusion coating	395	13	30	—	4§	—
Powder	—	8	—	—	—	—
Rotomoulding	60	—	35	—	—	—
Miscellaneous	140	4·2	55	1·9	>83‖	2·0

* Sheet and pipe.
† Film.
‡ Not subsequently formed into rigid and semi-rigid containers.
§ For use in packaging.
‖ Subtraction of categories shown from total consumption gives moulding + miscellaneous = 755 m. lb.

At present, consumption of high density polyethylene is growing faster than that of low density, as is to be expected for a newer product. The pattern of end-uses of polyethylene is shown in Table 10.3. The major outlet for low density material is in film, which is used mainly for packaging. Injection moulded products use large amounts of both forms of polyethylene in such products as buckets, washing-up bowls and plastic dustbins. Wire and cable insulation, which was the first use for polyethylene to be developed, is still important. (In 1938 the Telegraph Construction and Maintenance Company manufactured a mile of polyethylene covered cable and laid it between the Isle of Wight and the mainland.) The major use for high density polyethylene is for blow-moulding compound, used among other things for detergent and bleach bottles. Its rigidity makes it suitable for this use but the homopolymer is prone to stress-cracking. To overcome this defect about 0·7% of a comonomer such as propylene, butene-1 or hexene-1 is included, which reduces density somewhat and improves crack resistance.

10.2.2 Polypropylene

Polypropylene, like high-density polyethylene, is made by the Ziegler process. Propylene is polymerized at about 70°C and three atm. pressure in the presence of an aluminium trialkyl–titanium tetrachloride catalyst. The product is unbranched and stereoregular, i.e. the methyl groups are oriented to the backbone of the molecule in a regular way. In the isotactic form they all lie on the same side of the chain and in the syndiotactic form on alternate sides. Stereoregular polymers pack together more easily than non-stereoregular (atactic) polymers and are therefore more crystalline and more rigid. The Ziegler process is unique in that it invariably yields completely stereoregular products.

Polypropylene is similar in many ways to high-density polyethylene. It has high strength, low density, high resiliency in moulded parts, high softening temperature, excellent resistance to stress cracking, good electrical properties, and is moisture and solvent resistant. It has, however, certain inherent disadvantages. Like polyethylene it has no affinity for ordinary dyes and it must be modified or pigmented if it is to be coloured (section 12.3.3). Moreover, as a branched chain hydrocarbon polymer it is very liable to oxidation, which is promoted by heat, light or trace metals. In the first step the chain breaks to give alkyl radicals which react with oxygen to give peroxyalkyl radicals; the latter abstract the tertiary hydrogen atoms of the polymer chain to form new alkyl radicals and alkyl hydroperoxides which themselves decompose to other compounds. The products of these reactions include ketones, aldehydes and cross-linked polymer chains formed by the union of two free radicals. The carbonyl compounds themselves absorb ultraviolet light and act as sensitizers for further photolytic degradation. Polypropylene must therefore be used in conjunction with anti-oxidants and light stabilizers. The design of such systems is very complicated; well known antioxidants include butylated hydroxytoluene (BTH)

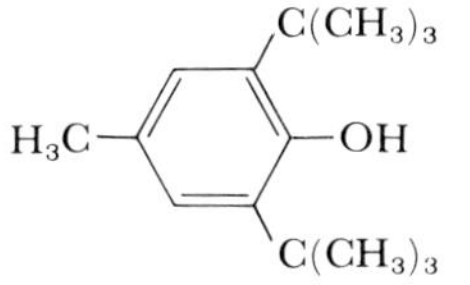

and AO—I

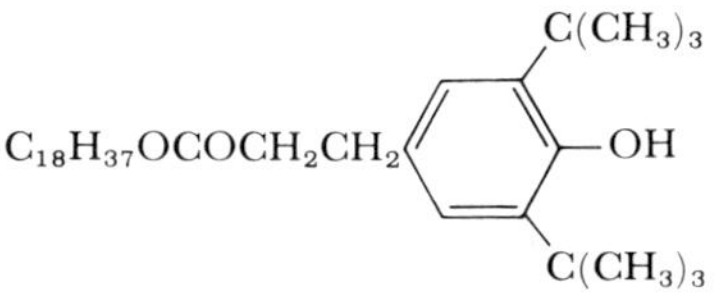

while UVA 1

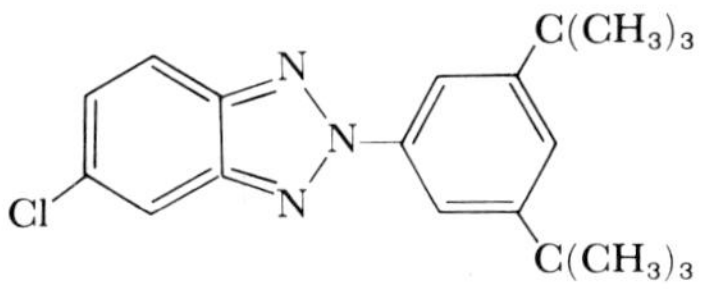

is an important light stabilizer.

While low-density polyethylene is used mainly in film and high-density polyethylene in injection and blow moulding, polypropylene is used in mouldings, film and fibres (Table 10.3). Oriented polypropylene film has many of the properties of cellophane and certain advantages. Polypropylene fibre is light and strong and resists corrosion and abrasion. It is increasingly replacing sisal in the manufacture of ropes.

10.2.3 Polystyrene

Polystyrene is made from styrene by conventional polymerization techniques (section 16.11). The styrene is made from ethylene and benzene in a two-stage process. First, ethylbenzene is made by the alkylation of benzene with ethylene and hydrogen chloride over aluminium chloride or phosphoric acid (Friedel–Crafts reaction) at 95°C

$$\text{C}_6\text{H}_6 + \text{C}_2\text{H}_4 \xrightarrow[\text{HCl}]{\text{AlCl}_3} \text{C}_6\text{H}_5\text{—CH}_2\text{—CH}_3$$

The ethylene does not need to be especially pure, but acetylene must be avoided as it spoils the catalyst. The alkylation is reversible and, under the above conditions, the equilibrium mixture contains 51% ethylbenzene, 31% polyalkylbenzenes and 18% benzene. The last two are recycled. Ethylbenzene is also extracted from the 'mixed xylenes' fraction produced by catalytic re-forming (section 7.8).

Dehydrogenation to styrene is brought about by passage of the ethylbenzene over zinc, magnesium, iron or aluminium oxide on activated charcoal at 630°C:

$$\text{C}_6\text{H}_5\text{—CH}_2\text{—CH}_3 \longrightarrow \text{C}_6\text{H}_5\text{—CH=CH}_2 + \text{H}_2$$

The forward reaction is favoured by operation at low pressures but if the reactor were run at less than atmospheric pressure air might be sucked in through a small leak and cause an explosion. The ethylbenzene is therefore mixed with 2·6 parts of steam per part of ethylbenzene and the reactor operated slightly above atmospheric pressure. In this way air is prevented from leaking into the reactor, but on the other hand the partial pressure of ethylbenzene is only about $\frac{1}{4}$ atm, and this is effectively the same as operating the reactor below atmospheric pressure. Addition of a greater proportion of steam would result in the equilibrium being pushed further to the right, but would reduce the throughput to an unacceptable level.

Removal of the diethylbenzene impurity from the ethylbenzene feedstock to the dehydrogenation unit is extremely important since, if this material were dehydrogenated, it would give divinylbenzene. This has two double bonds and would polymerize rapidly giving an infusible cross-linked polymer. Cost data for styrene manufacture are given in Fig. 10.4.

The importance of styrene dates back to the Second World War when it was a major constituent (about 25%) of synthetic rubber. This was the GR-S rubber, and it came into prominence because of the Japanese occupation of the principal areas which produced natural rubber. After the war, natural rubber failed to recapture its former markets but most of the GR-S market was taken by a modification called SBR rubber, a styrene–butadiene copolymer in the proportion of 1 mole styrene to 6 moles of butadiene (section 10.6.3). This too has passed its peak, and the main outlet for styrene now is as the homopolymer.

Polystyrene is a somewhat brittle material which makes a metallic noise when dropped. It has good moisture resistance and electrical properties. Its bulky phenyl groups cause it to have low crystallinity and impact strength, and also enable it to be obtained in a crystal clear glassy form. It is one of the cheapest thermoplastics.

Two major styrene copolymers are 'impact' polystyrene and acrylonitrile–butadiene–styrene resins (ABS). Impact polystyrene consists of styrene copolymerized with 3–10% of butadiene or styrene–butadiene rubber. The product is much tougher than styrene homopolymer but cannot be obtained in a transparent form. The reason for this became apparent in 1956 when it was shown by microscopy that impact polystyrene is a two-phase material, a sort of plastic 'alloy' in which rubber particles of about 1–5 μm in diameter are dispersed in a polystyrene matrix. ABS copolymers are also two-phase systems with the rubber element dispersed in the rigid styrene-acrylonitrile (SAN) copolymer matrix. SAN copolymers are plastics in their own right but are not widely used.

ABS resins, as befits their structure, were originally a blend of 60% SAN copolymer with 40% butadiene–acrylonitrile rubber. They are now copolymerized from the three separate components the proportions of which can be varied quite widely depending on the properties required. A possible formulation is 20% styrene, 20% acrylonitrile and 60% butadiene. Among other things, ABS resins are used in the heads of golf clubs.

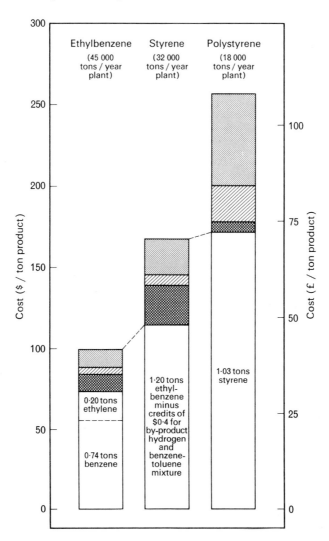

Figure 10.4 US costs for the production of polystyrene from benzene ($75/ton) and ethylene ($88/ton) via ethylbenzene and styrene.

Table 10.4 Polystyrene – End-use pattern 1971 for moulding and extrusion material

Polystyrene	USA (m. lb)	(%)	UK ('000 tons)	(%)
Packaging	} 600	26.7	41.0	31.5
Vending cups and cartons			21.5	16.5
Toys	290	12.9	13.4	10.3
Housewares	275	12.2	11.0	8.5
Furniture	100	4.4	1.8	1.4
Refrigerators	90	4.0	6.0	4.6
TV and radio	90	4.0	2.4	1.8
Other consumer durables	59	2.6	7.3	5.6
Combs, brushes, etc.	48	2.2	1.4	1.1
Light fittings and signs	44	1.9	5.2	4.0
Shoe heels and soles	n.a.*†	—	5.0	3.8
Flower pots	n.a.*	—	6.5	5.0
Miscellaneous	660	29.4	7.5	5.8
Total consumption: moulding	2 006		} 130	
extrusion	957			

* Included under miscellaneous uses.
† ABS is the preferred plastic in the USA and heels consumed 20 m. lb. of this.

The end-uses of polystyrene are shown in Table 10.4 and those of ABS resins in 10.5.

Table 10.5 ABS – End-use pattern 1971

	USA (m. lb)	(%)	UK ('000 tons)	(%)
Automobiles	105	16.8	4.4*	19.1
Sheet	n.a.†	—	5.0	21.7
Business machines and telephones	41	6.6	2.5‡	10.9
Consumer durables	115	18.4	3.8	16.5
Pipe and fittings	115	18.4	3.1	13.5
Miscellaneous	249	39.8	4.2	18.3
Total consumption (excluding exports)	625		23.0	

* Other than sheet material.
† Included under automobiles and miscellaneous.
‡ Telephones only.

10.2.4 Polymethyl methacrylate

Polymethyl methacrylate ('Perspex' in the UK, 'Lucite' in the USA) was one of the few polymers to be manufactured before the Second World War. UK production in 1939 was about 400 tons and during the war it was used for aeroplane cockpit covers.

It is made by reacting acetone with hydrogen cyanide to give acetone cyanhydrin. The hydrogen cyanide was made from methane, air and ammonia, but it is now generally available as a by-product of the ammoxidation route to acrylonitrile. Acetone cyanhydrin, on treatment with concentrated sulphuric acid, goes to methacrylamide sulphate and, without the latter being purified, it is treated with methanol and water to give methyl methacrylate monomer (Fig. 10.5). Polymerization can be brought about by free radical or anionic but not cationic initiators. Benzoyl peroxide is usually used.

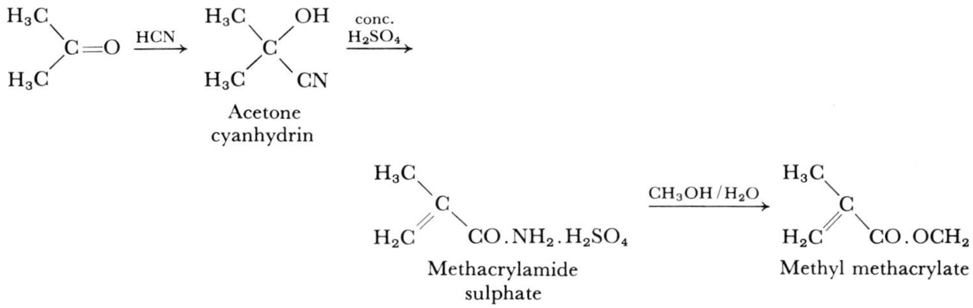

$$\begin{array}{c}H_3C\\ C=O\\ H_3C\end{array} \xrightarrow{HCN} \begin{array}{c}H_3COH\\ C\\ H_3CCN\end{array} \xrightarrow{\text{conc.}\ H_2SO_4}$$

Acetone cyanhydrin

$$\begin{array}{c}H_3C\\ C\\ H_2CCO.NH_2.H_2SO_4\end{array} \xrightarrow{CH_3OH/H_2O} \begin{array}{c}H_3C\\ C\\ H_2CCO.OCH_2\end{array}$$

Methacrylamide sulphate | Methyl methacrylate

Figure 10.5 Manufacture of methyl methacrylate.

Methyl methacrylate is a liquid and is cast into blocks or sheets of polymer. Its polymerization is highly exothermic and accompanied by a fairly large decrease in volume. It is important therefore to control the temperature to prevent the reaction from getting out of hand, and also to carry it out in a vessel with a gasket so that the decrease in volume does not result in the appearance of air holes (voids) in the block of polymer.

The larger the block of 'Perspex' it is desired to cast, the worse these problems become and, until the early 1950s, it was impossible to buy 'Perspex' in sheets more than an inch thick.

Unlike most polymers, its polymerization is reversible and the heating of polymethyl methacrylate scrap will regenerate the monomer. One firm in England makes a living doing this.

Polymethyl methacrylate is held together by moderate dipole forces but its molecule is very unwieldy, so it is a plastic and is amorphous. It is therefore transparent and, indeed, has outstanding optical properties. Its critical angle is very high (42°) and it can be used for 'piping' light round corners.

It is brittle and easily scratched but, interestingly, is harder than the isomeric polyethyl acrylate and ICI add about 4% of ethyl acrylate to their methyl methacrylate as a processing aid.

'Perspex' is a relatively expensive plastic since it is made by a multistage process and HCN is expensive. It finds mainly 'premium' uses where people are prepared to pay for its having better optical qualities than glass and being less brittle and more easily worked. Its main outlets are in such things as caravan skylights and windows, 'glasses' for car rear lights, light fittings and advertising signs generally and baths. Acrylic baths are expensive and fairly easily scratched; nonetheless they accounted for 146 000 out of the 800 000 baths sold in the UK in 1971.

The surface coating of General Motors cars in 1971 was still methacrylate based and this outlet accounted for 20% of the US market, though other manufacturers regard the formulation as obsolete.

It is possible that in future methacrylic acid might be made by a process similar to that available for acrylic acid (section 10.5.2) and this might ultimately bring down the price of polymethyl methacrylate.

244 *Industrial polymers*

10.2.5 Polyvinylidene chloride

Vinylidene chloride monomer is made by simultaneous chlorination and dehydrochlorination of ethane:

$$CH_3 \cdot CH_3 \xrightarrow[\substack{530-580°C \\ \sim 1\text{ s contact time}}]{Cl_2} \underset{\substack{\text{Vinylidene} \\ \text{Chloride}}}{CH_2 = CCl_2} \; (+ CH_2{=}CHCl + CHCl{=}CHCl)$$

It is an extremely reactive monomer and is copolymerized in the presence of peroxides with 5–17% vinyl chloride or acrylonitrile to give 'Saran', a material used in the form of thin film for wrapping because of its low water transmission, and also as a fibre. It can be extruded to give chemical and solvent resistant tubing. The polymer is highly crystalline when cooled slowly because of the two C—Cl dipoles on each mesomer, but can be cooled rapidly to give an amorphous material.

10.2.6 Polycarbonates

In formal terms, polycarbonates are polyesters of carbonic acid H_2CO_3 but neither their mode of preparation nor their properties bear any resemblance to inorganic carbonates. They are made by reaction between a dihydric alcohol or phenol and phosgene or an alkyl or aryl carbonate. The commonest raw materials are phosgene and Bisphenol 'A' (section 10.4.6)

$$n\left[HO{-}\phenyl{-}\underset{\substack{|\\CH_3}}{\overset{\substack{CH_3\\|}}{C}}{-}\phenyl{-}OH\right] + n\,COCl_2 \longrightarrow$$

$$\left[{-}\phenyl{-}\underset{\substack{|\\CH_3}}{\overset{\substack{CH_3\\|}}{C}}{-}\phenyl{-}O{-}\overset{\substack{O\\\|}}{C}{-}O{-}\right]_n + 2n\,HCl$$

Polycarbonates are transparent, have high impact strength, good heat, creep and chemical resistance, and good electrical properties. They are used in transparent electrical parts, cooling fans and propellors, safety helmets, lenses and babies' feeding bottles. Like polyacetals they are known as engineering plastics, and are more akin to metals than to general purpose thermoplastics.

10.2.7 Polyacetals

Commercial polyacetals are the result of combination of formaldehyde units in linear fashion to give the unit structure:

$$[-O-CH_2-O-CH_2-]_n$$

Of the two polyacetals currently on the market, one is made by polymerization of formaldehyde in hexane with an amine or cyclic nitrogen-containing catalyst. The upgrading of the formaldehyde to a satisfactory purity is expensive, and the other polyacetal is made by copolymerization of trioxane (a trimer of formaldehyde) and ethylene oxide.

Polyacetals, like polycarbonates, are high performance plastics which can be used to replace metals in gear wheels, structural units, etc., and they are therefore known colloquially as engineering plastics. ICI are importing polyacetals into the UK but have a plant scheduled to open in 1974.

10.2.8 Polytetrafluoroethylene

Polytetrafluoroethylene (PTFE or 'Teflon') is a speciality polymer made in relatively small quantities – of the order of 500–1 000 tons/year in the UK. It is expensive and difficult to fabricate because of its infusibility and toughness, and powder metallurgy techniques (preforming and sintering) have to be used. These qualities are also the reason for its commanding a premium price. It is stable and its properties are virtually unchanged between -70 and $350°C$. It has excellent low-friction and anti-stick properties, is hydrophobic and resistant to all corrosive chemicals. It is used in electrical insulators, wire covering, valve seats, seals and gaskets. Thin layers are used on non-stick pans and on the bottoms of the more expensive grades of ski.

Tetrafluoroethylene is made by the action of hydrogen fluoride (section 16.10.1) on chloroform (section 16.10.2) followed by cracking of the resultant difluorochloromethane:

$$CHCl_3 + 2HF \longrightarrow CHClF_2 + 2HCl$$

$$2CHClF_2 \xrightarrow[\text{Pt tubes} \atop \text{sub-atmospheric pressure}]{600-800°C} [2CF_2] \longrightarrow CF_2{=}CF_2 + 2HCl$$

As indicated, the reaction goes via the free radical CF_2. A contact time of only one second is used in the cracking, otherwise the CF_2 reacts with the product to give hexafluoropropylene:

$$CF_2{=}CF_2 + CF_2 \longrightarrow CF_3{-}CF{=}CF_2$$

Some hexafluoropropylene is always formed as a by-product and its copolymer with tetrafluoroethylene is a commercial product (FEP). It has poorer chemical resistance and lower melting point than PTFE but can be fabricated by conventional methods.

Tetrafluoroethylene is polymerized by a method similar to that for low density polyethylene. A pressure of 50–400 psi and a temperature of 40–80°C is used together with a peroxide catalyst.

10.3 Polyvinyl chloride, plasticizers and other additives

Polyvinyl chloride (PVC) is the most versatile of all plastics and vies with polyethylene as the most widely used. Production figures are given in Table 10.2. Every man, woman or child in developed countries consumed on an average about 15 lb. of PVC in 1971, the differences between the USA and elsewhere being unusually small. The reason for the great versatility of PVC is that it is capable of being plasticized – that is to say that the addition of certain materials called plasticizers to the polymer formulation can produce any desired degree of softness between rigidity and extreme pliability.

246 *Industrial polymers*

In addition to PVC resin and perhaps a plasticizer, a formulation will usually contain a stabilizer and may also contain a lubricant or processing aid, an extender (section 10.3.5) and a filler (section 10.3.6). Furthermore, the polymerization of vinyl chloride to give the original resin will have required peroxide initiators and emulsion or suspension stabilizers (section 16.11). The provision of these additives forms a branch of the plastics industry in itself.

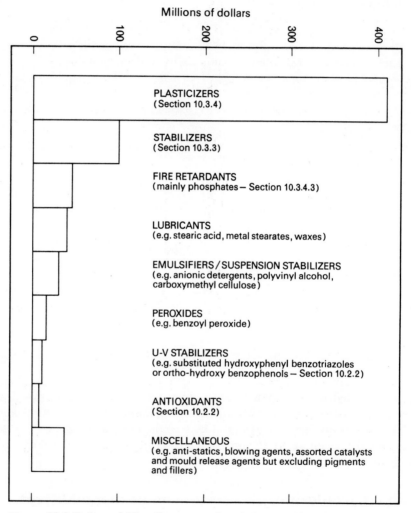

Figure 10.6 Estimated West European sales of plastic additives in 1970.

Plastics additives find their largest outlets into PVC but polypropylene, for example, relies heavily on anti-oxidants and UV stabilizers to prevent its degradation. Figure 10.6 shows estimated sales of plastics additives in Western Europe in 1970. Some of them will be discussed later in this chapter.

10.3.1 Manufacture of vinyl chloride

Vinyl chloride may be manufactured from acetylene by reaction with hydrogen chloride at 5 atm. in the presence of mercuric chloride catalyst. The reaction is exothermic and the system is cooled to keep its temperature between 160 and 250°C. Vinyl chloride boils at $-14°C$ and can be stored either under pressure or under refrigeration:

$$CH{\equiv}CH + HCl \xrightarrow{HgCl_2} CH_2{=}CHCl$$

It may also be made from ethylene in a two-stage process. Ethylene and chlorine are reacted directly together to give ethylene dichloride:

$$CH_2{=}CH_2 + Cl_2 \longrightarrow CH_2Cl.CH_2Cl$$

The ethylene dichloride is then pyrolysed at 500°C and 3 atm to give vinyl chloride:

$$CH_2Cl.CH_2Cl \longrightarrow CH_2{=}CHCl + HCl$$

One economic problem was the choice of raw material. The acetylene-based route consumes 1 mole of HCl for every mole of vinyl chloride produced, whereas the ethylene-based route produces 1 mole of HCl and consumes one of chlorine per mole of vinyl chloride. During the 1960s, many manufacturers had the idea of using both processes so that the HCl produced in the ethylene route was consumed in the acetylene route, resulting in complete utilization of the available chlorine. This was described as a 'balanced chlorine economy'.

Recently, with the drop in ethylene prices, it is doubtful whether the use of acetylene as raw material is economic even in such a system as this. It is probably better to use the ethylene-based process entirely and to convert the by-product HCl back into chlorine by some variation of the Deacon process (section 16.10.3). Various oxychlorination processes have also been developed in which the ethylene dichloride is produced from ethylene, hydrogen chloride and air, thus eliminating a whole stage from the process:

$$CH_2{=}CH_2 + \tfrac{1}{2}O_2 + 2HCl \longrightarrow CH_2Cl.CH_2Cl + H_2O$$

Finally, there are two processes in which naphtha is cracked to an equimolar mixture of ethylene and acetylene as the feedstock for a 'balanced chlorine economy'. A comparison of the economics of the processes is given in Fig. 10.7.

Vinyl chloride can be polymerized by any of the usual methods but emulsion polymerization under pressure in the presence of benzoyl peroxide or a redox system is the most usual:

$$CH_2{=}CHCl \longrightarrow {\left[CH_2{-}\underset{\underset{Cl}{|}}{CH} \right]}_n$$

A two-stage polymerization using azobisisobutyronitrile as initiator is reported to have certain advantages and to be gaining in popularity.

PVC is a white fluffy powder when manufactured by the above methods and it can be moulded into a rigid block. It is hard but not brittle and can be processed by almost all the methods developed for thermoplastics. Its C–Cl dipoles give it a moderate tendency towards crystallinity and it tends to be opaque or slightly translucent.

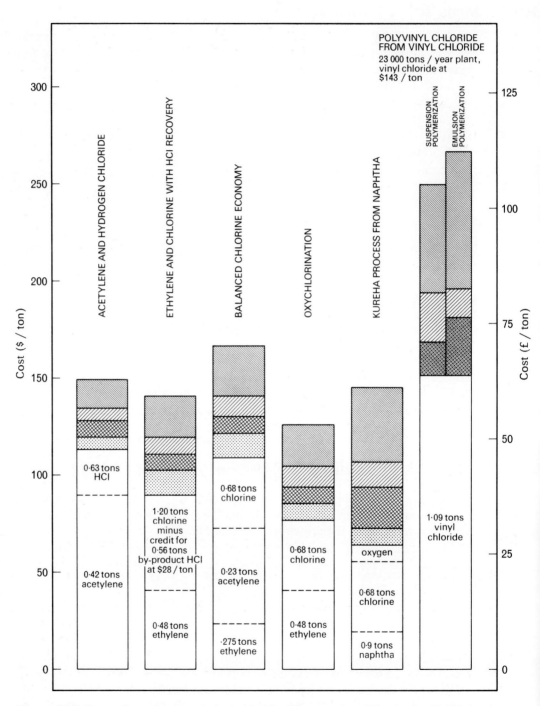

Figure 10.7 Comparison of routes to vinyl chloride. Diagram gives US costs for 45 000 tons/year plant assuming captive acetylene at $220/ton, ethylene at $88/ton, captive chlorine at $55/ton and hydrogen chloride at $30/ton.

10.3.2 End-uses

The uses to which PVC is put are shown in Table 10.6 together with the market shares. The table is subdivided into uses for plasticized or flexible PVC and for rigid, unplasticized material.

Table 10.6 PVC end-uses 1971

	USA (m. lb)	(%)	UK ('000 tons)	(%)
Plasticized (flexible) PVC				
Calendered sheet and film	600	18·3	47·8	13·5
Fabric and paper coating	227	6·9	27·4	7·7
Flooring	385	11·7	28	7·9
Coal board belting*	—		5·5	1·6
Wire and cable coating	370	11·3	44	12·4
General extrusions	35‡	1·1	15·9	4·5
Dipping, slush moulding, etc. (plastisols)	140	4·3	13·7	3·9
Footwear	115	3·5	13·5	3·8
Rigid PVC				
Sheet and film	130	4·0	16·3	4·6
Extrusions (pipe and conduit, etc.)	659	20·0	80·7	22·7
Gramophone records	125	3·8	14·4	4·2
Blow moulding (bottles)	70	2·0	7·5	2·2
Exports	179	5·4	19·2†	5·5
Other injection moulding	124§	3·8	10·9	3·2
Miscellaneous	127	3·9	8·3	2·3
Total consumption	3 286	100·0	355	100·0
Total production	3 320		335	

* The UK National Coal Board permits the use of PVC conveyor belting with a phosphate (non-inflammable) plasticizer for use in coal mines.
† Exported compound, i.e. material already blended.
‡ Hosepipe alone.
§ Rigids only.
Attempts to relate US to UK statistics showed certain internal inconsistencies in the former and data should be treated with caution.

None of the outlets for flexible PVC is growing very rapidly with the exception of footwear. Calendered sheet and coated paper and textiles find their way into packaging, upholstery in cars and in the home, vinyl wall-coverings, rainwear, shower curtains, PVC leathercloth etc.

Flooring is a major outlet for PVC. Cheap, brittle floor tiles are made from fillers (section 10.3.6) and asbestos bound together with a small quantity of PVC and plasticizer. The asbestos provides the strength. Expensive, flexible PVC tiles are only lightly filled and are quite resilient.

Continuous flooring is made by the laminating together of several base sheets of cheap reclaimed PVC and a top printed sheet. This is then covered with a 'wear' layer of crystal clear PVC. The layer is specially formulated to resist absorption of softening oils from black rubber heels etc. which would otherwise produce yellow/brown traffic stains. These laminates are quite cheap and compete with linoleum.

'Belting' covers the end-use in coal-mine conveyor belting. PVC belting is uneconomic compared with rubber, but was specified for use in coal mines by the National Coal Board following the Cresswell colliery disaster in 1950 when a rubber belt caught fire in a ventilation passage with considerable loss of life. Most pits were re-equipped in the early 1960s. Only replacement quantities are used now and with the run-down of the coal industry, the market is static or declining.

The main use for PVC in the electrical industry is wire and cable coating. Polyvinylchloride is extruded onto the wire with the aid of a cross-head die (section 16.12.2) and provides both electrical insulation and protection against corrosion.

'General extrusions' covers a multitude of uses of which garden hose and related PVC tubing is the most familiar. Plastisol uses include PVC gloves and children's toys.

'Footwear' is easily the most rapidly growing outlet for flexible PVC. The material is finding its way into the uppers, the insides and the soles of many cheap and middle priced shoes, and has even had an impact on the fashion trade where its shiny 'patent leather' look gives it a certain appeal, especially as it keeps cleaner and wears better than patent leather. A large amount of PVC footwear is also being exported to Africa where there is a large and scarcely tapped market.

The market for rigid PVC is growing much faster than that for the flexible material but still accounts for only about one-third of consumption. Rigid film and sheet goes into packaging and into credit cards. Rigid extrusions and injection moulding covers major outlets in construction, particularly in pipes (extruded) and the associated fittings (injection moulded), siding, windows, profiles (e.g. curtain rail) and drainpipes and guttering where these are permitted.

Gramophone records are an outlet for PVC which will be readily recognized. They are internally plasticized with about 10% vinyl acetate (section 10.3.4.6) and contain carbon black as a filler and for its anti-static properties. The PVC long-playing gramophone record burst on the world in the early 1950s and rapidly ousted the traditional 78 rpm shellac disc. It can thus be held at least partly responsible both for the 'teen-age' thing and the 'Hi-Fi' thing and execrated or praised according to one's taste. Classical $33\frac{1}{3}$ rpm records are normally made from 'virgin' PVC by pressing with an appropriately grooved matrix; 'pop' 45 rpm records are usually pressed from reprocessed scrap.

Blow moulding of bottles is a growing market in the UK as the throw-away container becomes more popular with manufacturers if not with ecologists. In the USA this market seems to have levelled out.

In the future it seems likely that the use of rigid PVC in construction industries will continue to be the fastest growing market, with the use in footwear helping to boost the figures for flexible PVC. Meanwhile, PVC is so widely used that any increase in economic activity will result in increased consumption of both the rigid and the flexible plastic.

10.3.3 Stabilizers for PVC

Polyvinyl chloride is not very stable to heat and light. It tends to degrade spontaneously, giving off HCl:

$$\{CH_2-CHCl\} \longrightarrow \{CH=CH\} + HCl$$

The process is a chain reaction and once a single C—Cl bond has been broken in a polymer chain, successive units of HCl will be eliminated to leave a polyene:

$$-CH_2-CH-CH_2-CH-CH_2-CH- \xrightarrow[\text{atom attack}]{\text{Initiation possibly by Cl}} -CH_2-\dot{C}H-CH_2-CH-CH_2-CH- + Cl \longrightarrow$$
$$\;\;|\;\;|\;\;|\phantom{HH- \xrightarrow[\text{atom attack}]{\text{Initiation possibly by Cl}} -CH_2-\dot{C}H-CH_2-}\;\;|\;\;|$$
$$ClClCl\phantom{H- \xrightarrow[\text{atom attack}]{\text{Initiation possibly by Cl}} -CH_2-\dot{C}H-CH_2-}ClCl$$

$$-CH_2-CH=CH-\dot{C}H-CH_2-CH- + HCl \longrightarrow -CH_2-CH=CH-CH=CH_2-\dot{C}H- \text{ and so on } + HCl$$
$$\phantom{-CH_2-CH=CH-\dot{C}H-CH_2-}\;\;|$$
$$\phantom{-CH_2-CH=CH-\dot{C}H-CH_2-}Cl$$

The evolution of even small quantities of hydrogen chloride from a domestic plastic is undesirable. In addition, the residual polyene is a chromophore (section 12.2) and absorbs strongly causing the plastic to yellow and eventually blacken.

The chlorine atom liberated at the initiation stage can attack another polymer chain and start the whole process over again. PVC is therefore always used with a stabilizer such as basic lead carbonate added at the 0·1–1% level. This absorbs any chlorine atoms liberated to give insoluble lead chloride, thus breaking the chains and inhibiting the decomposition. Spontaneous initiation, by itself, is negligible.

In many end-uses, such as toys which are liable to be sucked by the child, the presence of lead compounds is undesirable and non-toxic stabilizers are used instead. These are complex mixtures usually containing cadmium, barium or zinc 'soaps' (stearates, palmitates etc.) plus an epoxidized oil such as epoxidized soya bean oil. Such oils are the triglycerides of long chain fatty acids (linoleic acid, stearic acid etc.) which contain double bonds. In the epoxidation process, these are replaced by epoxy groups which also react readily with chlorine atoms. In addition to their stabilizing action, epoxidized oils act as plasticizers. Tin stabilizers are also used and the non-toxic 'package' may contain antioxidants and light stabilizers.

Sometimes lead and epoxy stabilizers are used together and are said to have a synergistic effect, i.e. the total stabilizing effect of the two additives is more than the sum of the individual effects.

In addition to stabilizers, metal soaps such as calcium stearate are often added at a level of about 0·5 per cent in order to improve processing. Lead stearate serves both purposes.

10.3.4 Plasticizers

A plasticizer is a material which tends to soften or decrease the brittleness of a material. Normally it does this by coming between the polymer chains and thus weakening the intermolecular forces of attraction between them. It also acts as a 'lubricant' so that they can slide more easily over one another. If, for example, 50 parts of dioctyl phthalate

is added to 100 parts of polyvinyl chloride (described as 50 phr of dioctyl phthalate; phr = parts per hundred resin) then a soft pliable material is obtained which can be processed further by any desired method. One can think of the plasticized polymer as a solution of the plasticizer in the solid.

In order for a plasticizer to dissolve in a polymer, certain conditions, which are similar to those which determine whether or not two liquids will be miscible, have to be satisfied. First, the polymer must be polar otherwise it will not interact with the plasticizer; it must not be too polar otherwise it will prefer to interact with itself. Thus polyethylene and polypropylene cannot be plasticized in this way; neither can nylon.

Second, the plasticizer must also be polar, but not too polar, for similar reasons. It may be thought of as solvating the polar groups in the polymer. The most suitable polarity appears to be that of the ester group. Phosphates are also suitable. Carbonyl compounds, ethers and nitriles appear suitable in theory but are rarely used; chlorinated hydrocarbons are used as secondary plasticizers (section 10.3.5).

Plasticizers which will indeed dissolve in resins are said to be compatible with them. In practice, they are normally compatible only in certain proportions and these must, of course, be known before a plasticizer can be used.

Certain qualities are looked for in a plasticizer. They are never found simultaneously in a single plasticizer and the relative importance attached to each of them depends on the end use to which the plasticized resin will be put.

These characteristics are:

1. Wide compatibility – the plasticizer should be miscible with the resin in as wide proportions as possible.
2. Low vapour pressure – otherwise the plasticizer will evaporate during milling (when it is heated) or during subsequent use of the plastic which will then go brittle.
3. Resistance to migration – for example, if a PVC tablecloth is placed on a table, the plasticizer in it must not migrate into the nitrocellulose lacquer on the table.
4. Chemical stability – the plasticizer should be chemically stable and resist corrosion.
5. High plasticizing efficiency – a given amount of plasticizer should have the greatest possible softening power. This requirement is a consequence of the fact that plasticizers are more expensive than resin. If the reverse were true, one would want to incorporate as much plasticizer as possible in a compound of given softness.
6. Retention of flexibility of compound at low temperatures – when a plasticized resin is cooled, it tends to go hard and brittle. Certain plasticizers, e.g. adipate esters, are less affected than others, and this is of value in stormwear, etc. Similarly, the viscosity of a plasticizer should vary as little as possible with temperature.
7. Light stability – many plasticizers go yellow on prolonged exposure to light which makes them useless in clear sheet.
8. Good electrical characteristics – as plasticizers are used in insulating material this means a high volume resistivity. For a typical sample of dioctyl phthalate this would be 10^{14} ohm cm but apparently trivial differences in processing can affect this by a factor of 10^3 or 10^4.
9. Ability to make the compound non-inflammable – this is of obvious importance in conveyor belting in coal mines, but is also important in building materials, etc.
10. Odourless, tasteless and of low toxicity.
11. Resistance to extraction by water or organic solvents – for example, the plasticizer in baby pants must remain in the material in spite of repeated use and subsequent washing in hot soapy water.
12. Low price – plasticizers can usually be made by a simple esterification process in relatively inexpensive plant. The market is therefore extremely competitive and customers are very cost conscious.

Polyvinyl chloride, plasticizers and other additives 253

A variety of plastics can be plasticized including cellulose nitrate (with camphor), cellulose acetate (with dimethyl or diethyl phthalate or aryl phosphates), some synthetic rubbers and polyvinyl acetate (section 10.5.1). About 90% of all plasticizers, however, go into PVC and only this use and the use in polyvinyl acetate will be discussed further.

10.3.4.1 General purpose PVC plasticizers and their molecular structure It was found fairly easily that the ester grouping resulted in the best plasticizers. As a general guide, one can say that the ester grouping confers the compatibility while the hydrocarbon groupings of the acid and alcohol involved reduce the volatility of the compound but also reduce its plasticizing efficiency. Esters of monobasic acids turn out to be almost useless as plasticizers because as their molecular weight increases, they become incompatible with PVC before their volatility has dropped to an acceptable value. This problem may be surmounted by the use of esters of di- and tribasic acids. The cheapest and most readily available source of such esters is phthalic anhydride which can readily be esterified to give esters of phthalic acid. About three-quarters of all plasticizers are in fact phthalates (referred to as higher phthalates to distinguish them from dimethyl and diethyl phthalates which are used with cellulose acetate) and Table 10.7 shows plasticizer consumption by type of ester.

Table 10.7 Plasticizers*

	USA 1969				UK	
	Production m. lb.	Sales m. lb.	Value ¢/lb.	Price ¢/lb 4/72	Production 1970 '000 tons	Price £/ton 4/72
Total plasticizers	1 382	1 275	21	—	—	—
Total organic phthalates	884	821	15	—	95·1	—
Diethyl	22·5	18·6	19	21	—	180·6
Dibutyl	34·5	30·0	18	22	—	159·9
Di-2-ethylhexyl	355	340	13	12	—	179
Di-isooctyl	83·0	70·4	18	12	—	138·8
Di-'mixed octyl'	—	—	—	—	—	139
Di-*n*-octyl-*n*-decyl	54·5	49·4	18	20	—	—
Di-isodecyl	137	126	14	12·5	—	—
Di-tridecyl	22·0	21·6	22	20	—	—
Total phosphates	18·9	14·9	46	34†	25·8	249†
Total epoxidized oils	104	94·7	26	—	—	—
Epoxy soya oils	73·5	67·1	26	—	—	—
Total polymeric plasticizers	54·0	48·7	38	—	—	—
Total adipates	65·8	60·5	25	—	—	—
Di-2-ethylhexyl	40·1	37·1	24	25	—	—
Chlorinated paraffin extenders	61·9	59·1	13	10·5	—	—

* Some items in this table are not used exclusively as plasticizers.
† Tricresyl phosphate.

The point made above about plasticizing efficiency against volatility can be illustrated by considering the plasticizing properties of dialkyl phthalates as the homologous series is ascended. Figure 10.8 shows the 'BS softness' and the 'AC volatility' for various phthalate esters added at a level of 67 phr. Data are plotted against the total number of

254 *Industrial polymers*

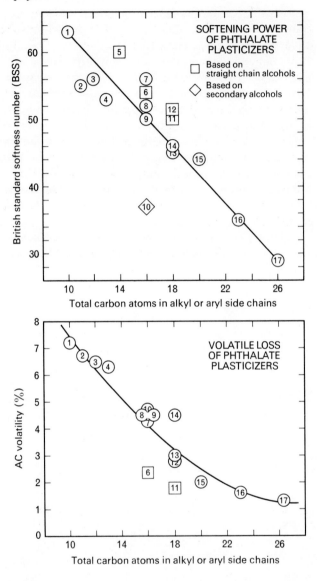

Key: Phthalate plasticizers

No.		No.	
1	Butyl benzyl	10	Di-capryl
2	Butyl octyl	11	Straight chain C_8/C_{10}
3	Di-isohexyl	12	n-octyl-n-decyl
4	Butyl decyl	13	Isooctyl decyl
5	Di-n-heptyl	14	Di-nonyl
6	Straight chain C_6/C_{10}	15	Di-isodecyl
7	C_7/C_9 mixed alcohols	16	Tridecyl isodecyl
8	Di-2-ethylhexyl	17	Di-tridecyl
9	Di-isooctyl		

Figure 10.8 Softening power and volatility of dialkyl phthalate plasticizers.

carbon atoms in the side chains. BS softness is calculated from the indentation made in a sample of compound by a 500 g. weighted, 2·38 mm diameter steel ball; AC volatility is the percentage of plasticizer lost when the compound is heated for 24 hours at 90°C surrounded by activated carbon.

The lower chain length esters, e.g. diethyl and dibutyl, have excellent solvating power but poor low temperature properties and high volatility. Long chain length esters, e.g. ditridecyl, have much lower efficiencies but also low volatilities. The medium chain length esters represent a compromise between these factors and the main general purpose plasticizers in the USA are di-(2-ethylhexyl) phthalate otherwise known as dioctyl phthalate (DOP) and di(2-methylheptyl) phthalate otherwise known as diisooctyl phthalate (DIOP). In the UK a mixed phthalate, di-alphanol phthalate, accounts for about two-thirds of the market with DIOP taking the other third and with the manufacture of DOP just beginning. Di-alphanol phthalate is made from a mixture of C_7–C_9 primary alcohols called Alphanol which is made by application of the OXO process to a refinery product called 'slack wax'. The other main plasticizer alcohols – isooctanol, isodecanol and tridecanol – are made similarly by the action of CO and H_2 on the appropriate olefines, but manufacture of 2-ethylhexanol differs somewhat (section 9.2.4).

The general purpose plasticizer market is currently being attacked by phthalates made from straight chain alcohols. These include Alfol 610 and 810 which are mixtures of C_6–C_{10} and C_8–C_{10} straight chain primary alcohols made by a variation of the Ziegler process, and the Linavols which are similar materials made by wax cracking. These materials appeared on the market as by-products of the processes which make straight chain olefines as feedstocks for biodegradable detergents (section 11.4.2). As can be seen from Fig. 10.8, straight chain phthalates have higher efficiencies and lower volatilities (and also better low temperature properties) than the traditional branched chain phthalates. They will stand a small premium in price since smaller quantities can be used to achieve a desired softness, but a premium of about 3¢/lb. (£28/ton) restricts their market. At the same price as traditional phthalates, they would undoubtedly capture a very large share of the general purpose plasticizer market, supposing adequate supplies to be available.

One interesting straight chain alcohol which is available in limited quantities and does not arise as a by-product from detergent manufacture is n-heptanol which is a by-product of the manufacture of nylon 11 from ricinoleic acid in France.

10.3.4.2 Specialized phthalate plasticizers Of the phthalates below di-octyl phthalate in the homologous series, only dibutyl and di-n-heptyl phthalates are of real commercial importance. The former used to be used as a plasticizer in polyvinyl acetate paint formulations, and is now used in some surface coatings and in conjunction with DOP in PVC plastisols (section 16.12.2), where very swift jelling of the mixture is required, as in the manufacture of foamed PVC. It is too volatile for use in formulations where hot milling takes place.

Butyl benzyl phthalate is the usual plasticizer for the top transparent layer of high quality floor tiles where resistance to staining by tar, asphalt, lipstick, black heel marks, etc., is required. Its resistance to staining appears in some way to be related to the presence of two aromatic rings in the molecule, since the property is also shown by diethylene glycol dibenzoate which competes very well with butyl benzyl phthalate in everything except price. The latter is one of the few plasticizers based on a monocarboxylic acid but, of course, has a dihydric alcohol to make up for it.

Butyl benzyl phthalate

Diethylene glycol dibenzoate

Long chain phthalates such as di-tridecyl phthalate are used where low volatility is required, e.g. in heating cables or car interiors where, on a hot day, volatilization of plasticizer and subsequent condensation on the windscreen can seriously reduce visibility.

10.3.4.3 Phosphate ester plasticizers Phosphate plasticizers are organic esters of phosphoric acid. They have excellent flame resistance and high efficiency and compatibility but poor low temperature properties and heat and light stability. Tricresyl and cresyl diphenyl phosphates are the two main esters at present but they may be replaced by alkylphenol–phenol mixtures.

10.3.4.4 Epoxidized oil and polymeric plasticizers Epoxidized oils are natural oils (soyabean oil, linseed oil, tall oil) in which a double bond has been replaced by an epoxy group, by the action of peroxyacetic acid. They have low volatility and good heat and light stability but are expensive and are more usually used as stabilizers (section 10.3.3).

Polymeric plasticizers are polyesters made from a dibasic acid (adipic, azelaic or sebacic) and a dihydric alcohol (propylene or ethylene glycol):

$$HOOC(CH_2)_4COOH + CH_3CHOH.CH_2OH \longrightarrow$$

Adipic acid　　　　　　Propylene glycol

$$HOOC(CH_2)_4CO\left[.O.\underset{CH_3}{CH}.CH_2O.OC(CH_2)_4CO.\right]_n O.\underset{CH_3}{CH}.CH_2OH + nH_2O \text{ etc.} \ldots$$

Polypropylene glycol adipate

The chains are 'stopped' and the chain length therefore controlled by addition of a small amount of a monofunctional acid or alcohol, e.g. butanol:

$$\ldots(CH_2)_4COOH + HO.C_4H_9 \longrightarrow \ldots(CH_2)_4COOC_4H_9 + H_2O$$

Butanol

The molecular weight is generally between 1 000 and 7 000. Polymeric plasticizers have very low volatilities and show low extractability and migration, but they have very high viscosities which cause processing difficulties and poor low temperature properties. They are usually used in conjunction with long chain phthalates or low temperature plasticizers.

10.3.4.5 Plasticizers based on aliphatic acids Ranges of plasticizers are available based on dibasic aliphatic acids. Adipic acid is easily the most significant, with azelaic acid second and sebacic acid a poor third. Esters with plasticizer alcohols (e.g. di-(2-ethylhexyl)adipate) have very good low temperature properties and compounds containing them remain flexible under conditions where compounds containing general purpose plasticizers would crack and split. They are frequently used in a blend with general purpose plasticizers. Esters of all three acids have very similar properties but adipic acid is the cheapest, being made on a large scale for the manufacture of nylon 66.

10.3.4.6 Internal and external plasticizers All the plasticizers so far considered are materials which are added to the resin. As additives, they can never be completely non-migratory and there must inevitably be a certain amount of volatile loss no matter how carefully the formulation is designed. As the chemistry of polymer systems became more fully understood, it was suggested that a polymer might be plasticized by the addition of another monomer at the time of polymerization so that a random copolymer might be obtained. By varying the proportions of the two monomers any desired degree of softness intermediate between the softnesses of the two homopolymers can be achieved.

A monomer which is copolymerized with another resin in order to give a softer copolymer is called an internal plasticizer, because the plasticizer molecule becomes an integral part of the polymer chain. Similarly the materials already discussed, which are not incorporated into the polymer molecule, are called external plasticizers.

The best example of these two types is in the case of vinyl acetate emulsion paints where dibutyl phthalate, an external plasticizer, was replaced by 2-ethylhexyl acrylate, an internal plasticizer (section 10.5.1).

Vinyl chloride itself can be copolymerized with vinyl acetate to give a copolymer which is softer than PVC and is used in gramophone records. On a molecular level, the bulky acetate groups of the vinyl acetate keep the C—Cl dipoles of the vinyl chloride at a distance and thus the intermolecular attraction is reduced and the material is softer. In this case, unlike the case of emulsion paint, vinyl acetate is the internal plasticizer rather than the resin.

10.3.5 Extenders

It was stressed in section 10.3.4 that a plasticizer has to be compatible with the resin it is to plasticize. It need not necessarily be compatible in all proportions, but the plasticizer must dissolve to some extent in the resin.

Other compounds exist which, though they are incompatible with PVC, *are* compatible with plasticized PVC and add a softening effect to that of the plasticizer already present. Such materials are called extenders or secondary plasticizers, the true plasticizers then being called primary plasticizers.

The best known extenders for PVC in the UK are the Cereclors, a range of chlorinated hydrocarbons marketed by ICI and made originally from paraffin wax but now from *n*-alkanes. They do not fit the definition of secondary plasticizers given above in that they *are* compatible with unplasticized PVC, but confer virtually no softness unless used with a primary plasticizer. Figure 10.9, for example, shows that addition of 33 phr of Cereclor S52 to unplasticized PVC does not even give a compound with British Standard Softness (B.S.S.) 10, whereas addition of the same amount to a compound containing 70 phr DIOP raises the B.S.S. from 60 to 80. Extenders are cheaper than plasticizers as they must be if anyone is to be persuaded to replace part of their plasticizer by them. Apart from cost their only advantage is that they are less inflammable than primary plasticizers and indeed their compatibility limits are possibly not quite as good as is claimed. It is said that if a plastic is heavily stressed over a long period, the extender may well exude from the resin. Nonetheless, in end-uses such as wire-coating where price is critical, extenders are almost universally used.

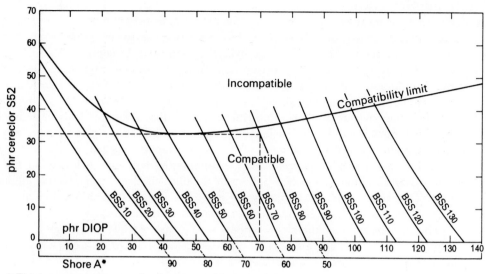

Figure 10.9 Compatability and softness of PVC compounds plasticized with blends of di-iso octyl phthalate and 'Cereclor' S52. A typical compatability curve – 'Cereclor' S52 used in conjunction with DIOP – is presented in the graph above in which the 'Cereclor' content (expressed as parts per hundred parts of resin, that is phr) of the plasticizer blend is plotted along the vertical axis and phthalate content is plotted along the horizontal axis. The series of parallel curves represents the loci of points of equal softness.

10.3.6 Fillers

Fillers are cheap inorganic materials such as asbestos, woodflour, precipitated chalk and glass fibre which are added to polymers to give them certain desirable tensile or electrical properties or merely to cheapen them. Carbon black is used in gramophone records partly as an anti-static and asbestos and chalk are used in cheap floor tiles, the former to confer strength and the latter cheapness. Fillers are used extensively with thermosetting resins.

10.4 Thermosetting plastics

All the polymers discussed in the previous section were thermoplastic. They could be melted and frozen repeatedly and consisted largely of linear chains. Thermosetting resins on the other hand degrade when heated and are made from polyfunctional monomers i.e. a unit of monomer is able to form more than two bonds with neighbouring units. The structure of thermosetting resins is elaborately cross-linked and the units of monomer form a three-dimensional network. Market data on the commoner thermosets are given in Table 10.8.

Table 10.8 Thermosetting plastics

	USA 1971 Production (m. lb.)	USA 1971 Sales (m. lb.)	USA 1971 Price (¢/lb.)	UK 1971 Production ('000 tons)	UK 1971 Consumption ('000 tons)	UK 1971 Price £/ton
Phenolics	1 067	862	22‡	74·0	38·1¶	196
Amino-plasts	679	622	19‡	169	145	
Urea–formaldehyde resins	n.a.	571†	—	—	—	225§
Melamine–formaldehyde resins	n.a.	51*	—	—	—	354
Unsaturated polyesters	810	639	320	37·6	37·5	340
Polyurethane foams	930	908	55‡	48·0	47·0	—
Rigid	n.a.	276	—	7·0	7·0	—
Flexible	n.a.	632	—	41·0	40·0	—
Epoxy resins	165	151	51‡	13·5	9·7	—
Alkyd resins‖	610	290	28‡	63·0	38·6¶	—

* Moulding powders only.
† Includes M/F impregnating resins for decorative laminates and improved particle board. In W. Europe these form 40% of M/F market, so by analogy they should account for about 34 m. lb. in USA.
‡ Unit values 1969.
§ Cellulose-filled moulding powder.
‖ See section 10.5.
¶ 1970 sales.

10.4.1 Phenoplasts

The name 'phenoplasts' covers the thermosetting resins made from phenols, or occasionally cresols, and aldehydes, usually formaldehyde. The manufacture of phenol was outlined in section 9.5.1.1 and its end-uses are shown in Table 10.9. Formaldehyde production is to be found in section 14.1.7.

Phenol–formaldehyde (P/F) resins were discovered by Baeyer in 1872 but were first studied thoroughly in 1909–26 by an American, Baekeland, hence the trade name Bakelite. At the outbreak of the Second World War, P/F resins accounted for about 60% of the synthetic plastics production of the UK – about 15 000 out of 25 000 tons, the balance being mainly UF resins and polymethyl methacrylate. In 1971 production of phenolics was 74 000 tons having risen from 53 800 in 1962. This represents a rather low growth rate of 3% per year, and is the pattern one would expect from a material which had reached economic maturity. More specific reasons why phenoplasts have not shared in the boom which has affected other plastics are that they are difficult to process and one cannot achieve the production speeds possible with injection moulded thermoplastics, their raw materials have been less affected by the petrochemical revolution, and their use is limited by their dark colour.

P/F resins divide into two main groups which are made by acid and alkaline catalysis.
10.4.1.1 Alkaline catalysed P/F resins When phenol and a concentrated aqueous solution of formaldehyde, known as formalin, are mixed in a molecular ratio of 1 : 1·5 or more, in

260 *Industrial polymers*

the presence of an alkaline catalyst such as sodium hydroxide, condensation polymerization takes place to give a resin known as a resol which is fusible, and soluble in ethanol or acetone. Further heating gives a resin which is fusible but insoluble, and eventually an insoluble, infusible resin is obtained.

The first stage of the reaction may be represented as the formation of the methylol derivatives of phenol. The hydroxyl group on the phenol activates the ortho and para positions on the ring and in this case the ortho positions are slightly preferred:

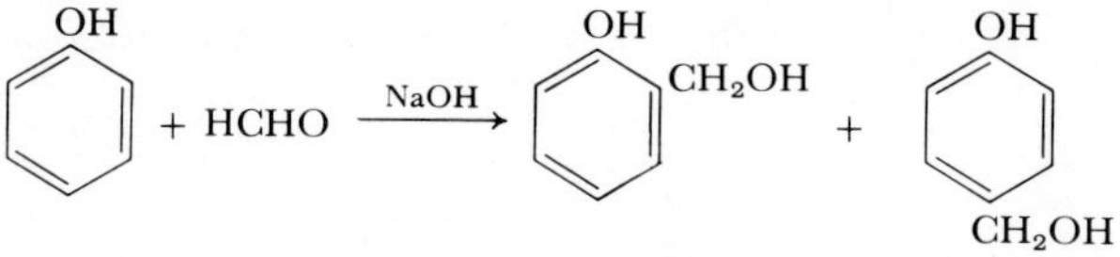

[Reaction scheme: phenol + HCHO, NaOH → saligenin (o-hydroxymethylphenol) + homosaligenin (p-hydroxymethylphenol)]

and also [2,6-bis(hydroxymethyl)phenol, 2,4-bis(hydroxymethyl)phenol, and 2,4,6-tris(hydroxymethyl)phenol structures shown]

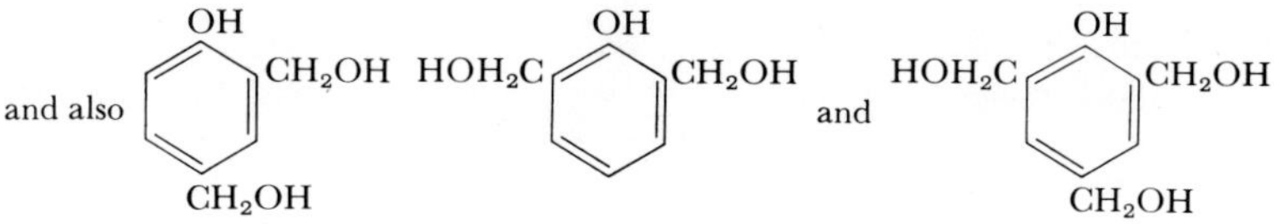

The methylol phenols condense with one another losing water to give low molecular weight linear polymers of somewhat random structure – these are resols:

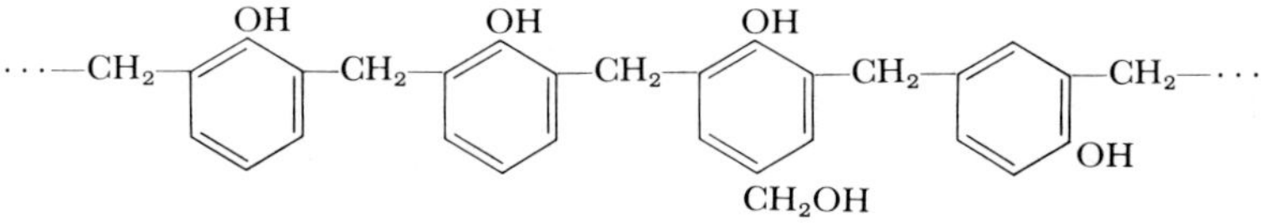

[Linear resol structure with phenol rings linked by CH₂ bridges, bearing OH and CH₂OH substituents]

Some oxygen bridges of structure

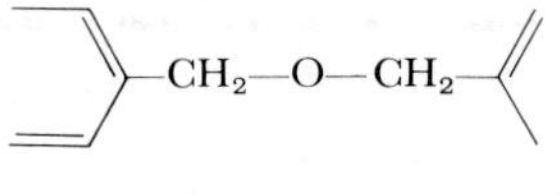

$$\rangle\!\!-\!CH_2\!-\!O\!-\!CH_2\!-\!\langle$$

also occur. The progress of the condensation can be followed by measurement of the amount of steam given off.

If the resols are heated further, they condense with one another giving a cross-linked polymer which at a certain point becomes insoluble and eventually cross-links further and becomes infusible:

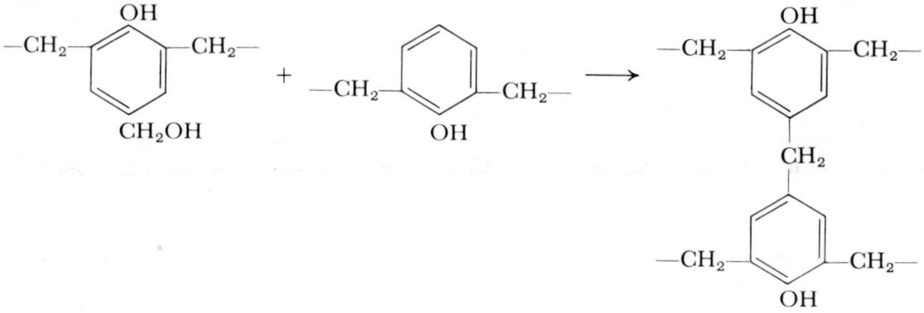

[Cross-linking reaction scheme showing methylol-bearing phenol unit + phenol unit → cross-linked network with CH₂ bridge linking rings at additional positions]

It should be noted that the hydroxyl groups of the phenol are not involved in bond formation – they simply activate the ortho and para positions on the benzene ring.

It will be seen that the final plastic has been made from a single mixture of phenol and formaldehyde. It is therefore called a one-stage resin. One-stage resins are used as plywood adhesives, for making laminates (e.g. Formica), for high shock resisting moulding materials and for stoving lacquers.

10.4.1.2 Acid catalysed P/F resins If the P/F resin is made using an acid catalyst and an excess of phenol (e.g. sulphuric acid; phenol : formaldehyde > 7:6) long straight chain polymers are obtained in which the phenol rings are joined only by methylene bridges. These are called novolacs. The simplest is dihydroxydiphenylmethane.

$$HO-\langle\bigcirc\rangle-CH_2-\langle\bigcirc\rangle-OH$$

Novolacs differ from resols in having no free methylol groups. This is a consequence of an excess of phenol in the reaction mixture. A typical structure is:

[Structure showing chain of phenol rings linked by CH_2 bridges with OH groups]

Novolacs will not cross-link by the action of heat alone because of their lack of methylol groups. They are fusible solids and because of their phenolic hydroxyl groups, they are soluble in alkali.

They are sold and stored in this form. When required, they are treated with more formaldehyde under alkaline conditions or with 'Hexa'. 'Hexa' is hexamethylene tetramine, a condensation product of formaldehyde and ammonia:

$$6HCHO + 4NH_3 \longrightarrow [\text{hexamethylene tetramine}] + 6H_2O$$

In the heated mould, this reaction goes in the opposite direction providing both formaldehyde and alkaline conditions.

The novolac, treated in one of these ways, reacts with the extra formaldehyde to give a polymer which contains free methylol groups:

$$-CH_2-\underset{}{\underset{}{C_6H_3(OH)}}-CH_2-\underset{}{\underset{}{C_6H_3(OH)}}-CH_2- \;+\; HCHO \longrightarrow \; -CH_2-\underset{}{\underset{}{C_6H_3(OH)}}-CH_2-\underset{}{\underset{CH_2OH}{C_6H_2(OH)}}-CH_2-$$

and under the influence of heat this material cross-links in exactly the same way as did the resol.

The resin resulting from the above process is called a two-stage resin. Two-stage resins are used as general purpose moulding materials. They are used with an equal weight of filler which can be such things as wood flour, disintegrated wood pulp, asbestos fibre, nylon fibre, powdered mica or graphite.

Though novolacs and resols are conventionally made by acid and alkaline catalysis respectively, it is the phenol:formaldehyde ratio which determines the product and it is possible to make novolacs by alkaline catalysis if there were any reason for doing so.

All phenolic resins have a dark colour and this can only be slightly modified by pigments to give the characteristic dark brown, dark green or black phenolic mouldings.

10.4.2 Aminoplasts

The term 'aminoplasts' embraces those thermosetting resins made from an amine and an aldehyde. The important aminoplasts are urea-formaldehyde (U/F) and melamine-formaldehyde (M/F) resins.

The U.K. production of aminoplasts rose from 91 000 tons in 1962 to 169 000 tons in 1971, about twice the size of the phenoplast production. The growth rate (about 7%) is more satisfactory but still below that of the newer thermoplastics (Fig. 2.3).

10.4.2.1 Urea-formaldehyde resins The manufacture of urea is described in section 14.1.6 and its end-uses compared with those of phenol and melamine in Table 10.9. It will react with formaldehyde under slightly alkaline conditions to give methylol ureas:

$$O=C(NH_2)_2 \;+\; HCHO \longrightarrow O=C(NH \cdot CH_2OH)(NH_2) \;+\; O=C(NH \cdot CH_2OH)_2$$

Mono and dimethylol ureas

Upon acidification and heating, condensation occurs to give linear polymers analogous to resols:

$$\underset{NH \cdot CH_2OH}{\underset{CO}{NH \cdot CH_2OH}} \;+\; \underset{NH \cdot CH_2OH}{\underset{CO}{NH \cdot CH_2OH}} \;+\; \cdots \longrightarrow \underset{NH \cdot CH_2OH}{\underset{CO}{NH}}-CH_2-\underset{NH \cdot CH_2OH}{\underset{CO}{N}}-CH_2-\underset{NH \cdot CH_2OH}{\underset{CO}{N}}-\cdots$$

Thermosetting plastics 263

Table 10.9 Uses of phenol, urea and melamine in Western Europe

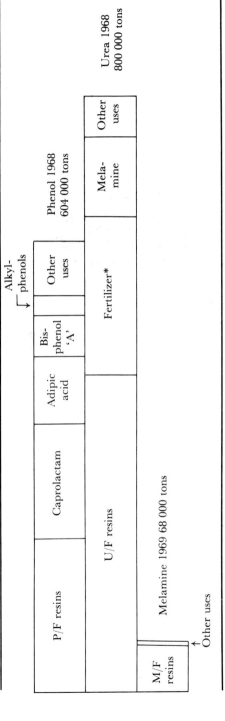

* Urea is much more widely used as a fertilizer in the USA. In 1969 production was almost 6 000 m.lb. of which about 5 000 m.lb. probably went into fertilizers.

264 *Industrial polymers*

and further heating causes cross linking between the methylol groups

$$\begin{array}{c} \sim\!\!N\!\!\sim \\ | \\ CO \\ | \\ NH.CH_2OH \end{array} \;+\; \begin{array}{c} NH.CH_2OH \\ | \\ CO \\ | \\ \sim\!\!N\!\!\sim \end{array} \;\longrightarrow\; \begin{array}{c} \sim\!\!N\!\!\sim \\ | \\ CO \\ | \\ NH \\ | \\ CH_2 \\ | \\ N\!\!-\!\!CH_2OH \\ | \\ CO \\ | \\ \sim\!\!N\!\!\sim \end{array}$$

The mechanism is not fully understood.

U/F resins are always used with fillers, usually bleached paper pulp or woodflour, as their tensile strength in their absence is too low. U/F mouldings are colourless or white, and translucent and light-stable and so, by addition of pigments, they can be fabricated in any desired colour. They are less water resistant than phenolics and prolonged contact with water causes cracking due to its alternate absorption and loss. The mouldings have high dielectric strength and non-tracking properties and are suitable for many electrical applications. They are not as heat resistant as P/F resins as their fillers are liable to combustion.

10.4.2.2 Melamine–formaldehyde resins Melamine is the only chemical, apart from acetylene, to be made from calcium carbide. The latter reacts with nitrogen in a process which is exothermic and self-sustaining but which requires an initiation temperature of 900°C. The product is calcium cyanamide. Fluxes such as calcium chloride or fluoride

Table 10.10 Phenoplasts and aminoplasts – end-uses 1971

	Aminoplasts		*Phenoplasts*	
	USA (m. lb.)	UK ('000 tons)	USA (m. lb.)	UK ('000 tons)
Bonding and adhesive resins for:				
plywood and fibrous and granulated wood	355	42·0	244	n.a.
Foundry or shell moulding	—	—	86	5·9
Friction materials (e.g. brake linings)	—	—	43	7·0
Laminating	47·0	3·7	107	6·2
Other bonding and adhesive resins	—	—	160	n.a.
Moulding powders	107	41·6	295	22·3
Paper treatment and coating resins	48·0	5·9	—	—
Surface coatings	62·0	10·5	35	n.a.
Textile treatment	60·0	4·2	—	—
Other uses	—	11·1	97	n.a.
Total	679	119*	1 067	74·0

* *Brit. Plastics* gives consumption as 119 on p. 65, but production figures differ on pp. 65, 67.

are added to reduce the initiation temperature of the reaction. Acidification of calcium cyanamide with aqueous carbon dioxide solution or dilute sulphuric acid gives cyanamide which dimerizes to dicyandiamide. Treatment with an ammonia catalyst at high temperature and pressure leads to melamine (Fig. 10.10).

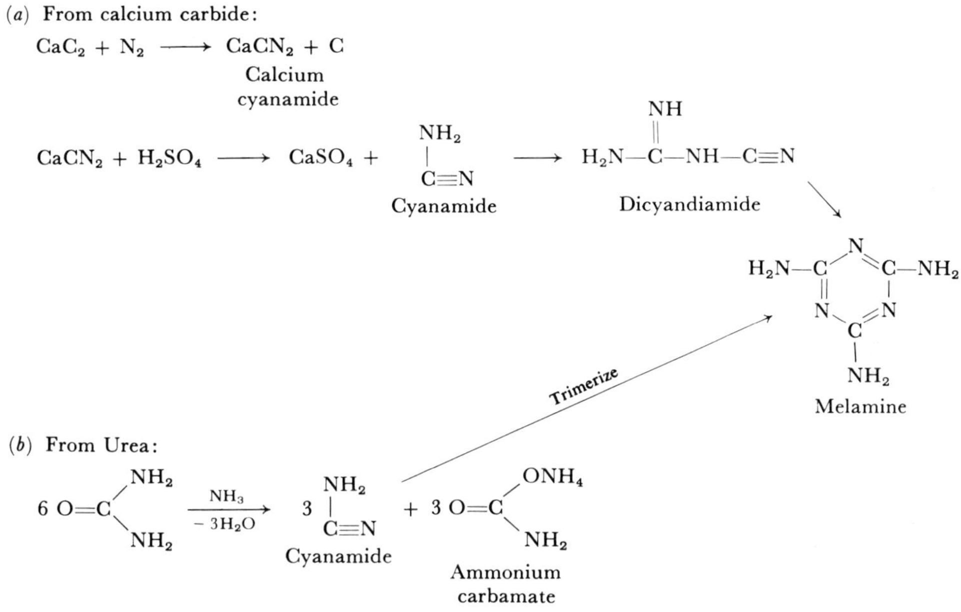

(a) From calcium carbide:

$$CaC_2 + N_2 \longrightarrow CaCN_2 + C$$
$$\text{Calcium cyanamide}$$

$$CaCN_2 + H_2SO_4 \longrightarrow CaSO_4 + \underset{\text{Cyanamide}}{\begin{array}{c} NH_2 \\ | \\ C\equiv N \end{array}} \longrightarrow \underset{\text{Dicyandiamide}}{H_2N-\overset{\overset{NH}{\|}}{C}-NH-C\equiv N}$$

$$\xrightarrow{\text{Trimerize}} \underset{\text{Melamine}}{\begin{array}{c} H_2N-C \overset{N}{\underset{N}{\diagup\diagdown}} C-NH_2 \\ \| \qquad | \\ N \diagdown \underset{C}{\diagup} N \\ | \\ NH_2 \end{array}}$$

(b) From Urea:

$$6\ O=C\begin{array}{c} NH_2 \\ \diagdown \\ NH_2 \end{array} \xrightarrow[-3H_2O]{NH_3} \underset{\text{Cyanamide}}{3\ \begin{array}{c} NH_2 \\ | \\ C\equiv N \end{array}} + 3\ O=C\begin{array}{c} ONH_4 \\ \diagup \\ \diagdown NH_2 \end{array}$$
$$\text{Ammonium carbamate}$$

Figure 10.10 Manufacture of melamine.

Another method starts with urea which is dehydrated in the presence of ammonia to cyanamide and ammonium carbamate. The former can be trimerized to melamine; the latter is reconverted to urea. Though the route is attractive on paper, the theoretical yield of cyanamide is only 35% by weight per cycle, and there are formidable corrosion problems arising from acid deamination products such as cyanuric acid. Allied Chemical operate the process in the USA, and Dutch States Mines do so in Europe, but there is as yet no UK operator.

Melamine reacts with formaldehyde in a manner analogous to urea. Under slightly alkaline conditions the hydrogen on the amine groups is replaced by methylol groups. On acidification and heating, the methylol melamines condense to give linear polymers which eventually cross link in the same way as U/F resins.

M/F resins combine the good qualities of P/F and U/F resins. Like P/F resins they are stable to heat and moisture and can be used continuously above the boiling point of water; like U/F resins they are colourless and stable to light. They can be used with or without a filler as desired.

The main outlet for M/F mouldings is in tableware where it is lighter and less fragile than conventional materials. M/F resins are also used in some decorative laminates, as adhesives in speciality chipboard and in certain automobile surface coatings (section 10.5.5).

266 Industrial polymers

10.4.3 Economic aspects of phenoplasts and aminoplasts

P/F and U/F resins were at one time used almost entirely in moulding powders, which were fabricated into electrical parts, toilet seats, etc. These end-uses are rapidly being overtaken by the employment of P/F and U/F resins as adhesives in plywood, blockboard, chipboard and the other wood-based materials which are coming to replace solid wood. The 1971 end-use pattern is shown in Table 10.10. U/F resins are much more important than P/F resins in the UK and Western Europe; the opposite applies to the USA.

P/F, U/F and M/F resins represent three plastics competing for similar markets. In each individual case, a choice has to be made between them by assessment of financial versus technical factors. Table 10.11 shows the published UK prices of raw materials and moulding powders in 1972:

Table 10.11

	£/ton		£/ton
Phenol	84·6	P/F moulding powder	181·7
Urea	32·2	U/F moulding powder (cellulose filled)	224·9
Melamine	220·8	M/F moulding powder	353·6
Formalin (37% aqueous formaldehyde)	27·5		

An apparent anomaly is that while urea is cheaper than phenol, U/F moulding powders are more expensive than P/F powders. The reason is that fillers of a quality to permit light-coloured U/F mouldings are relatively expensive. If no filler were required, as for example in an adhesive, the U/F resin would be, and is indeed, cheaper.

If price were the dominant factor, therefore, P/F resins would have the moulding powder market while U/F resins would be invariably used in chipboard. This is what actually happens until the situation arises where chipboard is required which will not be damaged by water, or a moulding is required in a pastel colour. At this point, the premium must be paid for P/F adhesives and U/F moulding powders.

If water resistance and pastel colours were required in a single article, a cup and saucer perhaps, then an M/F moulding powder would have to be used and a further premium would need to be paid.

10.4.4 Unsaturated polyester resins

An unsaturated polyester resin is made by reaction of a glycol with a mixture of an unsaturated dibasic acid or anhydride and saturated dibasic acid or anhydride. Usually propylene or ethylene glycol is used with maleic anhydride and phthalic anhydride or adipic or sebacic acids. As the reactants are all bifunctional, the resulting polymer is linear and contains occasional double bonds. Propylene glycol maleate phthalate, for example, would have the approximate structure:

$$-CH_2-\underset{\underset{CH_3}{|}}{CH}-O-CO-\underset{\text{(benzene)}}{\bigcirc}-CO.O-\underset{\underset{CH_3}{|}}{CH}-CH_2-O-CO-CH=CH-CO-O-CH_2-\underset{\underset{CH_3}{|}}{CH}-O.OC-\underset{\text{(benzene)}}{\bigcirc}-CO.O.\underset{\underset{CH_3}{|}}{CH}.CH_2O.OC-$$

Its molecular weight would be in the thousands and it would be a viscous liquid. If styrene or some other unsaturated monomer is now added to the polyester, along with a polymerization catalyst such as benzoyl peroxide, the styrene double bond opens and cross links neighbouring polyester chains to give an insoluble infusible compound:

$$\begin{array}{c}\sim\sim CH=CH\sim\sim\\+\\\sim\sim CH=CH\sim\sim\end{array} + \underset{\text{(benzene)}}{\bigcirc}CH=CH_2 \longrightarrow \begin{array}{c}CH-CH-CH-CH\\|\quad\quad\quad\quad|\\-HC\underset{\text{(benzene)}}{\bigcirc}CH-\end{array}$$

The advantage of polyester resins is that the cross-linking process is not accompanied by the evolution of water. P/F, U/F and M/F resins all have to be cured (i.e. cross-linked) at high temperatures in order that the aqueous by-product be driven off. Unsaturated polyesters, in contrast, will cure in the cold.

Unsaturated polyesters are usually used with a filler and the commonest is glass fibre. In this form, it is used in car bodies, boat hulls and roofing and electricals generally. A car or boat body may be built up fairly easily by the placing of polyester/glass fibre on a mould or former where it is allowed to cure. One side of the moulding has a finish dependent on the quality of the mould; the other is rough and random strands of glass fibre are visible. For short production runs, this is much cheaper than the purchase of and operation of equipment for stamping the bodies out of metal sheet. It is thus a low capital outlay – high labour cost method compared with steel pressing which is high capital outlay – low labour cost. Glass fibre bodies therefore tend to be found on specialized sports cars with a limited market.

A less labour-intensive method of processing polyester resin is dough-moulding. The polyester is mixed with $\frac{1}{4}$ in. sections of glass fibre and a curing agent; this 'dough' is then injection-moulded in the conventional manner (section 16.12.2).

Polyesters can be cross-linked with materials other than styrene. Diallyl phthalate gives a more flexible resin and triallyl cyanurate confers greater heat resistance.

Production of unsaturated polyesters has grown by a factor of three between 1962 and 1968. It is now comparable with aminoplast and phenoplast production.

10.4.5 Polyurethanes

When an isocyanate reacts with a compound containing a hydroxyl group, a urethane linkage is formed:

$$\underset{\text{Isocyanate}}{R-N=C=O} + \underset{\substack{\text{Hydroxylic}\\\text{compound}}}{R'OH} \longrightarrow \underset{\text{Urethane}}{R-NH-\underset{\underset{OR'}{|}}{C}=O}$$

Industrial polymers

If a di-isocyanate and a di-hydroxylic compound are used, a linear polymer results known as a polyurethane:

$$nR(OH)_2 + n\ R'(NCO)_2 \longrightarrow H(-O-R-O-\underset{\underset{O}{\|}}{C}-NH-R'-)_n NCO$$

Polyurethane

A polyisocyanate or a polyol will, of course, give rise to a cross-linked polyurethane so that thermosetting and thermoplastic polyurethanes are both possible and are indeed manufactured.

The commonest di-isocyanate is tolylene di-isocyanate which is sold as a mixture of isomers produced from toluene in a four-stage process:

Toluene → (HNO$_3$/H$_2$SO$_4$) → Dinitrotoluene → (H$_2$) → Diaminotoluene → (COCl$_2$ carbonyl chloride) → Tolylene di-isocyanate

Various diols are used including polyethers, polyesters with terminal hydroxyl groups and hydroxyl bearing oils such as castor oil. The usual practice is to base flexible foams on polypropylene glycol and rigid foams on polyols produced by the action of propylene oxide on glycerol, pentaerythritol etc.

Polyurethanes are used as lacquers for cloth, paper and leather treatment, in polyurethane rubbers and in surface coatings where they are very tough but tend to yellow. The greatest outlet, however, is in foams which can be produced either flexible or rigid and are used for thermal insulation, cushioning and to a small extent for electrical potting.

Polyurethane foams are made either by including in the formulation a conventional blowing agent (section 16.12.1) or by the following technique: if sebacic acid, phthalic anhydride, succinic acid and glycerol are mixed together and partially esterified, a polyester is obtained containing free hydroxyl and carboxyl groups. If tolylene di-isocyanate is now added, rapid reaction takes place and heat is evolved. Part of the di-isocyanate reacts with the hydroxyl groups and cross links the polyester chains by urethane bonds:

$$\{-OH + OCN.C_7H_6.NCO + HO-\} \longrightarrow \{-O-CO-NH-C_7H_6-NH-CO-O-\}$$

Owing to the increase in temperature, however, the carboxyl groups also react with the di-isocyanate leading to further cross linking and the production of carbon dioxide:

$$\{-COOH + OCN.C_7H_6.NCO + HOOC-\} \longrightarrow \{-CO-NH-C_7H_6-NH-CO-\}$$

The carbon dioxide causes the mass to bubble, and so a hard, rigid, polyurethane foam is obtained. This method, though instructive, is rarely economic.

Thermosetting plastics

Table 10.12 Polyurethane consumption

	USA 1971 (m. lb.)	UK 1971 ('000 tons)	W. Europe 1969 ('000 tons)
Non-rigid foams	632	41*	260
Rigid foams	276	7*	55
Paints	n.a.	n.a.	25
Elastomers	n.a.	n.a.	11
Fibres ('spandex')	n.a.	n.a.	6
Other uses	n.a.	n.a.	3

* Production.

The consumption figures for polyurethanes are shown in Table 10.12. The biggest outlets are in bedding and car upholstery. The growth of polyurethanes has paralleled that of polyesters and this area too is one of rapid expansion.

10.4.6 Epoxy resins

Epoxy resins have some claim to be thought of as the first cross-linked polymers to be developed on a basis of chemical theory rather than by trial and error. They were used during the Second World War as adhesives for sticking the wings on the 'Mosquito' fighter, the last wooden military aeroplane, and this provides ample testimony to their superiority over traditional glue.

A typical epoxy resin is made by reacting Bisphenol A (dimethyl, di-*p*-hydroxyphenyl-methane) with epichlorohydrin in the presence of alkali. Neither of these compounds is obviously an article of commerce, but both are made fairly simply, largely for this end-use. Bisphenol A is made by condensation of acetone with phenol:

$$\text{HO-C}_6\text{H}_4\text{-OH} + \text{O=C(CH}_3\text{)}_2 \xrightarrow{\text{acid}} \text{(HO-C}_6\text{H}_4\text{)}_2\text{C(CH}_3\text{)}_2 + \text{H}_2\text{O}$$

Since the cumene–phenol process (section 9.5.1.1) produces phenol and acetone, the process which uses them both is clearly attractive.

Epichlorohydrin is made by chlorination of propylene with chlorine, then hypochlorous acid, followed by its dehydrochlorination:

$$\underset{\text{Propylene}}{\text{CH}_3\text{—CH=CH}_2} \xrightarrow[\text{400–500°C}]{\text{Cl}_2} \underset{\text{Allyl chloride}}{\text{CH}_2\text{Cl—CH=CH}_2} \xrightarrow{\text{HOCl}}$$

$$\underset{\text{Glycerol dichlorohydrin}}{\text{CH}_2\text{Cl—CHCl—CH}_2\text{OH}} \xrightarrow[\text{Ca(OH)}_2]{-\text{HCl}} \underset{\text{Epichlorohydrin}}{\text{H}_2\overset{\overset{\displaystyle O}{\diagup\diagdown}}{\text{C—CH}}\text{—CH}_2\text{Cl}}$$

Industrial polymers

These react together to give an epoxy resin which is a type of polyether:

$$CH_2\!\!-\!\!\overset{O}{\overset{\diagup\,\diagdown}{CH}}\!\!-\!\!CH_2\!\!-\!\!\left[\!O\!-\!\!\bigcirc\!\!-\!\!\underset{CH_3}{\overset{CH_3}{\overset{|}{C}}}\!\!-\!\!\bigcirc\!\!-\!\!O\!-\!CH_2\overset{OH}{\overset{|}{CH}}.CH_2\!\right]_n\!\!-\!O\!-\!\!\bigcirc\!\!-\!\!\underset{CH_3}{\overset{CH_3}{\overset{|}{C}}}\!\!-\!\!\bigcirc\!\!-\!O\!-\!H_2C\!-\!\overset{O}{\overset{\diagup\,\diagdown}{CH}}\!\!-\!\!CH_2$$

The commercial resins end in epoxy groups. They are thermoplastic but will cross link via their terminal epoxy groups when heated with curing agents such as diethylene triamine (I), triethylene tetramine (II) or metaphenylene diamine (III). Certain acid anhydrides such as phthalic anhydride or pyromellitic dianhydride (IV) may also be used. Compounds (I) and (II) result from the action of ammonia on ethylene dichloride:

$$H_2N.CH_2CH_2NH.CH_2CH_2NH_2 \qquad H_2N.CH_2CH_2NH.CH_2CH_2NH.CH_2CH_2NH_2$$
$$\text{I} \hspace{8em} \text{II}$$

III (metaphenylene diamine), IV (pyromellitic dianhydride)

Cured epoxy resins tend to be brittle and modifiers such as polysulphide rubber are sometimes added. Substitution of a P/F novolac resin for the Bisphenol A results in greater cross linking due to the additional phenolic groups, and hence greater heat stability.

Epoxy resins are probably best known as 'Araldite' and similar adhesives. They are used as binders for fibreglass, when their superior properties justify their premium over polyesters, and are also used for 'potting' electrical components. Their largest use is in surface coatings and they are used to coat the much-battered skittles in bowling alleys. Combined with fillers they are used as chemically resistant floor coatings. They are also used as primers on cars and as chemical and water resistant finishes on such things as bridges. Their use is restricted by the high price – of the order of £500 ($1 200) per ton.

Production of epoxy resins and their raw materials is given in Table 10.8 and 10.13. Production doubled between 1966 and 1970, but as epoxies command a substantially higher price than the other materials in the table, the tonnage involved is correspondingly smaller.

Table 10.13 Raw materials for polyurethanes and epoxies

		USA 1969		
	Production (m. lb.)	Sales (m. lb.)	Value (¢/lb.)	Price (¢/lb.) 4/72
Isocyanic acid derivatives	421	354	33	—
Toluene 2,4- and 2,6-di-isocyanate (80/20 mixture)	258	252	30	33
Bisphenol 'A'	182	64	19	—
Epichlorohydrin	n.a.	n.a.	n.a.	27

10.5 Paints

The existence of coloured cave paintings dating from 20 000 to 10 000 BC shows that the paint industry is of a respectable antiquity. Mediaeval European painters decorated churches and palaces with a mixture of pigment (e.g. red lead) and animal glue ground into water, and in the thirteenth century linseed oil was added.

The UK paint industry proper was established in and around London in the eighteenth century, its products being based on linseed oil. This was mixed with whiting to form a sealer (putty), with Spanish Brown to form a primer and with white lead to form a topcoat. To increase its rate of drying, the oil was first heated with litharge but even so, drying took at least a couple of days. By 1900 there were 400 manufacturers producing £$8\frac{1}{2}$ m. of products. The paints were still based on linseed oil, pigmented with white lead or zinc oxide, and thinned with turpentine, but a large range of newly developed organic and inorganic pigments was incorporated. Natural resins such as shellac or rosin were also included to give a harder surface. Distempers were popular for interior use, being a composition of pigments ground in glue solution, later replaced by aqueous casein solution.

In 1927, ICI introduced the oil-modified alkyd paint (section 10.5.3) from America and within a few years this new paint medium dominated the market as it has done ever since. In spite of the presence of ICI in the field, paint manufacture remained a backwater of scientific progress until cheap synthetic monomers, of the type used in the plastics industry, became available in the mid-1950s. Since then, there has been extensive rationalization within the paint industry and it has become more and more a part of the chemical industry proper.

The paint industry manufactures few if any chemicals. It operates by buying the chemicals it needs, mixing them and packaging and distributing the resultant paint.

As the only plant required is equipment for mixing the paint formulation (usually a ball mill), it is possible to set up as a manufacturer with very little capital. The overwhelming majority of paint firms have fewer than 25 employees (Table 10.14), and although the 'giants' take a large share of the market, the small firms still manage to show satisfactory profits and it is the medium-sized firms that are being financially squeezed. In 1963 the top ten firms held nearly 60% of the market and this figure had probably risen to 70% by 1969. On the other hand, the two hundred 'minnows' still share some 7% of the market between them.

Table 10.14 Size of paint firms as measured by number of employees (1963)

Number of employees	Number of firms
<25	~200
25–99	94
100–199	39
200–499	12
500–999	7
1 000–1 999	6
>2 000	4

A consequence of this is that price competition is fierce. The larger manufacturers can charge a premium because of their advertising and superior quality control, but in the sort of situation where, for example, a local authority asks for tenders for paint for its hospitals, it is very likely that a local small manufacturer will win the contract.

A paint or surface coating must have some or all of the following properties. It must be easily applicable to a surface by brush, roller, spray-gun, dipping, etc., and must adhere to that surface in a uniform homogeneous film. It should be opaque (except for varnishes, etc.) and should protect the surface which it covers. Consequently, it should resist abrasion and corrosion. The colour should be stable to light and the paint should not yellow with age. In the case of domestic paints, speed of drying is important. There are, of course, specialized paints designed to have additional particular properties, e.g. heat-resistant, fungicidal or odourless paints.

A paint formulation consists of five items:

1. Resin – this is the paint binder which binds the pigment into a homogeneous film. Polyvinyl acetate is the binder in water-based emulsion paints, and oil-modified alkyd resins are used in gloss paints.
2. Pigment – this gives the paint its covering and protective power. Titanium dioxide is the most commonly used. White lead is used, especially by the army where its protective power is important and its toxicity is not. There are also ranges of coloured inorganic pigments (e.g. lead chromate) and organic pigments such as Hansa Yellow or Indanthrone Blue. The pigment is the most expensive single item in a pot of paint.
3. Solvent – This is used to facilitate the application of the paint. White spirit, or white spirit plus xylene are usually used, though trichloroethylene has certain specialized uses due to its rapid rate of evaporation and low inflammability. Since solvents evaporate into the air, they are something of a waste of money and many attempts have been made to devise paint formulations in which water is the solvent. This has been successful in the case of emulsion paints, but 80% of all paints are still organic-solvent based. The discovery of a satisfactory water-based gloss paint is the current aim of a number of research groups.
4. Extenders – these are added to cheapen paint and to improve its physical properties. They include barytes, blanc fixe, whiting and china clay.
5. Dryers – these are essential to accelerate the drying of air-drying paints. They are usually the naphthenates of heavy metals such as lead, cobalt or manganese, and are added at a level of tenths of a per cent. In water-based emulsion paints, a 'Redox' system is used to initiate polymerization (section 16.11.3) and this may be thought of as a drier. A typical formulation involves ammonium persulphate and sodium metabisulphite at a level of hundredths of a per cent.

Other additives may include surfactants to assist dispersion of the pigment, silicones to improve weathering properties, thixotropic agents, anti-settling agents and bactericides. Figure 10.11 gives the formulations of two representative decorative paints.

The UK paint market in 1968 was about 100 m. gallons and is growing at about 3% per annum, the same as the economy as a whole. This underlines the point that one can only paint something which someone else has made, and that the paint industry is limited in a way which does not apply, say, to plastics. In addition, the cost of applying paints is high and there is a modern tendency to make things out of more expensive materials so that they do not need painting. A good example is the aluminium carriages on the London Underground which are replacing the old, steel, red-painted ones.

The paint market may be divided into industrial and decorative paints. The former term refers to paints used on products manufactured by a particular industry, e.g. automobiles, domestic appliances, ships and bridges but not houses. All other paints are classified as 'decorative' and are sub-divided by outlet into 'retail and 'trade'. In 1968

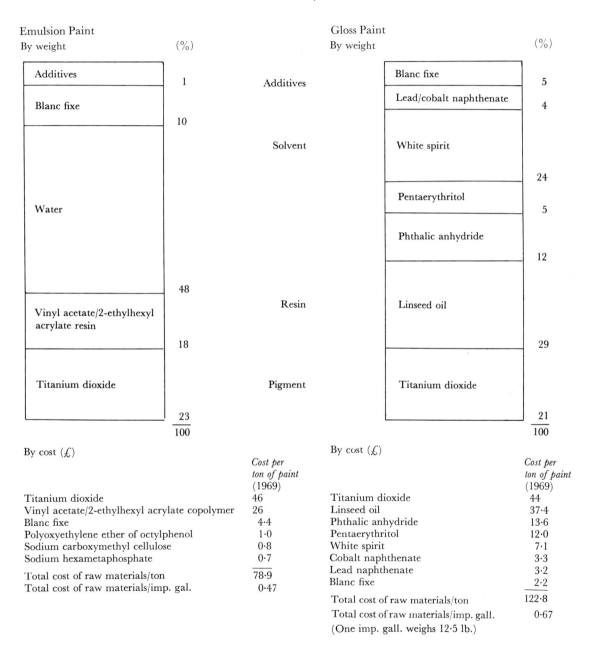

Figure 10.11 Formulations and UK costs of typical decorative paints.

274   *Industrial polymers*

paint consumption was about 50/50 industrial to decorative by volume and decorative paint was about 66/33 trade to retail by volume.

The retail market (i.e. sales to the do-it-yourself decorator) is currently the growth sector of the paint industry. This is due to the high cost of professional decorating, increased leisure time, the ease of application and improved quality of modern paints, and the increased standard of living. It is interesting that this last is sufficient to cover the cost of paint but not sufficient to permit the luxury of a decorator.

The main decorative paints are the oil-modified alkyd gloss paint and the vinyl acetate copolymer water-based emulsion paint. Industrial paints are nearly all 'custom made' for the user, but involve a much higher proportion of thermosetting cross-linked stoving finishes in which the surface coating is 'baked' onto the product.

10.5.1   Polyvinyl acetate and the water-based emulsion paint

Vinyl acetate monomer is made by passing acetylene and acetic acid vapour over a zinc acetate on charcoal catalyst at 175–200°C and atmospheric pressure. The reaction is

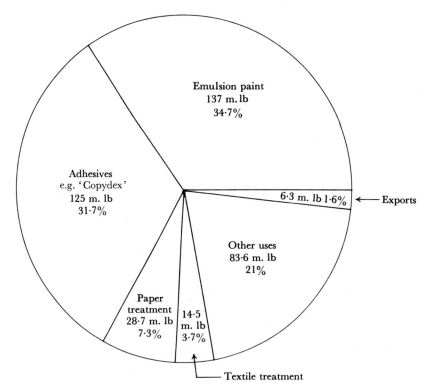

Figure 10.12 End-uses of polyvinyl acetate – USA 1969. Production = 426 m.lb. Sales and use = 395 m.lb divided as in diagram.

The above end-use pattern does not list material used for polyvinyl alcohol manufacture, amounting to 57m. lb which would require about 120 m. lb polyvinyl acetate, too much to be accounted for under 'other uses'. Vinyl acetate production, however, was 729 m. lb and presumably the material for the alcohol comes under this heading, as must the approximately 14 m. lb used in gramophone records, and other copolymer uses, etc.

exothermic and cooling is necessary. The monomer can be emulsion polymerized (section 16.11.3) by heating with a peroxide catalyst, and a latex containing about 50% resin is obtained.

$$HC\equiv CH + CH_3COOH \longrightarrow CH_2=CH\ OOC.CH_3 \longrightarrow (-CH_2-CH-)_n \mid CH_3.COO$$

It is also possible to make vinyl acetate by a variation of the Wacker process (section 9.1.4) in which ethylene and air at slightly above atmospheric pressure are passed into glacial acetic acid in the presence of a palladous chloride–cuprous chloride catalyst system. The overall reaction is

$$C_2H_4 + PdCl_2 + CH_3COOH \longrightarrow CH_2=CH.OOC.CH_3 + Pd + 2HCl$$

The cupric chloride then oxidizes the palladium back to palladous chloride and is itself reduced to cuprous chloride. The air oxidizes the cuprous ion back to cupric.

In 1965 or thereabouts, ICI started to build plant to operate some variant of this process and for this reason no further UK acetylene-based plant was built. Unfortunately

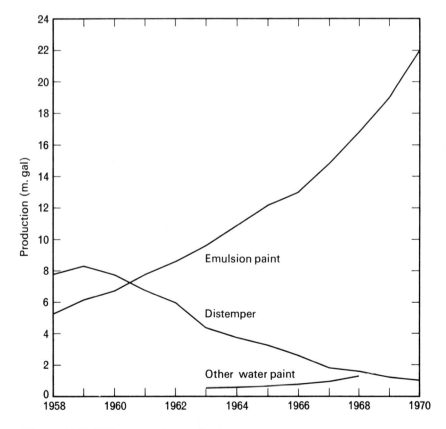

Figure 10.13 UK water paint production.

there were severe corrosion problems associated with the use of glacial acetic acid and in spite of large sections of the plant being rebuilt in titanium metal, it never operated near capacity and was closed down in 1969. There was thus a shortage of vinyl acetate capacity in the early 1970s. A vapour phase variation of the process is nonetheless viable and is being operated in other countries.

Annual UK polyvinyl acetate consumption rose from 9 000 tons in 1955, to 25 000 tons in 1968. The US end-uses are shown in Fig. 10.12.

The first three of these are related to the emulsion paint use. The copolymer use is mainly in combination with polyvinyl chloride in certain specialized uses. A somewhat different use is that polyvinyl acetate can be hydrolysed to polyvinyl alcohol:

$$(-CH_2-CH-)_n \xrightarrow{H_2O} (-CH_2-CH-)_n + nCH_3COOH$$
$$|\phantom{CH-)_n \xrightarrow{H_2O} (-CH_2-}|$$
$$CH_3.COO \phantom{-)_n \xrightarrow{H_2O} (-CH_2-}HO$$

Polyvinyl alcohol

The presence of a hydroxyl group on every mesomer results in polyvinyl alcohol being soluble in water and the presence of the polymer chains in solution causes a marked increase in viscosity. Polyvinyl alcohol is thus used as a thickening agent in certain foods and it is also used as a textile size and in grease-proofing paper. Another derivative, polyvinyl butyral (made from the alcohol plus butyraldehyde) makes an excellent interlayer in a safety glass 'sandwich'.

The main outlet for polyvinyl acetate, however, is in emulsion paints. These have now almost completely displaced the traditional water-based distemper. Data on the consumption of these are given in Fig. 10.13 and provide a good example of the displacement of one material by another in a market which as a whole is not expanding very rapidly.

Vinyl acetate polymer emulsions were introduced into the UK from Germany in the mid-1950s. The paints made from them however did not have a sufficiently low glass transition temperature, i.e. they gave a film which was too brittle (the glass temperature is the temperature below which the resin will break if subjected to applied pressure).

The polyvinyl acetate was therefore plasticized with dibutyl phthalate or less frequently with tricresyl phosphate. These are external plasticizers (section 10.3.4.6) and the paints, though better than those without plasticizer, tended to age because of evaporation of the plasticizer. In 1960, the externally plasticized PVA emulsion was replaced by an 80:20 vinyl acetate:dibutyl or di-2-ethylhexyl maleate copolymer emulsion. The maleate ester plays the part of an internal plasticizer (maleic anhydride is made by the oxidation of benzene (section 9.5.1)). Maleate esters were, however, used for only a short time. 2-Ethylhexyl and butyl acrylates both show very high plasticizing efficiency and give a finer particle sized emulsion than maleates, with greater binding power for cheap filler pigments. When the price of acrylates dropped in about 1962 maleates were ousted and most emulsion paints at present are vinyl acetate/acrylate copolymer emulsions.

Polyacrylate homopolymers are widely used in the USA, but do not appear likely to oust vinyl acetate in the UK. If emulsion paints do move away from vinyl acetate, it seems likely that the cheap end of the market will go over to a vinyl acetate–ethylene copolymer while the heavy duty end goes towards an acrylic homopolymer.

10.5.2 Acrylics

The term 'acrylics' is used rather confusingly to cover methacrylates, acrylates and polymers based on acrylonitrile. The manufacture of the first and last of these is discussed in sections 10.2.4 and 10.7.4.

Acrylic acid is made industrially in three different ways:

1. Ethylene oxide and hydrogen cyanide react together to give ethylene cyanohydrin. On treatment with concentrated sulphuric acid at 175°C this is both hydrolysed and dehydrated to give acrylic acid:

$$H_2C\overset{O}{-}CH_2 + HCN \longrightarrow \underset{\text{Ethylene cyanohydrin}}{CH_2OH.CH_2CN} \xrightarrow[175°]{H_2SO_4} \underset{\text{Acrylic acid}}{CH_2=CH.COOH}$$

2. Acetylene, carbon monoxide and water are passed over a catalyst of nickel or any other metal which will form a carbonyl (Reppe synthesis):

$$C_2H_2 + CO + H_2O \xrightarrow[HCl]{Ni(CO)_4} CH_2=CH.COOH$$

3. Ketene, obtained by the cracking of acetic acid, reacts with formaldehyde to give β-propiolactone and this rearranges to give acrylic acid.

$$CH_3.COOH \xrightarrow{-H_2O} \underset{\text{Ketene}}{CH_2CO} \xrightarrow{CH_2O} \underset{\text{β-Propiolactone}}{CH_2CH_2OC\!-\!\!\!-\!\!\!O} \longrightarrow CH_2=CH.COOH$$

The first and third of these methods involve a large number of stages between the olefine feedstock from a petrochemical plant and the final product; the second method is based on acetylene which is itself expensive. There is, however, another route to acrylic acid which is apparently more economic. It has already been noted that acrylonitrile is made by the reaction of propylene, air and ammonia over a stannic oxide–antimony oxide or a bismuth phosphomolybdate catalyst. If ammonia is omitted from this reaction mixture, then acrolein is obtained. The ammonia can then be added to part of the acrolein to give acrylonitrile, while another part can easily be oxidized to acrylic acid:

$$CH_2=CH.CH_3 \xrightarrow{O_2} \underset{\text{Acrolein}}{CH_2=CH.CHO} \begin{array}{c} \xrightarrow{NH_3} CH_2=CH.CN \quad \text{Acrylonitrile} \\ \xrightarrow{O_2} CH_2=CH.COOH \quad \text{Acrylic acid} \end{array}$$

Since acrylic acid is obtained from propylene in only two stages, one of which would be operated anyway, this would appear easily the best route to acrylic acid. It was developed by The Distillers Company in the 1950s but the market for acrylates was not large enough

Industrial polymers

Table 10.15 Manufacture of oil-modified alkyd resins

(*a*) Condensation of glycerol with phthalic anhydride gives the commonest alkyd resin: (glycerol units shown in bold)

[Structural diagram showing phthalic anhydride + glycerol (CH₂OH–CHOH–CH₂OH) reacting to give a polyester chain with phthalate and glycerol units]

(*b*) Oil modified alkyd resins are made by heating a drying oil (e.g. linseed oil) with a polyol (e.g. pentaerythritol) and esterifying the product with a dibasic acid or anhydride (e.g. phthalic anhydride).

Linseed oil is mainly linolenic acid triglyceride. Linolenic acid is

$$CH_3CH_2CH=CHCH_2CH=CHCH_2CH=CH(CH_2)_7COOH$$

If this is written L.COOH then the triglyceride is

$$\begin{array}{l} CH_2O.OC.L \\ | \\ CHO.OC.L \\ | \\ CH_2O.OC.L \end{array}$$

When heated with pentaerythritol, the triglyceride undergoes transesterification reactions to a mixture of partial linolenic acid esters of pentaerythritol $(HOCH_2)_3C(CH_2O.OC.L)$, $(HOCH_2)_2C(CH_2O.OC.L)_2$, etc., plus a mixture of linolenic acid mono- and diglycerides

$$\begin{array}{l} CH_2O.OC.L \\ | \\ CHOH \\ | \\ CH_2OH \end{array} \quad \text{and} \quad \begin{array}{l} CH_2O.OC.L \\ | \\ CHO.OC.L \\ | \\ CH_2OH \end{array}$$

These react with phthalic anhydride to give a more or less linear oil-modified alkyd resin polymer, the chain length of which depends on the ratio of mono- to diglycerides in the mixture, the latter acting as chain-stoppers: (monoglyceride units are shown in bold and diglyceride units in italic)

[Reaction scheme: phthalic anhydride + monoglyceride (CH₂OH–CHOH–CH₂O.OC.L) + diglyceride (CH₂OH–CHO.OC.L–CH₂O.OC.L) giving a linear alkyd polymer with phthalate linkages, monoglyceride units (bold) bearing pendant CH₂–O–CO–L groups, and terminal diglyceride units (italic) with two L ester groups]

These polymers are the paint resins. In the presence of an initiator, the linolenic acid groups cross link between their double bonds to give a three-dimensional polymer.

at the time to make it worthwhile. The market had grown to a satisfactory size in the mid-1960s, but just as Distillers were deciding to go ahead, their chemical interests were taken over by BP who have apparently been thinking about the project ever since. A Japanese variant of the process is possibly in operation in Japan and is due to come onstream in the UK in 1972. The UK consumption of acrylate esters in 1970 was about 20 000 tons, all of which was imported.

Acrylic acid is used almost entirely as its esters which are sold either as monomers for copolymerization or as an emulsion similar to polyvinyl acetate emulsion. The most important esters on the market are methyl (used as a comonomer in polyacrylonitrile fibres), ethyl (used as a processing aid in the production of 'Perspex', in leather treatment and to give 'wet strength' to paper products), and butyl and 2-ethylhexyl (used as comonomers in emulsion paints).

Polyacrylate homopolymer emulsion paints (i.e. the resin is entirely acrylic and polyvinyl acetate is omitted) are used in the USA for painting the outsides of houses where their stability to bright sunlight is an asset. In Britain, both wooden houses and bright sunlight are lacking and this market has never developed.

Apart from the above uses, two other 'acrylic' surface coatings are of importance. One, a thermoplastic acrylic, is used on General Motors cars (Vauxhall in the UK). It is essentially a polymethyl methacrylate resin with plasticizer and solvent. It dries by solvent evaporation and a gloss finish is achieved by polishing.

Thermosetting acrylics were pioneered by Pittsburgh Plate Glass who produced a range of 'Scopacron' resins in the early 1960s based on acrylamide, acrylonitrile and styrene. A UK version used on UK Ford cars is based on hydroxymethyl methacrylate and methyl methacrylate which is cross-linked with an amino-resin. This finish has the advantage over the alkyd–melamine resin (section 10.5.3), used in the UK by British Leyland, in that scratches incurred in production can be removed by polishing. The future trend will probably be towards thermosetting acrylics applied in the form of a powder.

10.5.3 Alkyd resins

In spite of technological change, the paint industry is still based, as it was forty years ago, on the oil-modified alkyd resin. In 1963 paint manufacturers bought 26 000 tons of these materials and although this figure is now more or less static it still accounts for about half of manufacturers' total expenditure on resins.

Alkyd resins are condensation polymers of dibasic acids or their derivatives, with polyfunctional alcohols. The resulting resins cross-link too slowly and are not sufficiently soluble to be of any use as moulding materials, but when modified with so-called drying oils they become an excellent basis for paints. As there are many different dibasic acids, polyfunctional alcohols and drying oils, there is an extraordinarily large number of alkyd resins possible and about 450 of these are actually available. We shall consider only the commonest members of each group.

When glycerol is heated with phthalic anhydride a polyester, known as an alkyd resin, is formed, which on prolonged heating would cross-link to give an unfusible mass. See Table 10.15(a).

In the manufacture of the oil-modified resin, see Table 10.15(b), the drying oil (e.g. linseed oil which is mainly the triglyceride of linolenic acid) is heated with an excess of the polyol, say pentaerythritol. Alcoholysis takes place and the product consists of a mixture of mono- and diglycerides. This is then esterified with phthalic anhydride in the presence of a little xylol which is recirculated. Water is distilled off as esterification proceeds, the oil-modified alkyd being the residue product. The introduction of the linseed oil serves two purposes. First, by reducing the average number of hydroxyl groups per molecule, it prevents any substantial degree of cross-linking and limits the molecular weight and hence the viscosity. Second, the linolenic acid residues contain three double bonds each which allows for the possibility of further cross-linking once the alkyd resin based coating has been applied. This is in fact brought about by addition to the paint formulation of a small amount of cobalt naphthenate which, in the presence of air, catalyses the cross-linking.

Apart from the wide range of alkyd resins available, oil-modified synthetic resins are also made, as are normal air-drying alkyds modified by synthetic monomers. The discussion of these is beyond the scope of this book but it is worth mentioning that British Leyland use an alkyd–melamine finish on their cars. Melamine, formaldehyde and butyl alcohol are condensed to give a partially butylated melamine–formaldehyde resin (section 10.4.2.2) and this is mixed with an oil-modified alkyd resin and used as a stoving finish. The formulation is the major general purpose industrial stoving finish for domestic appliances and other uses, and in the UK in 1970 acrylics and epoxy finishes had made little impression on this market.

10.5.4 Nitrocellulose

Nitrocellulose is a cheap natural resin made by nitration of cellulose. The degree of nitration required involves about 10–12% nitrogen in the final product, corresponding to about two nitro groups to every anhydro-glucose unit.

Consumption of nitrocellulose is static and its share of the paint market is declining (15% in 1963 to 13% in 1968). It was the traditional resin used in furniture finishing but is being displaced by melamine precatalyst, a complicated synthetic polymer cross-linked by means of a metal catalyst. It is more expensive than nitrocellulose but provides a superior finish.

Nitrocellulose finishes used to be used on cars and the painting process took several days. The only firm still using such finishes is Daimler which thus enhances its 'traditional' image.

10.5.5 Other resins

A variety of other resins are used in paints and their consumption may expand in the future. Epoxy resins and polyurethanes have already been discussed (sections 10.4.5 and 10.4.6) but are at present consumed in only small quantities.

Marine paints, where corrosion resistance is important, used to be based on red lead and linseed oil which are very cheap. These have been replaced by an expensive finish involving a zinc-rich primer based on an epoxy-polyamide. This reduces the drag of the water on the ship and the saving in propulsion costs more than outweighs the premium paid for the paint.

10.5.6 Titanium dioxide

Titanium dioxide is the major pigment used by the paint industry. End-uses in the UK are shown in Fig. 10.14. UK production for 1970 was about 170 000 tons and consumption 121 000 tons. UK prices were by the sulphate route £182/ton and by the chloride route £187/ton. US consumption for 1969 was 654 000 tons.

It has a high covering power and a paint based on it has six times the opacity of one based on zinc oxide. Over recent years, manufacturers have increased the pigments content of their paints in the hope of developing a truly 'one-coat' paint which would have a high appeal to the domestic market.

The majority of titanium dioxide in the United Kingdom is made by the sulphate process. Ilmenite ($FeO \cdot TiO_2$) is digested with sulphuric acid to give a solid dry reaction mass from which ferrous, ferric and titanium sulphates are leached with water. Ferric sulphate is reduced to ferrous sulphate to prevent coloration of the product, and the ferrous sulphate is then separated by fractional crystallization. Titanium sulphate is then converted to the hydrate (insoluble metatitanic acid H_2TiO_3) calcined, milled and dried. It is commonly extended by being blended with anhydrous calcium sulphate.

The overall process takes about two weeks, involves about thirty stages and is labour intensive. Each ton of titanium dioxide requires about four tons of sulphuric acid so that the pigment industry consumes some half a million tons of sulphuric acid per year and is thus the second largest user of this material (section 14.2).

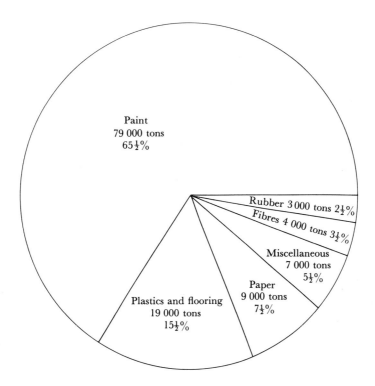

Figure 10.14 Titanium dioxide end-uses – UK 1970.

282 *Industrial polymers*

The most recently built titanium dioxide plants, however, use the much simpler chloride process. Rutile (95% TiO_2) is chlorinated in a furnace to give volatile titanium tetrachloride $TiCl_4$. This is purified by washing and distillation and burned in a specially designed and much patented burner to give rutile crystals which are treated to give optimum pigment properties. The chlorine is recycled with a loss of about 10% per pass. The process is said to be cheaper both in plant construction and operating costs, and the product is of a higher quality and is used in the whiter-than-white paints that appeared on the market in 1969. At that time, pigment made by the chloride process commanded a premium of about £5/ton. The main limitation of the process is that rutile is less common than ilmenite and supplies of it may be insufficient for the sulphate process to become obsolete. The chloride process can be applied to ilmenite, but the chlorine usage increases dramatically and, since chlorine is expensive, the economics of the process become less favourable.

In the near future, however, we may expect an increase in chloride-based titanium dioxide and a corresponding drop in the proportion of sulphuric acid consumed by the paint industry. This could be predicted more confidently were it not for the extreme difficulties experienced by Laporte in bringing the first UK chloride plant onstream and a world shortage of the rutile feedstock. British Titan Products, however, appear to have experienced no such difficulties.

10.6 Rubbers

Rubbers (or elastomers) differ from other plastics by virtue of the fact that they stretch; that is that they have high extensibilities and low moduli of elasticity. Furthermore, the

Table 10.16 US synthetic rubber production 1969 and world capacity 1970 (excluding planned economy countries)

	Production (m. lb)	USA Sales (m. lb)	Unit value (¢/lb)	World Capacity ('000 tons)	%
Styrene-butadiene rubber (SBR)	2 530†	2 130†*	24	3 486	60·5
Polybutadiene rubber (BR)	597	553	18	974	16·9
Butyl rubber	291	n.a.	n.a.	355	6·2
Polyisoprene rubber (IR)	244	200	20	213	3·7
Ethylene–propylene copolymer (EPM) and terpolymer (EPDM)	167	142	27	151	2·6
Nitrile rubber	154	137	47	259	4·5
Polyurethane rubbers	27	23	106	n.a.	<1
Silicone rubbers	14	13	319	n.a.	<1
Polychloroprene rubbers	n.a.	n.a.	n.a.	324	5·6
All other rubbers	500	720‡	38	—	—
Total	4 524	3 918	27	5 762	100

* Partly estimated.
† Excludes butadiene–styrene–vinylpyridine rubbers (34 m. lb production).
‡ Sales figure includes sales of butyl rubber.

extensibility is reversible – when the stress is removed the rubber reverts to its original shape. Intermolecular forces in rubber are low and the dividing line between them and plastics is not sharp.

The story of natural rubber was summarized in section 2.2. Synthetic rubbers were developed mainly under the stress of actual or impending war. During the First World War, faced with the allied blockade, Germany developed a poor-quality synthetic 'methyl rubber' based on 2,3-dimethylbutadiene, and manufactured some 2 500 tons of it. From 1933 onwards, Germany again prepared for war and, encouraged by Hitler, IG Farben developed two synthetic rubbers: one based on butadiene and styrene (Buna S) and one based on butadiene and acrylonitrile (Buna N). With the rubber-producing countries occupied by the Japanese, the UK, USA and USSR were also compelled to build synthetic rubber plant during the Second World War and the technology became fairly well understood.

The price of synthetic rubbers remained high, however, until the boom in petrochemicals reduced the price of the hydrocarbon feedstock. Since then the growth of synthetic rubber has paralleled that of plastics. Nonetheless, natural rubber has not been displaced completely by synthetics, though consumption is more or less static. Table 10.16 shows the 1970 world capacities for synthetic rubber according to type and illustrates which are the more important of the synthetics. Table 10.17 gives market data. Overall growth rate is expected to be about 10%/year, concentrated mainly in EPM/EPDM and IR and to a lesser extent in butadiene and polychloroprene rubbers. World consumption of natural rubber in 1970 was 32·3% of total rubber consumption.

The disposition of resources in a modern industry is well illustrated by a comparison which shows that a synthetic rubber factory produces rubber on an area of land 1/30 000 of the size required to produce the same amount of natural rubber and with 1/70 of the labour force. It requires very much more capital to build the factory, but this is readily available in a developed country. It can be seen from these data that the economic need for colonies to produce raw materials with relatively cheap labour and land has dropped sharply since the Second World War. It is not entirely impossible that the willingness of many great powers to grant independence to their colonies, and the parlous economic state of these colonies once independence has been granted, are related to this.

Table 10.17 UK rubber production and consumption*

	1960	1970
Production	('000 tons)	('000 tons)
Imports of natural rubber	210·4	198·6
Synthetic rubber production	90·4	301·3
Reclaimed rubber production	45·8	29·0
Consumption		
Total of natural and synthetic rubber	280·7	428·5
into tyres and tubes	155·7	229·6
into cables	9·2	5·2
into belting	7·2	10·7
into footwear	16·5	13·0
into cellular products (foam rubber, etc.)	16·4	19·5
other uses and unclassified uses	75·7	150·4

* Figures are obtained from returns completed by manufacturers. They add up to slightly less than total rubber consumed because of underestimates. Consumption figures omit a substantial tonnage of exports.

284 Industrial polymers

The processing of rubber is extremely complicated and involves a wide range of chemicals as vulcanization accelerators, stabilizers, plasticizing and reclaim agents, vulcanization retarders, vulcanization agents, fillers, pigments, plasticizers, binders, blowing agents, lubricants and other additives; a discussion of these will be found in standard books on rubber technology. Suffice to say that in 1969 when the USA produced 4 524 m. lb. of synthetic rubber at a unit value of 27¢/lb., she also needed to produce 303 m. lb. of rubber-processing chemicals with a unit value of 63¢/lb.

10.6.1 Natural rubber

Ninety-nine per cent of natural rubber is obtained from *hevea brasiliensis* and this variety is called hevea rubber. Like many natural products, it is stereoregular and is the *cis*-1,4 polymer of isoprene (I). The *trans*-1,4 isomer (II) is also obtained from certain trees and is called balata or gutta-percha:

$$\left[\begin{array}{c} CH_2 \\ | \\ CH_3 \end{array} \!\!\! C\!=\!C \!\!\! \begin{array}{c} CH_2 \\ | \\ H \end{array} \right]_n \qquad \left[\begin{array}{c} CH_2 \\ | \\ CH_3 \end{array} \!\!\! C\!=\!C \!\!\! \begin{array}{c} H \\ | \\ CH_2 \end{array} \right]_n$$

(I) Hevea rubber (II) Gutta-percha

These polymers have low intermolecular forces and can easily be permanently deformed because the polymer chains slip over one another. In order to confer the important property of reversible extensibility, the chains have to be fixed loosely in space relative to one another and this is done by heating the rubber with sulphur which reacts at the allylic positions and forms sulphur 'bridges' between the chains. In ordinary soft rubber about 4% of sulphur is used so that 5–10% of the double bonds are linked. If 45% of sulphur is used, a hard rigid totally cross-linked polymer is obtained. This process is called vulcanization. The vulcanized rubber has the approximate structure:

$$\begin{array}{c}
\cdots\!-\!CH\!-\!HC\!=\!C\!-\!CH\!-\!CH_2\!-\!CH\!=\!C\!\!\!\underset{\underset{S}{|}}{\overset{\overset{CH_3}{|}}{{}}}\!\!\!-\!CH_2\!-\!CH\!-\!CH\!=\!C\!\!\!\underset{\underset{S}{|}}{\overset{\overset{CH_3}{|}}{{}}}\!\!\!-\!CH_2\!-\cdots \\
\underset{\underset{\underset{\underset{}{CH_3}}{|}}{S}}{|} \qquad \underset{CH_3}{|} \\
\cdots\!-\!CH\!-\!C\!=\!CH\!-\!CH\!-\!CH_2\!-\!C\!=\!CH\!-\!CH_2\!-\!CH\!-\!C\!=\!CH\!-\!CH_2\!-\cdots \\
\qquad\qquad\qquad\qquad\qquad |\!\!\!\ CH_3
\end{array}$$

10.6.2 Synthetic polyisoprene

The development of stereospecific catalysts of the Ziegler or titanium type has made possible the manufacture of *cis*-1:4-polyisoprene from petroleum feedstocks. The products of thermal cracking of petroleum include 2-methylbutene-1 and 2-methylbutene-2,

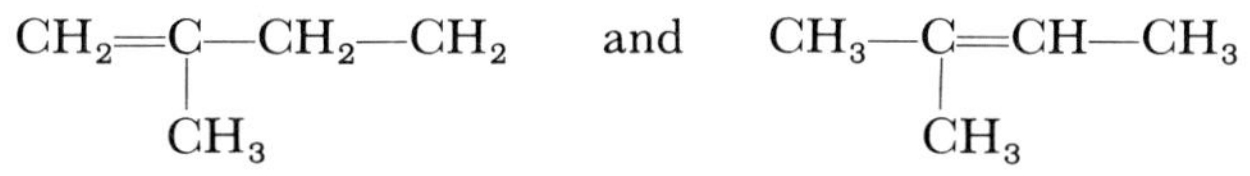

$$CH_2=C-CH_2-CH_2 \quad \text{and} \quad CH_3-C=CH-CH_3$$
$$|\phantom{-CH_2-CH_2 \quad \text{and} \quad CH_3-}|$$
$$CH_3 \phantom{-CH_2-CH_2 \quad \text{and} \quad CH_3-C}CH_3$$

and these can be extracted from refinery gases with 65% sulphuric acid. Catalytic dehydrogenation gives isoprene

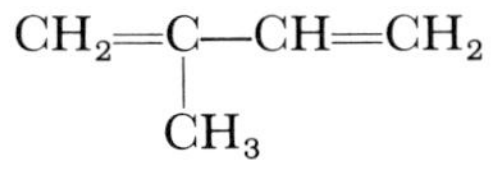

$$CH_2=C-CH=CH_2$$
$$|$$
$$CH_3$$

which is redistilled and dried, dissolved in *n*-pentane and polymerized at 50°C.

The product is becoming competitive in price with natural rubber and, if it does so, might well make inroads into its market. It is said to be easier to process than natural rubber. The USSR has invested heavily in polyisoprene rubber and hopes to stop the import of natural rubber but her motives are thought to be political rather than economic.

10.6.3 Styrene-butadiene

The original Buna S or GR-S rubber was based on a styrene–butadiene copolymer now slightly modified and known as SBR. About one mole of styrene is used to six of butadiene (about 1:3 by weight). Polymerization is brought about by a redox system (ferrous pyrophosphate/*p*-menthane hydroperoxide) at 5°C in an emulsion, or in solution by a stereospecific catalyst.

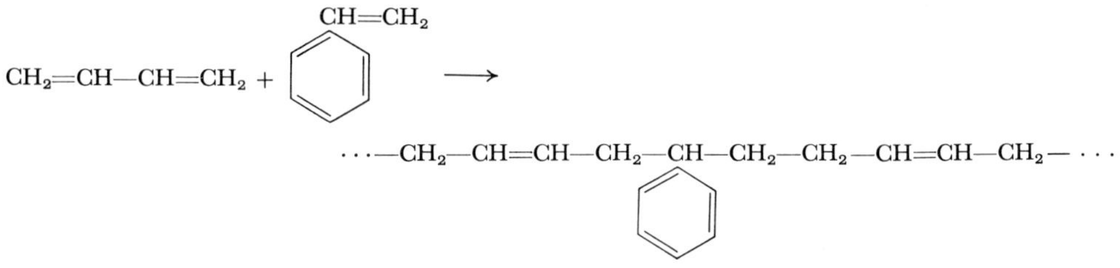

$$CH_2=CH-CH=CH_2 + \text{styrene} \longrightarrow$$
$$\cdots-CH_2-CH=CH-CH_2-CH-CH_2-CH_2-CH=CH-CH_2-\cdots$$

This can be vulcanized by similar methods to ordinary rubber.

SBR is easily the most important synthetic rubber, accounting in 1969 for some 56% by weight of the US market. An early competitor to SBR was butyl rubber which made a superior inner tube for a car tyre. This market disappeared with the advent of the tubeless tyre, although some butyl rubber is still used in non-squeal tyres. A much more serious threat to the SBR market comes from stereoregular polybutadiene although its advent will increase rather than decrease the market for butadiene. Polyisoprene has similar properties to polybutadiene but is more expensive and does not, at present, seem likely to compete.

10.6.4 Butyl rubber

Butyl rubber is a copolymer of isoprene and isobutylene. The reaction is carried out in methyl chloride solution in the presence of a Friedel–Crafts catalyst (aluminium chloride) at the unusually low temperature of $-95°C$.

$$CH_2{=}C{-}CH{=}CH_2 + CH_2{=}C{-}CH_3$$
$$\quad\quad |\quad\quad\quad\quad\quad\quad\quad\quad |$$
$$\quad\; CH_3 \quad\quad\quad\quad\quad\quad\quad CH_3$$
$$\text{Isoprene} \quad\quad\quad\quad\quad\quad \text{Isobutylene}$$

$$\cdots{-}CH_2{-}C{=}CH{-}CH_2{-}CH_2{-}C{-}CH_2{-}C{=}CH{-}CH_2{-}\cdots$$
$$\quad\quad\quad\quad |\quad\quad\quad\quad\quad\quad\quad\quad |\quad\quad\quad\quad |$$
$$\quad\quad\quad CH_3 \quad\quad\quad\quad\quad\quad CH_3 \quad\; CH_3$$
$$\text{Butyl rubber}$$

The proportion of isoprene used is only 1–3% so that the number of double bonds in the polymer is relatively small. They are all cross-linked in vulcanization and, unlike natural rubber, extensive vulcanization does not give a hard, rigid product. The final properties of the rubber can be controlled by varying the amount of isoprene. Heat resistance increases and chemical resistance decreases with increasing isoprene.

Butyl rubber has low gas permeability and was formerly used for tyre inner tubes. It is used for air cushions, bellows etc. It has good electrical properties and resistance to weathering and is cheap. It is therefore also used in convertible car tops, insulation of power cables and hose for steam and acids.

10.6.5 Polybutadiene rubbers

The original straight chain polybutadiene rubbers were extremely difficult to process and styrene had to be added as a copolymer to give a useable product. However, if the homopolymer is made using a Ziegler catalyst, a processable stereoregular polymer is obtained. The *cis*-isomer is enjoying considerable commercial success and is being used as an extender in SBR tyres, where it appears to increase abrasion resistance and tread life.

10.6.6 Nitrile rubbers

Nitrile rubbers date from the 1930s and are made by copolymerization of acrylonitrile and butadiene under much the same conditions as are used for SBR:

$$CH_2{=}CH{-}CH{=}CH_2 + CH_2{=}CH \longrightarrow$$
$$\quad\quad\quad\quad\quad\quad\quad\quad\quad\quad\quad\; |$$
$$\quad\quad\quad\quad\quad\quad\quad\quad\quad\quad\quad CN$$

$$\cdots{-}CH_2{-}CH{=}CH{-}CH_2{-}CH_2{-}CH{-}CH_2{-}CH{=}CH{-}CH_2{-}\cdots$$
$$\quad\quad\quad\quad\quad\quad\quad\quad\quad\quad\quad\quad\quad\quad\quad\quad\; |$$
$$\quad\quad\quad\quad\quad\quad\quad\quad\quad\quad\quad\quad\quad\quad\quad\; CN$$

Most commercial rubbers contain 20–40% acrylonitrile and have exceptional solvent resistance with minimum loss of low temperature flexibility. Increase of acrylonitrile to 60% gives a leathery product with better resistance to aromatic oils. Nitrile rubbers are used in petrol hose and belting.

Terpolymers of nitrile rubber with acrylonitrile–butadiene copolymers (ABS resins) have a very high impact resistance but lose all their other rubbery properties (section 10.2.3).

10.6.7 Polychloroprene rubber

Polychloroprene (neoprene) rubber contains chlorine and is made by the polymerization of 2-chloro-1,3-butadiene (chloroprene)

$$CH_2=CH-CCl=CH_2 \longrightarrow [CH_2-CH=C(Cl)-CH_2]_n$$

Chloroprene → Polychloroprene (Neoprene)

The chlorine content gives neoprene a high resistance to heat and flame and also to oil and chemicals. It is thus used for such things as petrol hose, speciality conveyor belts, balloons and protective gloves.

Chloroprene may be made by dimerization of acetylene to vinylacetylene followed by hydrochlorination of the latter:

$$2HC{\equiv}CH \longrightarrow H_2C=CH-C{\equiv}CH \xrightarrow{HCl} CH_2=CH-CCl=CH_2$$

Vinylacetylene is a dangerous material to make or handle and in a more recent process chloroprene is made by carefully controlled chlorination and dehydrochlorination of a very cheap mixed butene feedstock coming direct from a cracking unit. The major problem is to get the chlorine atom and the double bond in the correct places, and to prevent the whole mixture from polymerizing spontaneously. The acetylene route is still used by Dupont in Northern Ireland.

10.6.8 Ethylene–propylene rubbers

Copolymers of ethylene and propylene containing 20–80% ethylene can be made from the olefines using a Ziegler catalyst. They can be vulcanized (cross-linked) using dicumyl peroxide.

The rubber industry, however, is conservative and is geared to vulcanization with sulphur. It is therefore common for a terpolymer to be made from ethylene, propylene and about 3% of a diene, which can then be vulcanized with sulphur. Dicyclopentadiene used to be the favoured diene but the commonest is now ethylidene norbornene

$$\text{[bicyclic structure with CH}_2\text{ bridge and CH—CH}_3\text{ substituent]}$$

10.6.9 Miscellaneous rubbers

A wide variety of other rubbers is sold, but none in very large tonnages. They have nonetheless important specialized applications and some of them may become bulk chemicals in the future.

Polysulphide rubbers are made from organic dihalides and sodium polysulphides. Thiokol A is typical of the group:

$$Na_2S_4 + ClCH_2CH_2Cl \longrightarrow \cdots-\underset{\underset{S}{\|}}{S}-\underset{\underset{S}{\|}}{S}-CH_2-CH_2-\underset{\underset{S}{\|}}{S}-\underset{\underset{S}{\|}}{S}-CH_2\ CH_2-\cdots$$

Thiokol 'A' (before cross-linking)

Polysulphide rubbers have particularly good solvent resistance and are used for tank linings and for the taking of dental impressions.

Silicone rubbers are typified by polydimethylsiloxane

$$\begin{array}{c} CH_3 \quad\; CH_3 \quad\; CH_3 \\ | \qquad\; | \qquad\; | \\ -Si-O-Si-O-Si-O- \\ | \qquad\; | \qquad\; | \\ CH_3 \quad\; CH_3 \quad\; CH_3 \end{array}$$

They are expensive to make, the starting material being elementary silicon. The Si—O bond is about 20 kcal/mole (84 kJ/mole) more stable than the C—C bond so the polymers have great stability towards heat and chemical reagents and this is the area in which they are used.

Other rubbers are based on chlorosulphonated polyethylene, acrylate esters, epichlorhydrin, polyurethanes, and copolymers of vinylidene fluoride and hexafluoropropylene. Propylene–isoprene rubbers offer some hope of a very low cost rubber in the future.

10.7 Textile fibres

The origins of the textile industry are lost in biblical times when Adam delved and Eve span. The commercial revival of Western Europe in the eleventh and twelfth centuries was accompanied by a boom in the woollen industry which until the end of the thirteenth century was dominated by the Flemish weavers.

England had traditionally been an exporter of raw wool, but in 1347 Edward III put a mere 2% export duty on cloth exports compared with 33% on raw wool. This, plus the efforts of Flemish immigrants, raised cloth exports from 4 500 broadcloths/year in 1347–48 to 43 000/year in 1392–95, an annual growth rate of $5\frac{1}{2}$%, satisfactory even by the frenetic standards of the twentieth century. From then till the nineteenth century, England's prosperity rested primarily on wool and the Lord Chancellor sat and still sits on a woolsack in recognition of this fact. For most of the period, the industry was located in the South of England, in Gloucestershire, Norfolk and the Cotswolds. The increasing mechanization of the industry together with the gradual switch from woollen to worsted cloths made water power, access to coal, and a suitable atmospheric humidity essential and the industry slowly moved to Bradford and the Yorkshire Heavy Woollen district where it is still centred.

Cotton has also been known from time immemorial but did not become important in Western Europe till 1780–1800 when a series of discoveries dramatically reduced the cost of growing and processing it. Raw cotton was spun and woven in England and exported throughout the world. King Cotton was one of the pillars of Imperial prosperity.

The downfall of the British cotton industry began when indigenous cotton industries were set up in less developed countries and, since the First World War, it has declined dramatically. Between 1912, when they represented about one-quarter of total UK exports, and 1968, exports of cotton cloth dropped from 6 800 m. yd^2 to 224 m. yd^2.

The traditional textile industry was based on animal fibres (wool, hair, silk, etc.) with a long-chain protein structure, and vegetable fibres (cotton, linen, hemp, jute, flax, etc.) which are cellulose polymers. At the turn of the century, the industry drew on the chemical industry for synthetic dyestuffs and for bleach but the fibres themselves were all natural polymers.

The first semi-synthetic fibre was nitrocellulose which was developed by Sir Joseph Swan in England (1883–84) and Chardonnet in France (1884) as a by-product of the search for a strong carbon filament for lamps (see also section 2.2). Later, cuprammonium and viscose rayon fibres were developed but it was not until after the First World War that these semi-synthetic fibres based on cellulose made a serious impact on the textile market. In 1919, world production was about 11 000 tons; by 1924 it had risen sixfold to 61 000 tons and by 1929 it had trebled to 197 000 tons.

The development of fully synthetic fibres took place before the Second World War, but they did not make an impact on the textile market till the 1950s when the petro-

Table 10.18 World production of synthetic fibres in 1970 (excluding cellulosics) (m. lb)

	Acrylic	Nylon	Polyester	Polyolefine*	Other†	Total
Western Europe	885	1 315	1 016	102	41	3 359
Eastern Europe	158	398	203	19	43	821
North and South America { USA / other countries }	535	{ 1 355 / 254 }	1 625	287	16	{ 3 586 / 486 }
Japan / all other countries	634	860	778	144	203	{ 2 259 / 360 }
Totals	2 212	4 182	3 622	552	303	10 871

* Probably an underestimate; more likely figures are W. Europe 169, USA 259, others 328, total 756.
† Spandex elastomer, etc.

chemical revolution brought down the prices of raw materials and made the synthetic fibres comparable in price with traditional materials.

Synthetic fibres have been made from many polymers, the crystallinity of which can be increased to a sufficiently high value by drawing. The three main groups of synthetic

Table 10.17 Chemicals for synthetic fibres*

Materials	Production m. lb	US Tariff Comm. 1969 Sales m. lb	Unit value ¢/lb	US price ¢/lb 4/72	UK price £/ton 4/72
Acetic acid	1 770	426	6	4·3	73·5
Acetic anhydride	1 675	145	9	14	110·0
Carbon disulphide	799	509	4	4·8	—
Cellulose acetate	827	—	—	36	319·2
Other cellulose esters	1 026	229	34	—	—
Cellulose ethers	122	116	56	—	—
Cellulose plastics	193	187	59	—	—
Cyclohexane	2 232	2 062	3	24	—
Cyclohexanone	704	59	9	18	205
Cyclohexanol	—	5	22	28	220
Adipic acid	1 220	119	16	19	—
Hexamethylene diamine	663	—	—	—	—
Caprolactam	—	450	18	24·5	—
Nylon polymers for fibres / Polyamide resins for plastics	92	82	81	—	579·6
Chemicals for polyester fibres: See Tables 9.2 and 9.13					
Chemicals for acrylic fibres: See Table 9.4					

*Figures give total usage, not only that in fibres.

Table 10.20 UK prices of natural and synthetic fibres

	£/ton August 1972
Wool	
dry combed wool rope	1 325
Cotton	
US cotton c.i.f. Liverpool	300
Rayon	
ordinary, bright and matt rayon staple	290
167 decitex (150 denier) cuprammonium rayon	800
Nylon	
staple	770
continuous filament	900
Polyester	
staple	600
continuous filament	1 200
Acrylic	
staple	700
carpet grade	500

fibres, however, are the polyamides (e.g. nylon 6) the polyesters (e.g. Terylene and Dacron) and the newer acrylic fibres (e.g. Acrilan, Courtelle and Orlon). These polymers also have other non-textile uses such as polyester clear film (Melinex), nylon bearings for machinery and ABS copolymers. Other synthetic fibres are relatively unimportant, but mention should be made of polypropylene fibres which have displaced natural fibres in a large section of the rope market. Urethane fibres known variously as 'elastomerics' or 'spandex' have been used in garments requiring elastic properties (women's foundation garments, ski-wear, etc.) but their success is not yet assured. World production of synthetic fibres is shown in Table 10.18.

The growth of synthetic fibres in the UK relative to natural and semi-synthetic fibres is illustrated in Fig. 10.15 while the division of the market between the various man-made

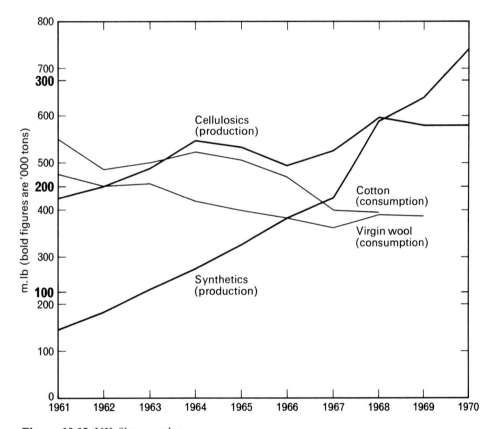

Figure 10.15 UK fibres market.

fibres is shown in Fig. 10.16. Tables 10.19–20 show the price and market structure of the fibres market and of their chemical precursors. Raw material prices of all synthetics dropped substantially between 1960 and 1970 (Fig. 3.21) but the price of polyester staple fibre dropped much more than that of polyamides or acrylics. There is obviously massive over-capacity for all three fibres. This suggests that there is now greater scope for price cuts in the last two fibres and future growth of polyesters may well be slow.

292 *Industrial polymers*

The most significant change which has taken place in the textile industry during the 1960s has been the erosion of the traditional weaving process by machine knitting techniques. The latter are particularly suitable for man-made fibres thus reinforcing other reasons for using them, and in addition, the knitting system is highly flexible and allows quick reaction to market needs. The process is simple and fast and the equipment requires little down-time. The trend towards knitting seems bound to continue, as do the other trends towards man-made and away from natural fibres.

A growing area with much potential for the future is non-woven fabrics. The fibres in these are matted together either under the influence of rolling, beating and pressure, as in traditional felt, or polyvinyl acetate or polyethyl acrylate adhesives as in the more modern materials. Fibres of nylon and rayon can be bonded to give a web-like structure, and fibres of paper can be bonded to give the materials used in disposable underwear, hospital sheets, etc.

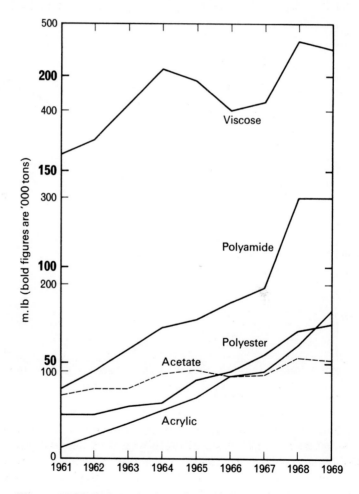

Figure 10.16 UK production of synthetic and semi-synthetic fibres 1961–69.

The end-use pattern for synthetic fibres is shown in Fig. 10.17.

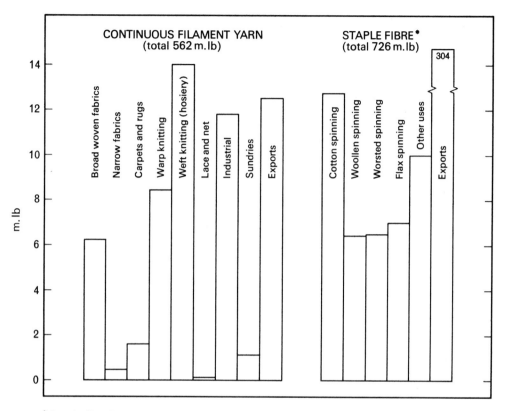

*Staple fibre is made by chopping continuous filament into short lengths. In this form, it is suitable for blending with other fibres before spinning

Figure 10.17 End-uses of man-made fibres (including cellulosics) – UK 1970.

10.7.1 Cellulosics

Rayon is a regenerated cellulose fibre in which up to 15% of the hydroxyl hydrogens have been replaced in a cross-linking operation. In the manufacture of viscose rayon, cellulose (wood pulp, straw, etc.) is dissolved in sodium hydroxide, and carbon disulphide is added to give cellulose xanthate. Cellulose has the formula $(C_6H_{10}O_5)_n$ and consists of long chains of glucose molecules linked by the elimination of water:

294   *Industrial polymers*

The formation of the xanthate may be written:

$$-\underset{|}{\overset{|}{C}}-OH \xrightarrow{NaOH} -\underset{|}{\overset{|}{C}}-ONa \xrightarrow{CS_2} -\underset{|}{\overset{|}{C}}-O-\underset{\parallel}{C}-SNa$$
$$\phantom{-\underset{|}{\overset{|}{C}}-O-\underset{}{C}-}S$$

0·5–0·6 xanthate groups are inserted per glucose nucleus and the resultant compound is dissolved in dilute sodium hydroxide to give the so-called 'viscose' solution. The latter may be extruded through very narrow holes (spinnarets) into a sulphuric acid bath which neutralizes the sodium hydroxide, hydrolyses the xanthate and regenerates the cellulose as a continuous filament. Its chain length is about 600 glucose units compared with 1 200–3 000 for the original cellulose. Alternatively, the viscose solution can be extruded through a narrow slot to give transparent regenerated cellulose sheet, known as cellophane, which is treated with a waterproofing agent and used as a packaging material.

In a related process, linters or wood pulp are dissolved in an ammonaical solution of copper oxide. This is spun similarly to viscose rayon by extrusion through spinnarets into dilute sulphuric acid. Cuprammonium rayon is chemically similar to viscose rayon but gives a finer yarn which is used in sheer fabrics and which commands a premium price.

Unlike rayon, cellulose acetate is an actual chemical compound of cellulose. It is made by partial hydrolysis of cellulose triacetate obtained by the action of equal quantities of glacial acetic acid and acetic anhydride on chemical cotton. The latter is cellulose obtained by purification and conversion of cotton linters:

Cellulose + 3n Acetic anhydride $\xrightarrow{H_2SO_4}$ Cellulose triacetate + $3n\,CH_3COOH$

Hydrolysis to between 51 and 62·5% acetyl content is carried out by dilution of the triacetate solution with water plus a little acetic and sulphuric acids. Addition of a large excess of water stops the hydrolysis at the appropriate stage and precipitates the acetate as flakes. These may be used (for example, as plastics moulding materials) or dissolved in acetone and ethyl alcohol and extruded through a perforated plate into a hot air stream which evaporates the solvent and leaves a continuous filament. The latter process is known as dry spinning.

Cellulose triacetate itself is more difficult to process than the acetate but can be heat-set to give 'wash-and-wear' fabrics. Other commercial cellulose esters include the nitrate (nitrocellulose) used in surface coatings, and the propionate and acetate–butyrate which are used as plastics materials.

The growth of the market for cellulose fibres has been hit by the increasing use of wholly synthetic fibres, and the market for cellulose plastics is more or less static. Cellulose acetate with triacetin (glyceryl triacetate) as an adhesive linking the separate fibres is still the preferred material for cigarette filter tips, but otherwise cellulosics have a reputation inferior to that of the more modern materials.

10.7.2 Nylon

Nylons are polyamides and are named according to the number of carbon atoms in the intermediates which react together to give the fibre. Thus the nylon from adipic acid and hexamethylene diamine:

$HOOC(CH_2)_4COOH$ + $H_2N(CH_2)_6NH_2$ ⟶

   Adipic acid   Hexamethylene diamine
(6 carbon atoms)   (6 carbon atoms)

$$-CO(CH_2)_4CO.NH(CH_2)_6NH.CO(CH_2)_4CO.NH(CH)_6NH-$$

is called nylon 66. Similarly the nylon from caprolactam:

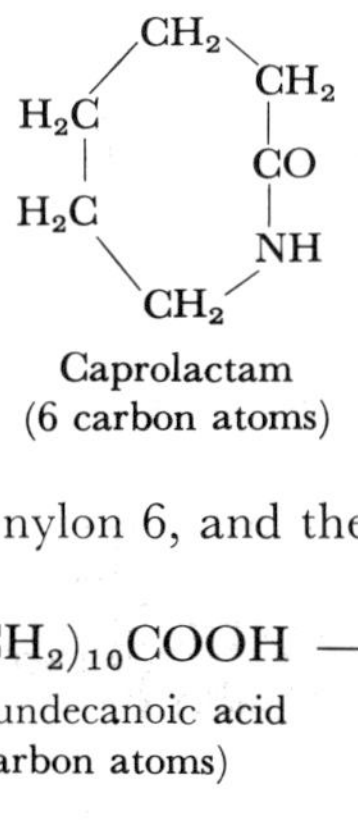

Caprolactam
(6 carbon atoms)

⟶ $-CO.NH(CH_2)_5CO.NH(CH_2)_5CO.NH(CH_2)_5CO-$

is called nylon 6, and the nylon from ω-aminoundecanoic acid:

$H_2N(CH_2)_{10}COOH$ ⟶ $-CO.NH(CH_2)_{10}CO.NH(CH_2)_{10}CO.NH(CH_2)_{10}-$
ω-Aminoundecanoic acid
(11 carbon atoms)

is called nylon 11.

Polyamide fibres were the first wholly synthetic fibres to be produced commercially. Although their structure is close to that of the animal fibres, they lack the moisture absorbency and softness of, for example, wool, but they are hard-wearing and make up into garments which are 'drip-dry'.

The routes by which nylons are manufactured are complicated and involve a larger number of steps than is usual in the manufacture of large-tonnage industrial chemicals. The reaction sequences leading to nylon 6 and nylon 66 are shown in Figs. 10.18 and 10.19. Both nylons can be made from benzene or cyclohexane feedstock which is air-oxidized by a variety of processes to give a mixture of cyclohexanol and cyclohexanone.

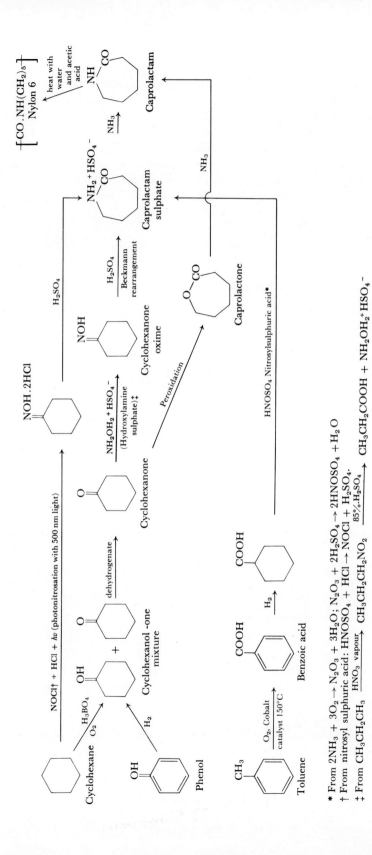

Figure 10.18 Routes to caprolactam and nylon 6.

* From $2NH_3 + 3O_2 \rightarrow N_2O_3 + 3H_2O$; $N_2O_3 + 2H_2SO_4 \rightarrow 2HNOSO_4 + H_2O$
† From nitrosyl sulphuric acid: $HNOSO_4 + HCl \rightarrow NOCl + H_2SO_4$.
‡ From $CH_3CH_2CH_3 \xrightarrow{HNO_3 \text{ vapour}} CH_3CH_2CH_2NO_2 \xrightarrow{85\%, H_2SO_4} CH_3CH_2COOH + NH_2OH_2^+HSO_4^-$

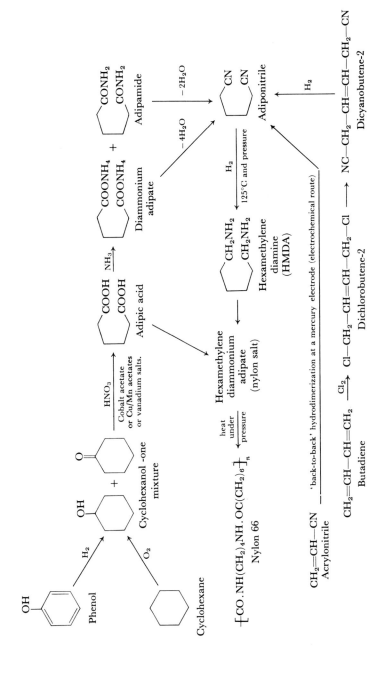

Figure 10.19 Routes to adipic acid, hexamethylene diamine and nylon 66.

298 *Industrial polymers*

In a widely used method for nylon 66 manufacture, this is oxidized with nitric acid in the presence of copper and manganese acetates to give adipic acid. Part of the adipic acid is set aside, and the rest converted to hexamethylene diamine. These are reacted together to give 'nylon salt' (hexamethylene diammonium adipate) and this, when heated under pressure in an autoclave in 60% aqueous solution at 280°C, gives nylon.

The production of nylon 6 is simpler in that it requires only a single monomer, caprolactam. In one route the cyclohexanol–cyclohexanone mixture is converted to an oxime by treatment with hydroxylamine, and this, on treatment with concentrated sulphuric acid, undergoes the classical Beckmann rearrangement to give caprolactam sulphate. Ammonia gives caprolactam and ammonium sulphate as a by-product. Heating at 250°C with 10% water and 1% acetic acid to control the chain length gives the polymer.

Nylon 11 is made only in France, Japan and West Germany and is derived from castor oil. This is hydrolysed to glycerol and ricinoleic acid and the latter is cleaved at the double bond to give ω-aminoundecanoic acid and n-heptyl alcohol. Nylon 11 is one of the few chemicals which still depends on an agricultural feedstock. The castor oil is grown mainly in France's former colonies in North Africa, and France has gone to great lengths to try to stabilize the raw material position. Other commercially available nylons include nylon 610 (based on sebacic acid) and nylon 12.

Bulk nylon is turned into a fibre by melt spinning. A molten mass of nylon is extruded through spinnarets and as the jets emerge they are cooled and solidified by a stream of cold air. At this stage, the nylon has not been stretched or drawn and does not show its natural strength and lustre as the polymer chains have not been aligned, that is, the polymer is insufficiently crystalline.

The filaments are therefore passed over two rollers, the second of which is rotating faster than the first. They are thus drawn to about four times their original length and acquire their characteristic properties.

About three-quarters of UK nylon consumption goes into clothing, and about half of this is textured yarn (stretch nylon). Other textile uses include tufted carpets from nylon, nylon–rayon or nylon–wool mixtures. Some nylon is also used in tyre cord and in gear wheels and machine bearings which make less noise than metal gears and do not require lubrication. The use in tyre cord is the largest UK industrial outlet for nylon, but it is still small compared with the USA where it accounts for about one-third of total nylon consumption. In the UK, rayon is still preferred.

Nylon 66 has greater strength, higher modulus, greater resistance to oxidation and a higher melting point than nylon 6. The latter, as it involves a single monomer, is marginally cheaper to manufacture. Meanwhile, nylon 6 has made few inroads into the nylon 66 market in the USA or UK but is the dominant fibre in the rest of Europe and Japan.

10.7.3 Polyesters

Polyethylene glycol terephthalate ('Terylene' and 'Dacron' when made by ICI and Dupont respectively) is a condensation polymer made by ester interchange of ethylene glycol with dimethyl terephthalate, or by direct polyesterification of ethylene glycol with terephthalic acid (Fig. 10.20). It is the only saturated polyester of any importance, although there is a small market for low molecular weight polyesters as non-migratory plasticizers (section 10.3.4.4).

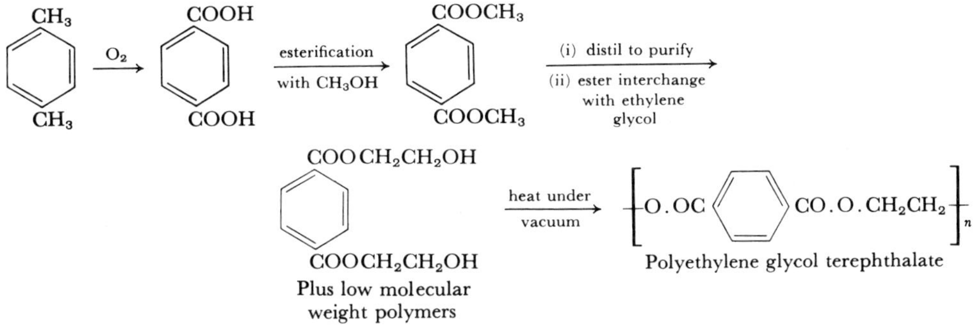

Figure 10.20 Manufacture of polyethylene glycol terephthalate.

Dimethyl terephthalate is sometimes used in preference to terephthalic acid for two reasons: the straightforward esterification reaction goes rather slowly, and there are problems associated with the purification of terephthalic acid. The latter is obtained from liquid-phase air oxidation of p-xylene (section 9.5.3) and older plant gives a product only about 77% pure. The easiest way to purify it is to esterify the acid with methyl alcohol and distil off the resulting dimethyl terephthalate which, having been obtained, might just as well be used in that form. The methyl alcohol is, of course, recovered after the ester interchange. Recent variations of the xylene oxidation process give a purer terephthalic acid, and in 1970 only 38% of polyester fibre production derived from installations producing only dimethyl terephthalate.

The polyethylene glycol terephthalate molecule contains the polar ester group. It has no side chains and the aromatic ring lends it rigidity. It can therefore be obtained in crystalline form and has fibre properties. The fibres are, in fact, spun from the melt in the same way as nylon.

Polyester fibres have great tensile strength and abrasion resistance being equal to polyamides and better than acrylics. On the other hand, they have a high specific gravity (polyesters 1·38, polyamides 1·14, polyacrylonitriles 1·15) so that the weight/unit area of textile fabric is high with a corresponding economic disadvantage. Polyester fibres are used by themselves or in a blend with cotton. The latter has proved successful in recent years as the blend has better moisture absorbency and opacity than the pure polyester and better 'wash and wear' properties than pure cotton.

The use in textiles is easily the largest outlet for polyethylene glycol terephthalate though a market also exists for the clear film 'Melinex'.

10.7.4 Acrylonitrile and acrylic fibres

Though acrylonitrile is now primarily used as a monomer for acrylic fibres, it first came into prominence during the Second World War as a copolymer in nitrile rubber which was used as a self-sealing agent in aircraft gasoline tanks.

Acrylonitrile was originally manufactured by the dehydration of ethylene cyanohydrin which in turn was made by the addition of hydrogen cyanide to ethylene oxide:

$$H_2C\underset{O}{\underset{\diagdown\diagup}{-}}CH_2 + HCN \longrightarrow \underset{OH\ \ CN}{H_2C-CH_2} \xrightarrow{-H_2O} CH_2=CH-CN$$

The process was superseded by one based on the addition of hydrogen cyanide to acetylene in the presence of cuprous chloride:

$$C_2H_2 + HCN \longrightarrow CH_2=CH-CN$$

This is apparently a cheap and simple process but has certain economic drawbacks. The price of acetylene is high compared with olefine feedstocks, and hydrogen cyanide is a dangerous reagent made by reaction of ammonia, air and methane over a platinum catalyst:

$$2NH_3 + 3O_2 + 2CH_4 \longrightarrow 2HCN + 6H_2O$$

Its use is almost entirely captive, so it is difficult to attach a price to it, but it is certainly expensive compared with ammonia.

These factors, combined with the underlying trend towards olefine feedstocks, have led to the replacement of the above process by the newer ammoxidation process based on propylene. Indeed, ICI 'mothballed' a plant of the above type the day it was completed.

The ammoxidation route to acrylonitrile is a good example of 'hot tube' chemistry. Propylene from the refinery, without further purification, is mixed with an equal amount of ammonia and twice as much oxygen and the mixture passed over a stannic oxide–antimony oxide catalyst at 450°C and 2–3 atm. Acrylonitrile is obtained in about 70% yield together with about 10% yields of acetonitrile and hydrogen cyanide. The hydrogen cyanide may be used for methyl methacrylate production, but few outlets for acetonitrile have been found.

In the British process, developed by the Distillers Company, propylene is converted first to acrolein via the relatively stable allyl radical:

$$CH_2=CH-CH_3 \longrightarrow \underset{\text{Allyl radical}}{[CH_2\cdots CH\cdots CH_2]^{\cdot}} \xrightarrow{O_2} \underset{\text{Acrolein}}{CH_2=CH-CHO}$$

and then, without any actual treatment of the acrolein, it is converted to acrylonitrile:

$$CH_2=CH-CHO + NH_3 + \tfrac{1}{2}O_2 \longrightarrow CH_2=CH-CN + H_2O$$

In the US variant of the process, developed by Standard Oil of Ohio (SOHIO), the whole reaction is carried out in a single step over a bismuth phosphomolybdate catalyst. There was a patent suit in the early 1960s between the Distillers Company and SOHIO which hinged on whether the SOHIO one stage process was an infringement of Distillers' patents on the two-stage process. Was acrolein an intermediate in the one stage process? This was one case where a reaction mechanism was of more than academic importance but the case was eventually settled out of court.

The main advantage of the SOHIO process is that it takes place in a fluidized bed and can therefore be operated on a greater scale than the DCL fixed bed process. The latter, however, has a potential advantage in that if some of the acrolein stream were diverted, it could be converted very simply to acrylic acid whose esters are important monomers in paints, etc. (section 10.5.2). The advent of the ammoxidation route to acrylonitrile brought down the UK price from £280 to £128/ton.

A process which may become important in future has been developed by Knapsack-Griesheim in Germany. Acetaldehyde, made from ethylene *via* the Wacker process, is reacted with hydrogen cyanide to give lactonitrile which is dehydrated to acrylonitrile:

$$CH_3CHO + HCN \longrightarrow CH_3CHOH \cdot CN \xrightarrow{-H_2O} CH_2{=}CH{-}CN$$

Yields are said to be higher than from ammoxidation and less purification of the product is required.

Acrylonitrile may be polymerized to polyacrylonitrile by free radical or anionic but not cationic polymerization:

$$CH_2{=}CH{-}CN \longrightarrow {-}CH_2{-}\underset{CN}{CH}{-}CH_2{-}\underset{CN}{CH}{-}CH_2{-}\underset{CN}{CH}{-}$$

Polyacrylonitrile chains are compact and the nitrile groups interact strongly with one another, so the polymer has fibre properties. It was impossible until the mid 1950s to use it as such because no solvent for it was commercially available. Since dimethylformamide has come on to the market, dry spinning of polyacrylonitrile has become feasible and the use of polyacrylonitrile in textiles has increased rapidly. 'Orlon' is a polyacrylonitrile homopolymer, while 'Acrilan' and 'Courtelle' contain small amounts of copolymers such as vinyl chloride.

Dimethylformamide is a highly polar solvent – it also dissolves acetylene – and is made by reaction of dimethylamine with methyl formate or carbon monoxide plus sodium methoxide:

$$H \cdot COOCH_3 + (CH_3)_2NH \longrightarrow (CH_3)_2N{-}\underset{H}{\overset{}{C}}{=}O \; ({+} \; CH_3OH)$$
$$\text{Dimethylformamide}$$

$$(CH_3)_2NH + CO \xrightarrow[CH_3OH]{CH_3ONa} \nearrow$$

Polyacrylonitrile fibres are about 50% cheaper than wool. Textiles based on them have a woolly feel and have made a substantial impact on the cheap end of the wool market. They have the great advantage of being machine-washable and non-shrink. In 1969, UK consumption of acrylonitrile was about 110 000 tons of which 77 000 tons went into fibres, compared with 58 000 tons into fibres in 1967.

Acrylonitrile is also used as a copolymer in plastics, rubbers and stoving finishes (sections 10.2.3, 10.5.5). A minor use is in the cyanoethylation of cotton to improve its properties.

Because polyacrylonitrile can be made from propylene in only two steps, it is potentially a cheaper fibre than nylon or polyesters and demand for it seems likely to continue to rise rapidly.

Notes and references

General Polymers are of such tremendous interest from academic, industrial and commercial points of view, that they have attracted a huge literature. The following biblio-

302   *Industrial Polymers*

graphy consists of books we ourselves have used, and we apologize for overlooking many others, no doubt of equal excellence.

The polymer equivalent of Kirk–Othmer is the massive *Encyclopaedia of Polymer Science and Technology,* Interscience, New York, 1965, and this provides a comprehensive source of information and data. Iliffe, part of the Butterworth Group (London), publishes a series of monographs on behalf of the British Plastics Institute dealing individually with a large number of polymers. Reinhold (New York) publishes a similar series on *Plastics Application.* The books in these two series could well fit into an overcoat pocket. A more comprehensive and academic treatment of the chemistry, physics and technology of high polymeric substances is given in *High Polymers,* Wiley–Interscience, New York, a series which has so far got through 24 volumes. Readers in search of information about an individual polymer are likely to find a volume on it in one of these series.

Other general books, less comprehensive but more compact, include the still valuable classic, B. Golding, *Polymers and Resins,* Van Nostrand, New York, 1959, and F. W. Billmeyer, Jr., *Textbook of Polymer Science,* Wiley, New York, 1962. An inexpensive hard-back on which we have drawn is L. K. Arnold, *Introduction to Plastics,* Allen and Unwin, London, 1969 (also published in USA in 1968 by Iowa State University Press). This gives a simple non-chemical approach to the subject. Two useful paperbacks are A. J. Gait and E. J. Hancock, *Plastics and Synthetic Rubbers,* Pergamon, London, 1970, and T. Alfrey and E. F. Gurnee, *Organic Polymers,* Prentice Hall, New Jersey, 1967.

Consumption and production of plastics are far better documented than those of other materials. In particular, the January editions of *Modern Plastics* and *British Plastics* give detailed analyses of the previous years' market figures in the USA and UK respectively. *US Tariff Commission* and *UK Business Monitor* and *Annual Abstract of Statistics* provide less information later, but with official blessing. All market and end-use data in this chapter, except where otherwise mentioned, come from these sources. Costings are taken from Hahn.

The physical and mechanical properties of plastics are quantified by standardized testing methods which in the UK are laid down in *B.S. 2782 Methods of Testing Plastics,* British Standards Inst., London, 1970. Those in use in the USA are given in *American Society for Testing Materials* ASTM-2240-68.

Section 10.1   Figure 10.2 (a) and (b) come from an article by A. Keller and F. M. Willmouth, *Materials Science Club, Bulletin No.* 20, December 1969, and 10.2 (c) is an unpublished electron micrograph by M. G. Bader of the University of Surrey.

Section 10.2.1   Polyethylene is treated comprehensively in a series of articles in the 781 page *Polythene* (ed. A. Renfrew and P. Morgan), 2nd ed., Iliffe, London, 1960. Market data on polyolefines are given by K. H. Römtz, *ECMRA Budapest,* 1970, Figure 10.3 is based on data from Hahn.

Section 10.2.2   Polypropylene is included with polyethylene in K. H. Römtz, op. cit. Polypropylene must be used with U.V. stabilizers and other additives. We have mentioned these only cursorily in this book but they are dealt with by H. E. Kinne, *ECMRA Amsterdam,* 1970.

Sections 10.2.6 and 10.2.7   Engineering plastics are discussed as such by Arnold, op. cit. and also in separate papers by A. D. Plaistowe and G. Battson, *ECMRA Amsterdam,* 1970.

Section 10.2.8 The world market for fluoroplastics is dealt with by E. Tarnell, *ECMRA Amsterdam*, 1970.

Section 10.3 M. Kaufman has written a *History of Polyvinyl Chloride*, McLaren, London, 1969. Technical aspects of PVC are discussed by H. A. Sarvetnick, *Polyvinyl Chloride*, Van Nostrand–Reinhold, New York, 1969, and European market aspects by H. Hurard, *ECMRA Budapest*, 1970. The role of plastics additives is discussed by G. D. Allen, *ECMRA Amsterdam*, 1970 and Fig. 10.6 is based on his paper. The same volume contains a paper by R. C. Corwin on fire retardant intermediates and additives.

Section 10.3.3 A much more detailed treatment is given in *Stabilization of Polymers and Stabilizer Processes*, Advances in Chemistry Series, No. 85, ACS, Washington, 1968.

Section 10.3.4 Books on PVC normally deal also with plasticizers. A specialized book is the classic D. N. Buttrey, *Plasticizers*, 2nd ed., Cleaver-Hume, London, 1957. Its approach is largely empirical and a science-based approach is to be found in *Plasticization and Plasticizer Processes*, Advances in Chemistry Series, No. 48, ACS, Washington, 1965.

Section 10.3.4.1 Data in Fig. 10.8 come from Sarvetnick, op. cit. but have been checked against B.P. Chemicals International trade literature.

Section 10.3.5 ICI produces an excellent glossy book on Cereclors from which Fig. 10.8 is taken.

Section 10.4 The development of U/F, P/F and M/F resins in Western Europe is discussed by M. L. M. Vermin, *ECMRA Amsterdam*, 1970 and Fig. 10.9 is based on his data.

Section 10.4.5 Urethane polymers in Western Europe are dealt with by G. D. Allen and Ch. W. Fryer, *ECMRA Budapest*, 1970.

Section 10.4.6 For a recent account of developments in the epoxy resin field, see *Epoxy Resins*, Advances in Chemistry Series, No. 92, ACS, Washington, 1970.

Section 10.5 The paint industry is one of the least documented branches of the chemical industry. Apart from the publications of the Oil and Colour Chemists Association, the only significant books we have seen are *The British Paint Industry*, F. Armitage, Pergamon, 1964, and D. H. Parker, *Principles of Surface Coating Technology*, Wiley, New York, 1965. A useful series of articles by J. M. Butler appeared in *Paint Manufacturer* in 1966 and, of course, we have drawn extensively on *A Study of the U.K. Paint Industry*, D. R. Palmer, B. G. Reuben and M. L. Burstall, University of Surrey, 1970. Stirling and Co.'s *Chemical Industry Handbook* has an excellent section on paint.

Section 10.5 Data from BOT *Census of production*, 1963, part 31, Paint and Printing Ink Industries and *Annual Abstracts of Statistics*.

Section 10.5.1 Data from DTI *Business Monitor*.

Section 10.5.6 Cost of acrylic acid manufacture is estimated in *European Chemical News*, 1 August 1969. The future of the chloride route is discussed in Stirling and Co's investment survey of Laporte though Laporte's problems with the new plant had not arisen at that time.

Section 10.6 A classic text-book is M. Merton, *Introduction to Rubber Technology*, Reinhold, New York, 1959, and a comprehensive modern one is C. M. Blow, *Rubber Technology and Manufacture*, Butterworth, London, 1971. An article on world rubber market trends and their interpretation by H. W. Kleinecke appears in *ECMRA Budapest*, 1970, and Table 10.14 is taken from it. UK end-use pattern in Fig. 10.15 comes from the *Annual Abstract of Statistics*.

Section 10.7 The textile industry has been a basic industry in the UK for centuries and the effects on it of the advent of man-made fibres have been discussed in a number of reports, for example, Clothing Economic Development Corporation, *Your Future in Clothing*, HMSO, London, 1970; Wool Textile Economic Development Corporation, *The Strategic Future of the Wool Textile Industry*, HMSO, London, 1969; *Cotton and Allied Textiles: A Report on Present Performance and Future Prospects*, Textile Council, 1969; and Hosiery and Knitwear Economic Development Corporation, *Hosiery and Knitwear in the 1970s*, HMSO, London, 1969.

The scientific and technological aspects of man-made fibres are dealt with in the two volume work, H. F. Mark, S. M. Atlas and E. Cernia (eds.), *Man-Made Fibres: Science and Technology*, Interscience, New York, 1966, in R. W. Moncrieff, *Man-Made Fibres*, 5th ed., Butterworth, London, 1970; in M. E. Carter, *Essential Fiber Chemistry*, Marcel Dekker, New York, 1971, and in a useful paperback, J. E. McIntyre, *The Chemistry of Fibres*, Edward Arnold, London, 1972.

Some statistical data appear in P. Hallam (compiler), *Man-Made Fibres in the U.K.*, Shirley Institute, Manchester, 1971, and a useful reference work is Lennox-Kerr (ed.), *Index to Man-Made Fibres of the World*, 3rd ed., Harlequin Press, 1967. Statistical material not from sources already cited may be found in the *Textile Council Quarterly Statistical Review* and *Textile Organon*. Each year's June issue of the latter is particularly useful. Data on the early 14th century wool trade came from the *Encyclopaedia Britannica*.

Data in Table 10.17 are from T. B. Kozlowsky, *ECMRA Budapest*, 1970. Prices in Table 10.19 come from the *Financial Times* and the British Man-Made Fibres Association. Prices paid for synthetics are likely to be lower than those quoted, especially in the case of polyester filament. Table 10.20 comes from *Textile Organon*, June 1971. UK wool prices rose sharply in 1972.

Section 10.7.3 The terephthalic acid versus dimethyl terephthalate controversy is discussed by T. B. Kozlowsky, *ECMRA Budapest*, 1970.

11 Soaps and detergents

11.1 The soap trade

Soap-making is the oldest and the most widespread of chemical operations. Soap was discovered in late Roman times; by the reign of Elizabeth I it was being manufactured on a substantial scale and, as Table 11.1 shows, the industry has now spread to almost all countries. So long as people wish to be clean they will want soap and experience shows that cleanliness increases with increasing national prosperity. In Britain, for example, the *per capita* consumption of soap increased by approximately four times during the nineteenth century and industrialization produced parallel changes in other nations. A high consumption of cleansing materials is an infallible sign of an advanced economy.

Over the past fifty years the industry has both expanded in size and changed in basic technology. In Western Europe and North America soaps have been increasingly displaced by cheaper and more effective synthetic equivalents. Cleansing products are now marketed in an increasing range of forms. Household soap (bar soap), once the staple of the trade, has lost much of its importance in advanced countries and has been largely replaced by more convenient products. Detergent powders and liquids, toilet soaps, shampoos and scourers have been developed for particular tasks where once yellow bar soap was used for all purposes.

Soaps and detergents are made by chemical processes but the soap trade stands apart from the rest of the chemical industry. To some extent the differences arise from differences in technology. The processes used in soap-making are in principle simple and have changed little during the past twenty years. The plant involved is notably less expensive than is usual in the chemical industry: whereas ICI or Du Pont expect annual sales to be equal to between 100 and 200% of total plant investment, the corresponding figure for Procter and Gamble or Colgate–Palmolive is between 400 and 600%.

Even more important is the nature of the market. Chemical companies sell their products to a limited number of sophisticated customers, whose decisions are based on rational economic grounds. Soap companies sell directly to housewives – to millions of unsophisticated customers, unresponsive to small differences in quality and price but powerfully influenced by social and psychological factors. In the broadest sense chemical companies are production-oriented while soap companies are preoccupied by the problems of marketing.

11.2 Surface-active agents and their properties

Components common to almost all cleansing materials are surface-active agents. A prominent characteristic of a liquid is its surface tension (μ) which is a measure of the

Table 11.1 World production of soaps, washing and cleansing mixtures 1968

	Soaps				Synthetic surfactants			Other	TOTAL	per capita consumption (kg/head)
	House-hold	Toilet	Flakes	Powders	Powders	Liquids	Scourers			
Western Europe	438*	285	31	238	2 172	346	324	291	4 281	12·7
Great Britain	31	93	8	146	273	116	89	12	798	12·7
EEC	222	145	14	48	1 523	172	178	232	2 534	13·0
Others	185	47	9	44	440	58	57	47	949	11·0
Eastern Europe	1 084	176	3	133	603	n.a.	n.a.	97	2 230	6·2
North America	35	327	77	126	1 905	1 008	231	592	4 338	19·2
USA	32	306	73	119	1 793	955	231	592	4 137	20·1
Central and South America	694	125	n.a.	66	305	20	23	n.a.	1 279	5·0
Asia	1 030	210	192	29	584	140	3	n.a.	2 416	1·3
Japan	92	32	n.a.	16	430	129	n.a.	n.a.	729	7·2
Australasia	22	28	3	28	50	48	2	n.a.	207	14·3
Africa	325	25	n.a.	2	124	n.a.	n.a.	24	710	2·4
TOTAL	4 190	1 800	410	995	10 202	2 992	1 138	1 887	24 608	

* Figures in units of '000 tons.

Surface-active agents and their properties 307

forces acting between the component molecules. In exact terms it is the work done in bringing a molecule from the interior of the liquid, where it is attracted by molecules on all sides, to the surface, in which it is attracted by adjacent molecules in the surface, in the interior of the liquid, and in the other phase (Fig. 11.1). The surface tensions of

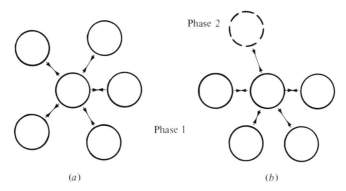

Figure 11.1 Forces acting on a molecule (a) in the interior of a liquid (b) at a phase interface.

some pure liquids are shown in Table 11.2, from which it is obvious that surface tensions are appreciable in magnitude and that they vary markedly with the nature of the liquid and of the interface. The surface tension of a liquid is greatest when the intermolecular forces are large and the attractive forces exerted by the other phase are small; thus the surface tension of water at the air–water or hexane–water interfaces is high because water is hydrogen-bonded and air or hexane molecules do not greatly attract water molecules. Conversely it is low at the *n*-butanol–water interface because of the ability of *n*-butanol to form hydrogen-bonds with water.

Table 11.2 Surface tensions (dynes/cm) of various liquids at 20°C

Liquid–Vapour		Liquid–Liquid	
H_2O	72.0	H_2O—$CH_3(CH_2)_2CH_2OH$	1.8
$(CH_3)_2S^+$—O^-	43.5	H_2O—C_6H_6	35.0
C_6H_6	28.9	H_2O—C_7H_{16}	50.2
CH_3OH	22.6		
C_7H_{16}	19.7	$(CF_2)_n$—C_6H_6	7.8
C_7F_{16}	11.0	$(CF_2)_n$—H_2O	57.0

In general the surface tensions of solutions differ from those of pure liquids. As increasing amounts of a solute are dissolved in a solvent, the surface tension of the solution may slowly increase, as in the case of sodium chloride–water; it may slowly decrease, as in the case of acetic acid–water; or it may fall rapidly to a low constant value (cf. Fig. 11.2). Compounds which show this last type of behaviour are termed surface-active agents or surfactants. They are of the structure R–X where R is a water-insoluble (hydrophobic) group – usually a hydrocarbon residue containing twelve or more carbon atoms – and X is a water-soluble (hydrophilic) group, which may be anionic, cationic or non-ionic

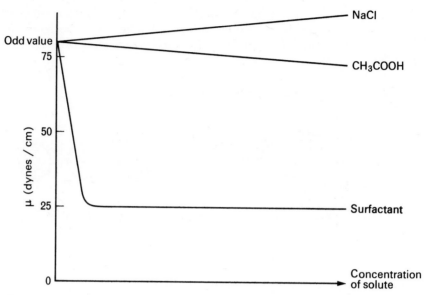

Figure 11.2 Surface tension of solutions of various solutes in water.

in nature (cf. Table 11.3). This combination of hydrophilic and hydrophobic properties reduces the attractive forces between surfactant and solvent and leads to the unusual properties of surfactant solutions. In dilute aqueous solution the surfactant forms a monomolecular layer at the surface of the liquid instead of being randomly scattered throughout its bulk. Moreover, above a certain critical concentration, surfactant molecules associate to form micelles, stable, highly-ordered spherical or rod-shaped aggregates (cf. Fig. 11.3) arranged with the hydrophilic ends outward and the hydrophobic ends inward.

Surfactants have many applications. They are wetting agents: they promote good contact between water and water-repellent surfaces. In isolation, liquids form spherical droplets since each molecule in the surface is pulled inwards by a force perpendicular to the surface (cf. Fig. 11.1). On a waxy surface, water forms comparatively large droplets because the attractive forces between solid and liquid are small compared with those within the liquid. In a surfactant solution the attractive forces between molecules in the surface and in the interior of the liquid are reduced; conversely, the presence of the hydrophobic ends of the surfactant molecules increases the forces between solid and liquid. Droplet formation is inhibited and the solution spreads smoothly over the solid surface. Such behaviour is desirable when aqueous solutions are to be applied to fatty or waxy surfaces as in the use of insecticidal sprays and dips or of printing ink. An interesting application of the same principle in reverse is the weather-proofing of fabrics, in which the fibres are given a waxy finish in order to promote the formation of large water droplets which cannot easily pass through the material under normal pressures.

Concentrated solutions of surfactants also act as solubilizing agents: they aid the solution of otherwise insoluble materials. This behaviour is only observed when micelles are present and is believed to involve the incorporation of the solubilized material into the micelle. Three modes of solubilization have been suggested: non-polar molecules may

Table 11.3 Structure, use and production of typical surfactants

Surfactant	Typical structure	1970 Production (000 tons) UK	Other Western Europe	USA	USA price ($/ton)	Uses*
Anionics						
Soaps		380	1 450	1 238	330	D
Synthetics		280	1 000	425	—	
Dodecylbenzene sulphonate	$C_{12}H_{25}$–⟨phenyl⟩–$SO_3^- M^+$	100	400	813	—	D
Alkyl sulphonate	$C_nH_{2n+1}SO_3^- M^+$	90	350	268	420	D
α-Alkenyl sulphonate	$C_nH_{2n+1}CH=CH(CH_2)SO_3^- M^+$	n.a.	20	—	—	D
Primary alkyl sulphate	$C_nH_{2n+1}CH_2OSO_2O^- M^+$	—	—	23	1 100	D, W
Alkyl ethoxysulphate	$C_nH_{2n+1}O(CH_2CH_2O)_xCH_2CH_2OSO_2O^- M^+$	—	—	49	1 200	D
Cationics						
Alkylamine salts	$C_nH_{2n+1}\overset{+}{N}H_3 X^-$	4	40	77	925	
Quaternary ammonium salts	$C_nH_{2n+1}\overset{+}{N}X(CH_3)_3 X^-$	—	—	22	820	G, H
Quaternary pyridinium salts	$C_nH_{2n+1}\overset{+}{N}$⟨pyridinium⟩ X^-	—	—	25	970	G, H
Amine oxides	$C_nH_{2n+1}\overset{+}{N}(CH_3)_2\text{—}O^-$	—	—	18	1 000	D, FS
Non-ionics						
Ethoxylated nonylphenols	C_9H_{19}–⟨phenyl⟩–$O(CH_2CH_2O)_xCH_2CH_2OH$	30	150	440	440	D, E, W
Ethoxylated primary alcohols	$C_nH_{2n+1}O(CH_2CH_2O)_xCH_2CH_2OH$	—	(150)	—	—	D, E, W
Fatty acid monoglycerides	$C_nH_{2n+1}COOCH_2CHOHCH_2OH$	—	—	—	—	E
Fatty acid diethanolamides	$C_nH_{2n+1}CON(CH_2CH_2OH)_2$	—	—	19	640	FS

* Uses: D = detergent, E = emulsifying agent, FS = foam-stabilizing agent, G = germicide, H = hair conditioner, W = wetting agent.

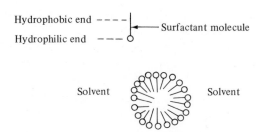

Figure 11.3 Cross-section of a micelle.

simply dissolve in the micellar core (Fig. 11.3), the hydrocarbon tails of the surfactant acting as the solvent; weakly polar molecules may be directly incorporated into the micelle structure; and hydrocarbon-insoluble materials may be adsorbed on the surface of the micelle. The solubilizing properties of surfactants are exploited in the dyeing of synthetic fibres (cf. Chapter 12), and in the so-called emulsion polymerization of alkenes (section 16.11.8).

A final major property of surfactants is their ability to stabilize emulsions. When two immiscible liquids are shaken together they form a dispersion but when the agitation is stopped phase separation follows at once. If, however, a small amount of a surfactant is added, a relatively stable emulsion is formed in which one phase (the discontinuous phase) is dispersed in the form of microscopic droplets in the other (continuous) phase. The surfactant forms a monomolecular layer around the droplets with the hydrophilic and hydrophobic ends of the molecules arranged appropriately (cf. Fig. 11.4). In

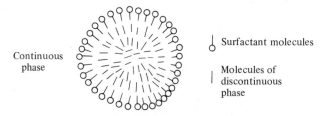

Figure 11.4 Partial cross-section of an emulsion.

contrast to micellar solubilization, the discontinuous phase constitutes a substantial part of the total volume and emulsions vary enormously in appearance. Emulsions in which an oil is dispersed in water include milk, shaving cream and burn ointments while butter, ice-cream and mayonnaise are typical of water-in-oil emulsions. Gas–liquid as well as liquid–liquid emulsions are readily formed in the presence of suitable surfactants: foams are air-in-liquid and aerosols liquid-in-air systems.

11.3 Detergency

11.3.1 Phenomenology

The removal of foreign material from solid surfaces is a complex process and most detergent compositions have been developed by empirical means. The efficacy of detergent mixtures may be measured by various practical tests from which general conclusions may sometimes be drawn. Many investigators have employed the launderometer, a miniature washing machine in which swatches of artificially soiled cloth are washed under controlled conditions, the extent of soil removal being estimated by changes in reflectance. Launderometer studies are cheap and easy to conduct and give consistent but often misleading results. The complexity of the phenomena involved makes it dangerous to rely upon conclusions drawn from simple systems; it is often more informative to employ tests which parallel practical washing conditions as closely as possible. Typical of the latter is the split-collar method devised by Procter and Gamble Ltd. Specially designed shirts are issued to a sample of male staff. Every day the detachable collars are collected and each is split in half; one pile of halves is then washed in the experimental mixture and the other in a reference composition; after drying, the individual halves are matched and their relative cleanness is assessed by a panel of judges. From a practical point of view this test scores heavily in employing naturally soiled material, realistic washing procedures and the subjective criteria of cleanness which potential customers would use.

On the basis of such experiments it is possible to make certain generalizations about detergents and detergency. Soils vary markedly in the ease with which they are removed. Food debris presents no problems, but the dirt found on clothes, typically a mixture of natural glycerides, free fatty acids, skin debris, carbon and silica, is far more tenacious. Proteins and inorganic material are particularly hard to remove and if retained lead to the irreversible greying of white fabrics. The nature of the surface to which the soil is attached is also of importance. Glass and ceramic surfaces are easily cleaned whereas fibres, especially synthetic fibres, retain dirt more strongly. A surfactant must be present if a mixture is to have detergent properties – although the amount needed may vary widely – but the efficacy of a particular surfactant is only vaguely connected with its efficacy as a wetting agent, the conditions under which it forms micelles, or its ability to stabilize foams. On the whole, detersive efficiency increases with an increase in the number of carbon atoms in the hydrophobic part of the surfactant molecule and, other things being equal, straight-chain compounds are superior to branched-chain compounds. The nature of the hydrophilic group is less significant. Anionic, cationic and non-ionic materials are of comparable efficiency, although anionic and non-ionic surfactants are preferred because of their greater compatibility with the other ingredients of detergent mixtures. The presence of calcium and magnesium ions in detergent solutions sharply reduces their cleansing powers, the effect being particularly marked with anionic materials with a hydrophilic carboxylate group; conversely these adverse effects may be suppressed by the addition of compounds which form soluble complexes with such ions. Finally, the physical conditions of washing are of significance: the best results are obtained when high temperatures and vigorous agitation are employed.

11.3.2 Theories

No completely satisfactory theory of detergency has yet been formulated but it is possible to give a qualitative account of some of the processes involved. A detergent mixture must both remove soil from a surface and prevent its redeposition. Consider an oily soil placed upon a basically hydrophobic surface immersed in water; the soil is more strongly attracted to the surface than to the water and therefore remains attached to it. When a surfactant is added to the aqueous phase, it will migrate to the soil interface and will increase the attractive forces between soil and solution; it will also migrate to the solid–water interface and so increase the attraction between solid and solution. The balance of forces will now be such that the soil may be readily detached by, for example, mechanical agitation (Fig. 11.5). The classic experiments of Adams and Stevenson show how

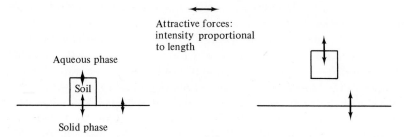

Figure 11.5 Schematic representation of soil removal by a surfactant.

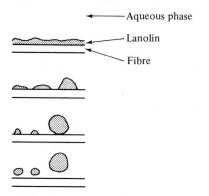

Figure 11.6 Stages in the removal of lanolin from a cotton fibre.

this may take place: when a surfactant is added to the aqueous medium, lanolin films adhering to wool fibres roll up into readily detachable sphere (Fig. 11.6). Under practical washing conditions the intermediate stages are much more complex and may involve the formation and separation of liquid crystals rather than of unchanged soil.

The ability of detergent compositions to prevent the redeposition of soil may also be explained in approximate terms. In a soil–water dispersion molecular bombardment keeps the particles in continuous motion provided that they do not agglomerate to such a size that this is no longer possible. Surfactants stabilize these dispersions by inhibiting coagulation in various ways. Anionic surfactants, for example, are adsorbed by soil particles to which they impart a negative charge; this is counter-balanced by diffuse layers (ionic atmospheres) of positively charged ions to produce an electrically neutral system. The ionic atmospheres of neighbouring particles repel one another and so coagulation is prevented. Another mechanism must clearly be involved in the case of non-ionic surfactants: it has been suggested that their already bulky hydrophilic ends become strongly hydrated and that the resulting steric interferences prevent soil particles from approaching sufficiently closely for agglomeration to occur. The adverse effect of calcium and magnesium ions is readily understood on the basis of these theories. The polarizing effect of an ion increases rapidly with increasing charge. A highly charged cation will be so strongly adsorbed by a negatively charged particle that the ionic atmosphere will be sharply reduced in size and in charge density. Such a cation will similarly compete for water molecules with a non-ionic surfactant, so reducing the effective bulk of the latter. In either case, soil particles are able to approach more closely and coagulation is facilitated.

11.3.3 Formulations

Detergent manufacturers wish to produce materials which are cheap, safe and effective. The history of the soap trade has been the history of the change from simple all-purpose products to relatively complex mixtures designed for specific ends. The formations in use today therefore vary widely in composition (cf. Table 11.4) but almost all contain both a surfactant and a builder.

It is instructive to examine the ingredients of a heavy-duty washing powder. The surfactant used is a synthetic anionic material or, less frequently, a soap; since it is relatively expensive, the formulation is such as to minimize the amount needed for a given performance. The builders fulfil several functions. Sodium tripolyphosphate (Fig. 11.7) is above all a cheap and efficient sequestering agent for calcium and magnesium ions with

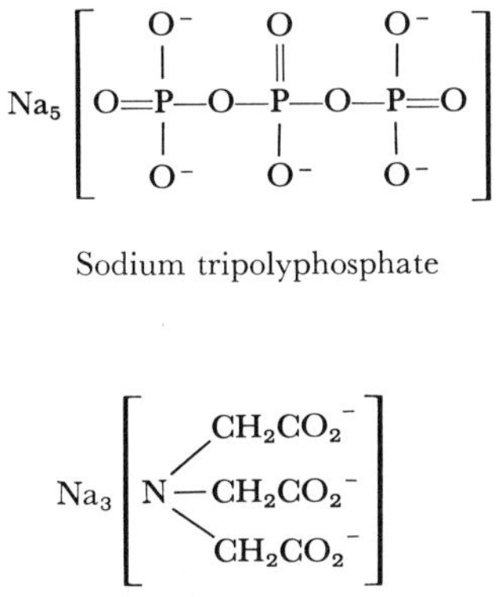

$$Na_5 \left[O=P(O^-)(O^-)-O-P(=O)(O^-)-O-P(O^-)(O^-)=O \right]$$

Sodium tripolyphosphate

$CH_3(CH_2)_{11}CONHCH_2CH_2OH$

Lauric acid ethanolamide

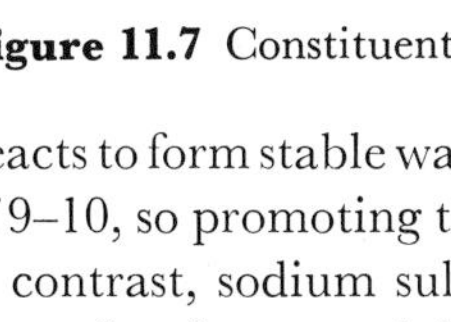

$$Na_3 \left[N(CH_2CO_2^-)_3 \right]$$

Trisodium nitrilo*iso*triacetate
(NTA)

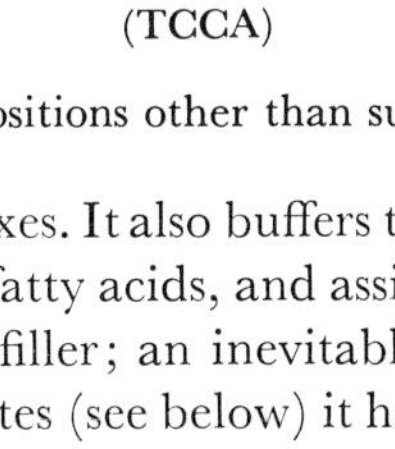

Trichloro*iso*cyanuric acid
(TCCA)

Figure 11.7 Constituents of detergent compositions other than surfactants.

which it reacts to form stable water-soluble complexes. It also buffers the aqueous medium to a pH of 9–10, so promoting the solution of free fatty acids, and assists in the suspension of soil. In contrast, sodium sulphate is merely a filler; an inevitable by-product of the manufacture of surfactant sulphates and sulphonates (see below) it has no objective function other than to make the powder bulky and smooth-flowing. A bleach is usually included to destroy soil that cannot be readily removed by the usual mechanisms. Sodium perborate ($NaBO_3 \cdot 4H_2O$), a compound made from sodium metaborate and hydrogen peroxide, is most commonly employed; safe and effective in hot water, it shows little

Table 11.4 Composition of some commercial detergent mixtures

	Heavy-duty powder	%	Low-foaming powder for automatic machines	%	Light-duty liquid	%	Powder for dishwashing machines	%	Scouring powder	%
Surfactant	NaDDBS*	14–18	Non-ionic	15	NaDDBS (HOCH$_2$CH$_2$)$_3$NDDBS	7 7	Non-ionic	3	NaDDBS	3
Builders	Na$_5$P$_3$O$_{10}$ Na silicate	30–45 5–9	Na$_5$P$_3$O$_{10}$ Na$_4$P$_2$O$_7$ Na silicate	15 10 15			Na$_5$P$_3$O$_{10}$ Na$_3$PO$_4$ Na silicate	50 15 25	Na$_5$P$_3$O$_{10}$ Na$_4$P$_2$O$_7$	10 5
Bleach	Na perborate or Enzyme	15–20 0·3–0·7	Na perborate	10					TCCA†	1
Other Ingredients	Foam stabilizer CMC‡ Optical bleach CH$_3$C$_6$H$_4$SO$_3$Na	1–2 0·5–1 0·1 2–3	CMC Optical bleach CH$_3$C$_6$H$_4$SO$_3$Na	1 0·1 2–3	Foam stabilizer	1–2			Abrasive	81
Fillers	Water Na$_2$SO$_4$	8–14 balance	Water Na$_2$SO$_4$ } balance		Water	84–85	Water	balance		

* DDBS = dodecylbenzene sulphonate   † TCCA = trichloroisocyanuric acid   ‡ CMC = sodium carboxymethylcellulose bleaching action in the cold. The more recently introduced proteolytic enzymes on the other hand work most effectively under ambient conditions in which they catalyse the hydrolysis of proteins to water-soluble peptides and amino acids. Optical bleaches, despite their name, have nothing in common with these materials: they are colourless dyes which fluoresce blue when adsorbed on fabrics, so offsetting any yellowness and producing a genuine 'whiter than white' appearance. Other minor ingredients include foam-stabilizers – usually lauric acid ethanolamides – soil suspending agents such as sodium carboxymethyl cellulose (CMC), anti-caking agents like sodium *p*-toluene sulphonate, and fungistats.

Other compositions call for less attention. Automatic washing machines need a low-foaming material because of the vigorous agitation employed; these usually incorporate a non-ionic surfactant or a mixture of soap with a synthetic anionic. Heavy-duty liquid detergents have also been introduced for household laundering in the past decade. Here the major problem is to incorporate the necessary ingredients in a bright clear solution: most formulations use the diethanolamine salt of an alkyl aryl sulphonate as the surfactant, silicates or pyrophosphates as builders, and sodium toluene sulphonate as a solvent aid. More important are the light-duty liquids which are mainly used for washing dishes. Food is easy to remove from china and glass surfaces – only a weak detergent action is required – but customers dislike the sight of the dirty solution resulting from their labours. The ability to produce a high, stable head of foam is a *sine qua non* for a successful product. Most dishwashing mixtures consist of an anionic surfactant – usually a mixture of sodium and ethanolamine salts – a foam stabilizer and a great deal of water. Products intended for dishwashing machines are quite different. As in washing-machines foam formation is undesirable and most formulations consist very largely of builders which assist what is a mainly mechanical cleaning process. It is customary to include an oxidizing agent to remove firmly held food particles which may cause streaking when the dishes dry. A particularly convenient compound is trichloro*iso*cyanuric acid (TCCA) which is rapidly hydrolysed to sodium hypochlorite in solution. A powerful bleach even at room temperature, it is now incorporated in most scouring powders, but is too reactive for other purposes.

11.4 Anionic surfactants

11.4.1 Soap

Soaps are the sodium salts of straight-chain carboxylic acids containing between 12 and 18 carbon atoms. They are made by the hydrolysis of natural fats and oils, which are the esters of such acids with glycerol. In practice tallow and coconut oil are the favoured materials: soaps made from the former have excellent detergent properties, while those made from the latter dissolve rapidly and excel in stabilizing foams. Mixtures of these fats are often used to obtain a product of optimum properties.

Most of the world's soap production is made in batches by the traditional kettle process in which the mixture of liquid fats is hydrolysed with aqueous sodium hydroxide at 80–90°C. Because of the mutual insolubility of oil and water, the initial reaction is slow, but when enough soap has been formed to produce emulsification, hydrolysis becomes rapid. At the end of the reaction the composition of the mixture is systematically adjusted (cf. Fig. 11.8) so as to permit the successive removal of aqueous lye and nigre

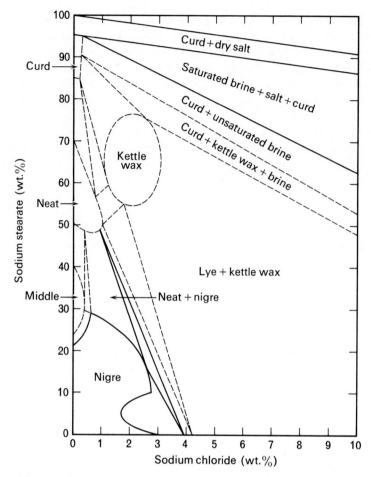

Figure 11.8 Phase diagram of the ternary system sodium stearate–sodium chloride–water at 90°C. *Note*: The triangles indicate heterogeneous equilibria between the three condensed phases at the apexes of the triangle. The regions between the triangles should be filled with tie lines, each connecting two condensed phases, the tie line nearest any triangle being almost parallel to the nearest boundary of that triangle.

phases, the product remaining as a separate liquid phase known as neat soap. The lye and wash solutions are worked up for glycerol; the nigre phase, which includes soap and coloured by-products, is purified and recycled; and the neat soap, which contains approximately 30% water, is dried and formed into bars or flakes. Where demand is high, continuous hydrolysis becomes attractive. In the Procter and Gamble process, a mixture of fats and zinc oxide is reacted by counter-current flow with water at 250°C and 40–50 atm pressure to give a continuous output of fatty acid from the top of the column and glycerol solution from the bottom. The fatty acids are purified by distillation and then continuously neutralized to neat soap in a high-speed mixer.

Soaps have excellent detergent properties, are easily incorporated in most formulations, and are completely biodegradable. On the other hand they form insoluble calcium and magnesium salts which sharply reduce their efficacy in hard water. Moreover, the cost of soap is critically dependent on the costs of natural fats which, though liable to sharp fluctuations, have in general increased substantially since the Second World War. For these reasons the production of soap in the advanced countries is declining (cf. Table 11.5) and soaps have been replaced for many purposes by synthetic surfactants.

Table 11.5 Production of soap and synthetic detergent powders 1960–69

	UK			EEC			USA		
	1960	1965	1969	1960	1965	1969	1960	1965	1969
Soap – all forms	447*	354	280	689	572	579	470	495	428
Synthetic detergent powders†	341	390	531	691	1 077	1 413	1 209	2 921	2 274

* Figures in units of '000 tons.
† Finished powders.

11.4.2 Synthetic anionics

The anionics are the most important class of synthetic surfactant and make up 60–70% of the total production of such compounds. They are the sodium salts of organic sulphonates or, less frequently, sulphates. As a group they are superior to soaps in that they form relatively soluble calcium and magnesium salts and are completely ionized under all conditions of acidity. Ultimately derived from petroleum in most cases, they are also cheaper than soaps in many applications.

The most important anionic surfactants are the alkylbenzene sulphonates. The manufacture of these compounds is a straightforward process. The Friedel–Crafts reaction of a C_{12} alkene mixture with benzene in the presence of hydrogen and aluminium chlorides gives an akylbenzene which is then sulphonated with a 4% SO_3–air mixture at 40–50°C. If a solid detergent is to be produced, the sulphonic acid is neutralized with sodium hydroxide and the resulting slurry is mixed with the builders and spray-dried to a smooth-running powder to which heat-sensitive ingredients such as bleaches and enzymes may then be added. In the production of liquid formulations the neutralization is effected with potassium hydroxide or with alkanolamines (cf. Table 11.4).

$$\text{C}_6\text{H}_6 + \text{C}_{12}\text{H}_{24} \xrightarrow[\text{AlCl}_3]{\text{HCl}} \text{C}_6\text{H}_5\text{-C}_{12}\text{H}_{25} \xrightarrow{\text{SO}_3} \text{HO}_3\text{S-C}_6\text{H}_4\text{-C}_{12}\text{H}_{25} \xrightarrow{\text{NaOH}} \text{NaO}_3\text{S-C}_6\text{H}_4\text{-C}_{12}\text{H}_{25}$$

Until the mid-1960s the C_{12} alkene used was a tetramer of propylene produced by cationic polymerization. Although cheap, it has a highly branched structure and leads to the formation of a surfactant which cannot be metabolized by bacteria. Such material therefore accumulates in rivers and may cause them to foam. The use of straight-chain alkenes, however, yields largely biodegradable surfactants and these hydrocarbons are

318 *Soaps and detergents*

now the required starting material. Straight-chain alkanes occur in petroleum from which they may be extracted by selective adsorption in the cavities of molecular sieves which are of such dimensions as to permit the entry of unbranched isomers only. In the widely used Molex process, the *n*-alkanes present in a C_{11}–C_{14} fraction are adsorbed in a fixed-bed zeolite sieve from which they are then displaced by a low-boiling straight-chain hydrocarbon. Catalytic dehydrogenation or chlorination-dehydrochlorination converts the *n*-alkanes to *n*-alkenes.

The other anionic surfactants are of less importance. Alkyl sulphonates have been little used, although the alkenyl sulphonates formed by the sulphonation of straight-chain 1-alkenes have attracted some attention recently. Sulphated alcohols are of greater significance. The secondary alcohol sulphate Teepol, formed by the addition of sulphuric acid to petroleum-derived alkenes, was indeed the first anionic to be widely used in the UK; it is, however, better as a wetting agent than as a detergent. A similar compound is the textile auxiliary Turkey Red Oil, which is a sulphated derivative of castor oil. The sulphates of straight-chain primary alcohols are excellent surfactants from every point of view except that of cost: in consequence they have been generally used in relative high-priced formulation where their remarkable detergent and foam-stabilizing properties are at a premium. A typical example is sodium lauroyl glyceryl sulphate ($C_{12}H_{23}CO_2CH_2CHOHCH_2OSO_3Na$), prepared by the sulphation of partly hydrolysed coconut oil: it is a constituent of shampoos, toothpastes – as Miracle GL-70 in *Gleem* – and contraceptive pastes.

11.5 Cationic surfactants

The most widely used cationic surfactants are quaternary alkyl-ammonium or alkyl-pyridinium salts, generally prepared by routine methods from the corresponding fatty acids: thus cetyltrimethyl-ammonium bromide ($C_{16}H_{33}\overset{+}{N}(CH_3)_3Br$; CETAB) is made from palmitic acid by conversion to the nitrile, reduction and quaternization. The cationics are stable in acid solution but undergo hydrolysis or the Hofmann elimination reaction in warm alkaline solution; they are also incompatible with many of the ingredients in common detergent mixtures. They are therefore confined to special uses – as germicides, as textile-softening agents, and in drilling fluids. Production amounts to no more than 3–5% of synthetic surfactant production.

$$C_{15}H_{31}CO_2H \xrightarrow[\text{heat}]{NH_3} C_{15}H_{31}CONH_2 \xrightarrow{-H_2O} C_{15}H_{31}C\equiv N \xrightarrow{H_2\text{-Ni}}$$
Palmitic acid

$$C_{15}H_{31}CH_2NH_2 \xrightarrow{CH_3Cl} C_{15}H_{31}CH_2\overset{+}{N}(CH_3)_3Cl^- \xrightarrow[\text{resin}]{\text{Ion-exchange}} C_{15}H_{31}CH_2\overset{+}{N}(CH_3)_3Br^-$$
Cetyltrimethyl-ammonium bromide

11.6 Non-ionic surfactants

In non-ionic surfactants the hydrophilic group is a polyoxyethylene chain –$(OCH_2CH_2O)_nCH_2CH_2OH$, which confers water solubility through the ability of oxygen atoms and hydroxyl groups to form hydrogen bonds with water. Such compounds are made by the reaction of an alcohol or amine with ethylene oxide in what is, in effect, an intramolecular S_N2 displacement by which the highly strained ring is opened to give a product which is itself nucleophilic and therefore capable of further reaction. About

60–75% by weight of polyoxyethylene content (5–10 ethylene oxide residues) must be incorporated if the product is to have surfactant properties.

$$\text{ROH} + \text{H}_2\text{C}\overset{\text{O}}{-}\text{CH}_2 \longrightarrow \text{ROCH}_2\text{CH}_2\text{OH} \xrightarrow{\text{H}_2\text{C}\overset{\text{O}}{-}\text{CH}_2} \text{ROCH}_2\text{CH}_2\text{OCH}_2\text{CH}_2\text{OH, etc.}$$

Non-ionic surfactants have many desirable properties. They are excellent wetting and solubilizing agents; they are compatible with both cationic and anionic surfactants; and their properties are largely unaffected by the presence of calcium and magnesium ions. They are often, however, low-melting waxes or liquids and are therefore difficult to incorporate in powders. A further disadvantage is that unlike the ionics they tend to separate from aqueous solution when the hydration of the polyoxyethylene tail is reduced, as is the case when the temperature or ionic strength of the medium is high. They are in practice used mainly in liquid and industrial washing products.

The production of non-ionic surfactants has increased rapidly in recent years as liquid detergent formulations have become popular and is now 30–35% of synthetic surfactant production. For some years the ethoxylated nonylphenols, prepared by the Friedel–Crafts alkylation of phenol with a C_8–C_{10} alkene mixture and ethoxylation of the product, were the most popular materials, but the ethoxylated linear alcohols have now come into prominence. Although more expensive – the necessary alcohols are made by the reduction of fatty acids or, better, by the Ziegler reaction (cf. section 16.11.5) – they are both excellent detergents and totally biodegradable.

11.7 Effluent problems and the detergent industry

From what has been said above it is clear that the use of detergents may give rise to substantial effluent problems. The crisis produced by the tetrapropylenebenzene sulphonates was largely overcome by the introduction of linear alkylbenzene sulphonates (cf. section 11.4.2 above) but even these compounds are not entirely satisfactory. Not only are they not completely biodegradable but they are poisonous to fish, especially if chains containing more than 14 carbon atoms are present.

Italy and France are at present contemplating legislation to control fish toxicity in commercial surfactants. Although other nations continue to tolerate linear alkylbenzene sulphonates, the trend of regulation in all countries is towards requiring complete biodegradability and non-toxicity. The surfactants of the future may well be alkyl sulphates or alkenyl sulphonates based upon linear C_{12} hydrocarbons.

The environmental effects of phosphate builders have also attracted much attention. In the presence of phosphates and nitrates blue-green algae may grow at an explosive rate in fresh water, covering large areas of the surface, killing fish and seriously upsetting the ecological balance of the river or lake. Wastes from the domestic use of detergent powders are a major source of phosphates in fresh water and much time and money has therefore been devoted to the development of alternate builders which do not have this disagreeable side-effect. These efforts have been unsuccessful. (Trisodium nitrilotriacetate (NTA: Fig. 11.7) is effective but may cause cancer and gene damage; moreover it is biodegraded to nitrate, itself an obviously undesirable product. Sodium citrate is harmless but not very effective. Other possible builders such as the benzene polycarboxylates are excessively expensive. In all probability the optimum solution to the problem is the installation of tertiary sewage treatment plants in which all phosphates would be recovered by precipitation at the sedimentation stage with iron alum (section 17.2.2), or a similar material. Both Sweden and Switzerland have adopted tertiary treatment extensively and Lake Zurich has been saved by this action.

11.8 Marketing

The soap and detergent industry sells most of its products direct to the final consumer. The problems of retail marketing therefore loom large to the manufacturer and it is hardly surprising that the more successful firms are characterized by extreme skill in this most difficult type of work. The products must 'walk off the supermarket shelf' not once, but repeatedly, in thousands of stores in all parts of the country. New products must be introduced and old products must be given a new face if sales are to be maintained. A heavy expenditure on market research, on advertising and on the sales and distribution force is inevitable.

The retail soap market is relatively price-inelastic. This has been shown in the field. In 1954–55 Lever Brothers sharply reduced the UK prices of their synthetic detergent powders in order to regain ground lost to Procter and Gamble. The latter did not respond in kind but, against all expectation, actually increased their share of that market during the period in question. More recently the success of the rather expensive liquid products and the failure of powders such as *Surf* which are sold at low prices because they are not advertised extensively, has confirmed this point. Differences in the performance or convenience of use of rival products are more important. Each new technical improvement – optical bleaches in 1952, perborates in 1954, enzymes in 1968 – has been rewarded by an increase in the sales of the innovator. When all products are much the same, small differences in colour, odour or packaging may help or hinder, but advertising becomes of central importance.

Advertising has three functions. First, it informs: it draws the customer's attention to the product. The introduction of a new brand is invariably accompanied by a massive advertising campaign designed to familiarize the populace with its name and characteristics. Second, it produces a halo of agreeable associations around the product: it adds, as it were, a psychological ingredient. The product may come to be seen as particularly mild or particularly effective. It may be associated with images of babies in spotless diapers or mothers with skins as perfect as those of their daughters. In extreme cases strong brand loyalties develop, and may help to maintain the position of an essentially inferior product, as was the case with *Persil* in the 1950s. Third, advertising is a form of pressure, intended to make the customer buy, and to buy *Ariel* rather than *Radiant*. In practice, advertisements combine these functions. Soap companies are the largest advertisers in most countries, especially on television, a medium which reaches the greatest audience with the greatest impact. Their campaigns are hardly an inspiring spectacle – the Western Front barrage laid down by Procter and Gamble is particularly offensive to the educated – but their efficacy is not open to doubt.

The commercial lifetime of a retail detergent is quite short. Whereas most bulk chemicals have been available for at least thirty years, few brands of detergent are more than twenty years old. When a major innovation – such as enzyme bleaches – is introduced, it is generally incorporated in a new product so as to maximize its psychological impact; conversely, a brand which has been sold for many years may eventually be seen as old-fashioned and therefore less desirable. Soap companies must innovate even though the changes may be apparent rather than real. The introduction of a new product is a complex and expensive process. Not merely buying habits, but attitudes to washing, are powerfully influenced by age, social class and geographical location; extensive research is necessary to define the extent and nature of potential markets. Elaborate testing procedures must be used to determine whether or not the product will perform adequately and, even more important, whether it will sell in a full competitive situation. These operations call for skills which are not found in the average chemical firm and emphasize yet again the singular nature of the soap industry.

11.9 The economics of soap

Little detailed information is available about the costs of soap and detergent manufacture. In the mid-1960s the Prices and Incomes Board estimated that raw materials probably amounted to about 40–50% of the price at which products were sold to retailers and that advertising and marketing costs made up another 20%. Capital-related charges are a relatively small part of the whole; as has already been noted (section 11.1), sales per unit of plant investment are much higher than in the general chemical industry. A rough estimate based on recent data suggests that depreciation and maintenance charges are no more than 2–3% of sales. Process costs are moderate since manufacturers buy in fats, alkylbenzene, tripolyphosphate and most other ingredients; the operations actually carried out in a typical detergent plant, for example, are sulphonation, mixing, spray-drying and packaging. Pre-tax profit margins are, however, far from excessive, and in the UK probably average about 8–10% of sales. It is believed that margins are highest on toilet soaps and liquid products and lowest on heavy-duty powders. The combination of reasonable profit margins with relatively low capital requirements makes soap and detergent production an exceptionally attractive business and indeed Procter and Gamble's British subsidiary has been for many years one of the most profitable firms in the country.

Yet if soap-making is both profitable and needs relatively little capital, how is it that in all advanced countries the industry is dominated by two or three giant firms – Lever Bros. and Procter and Gamble in Britain; Lever, Procter and Gamble and Colgate–Palmolive in the USA; Lever and Henkel in West Germany? There is no contradiction here. It is easy to enter the industry in a small way: in all countries there are many little firms making toilet soaps and liquid cleaners in particular. To launch a new product on a large scale, however, requires resources both of skill and money that are beyond all but very large companies. General chemical firms such as ICI or Shell have the money but lack the necessary expertise in retail marketing. Soap-making is a profitable business but only to soap-makers.

Notes and references

General   No book covers the whole field. In many ways the most useful are: A. M. Schwartz and J. W. Perry *Surface Active Agents – Their Chemistry and Technology* (vol. 1) and A. M. Schwartz, J. W. Perry and J. Berch, *Surface Active Agents and Detergents* (vol. 2), Interscience, New York, 1949 and 1958. The first volume covers the period before 1947 and the second reviews developments between 1947 and 1956. The best journal is *Soap, Cosmetic and Chemical Specialities*, formerly *Soap and Chemical Specialities*.

Statistics   Henkel A.G. produce a detailed annual survey of world production which is published in *Soap, Cosmetic and Chemical Specialities*. The latest refers to 1968 and was published in September 1970. The data in the OECD *Annual Report* is much less complete. The Department of Trade and Industry publishes detailed accounts of UK soap and detergent production classified by end-use in the *Business Monitor* series. US production classified by chemical constitution is given in US Tariff Commission, *Synthetic Organic Chemicals*.

Section 11.1   Table 11.1 is from the Henkel Survey, 1968 op. cit.

Section 11.2   There are many accounts of surfactants and their properties. Particularly useful are N. K. Adam, *The Physics and Chemistry of Surfaces*, Oxford, 1949; A. W. Adamson, *The Physical Chemistry of Surfaces*, 2nd ed., Interscience, New York, 1967; and A. M. Schwartz *et al.*, op. cit. The *Surfactant Science* series published by Marcel Dekker, New York, gives a very detailed account of the organic and physical chemistry of surfactants; vol. 1, *Non-Ionic Surfactants*, ed. M. J. Schick, 1967, and vol. 4, *Cationic Surfactants*, ed. E. Jungermann, 1970, are relevant here. C. E. Stevens, *Surfactants*, in Kirk-Othmer, 2nd edn., vol. 19, 1969, gives a good picture of current trends in usage.

Section 11.3.1   J. C. Harris, *Detergency Evaluation and Testing*, Interscience, New York, 1954, is the only book on this subject, but there are some useful papers in *Chemistry, Physics and Application of Surface-Active Substances*. vol. 3, Gordon and Breach, London, 1967.

Section 11.3.2   The best review of this complex subject is A. M. Schwartz, *Detergency*, in Kirk-Othmer, 2nd edn., vol. 6, 1965, although almost all the works cited in Section 11.2 above contain some comments about it. Figure 11.6 is based on N. K. Adam and D. G. Stevenson, *Endeavour*, 1953, **12**, 25.

Section 11.3.3   The largest collection of formulations is in A. Davidsohn and B. L. Milwidsky, *Synthetic Detergents*, 4th edn., Leonard Hill, London, 1967, on which Table 11.4 is based.

Sections 11.4–11.6   Details of preparation and uses are given in A. M. Schwartz *et al.* and C. E. Stevens op. cit. The manufacture of synthetic anionic powders is described by R. A. Duckworth, *CPE*, April 1970, p. 63, who gives some cost data. Recent developments are discussed by C. R. Hutchison, *Detergents Markets in Europe*, and D. W. Lawson, *Alkylates and Related Detergent Raw Materials*, summarized in *ECN*, Nos. 522 and 520, 1972. Figure 11.8 is from J. W. McBain and W. W. Lee, *Ind. Eng. Chem.*, **35**, 1943, 917. Table 11.5 is based on the OECD *Annual Reports*, 1960–61 to 1969–70.

Section 11.7   The *Surfactant Science* series contains a volume on *Surfactant Biodegradation* edited by R. D. Swisher, 1970. The ups and downs of phosphate substitutes are best followed in the trade press – Lawson op. cit. has some useful comments – and in the *13th Progress Report of the Standing Technical Committee on Synthetic Detergents*, London, HMSO, 1972.

Section 11.8   There is no good book on detergent marketing as such. The most informative single work on advertising is M. Mayer, *Madison Avenue*, Penguin, Harmondsworth, 1961, which gives a brilliant description of the American scene in the 1950s. J. L. Simon, *Issues in the Economics of Advertising*, University of Illinois, Urbana, 1970, is the best book on the subject and contains an excellent bibliography. *The Economics of Advertising*, The Advertising Association, London, 1967, is rather thin. British advertising expenditure on particular brands and by particular firms is given in the Legion Press monthly *Advertising Statistical Review*. Sources of information about American expenditure are discussed by Simon.

Section 11.9 The approximate cost structure is given in Report No. 4 of the National Board for Prices and Incomes, Cmnd. 2791, HMSO 1965. Depreciation, maintenance costs and profit margins were calculated from the Annual Report of Procter and Gamble Ltd.

12 Dyes and pigments

12.1 The colour industry

A dye is a compound which, when applied as a solution or dispersion, imparts to a solid substrate a more or less permanent colour. The substrate is usually a textile fibre but may also be paper, leather, a plastic material or a foodstuff; it normally possesses an affinity for particular dyes and absorbs them from solution under appropriate conditions. A pigment on the other hand is an insoluble coloured material which is applied as a suspension of particles in a continuous medium such as an oil or a synthetic resin.

Almost all dyes and an increasing proportion of pigments are synthetic organic compounds. Dyestuffs are the oldest product of the organic chemical industry; indeed from Perkin's discovery of mauve in 1856 until the Second World War, organic chemicals and dyes were almost synonymous terms (cf. Chapter 1). Although now dwarfed by the huge growth in plastic production during the past 25 years, the manufacture of dyestuffs remains of great importance. World production of dyes has grown steadily if not spectacularly and by 1970 amounted to about 500 000 tons per year (cf. Table 12.1). Moreover,

Table 12.1 Production of Organic Dyes and Pigments 1960–69

	1960	1965	1966	1967	1968	1969
UK organic colouring						
materials	37*	43	40	39	42	45
dyes	31	36	32	30	32	34
organic pigments	6	7	8	9	10	11
inorganic pigments	64	64	59	59	63	66
EEC	84	107	116	112	120	138
USA	71	107	112	115	113	108
Japan	35	45	49	54	59	66

* Figures in units of '000 tons.

they are of greater commercial significance than their tonnage would suggest. They are costly materials; their prices are in the range £500–£2 000 ($1 200–$5 000) per ton compared to £100–£500 ($250–$1 200) per ton for synthetic polymers. Accordingly it is understandable that, for example, UK dyestuff and pigment production in 1968 amounted to well under one per cent by weight of all chemical production, but to 4·4% by value of the sales, and almost 5% of the value added by the chemical and allied industries.

The high cost of dyes is due to the complexity of their structure and to the small scale on which they are made. Many of the most desirable dyes can only be made by laborious multistage syntheses of the type familiar to the classical organic chemist. Again, the demand for dyestuffs is a demand for many products for specific uses rather than for a few general purpose materials. No single dye accounts for more than perhaps 2–3% of the total production. In practice, dyes are therefore made by batch processes which use what is fundamentally scaled-up laboratory equipment. Even to the untutored eye, the manufacture of dyestuffs is recognizable as a branch of organic chemistry.

12.2 Colour and constitution

The human eye is sensitive to light of wavelengths between 380 and 750 nm and within this range is able to distinguish between about 150 distinct hues. The relationship between observed colour and wavelength of light is shown in Table 12.2. A material appears to be coloured if it selectively absorbs visible light of a particular wavelength; the transmitted or reflected light will produce an effect complementary to that absorbed. Thus a fabric which absorbs light of wavelength 435–480 nm appears yellow when viewed in white light. Substances which do not absorb light are white or colourless, and those which absorb light of all wavelengths are black.

The absorption of light by a compound involves a momentary increase in the electronic energy of its molecules. In any molecule the electrons are distributed among bonding

Table 12.2 Relationship of Colours Observed to Wavelength of Light Absorbed

Light absorbed		Colour seen
Wavelength (nm)	Colour	
400–435	Violet	↑ Green-Yellow
435–480	Blue	Yellow
480–490	Green-Blue	Orange
490–500	Blue-Green	Red
500–560	Green	Purple
560–580	Yellow-Green	Violet
580–595	Yellow	Blue
595–605	Orange	Green-Blue
605–750	Red	Blue-Green ↓

(Colour lightens ↑ / Colour deepens ↓)

σ and π orbitals, to which there correspond σ^* and π^* antibonding orbitals, and non-bonding (n) orbitals. In the ground state the bonding and non-bonding orbitals are full and the antibonding orbitals are empty or, at most, partly filled. The absorption of a quantum of light of appropriate energy will, however, promote a σ, π or n electron to an antibonding orbital; the energy E required is related to the wavelength λ of the light absorbed by the equation $E = hc/\lambda$ where h is Planck's constant and c is the velocity of light. In the ground state bonding electrons have less energy than non-bonding electrons and σ electrons have less than π electrons whereas the reverse is the case in antibonding orbitals. Hence it is not surprising to find that $\sigma \rightarrow \sigma^*$ transitions correspond to the absorption of light in the far ultra-violet (170 nm) region and that the colour of commercial dyes arises from $\pi \rightarrow \pi^*$ and $n \rightarrow \pi^*$ transitions. Finally it is obvious that the energy derived from the absorption of light must be dissipated by the excited molecules if absorption is to continue. It may be transformed into heat, used to effect photochemical processes, or it may be partly converted to rotational-vibrational energy and the balance re-emitted as fluorescent light.

Table 12.3 Chromophore systems of major importance

ethylenic chromophores				Aromatic chromophores
$C_6H_5(CH=CH)_nC_6H_5$				Triphenylmethyl cation
n	λ_{max} (nm)	ϵ_{max}	Colour	
1	319	24 350	none	
2	352	40 000	none	
3	377	75 000	pale yellow	
4	404	86 000	greenish-yellow	
5	424	94 000	orange	
6	445	114 000	orange-brown	
7	465	135 000	copper bronze	

Triphenylmethyl cation:

X	Y	Z	Colour
H	H	H	Orange
H	H	←(CH$_3$)$_2$N→	Green
←(CH$_3$)$_2$N—	—	(CH$_3$)$_2$N→	Violet

Polycyclic aromatics:

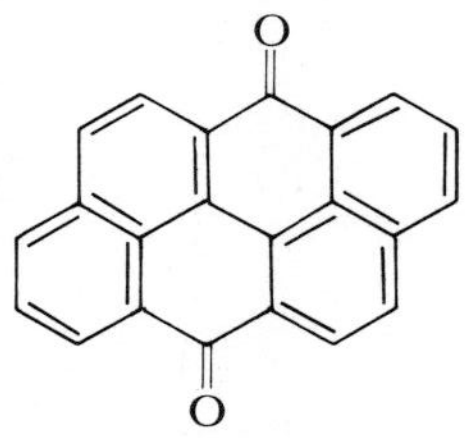

Anthanthrone — Orange
Pyranthrone — Golden
Dibenzanthrone — Blue-green

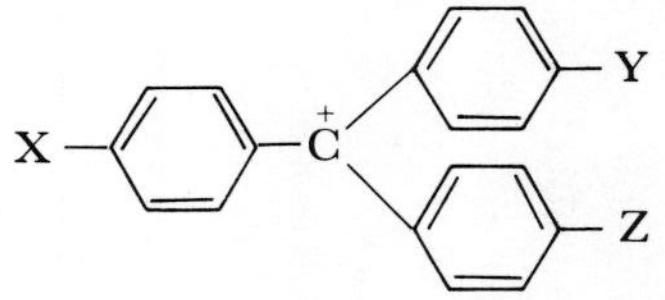

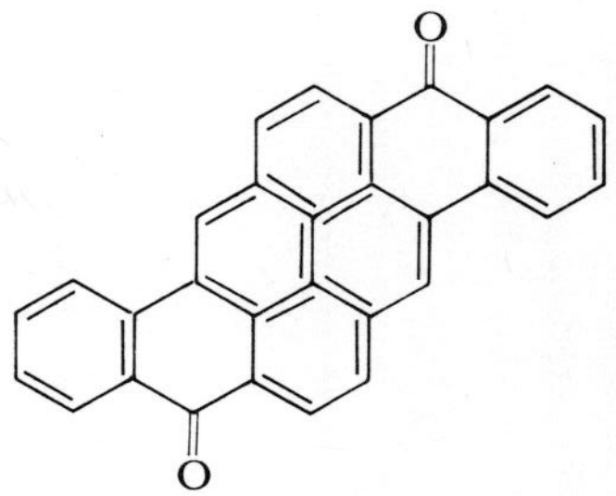

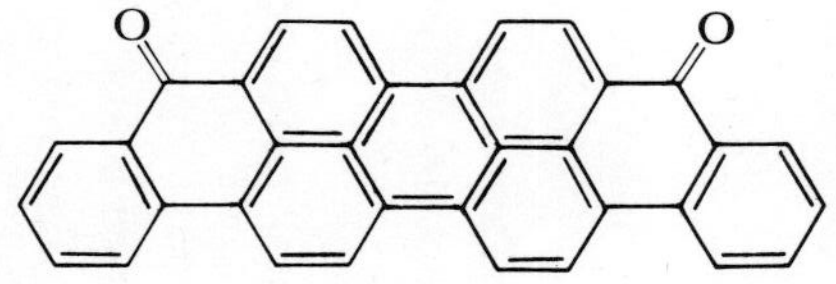

Azo chromophores

	Colour
Monoazo	
Alkyl–N=N–Alkyl	Pale orange
Aryl–N=N–Aryl	Yellow
Aryl–N=N–Alkyl	Orange–Red

Anthraquinones

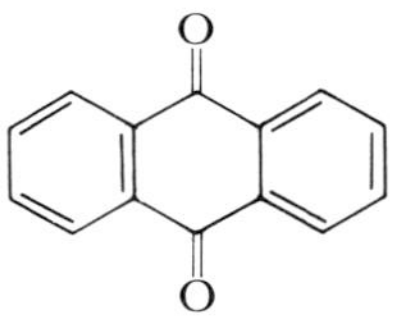

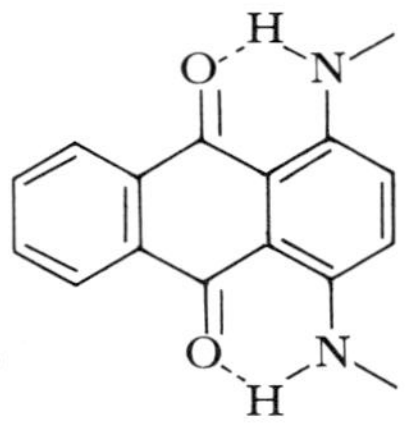

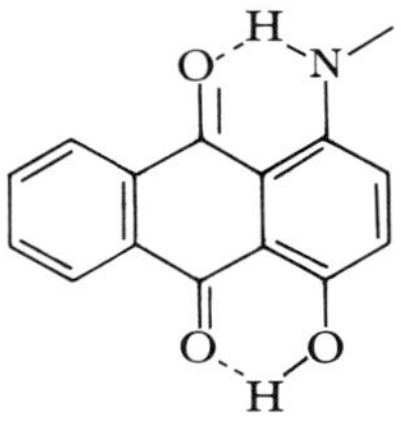

Pale yellow

Yellow–blue

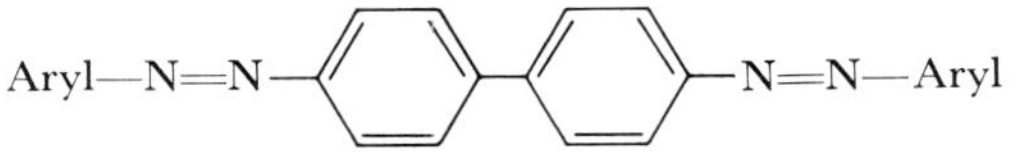

Blue–violet

Phthalocyanines

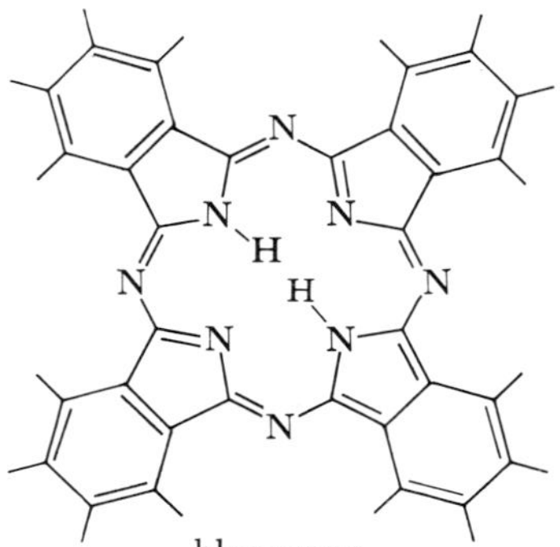

blue-green

Disazo

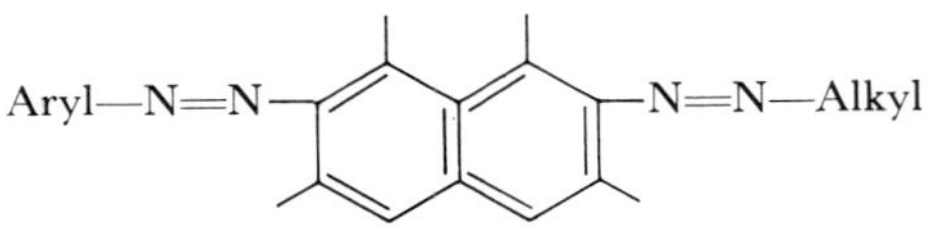

Aryl—N=N—Alkyl $\begin{cases} \text{Brown} \\ \text{Blue} \\ \text{Black} \end{cases}$

Aryl—N=N— $\begin{cases} \text{Yellow} \\ \text{Orange} \\ \text{Red} \end{cases}$ —N=N—Aryl

Trisazo

Aryl—N=N—Aryl—N=N—Aryl—N=N—Aryl

Brown–blue–green

In suitable cases it is possible to calculate the absorption spectrum from first principles but this is generally impracticable for dye molecules because the number of electrons is excessively large. Some success has been achieved by, for example, treating the π-electrons as if they were free to move within a conjugated system but most colour chemists have been forced to adopt a more qualitative approach. Almost a century ago Witt pointed out that the colour of dyes is associated with the presence of such features as carbon–carbon and carbon–oxygen double bonds, aromatic rings and nitro, nitroso and azo groups, which he termed *Chromophores*. The presence of certain of these groups, most notably the nitroso and azo groups, invariably confers colour on a compound, but others only do so under certain conditions. All chromophores contain π and often n-electrons and it is clear that they function by permitting $\pi \to \pi^*$ and $n \to \pi^*$ transitions which correspond to the absorption of visible light. The energy needed to excite an electron is critically dependent on the environment of the electron: if the latter forms part of an extended conjugated system the excitation energy is notably reduced because the excited state is stabilized by mesomeric electron delocalization. This effect increases with increasing conjugation, as has been confirmed with polyenes, where both the position and intensity of absorption vary, directly with the length of the conjugated system, with polycyclic aromatic systems and with azo compounds. Some typical chromophoric systems of commercial importance are shown in Table 12.3. It should be noted that in many cases the simplest examples are colourless or only faintly coloured; thus anthraquinone is very pale yellow whereas the 1,4-diamino and hydroxyamino compounds are blue. In this case it is probable that the O—H---O and O—H---N hydrogen bonds in the latter are so strong as to add two quasi-aromatic rings to the system, so altering profoundly its electronic structure. As this example shows, the structural features which confer desirable absorption characteristics are often quite specific and result from the combination of particular groups in particular ways.

Certain other groups containing lone pairs of electrons, notably amino and hydroxyl groups and halogen atoms, are able to influence the behaviour of the chromophores to which they may be attached, even though unable to confer colour themselves. Such groups are called *Auxochromes*; once more they stabilize the excited state by assisting electron delocalization. Their effect is generally to deepen colour by moving the position of maximum absorption to longer wavelengths.

The colour of a dye is not, however, entirely determined by the wavelength or the intensity of the principal absorption bonds. Purity of colour is also important. Bright colours result from narrow absorption bands with sharp peaks, and dullness from broad diffuse bands. In brown dyes, absorption is spread evenly over a wide range of wavelengths while black is the result of absorption throughout the whole visible spectrum. Particle size and shape effects play a considerable part in determining the colour of pigments, which are typically employed as molecular aggregates of about 10 μm diameter. Larger particles give notably duller and weaker colours and their formation must therefore be avoided. Moreover, many pigments are polymorphic and careful control is needed if the right crystalline form is to be produced.

12.3 Dyes and fibres

12.3.1 Acid and basic dyes

To be of commercial significance a dye must have both a suitable colour and an affinity for the substrate to which it is to be applied. In textile dyeing the dye is applied in an

Table 12.4 Acid and basic dyes for wool (D = chromophoric group)

(a) *Acid Dyes:* Wool-NH_2 + HX $\rightleftarrows$ Wool-$\overset{+}{N}H_3X^-$

Wool-$\overset{+}{N}H_3X^-$ + DSO_3^- $\rightleftarrows$ Wool-$\overset{+}{N}H_3DSO_3^-$ + X^-

CI name	Reference number	Structure	Colour
Acid orange 7	15 510	NaO_3S—C$_6$H$_4$—N=N—(naphthyl-OH)	Bright reddish-orange
Acid black 1	20 470	Ph—N=N—(naphthyl with HO, NH_2, NaO_3S, SO_3Na)—N=N—Ph	Bluish-black
Acid red 138	18 073	CH_3CONH, OH on naphthyl with NaO_3S, SO_3Na; —N=N—C$_6$H$_4$—$C_{12}H_{25}$	Bluish-red; abnormally fast
Acid blue 25	62 055	anthraquinone with NH_2, SO_3Na, NH—Ph	Blue

(b) *Acid mordant dyes:* Wool$\begin{matrix}NH_2\\CO_2H\end{matrix}$ + Cr^{3+} $\longrightarrow$ Wool$\begin{matrix}NH_2\\CO_2\end{matrix}$$Cr^{2+}$ $\xrightarrow{\begin{matrix}HO_2C\\HO\end{matrix}D}$ Wool$\begin{matrix}NH_2\\CO_2\end{matrix}\overset{+}{Cr}\begin{matrix}O_2C\\O\end{matrix}D$

CI name	Reference number	Structure	Colour
Mordant orange 6	26 520	NaO_3S—C$_6$H$_4$—N=N—C$_6$H$_4$—N=N—C$_6$H$_3$(OH)(CO_2Na)	Dull reddish-orange with Cr^{3+}
Acid blue 158	14 880	NaO_3S—(naphthyl-O)—N=N—(naphthyl-O)—SO_3^-, with $\overset{+}{Cr}$ coordinated to both O, plus H_2O, OH_2, OH_2	Greenish-blue premetallized dye

Table 12.4 – continued

(c) *Basic Dyes:* Wool-CO$_2$H + NaOH $\rightleftharpoons$ Wool-CO$_2^-$Na$^+$ + H$_2$O
Wool-CO$_2^-$Na$^+$ + D$^+$ $\rightleftharpoons$ Wool-CO$_2^-$D$^+$ + Na$^+$

CI name	Reference number	Structure	Colour
Basic green 4 (malachite green)	42 000		Bright bluish-green
Basic blue 26 (Victoria blue B)	44 045		Bright blue
Basic orange 21	48 035		Bright yellowish-orange
Basic red 18	11 085		Dull red

aqueous bath from which it must be taken up by the fibres as rapidly and irreversibly as possible. The processes used in particular cases depend on the nature of both the dye and the fibre.

Dyes may be attached to fibres by electrostatic or covalent bonds or by secondary valence forces such as hydrogen bonds or dispersion forces. Electrostatic bonding is involved in the dyeing of wool with *acid* dyes (Table 12.4). Wool is composed of the protein keratin which consists of polypeptide chains built up from 18 different α-aminoacids RCH(NH$_2$)CO$_2$H. Some of the R groups end in amino or carboxyl groups; in acid solution these amino groups are converted into ammonium ions which form salts with the anions of acid dyes. Dyeing of this type is an ion-exchange process and therefore reversible; this is an advantage in that level dyeing is readily attained but a disadvantage in that fastness to wet treatments is often low. Wet fastness may be improved by structural features which enhance the attractive forces between dye and fibre or reduce those between dye and solution. Dyes of relatively high molecular weight and those which contain strongly hydrophobic groups are more firmly held than are others. Alternatively the wet fastness of acid dyes capable of forming chelate metal complexes may be improved by treatment with chromic ion to give a derivative linked to the fibre through the metal. Chromium may be applied to wool as a *mordant* either before dyeing, simultaneously with the dye, or as an after-treatment. Best of all it may be incorporated in the dye itself. Other protein fibres such as silk and leather may be dyed in the same way.

Protein fibres also contain free carboxyl groups which in alkaline solution form salts with cationic or *basic* dyes. Many of the early synthetic dyes were cationics of the triarylmethane type, brilliant and intense in colour but unstable to light and to washing. Such compounds are now used only where durability is unnecessary, as in coloured papers, printing inks and crayons. The most important application of basic dyes in the textile industry is in the dyeing of polyacrylonitrile fibres such as Courtelle and Acrilan. Polyacrylonitrile homopolymers are extremely difficult to dye but copolymers containing carboxylic or sulphonic acid groups combine very readily with cationic dyes. The older basic dyes have been used to some extent for this purpose but cationic azo dyes have proved more popular.

12.3.2 Dyes and cellulose

Unlike the protein fibres, the cellulosic fibres, cotton, linen and rayon, contain no acidic or basic groups and are therefore unable to form electrostatic bonds with dyes. Since it is possible to dye cellulose the dye must be held to the fibre by other means. The simplest method of dyeing cellulosic fibres is by the use of a *direct* dye: the fibre is placed in a hot neutral aqueous solution of the dye to which an electrolyte has been added. The dye is rapidly adsorbed by the fibre to give moderately wash-fast dyeings. Most direct dyes are disazodyes based on benzidine, stilbene or naphthalene (cf. Table 12.5) which form long straight molecules with coplanar aromatic nuclei. It is believed that such molecules are attached to cellulose by multiple hydrogen bonds; many studies have shown that if a dye is to behave in this way, it must contain groups capable of forming hydrogen bonds and arranged so that they are adjacent to hydroxyl groups in the cellulose chain. High substantivity is also favoured by high molecular weight but decreased by the presence of hydrophilic groups.

Several other methods of dyeing cellulose depend on the formation of insoluble pigment particles within the fibre. A classical example is *vat* dyeing: the dye is reduced to a water-soluble derivative which is adsorbed by the fibre and then reoxidized *in situ* to the original compound. An after-treatment with hot soap solution assists crystallization and removes loose particles of pigment. The first vat dye to be used was indigo (cf. Table 12.6) which is reversibly reduced to the phenol indoxyl in alkaline solution; the glucoside of indoxyl occurs naturally in the indigo plant from which it was obtained until BASF discovered a practicable synthetic route in 1897. Today the most important examples are those based on anthraquinone and dibenzanthrone, which are used mainly for dyeing and printing cotton. They are extremely fast both to washing and to light but often give rather dull shades. Wool can be dyed in this way if protective agents such as glue are added to prevent hydrolytic degradation of the fibre under the alkaline conditions used.

Sulphur dyes are a related group. They are compounds of uncertain, probably polymeric, structure made by heating aromatic amines with sulphur or sodium polysulphide. They are applied from a sodium sulphide bath and then oxidized on the fibre. It is believed that they contain heterocyclic nuclei linked by disulphide groups $>$C—S—S—C$<$ which are reduced to mercaptide groups $>$C—SH by sodium sulphide and reformed by oxidation. Sulphur dyes are dull and not very fast to light, but they are very cheap.

Another process in which the fibre is pigmented is *ingrain* dyeing in which an insoluble azoic dye is produced by an *in situ* coupling reaction. The fibre is treated with an alkaline

332 Dyes and pigments

Table 12.5 Direct dyes for cotton

CI name	Reference number	Structure	Colour
Direct red 39	23 630		Bluish-red
Direct blue 67	27 925		Reddish-blue
Direct brown 57	31 705		Dull reddish-brown
Fluorescent brightener 32	40 620		Colourless-bluish fluorescence

Dyes and fibres 333

TABLE 12.8 Vat dyes

CI name	Reference number	Pigment Form	Reduced form	Colour
Vat blue 1 (indigo)	73 000	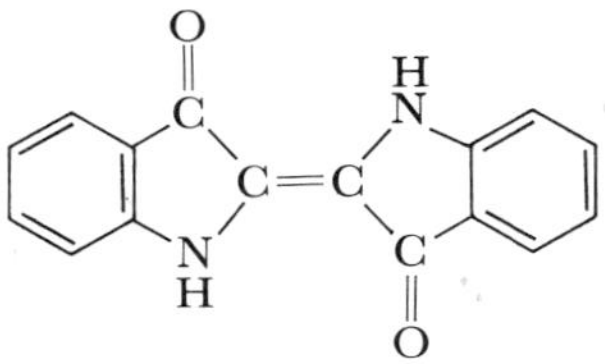	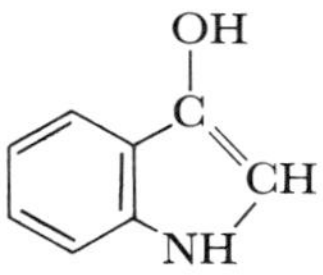	Navy-blue
Vat red 41 (thioindigo)	73 300	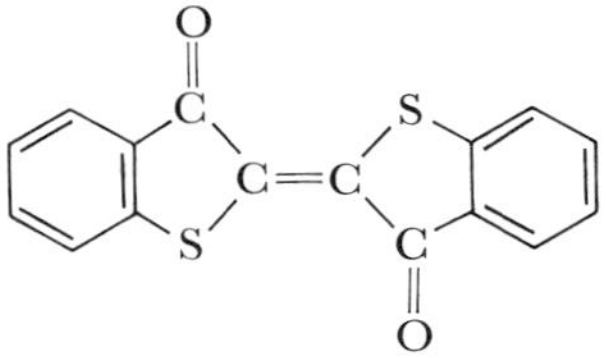	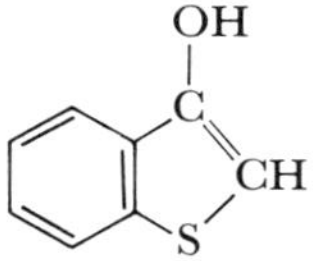	Bluish-red
Vat yellow 3	61 725	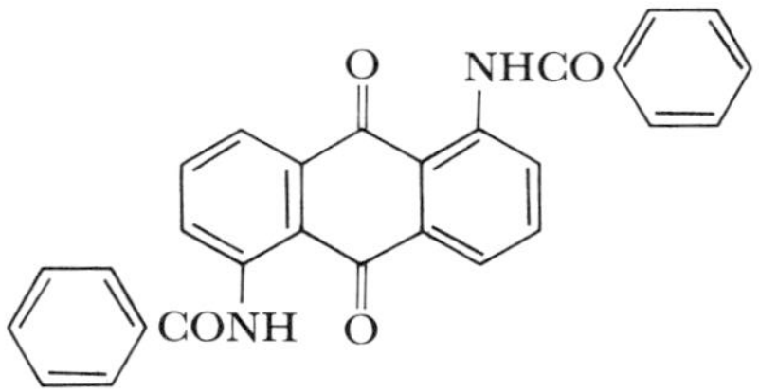	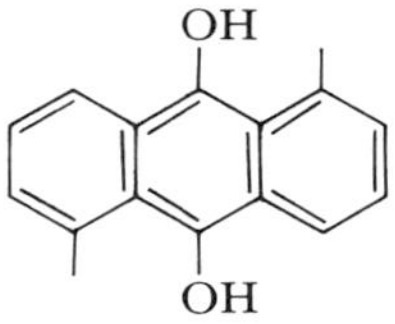	Yellow
Vat blue 4 (indanthrene blue)	69 800	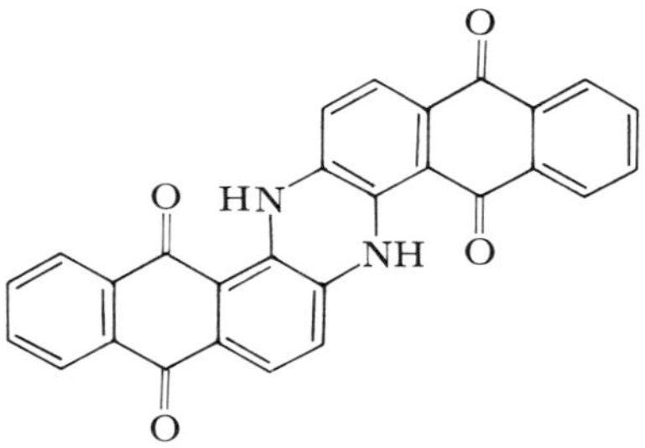	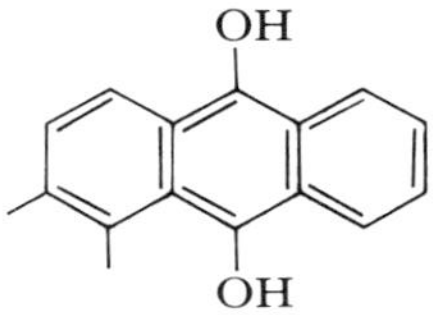	Bright reddish-blue
Vat green 1 (Caledon jade green)	59 825	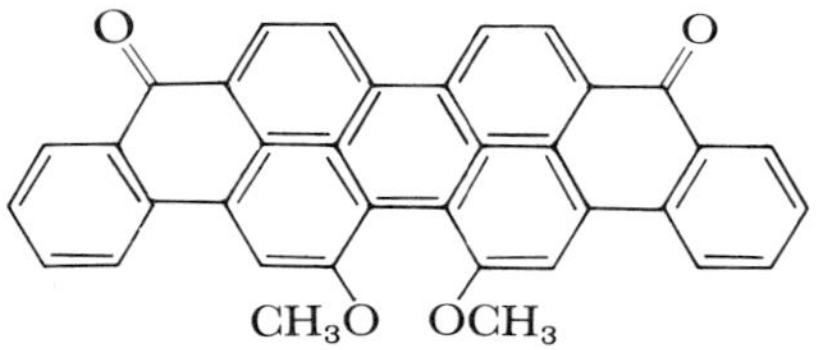	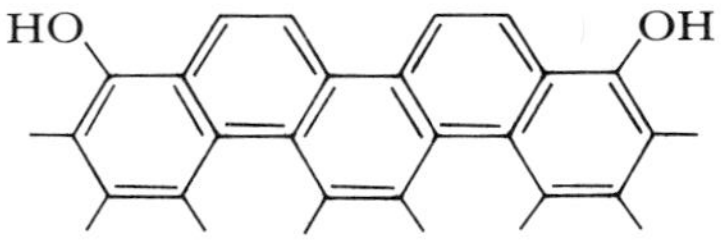	Bright green

334 Dyes and pigments

solution of a phenol or an amine which has an affinity for cellulose, such as the anilide of 2-hydroxy-3-naphthoic acid; the prepared fabric is then reacted with a diazonium salt. This process is extensively used for dyeing and printing cellulosic material since it is both convenient and effective. The prepared fabric may be stored before development and the diazonium salts are available as stable complex salts such as $(ArN_2)_2{}^+(ZnCl_4)^=$. The dyeing of cotton by basic dyes in the presence of a tannic acid mordant also appears to be a pigmentation process.

12.3.3 Dyes and synthetic fibres

The dyeing of synthetic fibres presents considerable problems. Cellulose acetate and triacetate have no affinity for direct cellulose dyes and *disperse* dyes must be used. Such dyes (cf. Table 12.7) are sparingly soluble and are applied as an aqueous dispersion stabilized by an anionic or non-ionic surfactant (cf. Chapter 11). The dye molecules form a solid solution within the fibre to which they are attached by secondary valence forces. Although

Table 12.7 Disperse dyes

CI name	Reference number	Structure	Colour
Disperse orange 1	11 080	$O_2N{-}C_6H_4{-}N{=}N{-}C_6H_4{-}NH{-}C_6H_5$	Dull orange
Disperse red 1	11 110	$O_2N{-}C_6H_4{-}N{=}N{-}C_6H_4{-}N(CH_2CH_2OH)(CH_2CH_3)$	Red
Disperse violet 1	61 100	1,4-diamino-anthraquinone	Bright violet on acetate; blue-violet on nylon
Disperse blue 1	64 500	1,4,5,8-tetraamino-anthraquinone	Blue

more reactive than the cellulose acetates, nylon is best dyed with disperse dyes as it is very difficult to produce nylon fibre with uniform dyeing characteristics. Disperse dyes have good levelling properties which conceal the effect of fibre irregularities. Polyester fibres such as terylene are markedly hydrophobic in character and are much more difficult to dye. In practice, if satisfactory dyeings are to be obtained, disperse dyes must either be applied in the solid state at 175–200°C, or as dispersions at 120°C and under increased pressure, or in the presence of a carrier such as *o*-hydroxydiphenyl. The carriers appear to function both by swelling the fibres and by providing a transfer phase compatible both with the solution and the fibres. Polypropylene is also highly resistant to dyeing but, as with polyacrylonitrile, the polymer chain may be modified to incorporate acidic or basic groups; alternatively, the molten polymer may be pigmented before it is spun into fibres.

12.3.4 Fibre-reactive dyes

The most important development in dyeing in recent years was the introduction of the fibre-reactive Procion dyes by ICI in 1955. These compounds are directly attached to cellulose fibre by covalent bonds and therefore show very high fastness to wet treatments. The Procion M dyes are based on symmetrical trichlortriazine (cyanuric chloride) in which the chlorine atoms are readily replaced by nucleophilic reagents. A conventional dye containing an amino or hydroxyl group is reacted with cyanuric chloride to give a substituted triazine derivative (cf. Table 12.8) which is then applied to cellulose in a mildly alkaline solution. The hydroxyl groups of the cellulose are largely ionized under these conditions and react smoothly to form cellulose–triazine–dye compounds. The Procion H dyes employ the less reactive dichlorotriazines which are more suitable for use in printing pastes. It is worth noting that the intermediate dye–triazine compound must have a substantial affinity for cellulose if reaction with the fibre is to predominate over alkaline hydrolysis. The commercial dyes of this type are based mainly on azo, anthraquinone and phthalocyanine compounds.

Other manufacturers have subsequently introduced new types of fibre-reactive dye for cellulose. Some are based on the nucleophilic substitution of dichloro- or trichloropyrimidines or of 2,3-dichloroquinoxaline residues. More interesting are those which exploit nucleophilic addition to vinyl sulphones or derivatives of acrylic acid; the Procilan range of fibre-reactive dyes for wool are believed to be reactive acrylamides. A

Table 12.8 Mechanisms of Attachment of Fibre-Substantive Dyes

D—NH$_2$ + [cyanuric chloride] ⟶ [D—NH-dichlorotriazine] ⟶ [D—NH-triazine-O—Cell]   Procion M (ICI)

Cell—OH + OH$^-$ ⇌ H$_2$O + Cell—O$^-$

D—SO$_2$CH$_2$CH$_2$OSO$_3$Na $\xrightarrow{OH^-}$ D—SO$_2$—CH=CH$_2$ + Na$_2$SO$_4$ $\xrightarrow{Cell-O^-}$ D—SO$_2$—CH$_2$CH$_2$O—Cell   Remazol (Hoechst)

D—NH—CO—CH=CH$_2$ + Wool—NH$_2$ ⟶ D—NH—CO—CH$_2$CH$_2$NH—Wool   Procilan (ICI)

D—CH$_2$CHOHCH$_2$Cl ⟶ D—CH$_2$HC—CH$_2$ (epoxide) $\xrightarrow{-NH-C(=O)-}$ D—CH$_2$CHOH—CH$_2$—N(—C(=O)—)   Procinyl H (ICI)

D = chromophoric group   Cell = cellulose chain   —NH—C(=O)— = polyamide chain similar reaction is apparently involved in the Procinyl dyes for nylon which combine the properties of disperse and fibre-reactive dyes. They are applied in weakly acid conditions which prevent reaction with the fibre and ensure level dyeing; the dyebath is then made alkaline and reaction with the amide groups in the fibre then takes place.

12.4 Miscellaneous applications of dyes

A significant use of dyes is as fluorescent brighteners or optical bleaches. These compounds are themselves colourless and non-fluorescent but when adsorbed onto cotton and other fabrics emit a strong blue fluorescence. This imparts a white appearance to yellowed fabrics by replacing the blue light absorbed and such compounds are universally used in detergent powders for this purpose (cf. Chapter 11). The structural requirements for optical bleaches are similar to those of direct dyes – they usually have linear molecules with coplanar ring systems – but they must of course be colourless. The most important group are the bistriazinyl derivatives of 4,4'-diaminostilbene-2,2-disulphonic acid, although other stilbenes, coumarins and pyrazoline derivatives are also used.

Dyes are often used in organic solvents; applications include wood stains, lacquers, polishes, candles, soap and foodstuffs. For these purposes they must be soluble in alcohol and hydrocarbon solvents, fast to light and generally non-toxic. The presence of long alkyl chains is desirable but that of hydrophilic groups is not. The bulk of solvent dyes are azo, anthraquinone or basic dyes of suitable structure.

12.5 Pigments

Pigments are used as suspensions of insoluble particles in a continuous medium which may be a solid – as in the mass coloration of plastics – a liquid, as in a paint, or a gas – as in smoke. To be of use as a pigment, a solid must not be merely coloured: it must form particles of reproducible shape and size which do not melt, vaporize, agglomerate or dissolve in the medium. Most early pigments were mineral in origin and inorganic compounds still account for the large majority of production. The most important are titanium dioxide (cf. section 10.5.6), the lead chromes, zinc oxide and the iron oxide pigments; although limited in range of colour, they are cheap and durable. The use of organic pigments has, however, increased steadily during the past decade. The earliest examples were insoluble salts of cationic and anionic dyes: examples are the *lakes*, insoluble aluminium or barium salts of hydroxyanthraquinones, and the insoluble phosphotungstates and phosphomolybdates of the cationic dyes. Azo pigments based on the anilides of 2-hydroxy-3-naphthoic acid and of acetoacetic acid are now widely used, as are some polycyclic vat dyes. Of outstanding importance is copper phthalocyanine, made by fusing phthalic anhydride, urea and cupric chloride, which combines a magnificent royal blue colour with extraordinary stability to light, heat and chemical reagents. Polychlorinated copper phthalocyanines are equally useful as green and yellowish-green pigments. Other commercially significant organic pigments include Pigment Green B, the iron complex of 1-nitroso-2-naphthol; aniline black, formed by the oxidation of aniline in the presence of vanadium salts; and carbon black itself.

Table 12.9 Production of Dyestuffs in UK (1968) and in USA (1969)

Class of application	By class of application					By chemical class USA only		
	UK		USA					
	Production (tons)	Value (£/ton)	Production (tons)	Value ($/ton)			Production (tons)	Value ($/ton)
Acid	8 077	924	11 133	5 311	Azo, total		34 295	4 408
Basic	2 657	869	6 754	6 303	monoazo		14 803	4 303
Direct	7 874	536	17 169	3 482	disazo		11 120	4 298
Disperse	2 565	2 392	11 552	5 510	trisazo		4 853	2 358
Mordant and chrome	3 155	469	1 023	3 570	other		3 519	4 633
Azoic	559	771	3 491	3 222	Azoic		3 491	3 222
Vat	n.a.	n.a.	23 043	2 579	Anthraquinone		23 520	4 453
Fluorescent Brightener } Fibre-reactive	7 264	1 365	18 046	3 659	Indigoid		2 369	1 410
			1 102	9 653	Stilbene		18 600	2 997
Sulphur	n.a.	n.a.	8 413	1 234	Sulphur		8 413	1 234
Solvent	2 520	657	5 078	3 835	Triphenylmethane		3 364	5 047
Other	n.a.	n.a.	2 183	7 083	Other		14 935	4 228
Total	34 671*	972	108 987	3 835	Total		108 987	3 835

12.6 Production and trends

The current pattern of production of dyestuffs in the UK and USA is shown in Table 12.9. As classified by application there are similarities between the two countries although certain differences are noticeable. Acid, basic, direct and disperse dyes comprise 42% of American and 62% of British production; fibre-reactive and mordant dyes are also of great importance in Britain and vat and sulphur dyes in the USA. Prices show greater divergences although it is clear that direct and sulphur dyes are relatively cheap and fibre-reactive dyes relatively expensive in both countries.

The US Tariff Commission also provides a breakdown of production by chemical constitution from which it is apparent that azo and anthraquinone compounds account for more than half the total. It is probable that this is also true of the British industry. This predominance is easily understood. Azo dyes are the most versatile of compounds: by suitable coupling reactions, compounds of almost any colour may be obtained at reasonable prices, while they may be applied to fabrics by every method except vat dyeing. The anthraquinone dyes include not merely the numerous substituted amino anthraquinones but also polycyclic derivatives such as the dibenzanthrones and anthraquinoneazines. Most of these compounds were originally developed as vat dyes but many of the simpler examples have now been adapted to fibre-reactive dyeing.

The current trend in Britain is towards the increased use of fibre-reactive dyes for cellulosic fibres in particular, and away from vat and azoic colours. Both the latter types give fast dyeings but vat dyes tend to produce dull shades while the azoics are relatively complex to apply: neither limitation applies to the fibre-reactive dyes. In the USA vat dyes still retain most of their importance and fibre-reactive dyes have as yet made little headway, perhaps because of the 'not-invented-here' syndrome. As the production of synthetic fibres continues to increase in both countries so does the output of disperse dyes, while the consumption of fluorescent brighteners is similarly tied to that of detergent powders. The use of organic pigments in paints and plastics is growing steadily and now amounts to about 20% of the total output of synthetic organic colouring materials (Table 12.10).

Table 12.10 Production of Pigments in UK (1968) and in USA (1969)

	UK		USA	
	Production (tons)	Value (£/ton)	Production (tons)	Value ($/ton)
Titanium dioxide	148 000	176	513 000	550
Zinc oxide	33 000	100	182 000	330
Iron oxide	24 000	71	109 000	310
Other inorganic pigments	20 000	n.a.	n.a.	n.a.
Organic pigments	10 000	1 290	26 000	4 730

Notes and references

General Many books deal with dyes from the standpoint of the organic chemist. A comprehensive treatise is K. Venkataraman, *The Chemistry of Synthetic Dyes*, 2 vols., Academic Press, New York and London, 1950, 1952; the same author is editing a series of supplementary volumes covering advances since 1950, of which vols. 3 and 4 appeared in 1970 and 1971. H. A. Lubs, *The Chemistry of Synthetic Dyes and Pigments*, Reinhold,

New York, 1955, is also useful. An excellent shorter book by a leading research worker is R. L. M. Allen, *Colour Chemistry*, Nelson, London, 1971. All these books are very fully referenced. The best all-round journal is the *American Dyestuff Reporter*, though the more technically oriented *Journal of the Society of Dyers and Colourists* and *Melliand Textilberichte International* are also useful.

The colour index The indispensable work of reference in the field is the *Colour Index*, compiled and published by the Society of Dyers and Colourists, 3rd edn., 5 vols., 1971. Approximately 7 900 dyes and pigments are listed, each of which is assigned a generic name which indicates its colour and usage pattern and a five-digit number which indicates its chemical structure; thus Acid Orange 7. no. 15 510, is an orange acid dye of structure

$$\text{C}_6\text{H}_5\text{-N=N-}\underset{\text{NaO}_3\text{S}}{\overset{\text{HO}}{\text{C}_{10}\text{H}_4}}\underset{\text{SO}_3\text{Na}}{\overset{\text{NH}_2}{}}\text{-N=N-C}_6\text{H}_5$$

In vols. 1 to 3 dyes are classified by generic name, with full technical data about methods of application, fastness, etc. In vol. 4 dyes are classified by chemical structure, with notes on their preparation, while vol. 5 gives trade names and suppliers. The whole work is lavishly cross-indexed and remarkably easy to use.

Statistics There are no readily available world-wide statistics. The OECD *Annual Report* gives overall production figures for dyes and – from 1969 – inorganic pigments. The *Annual Abstract of Statistics* contains UK production statistics for dyes, inorganic pigments and organic pigments, and the various *Census of Production* reports contain rather crude breakdowns of the total by class of application. The US Tariff Commission, *Synthetic Organic Chemicals*, gives detailed annual statistics classified both by class of application and by chemical structure.

Section 12.1 The data of Table 12.1 is from the OECD *Annual Reports*. UK production statistics are from *Annual Abstract of Statistics*, 1955 to date.

Section 12.2 Two classic short accounts are given by A. MacColl, 'Colour and Constitution', *Quart. Rev. Chem. Soc.*, 1947, **1**, 16 and M. J. S. Dewar, 'Modern Theories of Colour', *Chem. Soc. Special Publication No. 4*, 1956, p. 64. A more recent discussion is, that by S. F. Mason, 'Colour and the Electronic States of Organic Molecules', in Venkataraman, vol. 3, p. 169.

Section 12.3 The best book on the molecular processes involved in dyeing is T. Vickerstaff, *The Physical Chemistry of Dyeing*, 2nd edn., Oliver and Boyd for ICI Ltd., London, 1954. Good accounts of modern dyeing and printing technology are given by O. Glenz and K. Neufang respectively in Venkataraman, vol. 4, pp. 1, 75. W. F. Beech has written on *Fibre-Reactive Dyes*, Logos Press, London, 1970.

Section 12.5 Two useful sources are D. Patterson, *Pigments: An Introduction To Their Physical Chemistry*, Elsevier, Amsterdam, 1967, and J. S. Remington and W. Francis, *Pigments: Their Manufacture, Properties and Uses*, 3rd edn., Leonard Hill, London, 1954.

Section 12.6 UK production classified by method of application is given in the *Census of Production 1968*, part 35, HMSO, 1972. US Production statistics are from US Tariff Commission, *Synthetic Organic Chemicals*, 1969. Data on inorganic pigments is from the OECD *Annual Report* 1969.

13 Pharmaceuticals

Like the organic chemical industry, the pharmaceutical industry in its present form dates from the period of the Second World War. Before 1935, it was unusual for a doctor to be able to prescribe a drug to cure a specific disease, and virtually no drugs were known which would counter microbial infections. The two main exceptions were the use of arsenobenzyl and other arsenic compounds in the treatment of anthrax and syphilis, and the use of quinine to counter the malaria parasite.

Nonetheless, the second half of the nineteenth and first half of the twentieth century had seen a great increase in the expectation of life. In 1841, the expectation of life at birth in England and Wales was 40·2 years for men and 42·2 years for women. By 1951 these figures had risen to 66·4 and 71·5 years respectively and now (1967–69 data) stand

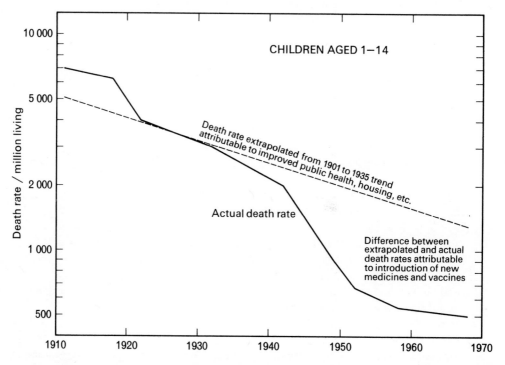

Figure 13.1 England and Wales – Registrar General's returns: 5 yearly averages 1911/15–1961/65 and the annual rate for 1966–68. Infant mortality rates show a steady drop due to improved living conditions and additional drop due to new medicines and vaccines. There were about 227 000 people alive in 1968 in England and Wales who would have died in childhood had the abrupt change in mortality not occurred.

at 68·7 and 74·9. Mortality in all age groups has declined but the drop in infant mortality has been spectacular (Fig. 13.1). This decline was due mainly to better hygiene and improved housing and living conditions, but one should also stress the role of vaccination in eliminating smallpox, and that of disinfection in preventing the spread of diseases, and the associated realization of the nature of microbial infection. Better and more food helped eliminate such diseases as rickets and also, for reasons which are unknown, certain diseases became less virulent. Bubonic plague declined at the end of the seventeenth century but this was not obviously the result of any human endeavour. It is thus only

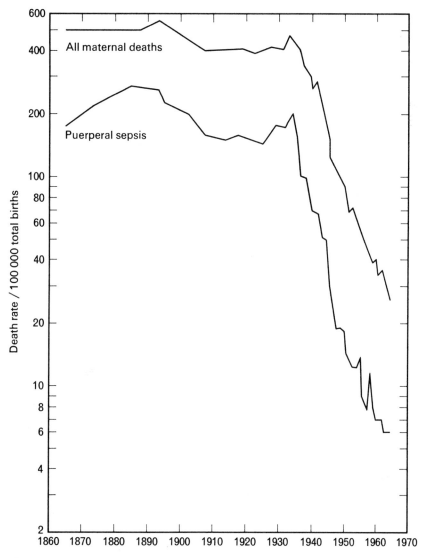

Figure 13.2 Maternal mortality death rate per 100 000 total births, 1860–1964, in England and Wales. *Note*: (i) Ten year averages 1861–90, five year averages 1891–1930, annual rates 1931–64. (ii) Logarithmic scale.

342 *Pharmaceuticals*

slightly unfair to the medical profession to attribute the increase in life expectancy up to 1935 to public health measures rather than to the ability of a doctor to cure disease. Indeed, if, in 1934, a patient became ill with a bacterial infection, there was very little a doctor could do except to advise him to stay in bed and drink plenty of fluids and to prescribe him a bottle of harmless medicine or a few aspirins in order to alleviate his symptoms. Nature would then take its course, and in cases of pneumonia or tuberculosis it was frequently a tragic one.

Diseases due to dietary and environmental deficiencies were largely conquered in the late nineteenth and early twentieth centuries (e.g. rickets, and cholera) and vaccination freed England from smallpox. Since 1935 generations of antibacterials have meant that

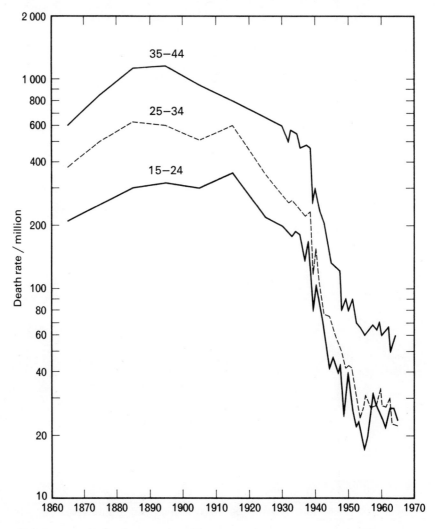

Figure 13.3 Pneumonia death rate per million males living by age, 1961–1964, England and Wales. *Note*: (i) Females rates are similar and in order to facilitate graphic presentation are not shown. (ii) Ten year averages 1861–1930, annual rates 1931–64. (iii) Logarithmic scale.

bacterial diseases (e.g. tuberculosis, pneumonia, puerperal fever) are no longer a serious threat to life in this country (Figs. 13.2, 13.3, 13.4). This is not to say that bacteria do not develop antibiotic-resistant strains – penicillin-resistant staphylococcus is a current problem in hospitals – but that the rate of development of antibiotics is keeping up with the rate of evolution of bacteria.

The main group of infectious diseases we are still unable to cure are those due to viruses – measles, scarlet fever, influenza and the common cold. In addition to virus diseases we are faced increasingly with the problems of ageing, nervous strain, and, ironically when two-thirds of the world is starving, overeating. The diseases whose incidence is increasing are heart disease and lung cancer. Table 13.1 shows the changing pattern of death.

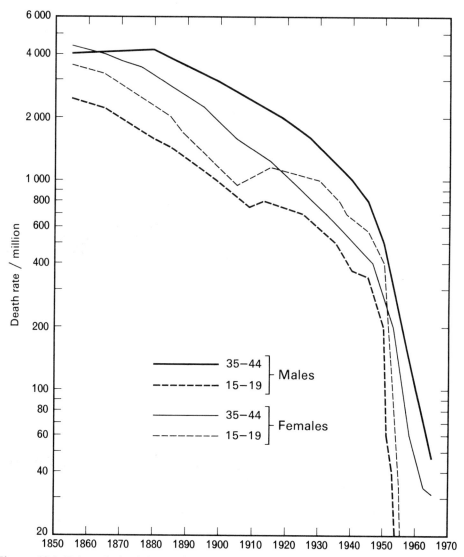

Figure 13.4 Tuberculosis of the respiratory system. Death rates per million living by sex and age. England and Wales 1851–1967.

344 *Pharmaceuticals*

The pharmaceutical industry is thus called upon to provide antibiotics (against bacterial infections), vitamins (against dietary deficiencies), cortisone and other hormones (against rheumatism and in birth control pills), sedatives and tranquillizers for people who cannot go to sleep and stimulants for those who cannot wake up. Drugs to cut blood pressure are also much in demand, and, of course there is a steady high level of

Table 13.1 Causes of death per million in certain age groups 1848–1965

	1848–72			1965		
Causes (groups)	Under 1 yr	1–4	45–64	Under 1 yr	1–4	45–64
Males:						
1. Infectious dis.	36 103	19 807	5 549	231	48	172
2. Cancer	20	17	715	117	109	3 979
3. Nervous dis.	42 636	3 079	3 014	483	62	1 204
4. Circulatory dis.	809	291	2 960	37	11	5 670
5. Respiratory dis.	30 446	6 262	4 722	3 065	188	1 572
6. Digestive dis.	23 554	2 309	2 499	815	59	353
Sub-total	133 568	31 765	19 459	4 748	477	12 950
All causes	202 655	36 383	23 936	21 785	868	14 060
Females:						
1. Infectious dis.	31 289	19 761	4 374	164	57	72
2. Cancer	23	17	1 701	65	94	2 806
3. Nervous dis.	31 762	2 864	2 665	379	53	977
4. Circulatory dis.	690	250	3 167	53	9	1 944
5. Respiratory dis.	22 832	6 065	3 497	2 341	186	476
6. Digestive dis.	18 990	2 284	2 454	569	51	217
Sub-total	105 586	31 241	17 858	3 571	450	6 480
All causes	162 281	35 457	20 618	16 595	763	7 295

demand for aspirin-containing preparations (q.v.) and harmless placeboes. Perhaps fortunately not all of those are consumed. In 1969 a sample of 500 households had between them 43 000 unwanted tablets of which 1 600 were psychotropic.

Table 13.2 (see pp. 346–347) shows USA and UK data on the contents and costs of prescriptions analysed according to the diseases they are intended to cure.

13.1 Sources of drugs

As we have already seen, the early organic chemical industry relied on fermentation processes to provide its main feedstock. The early pharmaceutical industry relied on alkaloids extracted from vegetable matter – principally opium, laudanum, morphine and quinine. Indeed, a well-known encyclopaedia published in 1933 considered all other sources of pharmaceuticals to be insignificant.

There are nowadays five sources of pharmaceuticals. These are shown in Table 13.3 listed in order of commercial importance.

Certain drugs can only be provided by a chemist by order on a doctor's prescription. These are known as ethical drugs and include all those the unsupervised use of which might be dangerous. Proprietary drugs, on the other hand, can be sold over the counter on demand. This chapter deals mainly with ethical drugs.

Standards for the purity of drugs and indeed for many other chemicals are laid down in the British Pharmacopoeia and the letters BPC on a jar or bottle indicate that these requirements have been satisfied. The US Pharmacopoeia fulfils a similar function in the USA.

Switzerland is the world's pharmaceuticals specialist, with the USA an important producer. In 1966, the USA ethical pharmaceutical industry made sales to the value of $4 660 m. of which $1 206 m. were exports. The pharmaceutical industry in the UK made sales of value £330·9 m. in 1968, of which £127·4 m. were to the health service and £45·3 m. were household medicines. £61·7 m. sales came outside those categories and £96·5 m. were exports. On a world scale, USA, West Germany, UK and Switzerland are the major drug exporters and Belgium–Luxemburg, Canada, Japan, Greece, Finland and Sweden are the major importers.

Table 13.3 Sources of pharmaceuticals

Source	Examples	Uses of examples
1. Chemical synthesis	Antihistamines, hypnotics, sulphonamides	Allergies or hay fever, insomnia, pain killing
2. Fermentation	Penicillin, streptomycin, tetracyclines	Active against microbial infections
3. Animal extracts	Hormones, liver extract, cortisone	Birth control, anaemia, rheumatoid arthritis
4. Biological sources	Vaccines, serums	Smallpox, diphtheria
5. Vegetable extracts	Alkaloids	Sedatives, antispasmodics, malaria

13.2 Characteristics of the pharmaceutical industry

The pharmaceutical industry differs from the rest of the chemical industry in a number of ways and these dictate its structure, and investment and sales policies. These differences are:

1. The speed with which new pharmaceuticals must be marketed. This is because the results of clinical trials tend to be published in the medical literature, indeed they have to be if doctors are to be persuaded to use the new material. On the other hand, clinical tests take a long time, and so a drug company must be ready to go into full-scale production and mount a massive advertising campaign the moment one of its new products is clinically approved.

2. The pharmaceutical industry produces a wide diversity of products in relatively small amounts by a large number of processes and operations. Furthermore, purity is of greater importance than high yield. Consequently, the processes of the pharmaceutical industry are much closer to conventional laboratory organic chemistry than are the pro-

Table 13.2 Number and cost of prescriptions dispensed through retail chemists, showing

Therapeutic group	Examples of group	UNITED	
		No. of prescriptions (m.)	%
Anti-infectives	Penicillins, tetracyclines, sulphonamides	35·7	15·0
Sedatives and tranquillizers	Barbiturates, phenothiazines, antidepressants	43·9	18·4
Metabolics	Sex hormones, other hormones, insulin	7·9	3·3
Cardiovasculars and diuretics	Digitalis, propranolol, methyl dopa, adrenaline derivatives, spironolactone, thiazide diuretics	19·6	8·2
Antacids, antispasmodics, laxatives, etc.	Magnesia and alumina formulations, belladonna derivatives	19·4	8·1
Analgesics and antipyretics	Aspirin, paracetamol, codeine	17·6	7·4
Anti-allergics	Antihistamines	6·5	2·7
Dermatologics	Antiseptics, cortisone and other ointments	16·8	7·0
Antirheumatics and anti-arthritics	Steroid hormones, indomethacin, ibuprofen, phenylbutazone	6·4	2·7
Cough preparations	Based on glycerine, menthol, etc., perhaps with ephedrine or morphine derivatives	19·3	8·1
Other groups	Stimulants, eye preparations, vitamins, vaccines	46·1	19·1
Totals		239·2	100

Drug classifications differ between the UK and USA, so data are only roughly comparable. USA figures show cost to consumer of prescription rather than net ingredient cost. We estimate the latter as 55% of the former, on average. The most expensive drugs are diuretics, hormone based preparations, anti-arthritics and antibiotics; the least expensive, aspirin, barbiturates, antacids and cough mixture. Note the high incidence of coughs in UK (see Section 17.4.2).

therapeutic group distribution

KINGDOM 1969				UNITED STATES 1966				
Total net ingredient cost (£m.)	%	Average net ingredient cost/ prescription (p)	No. of prescriptions (m.)	%	Value of prescriptions ($m.)	%	Average value of prescription ($)
19·6	18·4	55	152	17·1	611	20·1	4·00
18·8	17·7	43	124	14·0	430	14·1	3·47
6·2	5·8	78	118	13·3	389	12·8	3·29
16·8	15·8	86	103	11·5	392	12·9	3·81
6·4	6·0	33	77	8·8	250	8·2	3·25
4·2	4·0	24	62	7·0	176	5·8	2·84
1·8	1·7	28	43	4·9	125	4·1	2·91
7·0	6·6	42	42	4·7	122	4·0	2·91
6·0	5·7	94	33	3·7	135	4·5	4·11
3·8	3·1	17	31	3·5	73	2·4	2·35
16·0	15·2	35	103	11·5	344	11·1	3·34
106·1	100	45	888	100	3 047	100	3·43

cesses of the heavy chemical industry. The equipment used must be versatile. Batch processes are the rule and they are frequently carried out *in vitro*.

3. High unit cost. Pharmaceuticals are at least one order of magnitude more expensive than industrial organic chemicals. Thus, processes which would otherwise be prohibitively expensive can be used. On one occasion, Merck adopted a partial synthesis for cortisone which yielded 200 g of product per month which was sold at $200 per gram. This price dropped fairly quickly but, nonetheless, at the time the quantity and the price gave a viable commercial proposition for the company concerned.

4. The demand for pharmaceuticals is highly price inelastic. This is particularly true in the UK where the existence of the National Health Service means that neither doctor nor patient will normally know the exact cost of the drug which the one prescribes for the other. Furthermore, professional ethics demand that a doctor should prescribe what he thinks is the best drug regardless of price. Even when the patient pays for a drug, however, he is unlikely to refuse to buy a particular pharmaceutical because the price is too high. Indeed, he may feel that the efficacy of the drug is directly related to the amount he has had to pay for it (i.e. the demand curve has a positive gradient).

Competition in the pharmaceutical industry between patent protected ethical products therefore hardly ever takes the form of a price war. Instead, firms compete in innovation, that is by developing a new drug which no one else offers, and in trying to increase their share of the market for a particular pharmaceutical which several of them offer.

The importance of innovation is underlined by a 1969 estimate that 92% of ethical pharmaceuticals produced in that year originated after 1945, 45% after 1960 and 20% after 1965. The development of new products and the persuading of doctors to prescribe them demands high expenditure on research and development on the one hand and advertising and promotion on the other. In 1966 the UK ethical pharmaceutical industry spent £13·3 m. on advertising (13·2% of sales) and £13·0 m. on research (13·0% of sales) the latter figure rising to £22 m. in 1970. The analysis of 1966 advertising expenditure is as follows:

	(£m.)
Direct mail advertising	2·4
Advertising in medical journals, etc.	2·0
Samples	1·4
Representatives' calls on general practitioners	6·2
Other activities	1·3
	13·3

The magnitude of this figure may be gauged from the fact that there are some 20 000 general practitioners in the UK who consequently have about £650 per head spent on attempts to persuade them to prescribe particular medicines. The general practitioners for their part complain bitterly at the number of glossy brochures which arrive in the mail and the time which is wasted seeing representatives of the drug firms. They are criticized in turn for their conservatism and widespread failure to read even the most important medical journals, let alone the glossy literature put out by advertisers.

The above data apply to ethical products which are advertised only to the medical profession. Proprietary drugs are advertised to the general public through newspapers, television, etc. Research expenditure is proportionately lower and, as the consumer prescribes and pays for the drug himself, the market is slightly though not markedly more price-sensitive.

Table 13.4 Tonnages and unit values of selected pharmaceuticals and pharmaceutical groups

Pharmaceutical or group	Closest therapeutic group in Table 13.2	United States 1967			United Kingdom
		Production (000 lb)	Sales (000 lb)	Unit value ($/lb)	
Total		180 070	126 924	3·04	
Penicillins*	Anti-infectives	4 113	1 775†	18·8	1971 production 786 tons.
Tetracyclines		3 348	1 357	23·5	
Sulphonamides		4 794*	1 500	4·4‡	1958 production 859 000 lb
Streptomycins		795‡	n.a.	9·1	
Aspirin*	Analgesics and antipyretics	30 902	n.a.	0·61	
Barbiturates*		802	497	4·9	1958 production 855 000 lb
Meprobamate tranquillizers		1 260	913	2·55	
Other tranquillizers		444	85	27·4	
Antidepressants		116	—	—	
Hormones and synthetic substitutes	Metabolics	1 783	328	52·45	
Cardiovascular drugs	Cardiovasculars and diuretics	723	519	31·39	
Anticoagulants (total)		9	4	273	
Sodium heparin		3	2	504	
Gastrointestinals	Antacids, antispasmodics, laxatives, etc.	52 237	48 014	0·37	
Antihistamines*	Anti-allergics	442	265	1·67	
Dermatological agents	Dermatologics	12 996	9 388	0·44	
Amphetamines	Other groups	86	74	7·84	1960 production 18 000 lb
Vitamins		17 568	11 108	5·93	

* 1968 data   † Excludes veterinary uses   ‡ 1966 data

Note: These data are for production compared with Table 13.1 which relates to consumption on prescription. The drug classes are not always identical, but approximate correspondences are shown. Note the high production of vitamins, most of which are self-prescribed, not recommended by a physician.

350   Pharmaceuticals

5. The inapplicability of conventional ideas of demand and elasticity to the pharmaceutical industry means that its price structure is somewhat arbitrary, and subject to political pressures.

Whereas in the USA the drug firms compete to persuade doctors to prescribe their products to patients who pay for them, in the United Kingdom the National Health Service is overwhelmingly the largest purchaser of ethical pharmaceuticals and has all the advantages of a monopoly buyer. The cost of the health service in 1968 was £1 741 m. (5·23% of the national income). Of this £177 m.* (about 10%) was spent on ethical pharmaceuticals. 267·4 m. prescriptions (5·50 per head of the population of the UK) were filled at an average cost of 57p.* The corresponding figures for 1949 were 202 m., 4·75 per head at 15p each.

In the USA, in 1968, total expenditure on health was $57 103 m. (6·6% of gross national product). Of this $6 149 m. (10·8%) was spent on ethical pharmaceuticals plus about 40% as much again on proprietary pharmaceuticals. In 1966, 888·2 m. prescriptions were filled at an average cost of $3·43. This has risen from $0·91 in 1939.

The *per capita* cost of prescriptions in the UK in 1969 was £3·34 compared with a *per capita* expenditure of $28·66 in the USA.

Criticisms of rising costs are met by the industries in both countries by the claim that most of the cost increases are due to inflation and the rest are due to the fact that the consumer is getting a better (i.e. a more effective) product. Nonetheless, political pressures are directed at the industry and in the UK the government has used its dominant position to get the industry to agree to 'a voluntary price regulation scheme' by which prices are calculated on a 'cost plus' basis.

Pharmaceuticals are typical only in their diversity. The broad production picture is given in Table 13.4. A few of the more important products are considered below.

13.3   Aspirin (acetylsalicylic acid)

Aspirin is the most widely used pharmaceutical. It is a mild sedative, an antineuralgic (pain-killer) and an antipyretic (tends to bring down the temperature). It can be used either by itself or mixed with other compounds of complementary properties such as codeine.

Commercial manufacture began in 1899, and in 1968 in the USA 14 000 tons of aspirin were manufactured. The normal dosage is 0·3–1·0 g so it appears that in 1968 the average American bought, even if he did not consume, some 100–200 tablets of aspirin. The UK figures for 1958 were 2 359 m. aspirin and 3 249 m. aspirin-containing tablets purchased, which leads to a *per capita* consumption similar to that in the USA. The figures are astonishingly high and suggest an unexpected degree of international hypochondria.

Manufactured on a scale of thousands of tons per year, aspirin is the only pharmaceutical which is truly an industrial chemical and where questions of cost are paramount. It is made by the action of heat on dry sodium phenoxide with carbon dioxide under pressure (Kolbe–Schmitt process) and decomposition of the resultant sodium salicylate

---

* These are higher than figures given earlier in the chapter because they include various wholesale and retail margins.

to give salicylic acid. This is heated in turn with a slight excess of acetic anhydride and the resulting acetylsalicylic acid cooled, centrifuged, recrystallized and dried (Fig. 13.5).

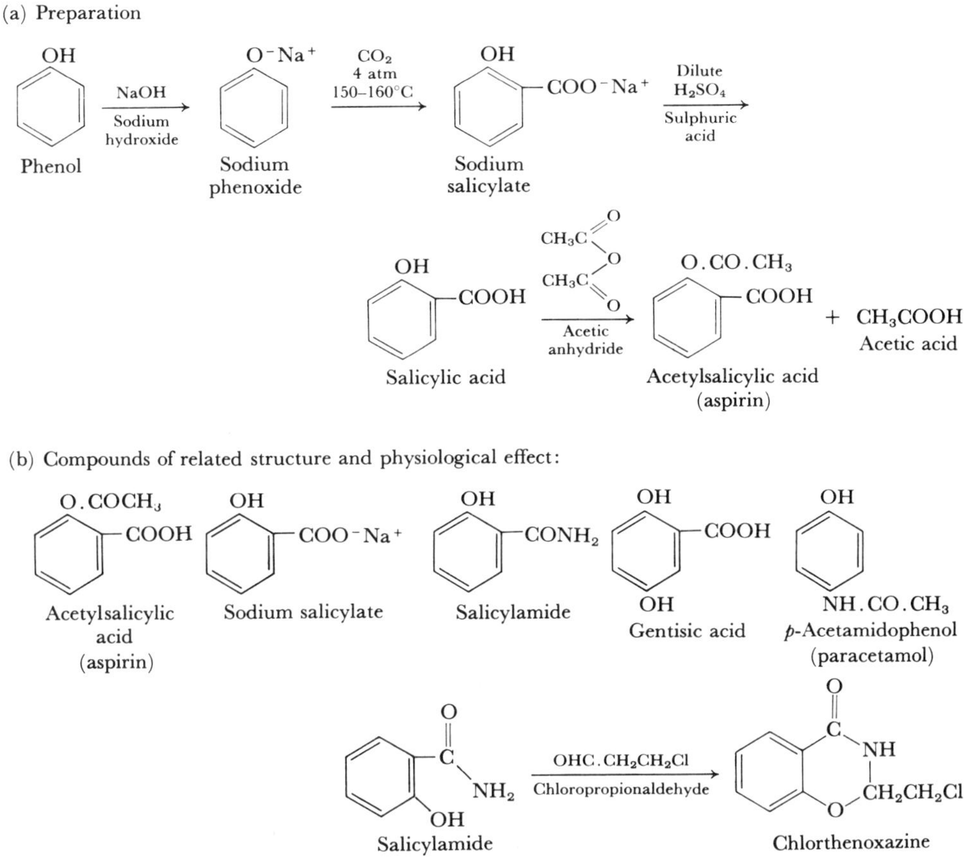

(a) Preparation

(b) Compounds of related structure and physiological effect:

Figure 13.5 Aspirin.

Aspirin has few side effects, is not habit-forming and is scarcely poisonous. It is a mild stomach irritant and people using aspirin in suicide attempts are usually sick before they have managed to ingest a fatal dose.

The irritant effect on the stomach is thought to be due to the insolubility of aspirin. If taken on an empty stomach, the tablet lies on the stomach lining without dissolving and may cause slight localized ulceration and bleeding. To avoid this, various companies market a soluble form of aspirin – sodium acetylsalicylate – which enters the stomach as a dilute solution. On mixing with hydrochloric acid in the stomach, insoluble acetylsalicylic acid is immediately precipitated so that a widespread mild inflammation replaces the localized bleeding. This is felt by the many purchasers of soluble aspirin to be an advantage.

Various compounds related to aspirin have a similar physiological effect. These include sodium salicylate, salicylamide, gentisic acid and chlorthenoxazine (Fig. 13.5). The last

352 *Pharmaceuticals*

of these is made from salicylamide and is claimed to be better than aspirin as an antipyretic and in relieving gout and rheumatism.

Para-acetamidophenol (paracetamol) has also come into the market in recent years as a substitute for aspirin and is said not to irritate the stomach (Fig. 13.5).

13.4 Sulphonamides

Forty years ago pneumonia and tuberculosis were major causes of death and puerperal fever was still a problem in maternity cases. These diseases are all due to infection by the haemolytic streptococcus. Puerperal or childbed fever has been included as a cause of death in mortality statistics for several centuries and is therefore a useful indication of the threat posed by the haemolytic streptococcus.

Before about 1850, the mother or child or both died of puerperal fever in between 10 and 30% of all births in hospitals. The source of the disease was not recognized, and midwives would go from confinement to confinement without even washing their hands. A Hungarian physician, Ignaz Semmelweis, noticed that many of the women who con-

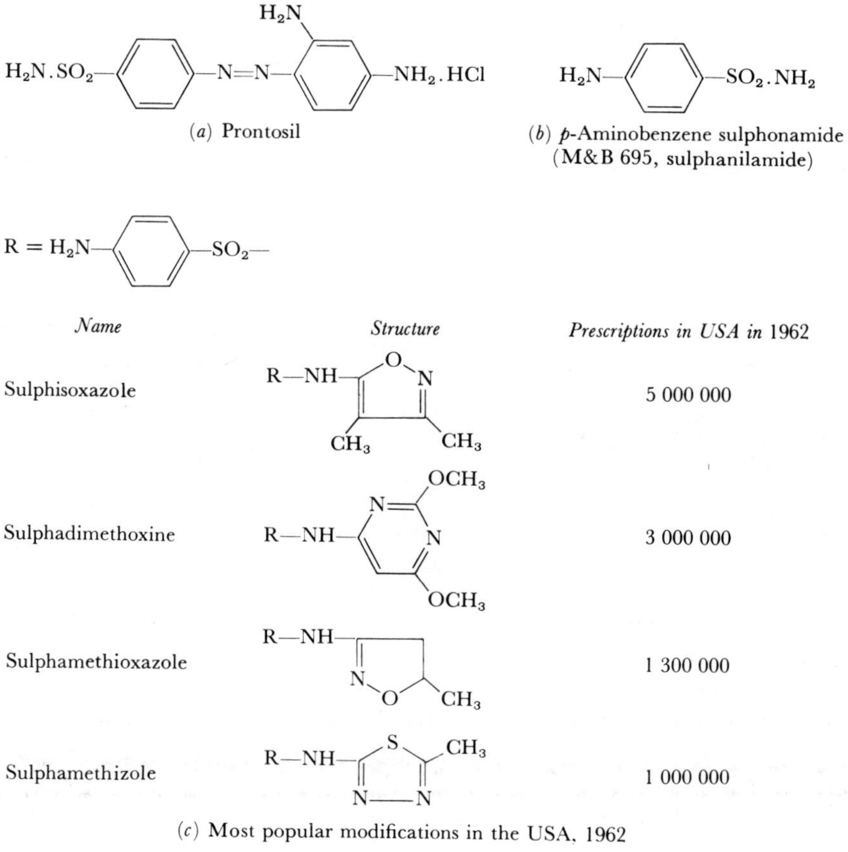

(a) Prontosil

(b) *p*-Aminobenzene sulphonamide (M&B 695, sulphanilamide)

$R = H_2N-C_6H_4-SO_2-$

Name	Structure	Prescriptions in USA in 1962
Sulphisoxazole	R—NH—(isoxazole with two CH$_3$)	5 000 000
Sulphadimethoxine	R—NH—(pyrimidine with two OCH$_3$)	3 000 000
Sulphamethioxazole	R—NH—(isoxazole with CH$_3$)	1 300 000
Sulphamethizole	R—NH—(thiadiazole with CH$_3$)	1 000 000

(c) Most popular modifications in the USA, 1962

Figure 13.6 Sulphonamides.

tracted the disease did so after being examined by medical students who had come straight from the dissecting room, or by doctors and midwives who had just attended other fever cases. Semmelweiss tried to persuade the medical staff to wash their hands in carbolic acid or dilute hypochlorite solution inbetween cases, and denounced any doctor who would not do so as a murderer. The medical establishment reacted characteristically, and Semmelweiss died a broken man in 1865, a victim himself of the haemolytic streptococcus. His methods eventually made headway and the death rate from puerperal fever dropped to 1% of all hospital births.

In 1935 Horlein and Elberfield announced the discovery of Prontosil (Fig. 13.6 (a)) which was an azo dye and which appeared to be specifically active against the haemolytic streptococcus. Further investigation showed that the therapeutically active agent was sulphanilamide (p-aminobenzene sulphonamide Fig. 13.6 (b)). Since then over 10 000 modifications of this material have been studied and Fig. 13.6 (c) shows the four most frequently prescribed 'sulphonamides' or 'sulpha' drugs.

The effect of sulphanilamide on puerperal fever has been dramatic. Figure 13.2 shows the deaths from this and related diseases from 1860 onwards. Deaths from pneumonia and tuberculosis have shown a similar decline, though in all cases the effect of the sulphonamides has been reinforced by subsequently discovered antibiotics.

Most sulphonamides are manufactured by condensation of p-acetamidobenzenesulphonyl chloride with the appropriate amino derivative of a heterocyclic compound. The steps in the production of sulphamerazine are shown in Fig. 13.7.

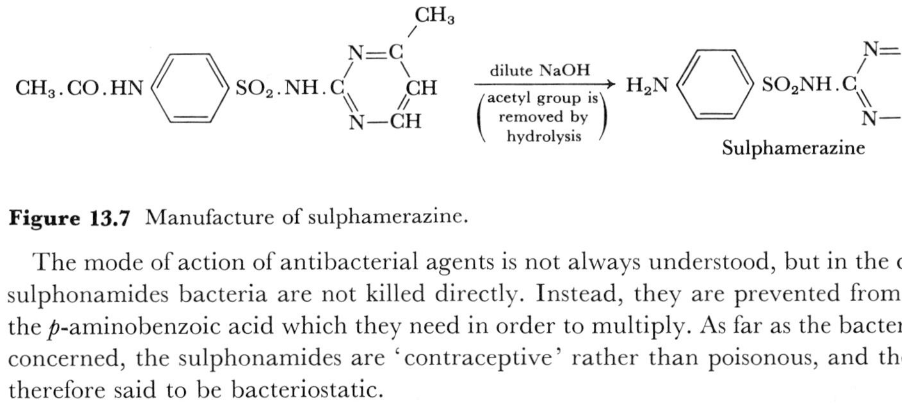

Figure 13.7 Manufacture of sulphamerazine.

The mode of action of antibacterial agents is not always understood, but in the case of sulphonamides bacteria are not killed directly. Instead, they are prevented from using the p-aminobenzoic acid which they need in order to multiply. As far as the bacteria are concerned, the sulphonamides are 'contraceptive' rather than poisonous, and they are therefore said to be bacteriostatic.

The sulphonamides have low toxicity. Their major disadvantage is that they are sparingly soluble, and when taken in large enough doses to be effective, they sometimes crystallize out in the urine which can be extremely painful. They are also potentially dangerous to the kidneys.

354 Pharmaceuticals

13.5 Barbiturates

The barbiturates are a class of heterocyclic compound used as hypnotics, i.e. sleep inducing drugs. They are white crystalline solids, insoluble in water but giving soluble sodium salts.

The first barbiturate sleeping tablet – Veronal – was introduced in 1903. Chemically it is diethylbarbituric acid. Hundreds of similar barbiturates are made with different substituents in place of the ethyl groups. The commonest are shown in Fig. 13.8. They are manufactured by a malonic acid type of synthesis. The route, starting with commercially available raw materials is shown in Fig. 13.9.

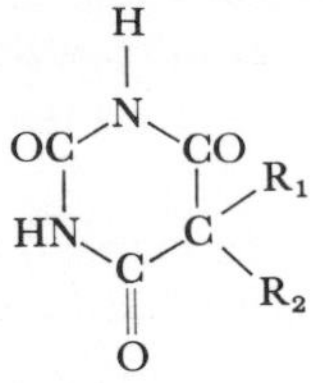

Compound		R_1	R_2
Veronal	(diethylbarbituric acid)	ethyl	ethyl
Luminal	(phenobarbitone)	ethyl	phenyl
Soneryl	(butobarbitone)	ethyl	n-butyl
Nembutal	(pentobarbitone sodium)	ethyl	1-methylbutyl
Tuinal	quinalbarbitone sodium	allyl	1-methylbutyl
	+amylobarbitone sodium	ethyl	isoamyl

Basic structure

Figure 13.8 Barbiturates. The most heavily prescribed barbiturates in the UK in 1970 were Soneryl, Tuinal, Nembutal and Sonalgin (a mixture of butobarbitone, codeine phosphate and phenacetin).

The UK market for barbiturates in 1958 was about 400 tons – considerably more per head of the population than in the USA. The typical dose is about 0·1 g so the production allows for 50–100 doses/year per head of the population.

Barbiturates have two major drawbacks. First, they are habit forming and patients develop a tolerance for them so that dosages have to be increased. Second, the fatal dose is only about ten to fifteen times the normal dose. Barbiturates therefore are truly dangerous drugs. They provide not only a simple means of suicide, but are a constant danger in a house containing inquisitive children. The present UK consumption must be viewed with misgiving but is steadily declining as the problem is publicized. It was for this reason that the apparently much less dangerous hypnotic, thalidomide, was so warmly welcomed when it first appeared on the market and why its use rapidly became widespread, thus increasing the magnitude of the eventual tragedy.

Meanwhile, in 1969 in England and Wales there were 2 934 deaths from poisoning by sedatives, hypnotics and anticonvulsants of which 329 were accidental and the remainder mainly suicides. About three-quarters of these were due to barbiturates and some less poisonous hypnotic would be a valuable addition to the medical armoury.

13.6 Antibiotics

Antibiotics are chemical compounds derived from micro-organisms which have the capacity even in dilute solution of inhibiting the growth of or even destroying other micro-organisms.

The first demonstration of microbial antagonism was performed by Pasteur and Joubert in 1877. They showed that many common bacteria inhibited the growth of anthrax organisms. In 1928 Fleming noticed that certain bacteria which he was cultivating on

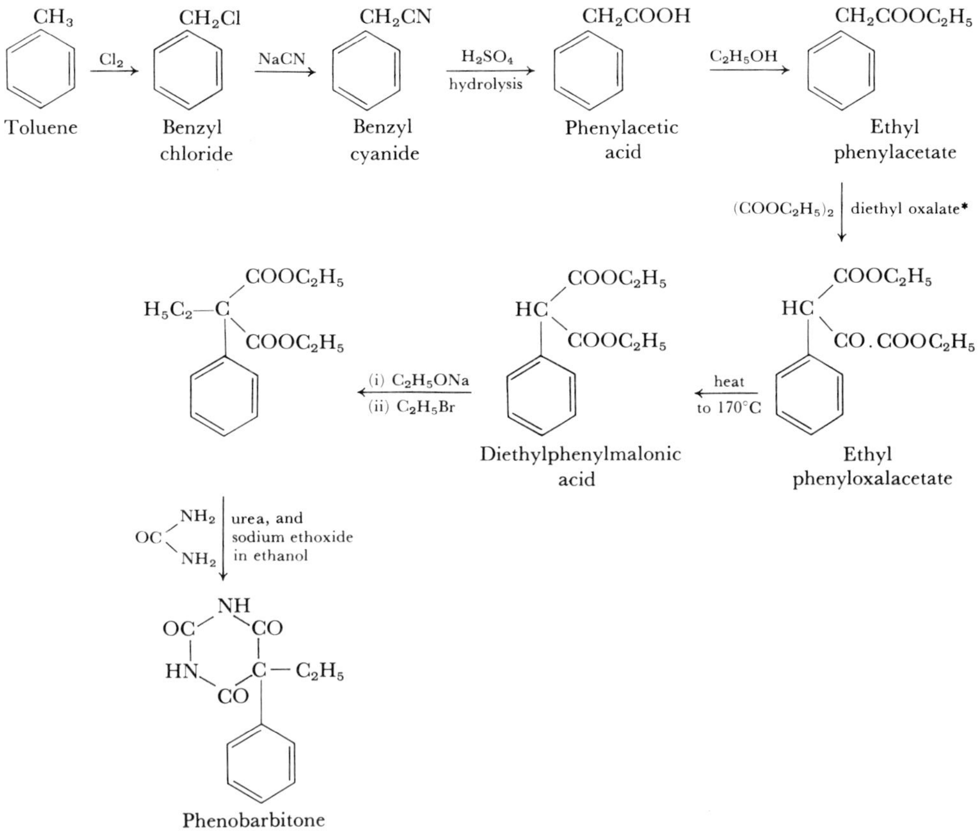

*Diethyl oxalate is made from sodium formate which, on alkali fusion, gives sodium oxalate, whence oxalic acid and its diethyl ester.

Figure 13.9 Synthesis of phenobarbitone.

the surface of a nutrient were killed by the action of the mould *Penicillium notatum* which had accidentally settled on the nutrient. The significance of the observation was unrecognized until the medical world had accustomed itself to the idea of sulphonamides, and it was not until 1939 that Florey and Chain re-started work on the material. The technology of large-scale production was worked out in the United States and by 1944 tonnage quantities were being produced on both sides of the Atlantic.

Since then, the penicillin molecule has been extensively modified and two other major groups of antibiotics have been discovered – the streptomycins and the tetracyclines.

Production of antibiotics is shown in Table 13.3. All antibiotics except chloromycetin, which is now used in the USA only as an animal feed additive, are made by fermentation processes rather than by chemical synthesis.

13.6.1 Penicillin

The penicillins, at least six of which have been isolated, are compounds produced during

356 *Pharmaceuticals*

the growth of *Penicillium notatum*, *Penicillium chrysogenum* and related moulds. They are all based on the 6-aminopenicillanic acid nucleus and differ only in the nature of the side chain R. The structures and activities of the six main naturally-occurring penicillins are shown in Fig. 13.10. They are quite strong acids ($p_K \approx 2 \cdot 8$) and easily undergo hydrolysis and rearrangement. This was the reason why penicillin had for so long to be administered by injection before an oral penicillin, stable in the stomach, was developed.

Pencillin is prepared by a variety of fermentation processes which differ slightly in the fermentation stages and widely in the extraction stages. A typical process is described in section 8.5.2.

A total synthesis of penicillin has been performed but is uneconomic. Penicillin can, however, be modified by chemical means, and this is important, because certain strains of the bacterium *Staphylococcus aureus* have acquired the ability to generate the enzyme penicillase which rapidly destroys natural pencillins. At one time it appeared that penicillins would no longer be useful antibiotics but they were given a fresh lease of life in 1963 when Beecham's, the UK firm, managed to produce a chemically modified penicillin – ampicillin – still based on 6-aminopenicillanic acid, which was resistant to penicillase. The initial discovery required a £3·5 m. research budget and since then many other chemically modified penicillins have been produced, especially the 2,6-dialkoxyphenyl penicillins. The most important modified penicillins which were available commercially in recent years are shown in Fig. 13.11.

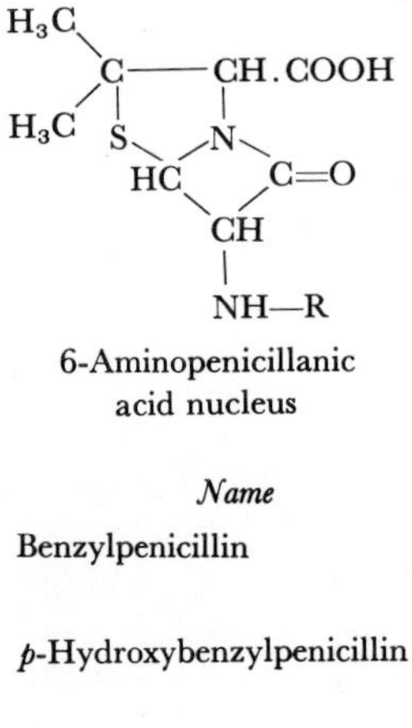

6-Aminopenicillanic acid nucleus

Name	Structure of side chain R	Antibiotic activity (units/mg sodium salt)
Benzylpenicillin	—CO.CH$_2$C$_6$H$_5$	1 667 (by definition)
p-Hydroxybenzylpenicillin	—CO.CH$_2$—C$_6$H$_4$—OH 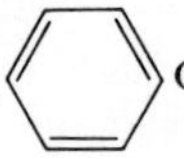	900
Penicillin F 2-pentenyl penicillin	—CO.CH$_2$CH=CH.CH$_2$CH$_3$	1 600
Penicillin dihydro F *n*-amyl penicillin	—CO.CH$_2$CH$_2$CH$_2$CH$_2$CH$_3$	1 500
Pencillin K *n*-heptyl penicillin	—CO.CH$_2$(CH$_2$)$_5$CH$_3$	2 300

Cephalosporin N
D-4-amino-4-carboxy butyl penicillin
	—CH$_2$CH$_2$CH$_2$CH·NH$_2$
	          |
	         COOH
		?

Figure 13.10 Naturally-occurring penicillins.

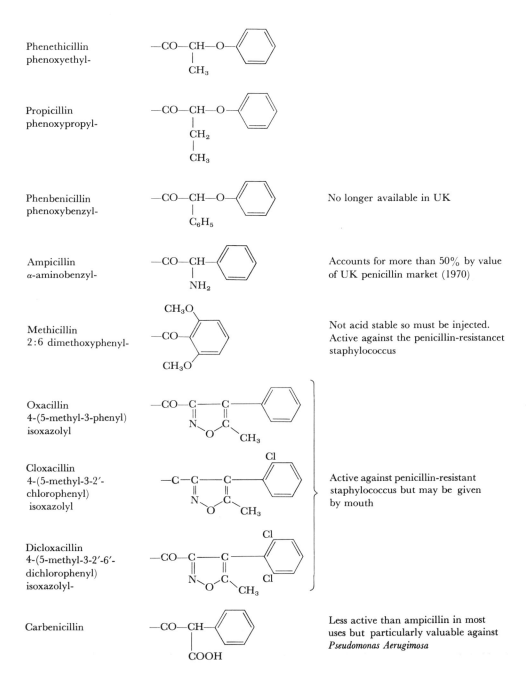

Figure 13.11 Chemically modified penicillins (based on the 6-aminopenicillanic acid nucleus. Structures of side chain —R only given.

358 *Pharmaceuticals*

13.6.2 Streptomyces antibiotics

The penicillin antibiotics are produced by the penicillium group of moulds; another large group, the streptomyces antibiotics, are made by the streptomyces moulds. Again a fermentation process is used and the products are complex organic molecules involving modified sugar nuclei. The two main antibiotics in this group are streptomycin and dihydrostreptomycin. The former consists of molecules of streptidine, streptose and N-methyl-L-glucosamine linked by oxygen bridges. Its structure is shown in Fig. 13.12.

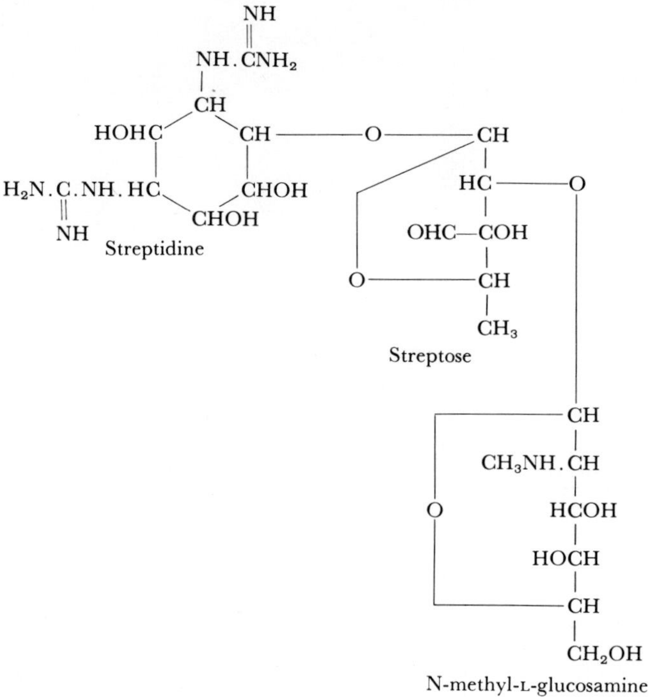

Figure 13.12 Streptomycin. In dihydrostreptomycin, the —C(H)=O group in the streptose molecule is replaced by —CH$_2$OH.

13.6.3 Tetracyclines

The tetracyclines are a further group of antibiotics prepared by a fermentation process. The structures are shown in Fig. 13.13. The tetracyclines are remarkable in that extensive modification of the molecular skeleton is possible without antibiotic activity being lost.

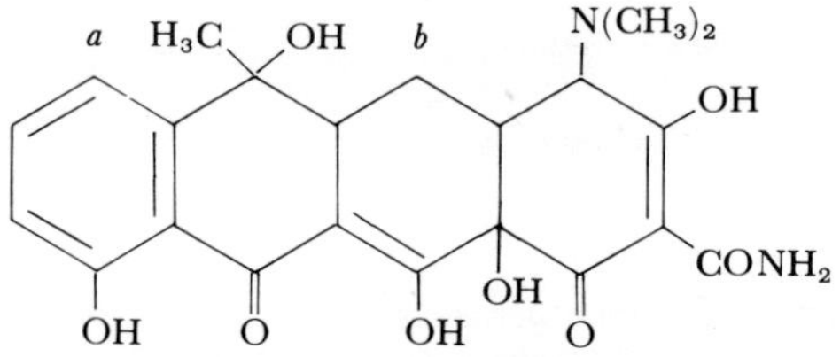

Figure 13.13 Tetracyclines. Chlorotetracycline has a chlorine atom at *a*. Oxytetracycline has a hydroxyl group at *b*.

13.7 Psychotropic drugs

Psychotropic drugs are those affecting the higher centres of the brain. They are used in treatment of psychiatric disorders and may be divided into five groups – hypnotics, tranquillizers, stimulants and appetite suppressants, antidepressants and psychotomimetics. Some structures are shown in Fig. 13.14.

Hypnotics

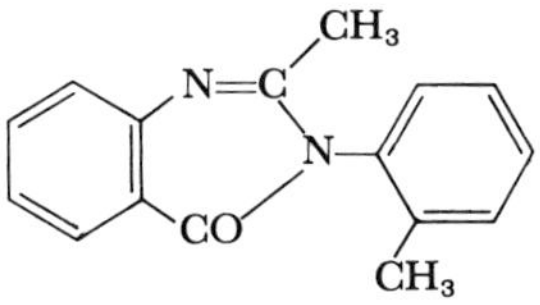

Methaqualone
(major constituent of Mandrax)

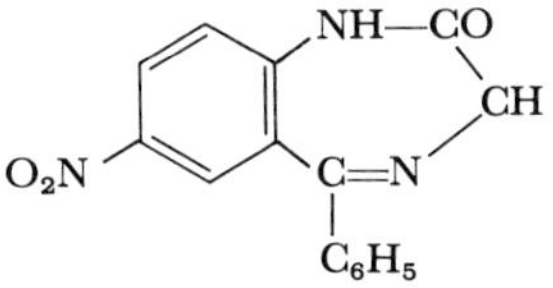

Nitrazepam
(Mogadon)

Tranquillizers

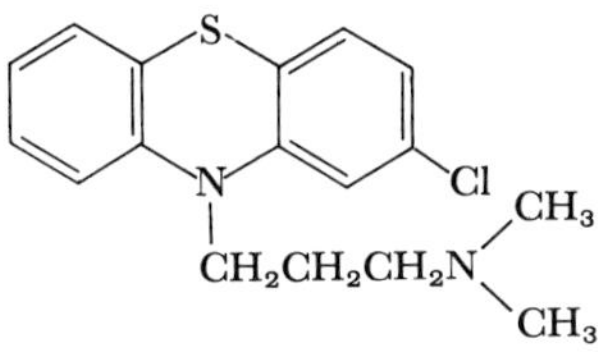

Chlorpromazine
(Largactil)

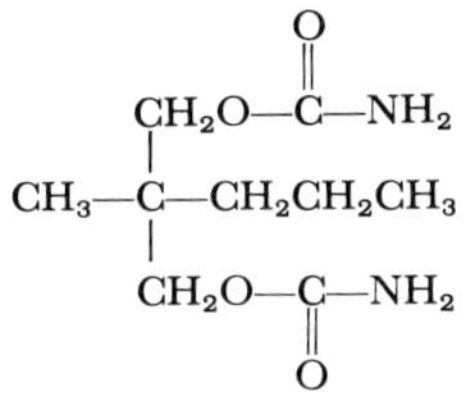

Meprobamate
(Miltown)

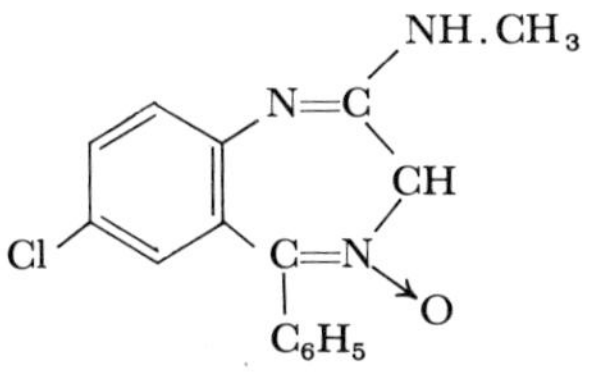

Chlordiazepoxide
(Librium)

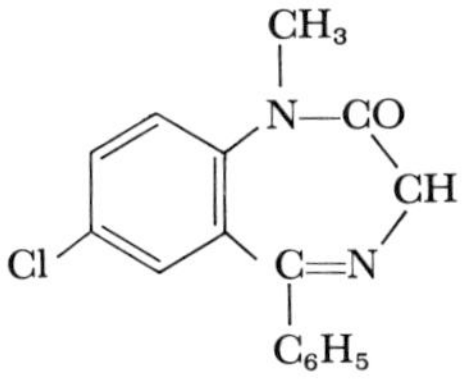

Diazepam
(Valium)

Stimulants and appetite suppressants

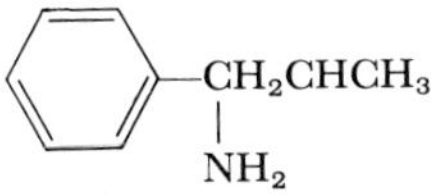

Amphetamine
(Benzedrine)

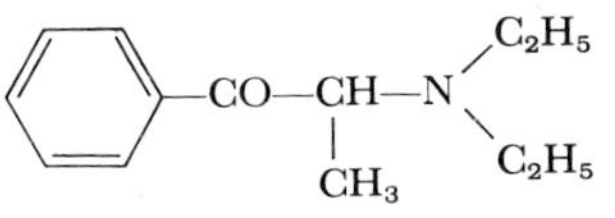

Diethylpropion
(Tenuate, Apisate)

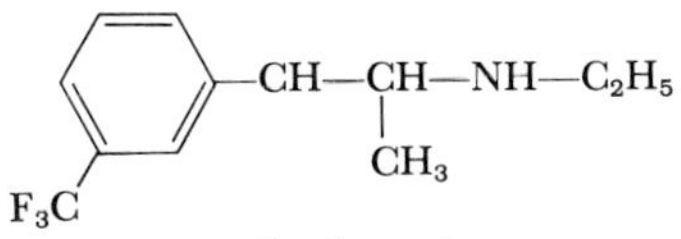

Fenfluramine
(Ponderax)

Figure 13.14 (continued overleaf)

Antidepressants

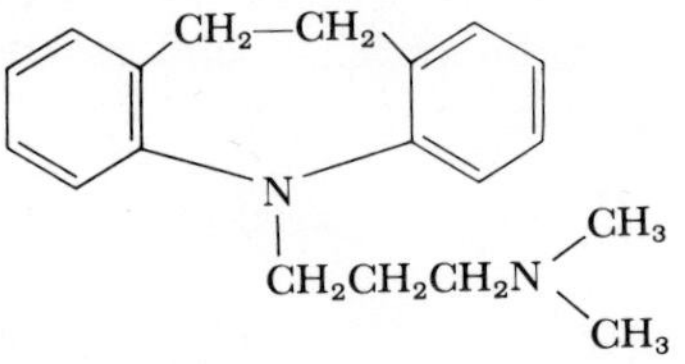

Imipramine (Tofranil)

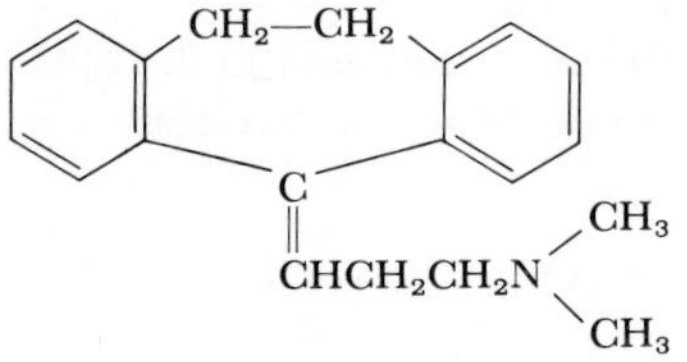

Amitriptyline (Tryptizol)

Psychotomimetics

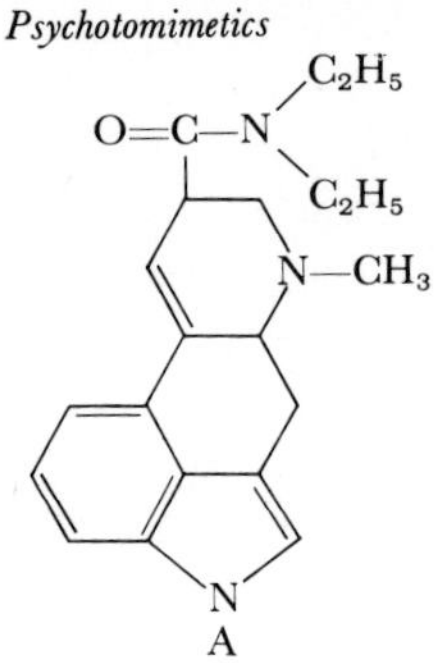

Lysergic acid (diethylamide)

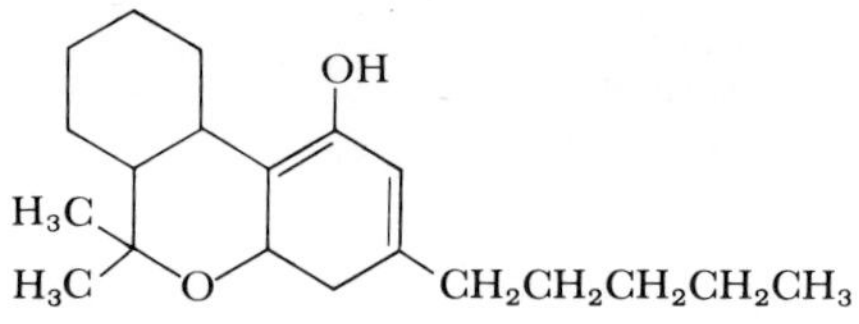

Tetrahydrocannabinol (constituent of cannabis)

Figure 13.14 Psychotropic drugs. The names shown directly below these structures are the abbreviated chemical names by which they are known; the names in parentheses are the better-known proprietary names. Some of these materials are marketed as their hydrochlorides.

Note the structural similarities: nitrazepam, chlordiazepoxide and diazepam are benzodiazepines, and chlorpromazine, imipramine and amitriptyline are tricyclic compounds.

Hypnotics are sleep-inducing drugs, the best known being the barbiturates, which have already been mentioned. The only important non-barbiturate hypnotics in the UK are Mandrax, a mixture of methaqualone with 10% of an antihistamine, diphenhydramine, and Mogadon, a benzodiazepine. They have some advantages and some disadvantages compared with barbiturates. In particular Mogadon is less dangerous when an overdose is taken.

Tranquillizers allay anxiety and nervous tension without impairing consciousness as do the barbiturates. The first tranquillizer, which came on the market in 1954, was chlorpromazine (Largactil). It is a phenothiazine derivative and a number of similar drugs exist in which the chlorine atom and the dimethylaminopropyl side chain are replaced by other substituents. Another early tranquillizer was meprobamate. The two tranquillizers which now dominate the UK market are chlordiazepoxide (Librium) and diazepam (Valium) and, like Mogadon, both are benzodiazepines. The synthesis of the former is shown in Fig. 13.15.

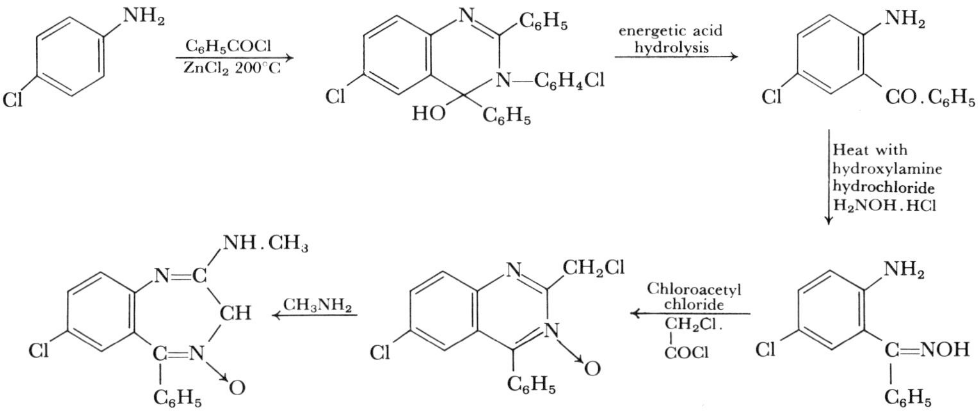

Figure 13.15 Synthesis of Librium.

Amphetamines were the most frequently prescribed stimulant and appetite suppressant drugs up to 1966, but they have since been overtaken, in a declining market, by diethylpropion derivatives (Apisate and Tenuate) and fenfluramine (Ponderax). These drugs all carry risks of addiction and psychosis, but public demand for slimming aids and for a feeling of well-being have kept them on the market. The British Medical Association nonetheless recommends that they be prescribed only for obesity and never for depression.

Antidepressants are related to tranquillizers but their effect is antidepressant rather than sedative. The two most commonly prescribed drugs in this class are imipramine (mainly sold as Tofranil) and amitriptyline (mainly sold as Tryptizol). They are both tricyclic compounds, related to chlorpromazine. The synthesis of imipramine is shown in Fig. 13.17.

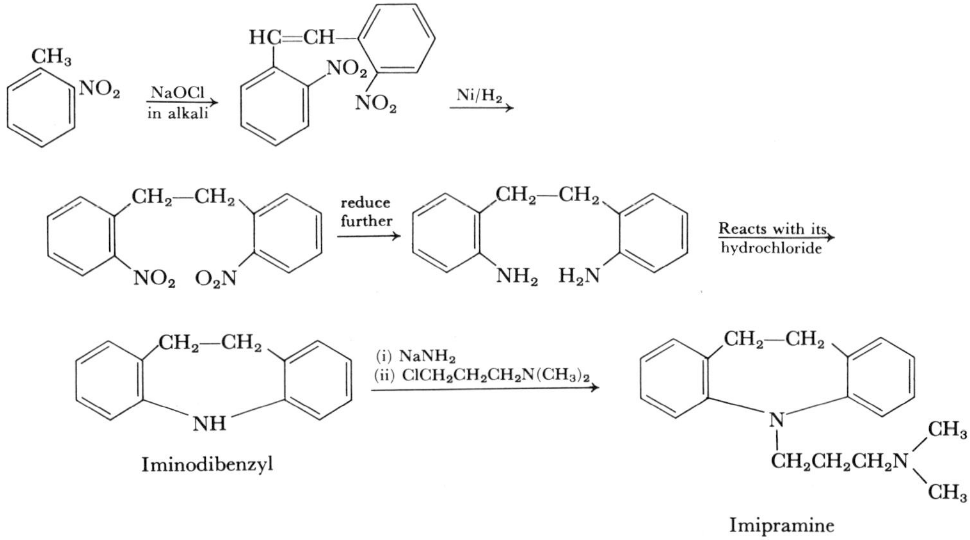

Figure 13.16 Synthesis of imipramine.

362 *Pharmaceuticals*

Psychotomimetic drugs cause mental and physical disturbances – rapid heart-beat, dilation of the pupil of the eye, tightness in the chest and abdomen and nausea, followed by vivid visual hallucinations, anxiety and delusions. They are structurally similar to 5-hydroxytryptamine and noradrenaline, two important amines occurring in the body, and in some way they replace these materials in bodily processes. The best known member of the group is LSD, lysergic acid diethylamide. The use of psychotomimetic drugs has been much abused and they find few medical applications.

UK prescribing trends, 1965 – 70, for the various groups of psychotropic drugs are shown in Fig. 13.17. Prescriptions for barbiturates and for stimulants and appetite suppressants decreased by 24 and 36% respectively but the drop in barbiturate hypnotics was more than wiped out by a massive 145% increase in non-barbiturate hypnotics, mainly Mandrax and Mogadon. The 59% increase in tranquillizers was mainly due to Librium and Valium and the 83% increase in antidepressants to Tryptizol and Tofranol. UK prescriptions for psychotropic drugs (including barbiturates) rose from

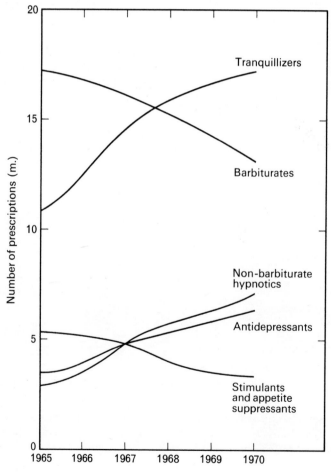

Figure 13.17 Prescribing trends for psychotropic drugs in England and Wales, 1965–70.

39·7 m. in 1965 to 47·2 m. in 1970. Approximately half of these were for the above six drugs only two of which (Librium, 2 m. prescriptions, and impramine 0·7 m. prescriptions) were of any importance in 1965.

The boom in these materials was economically the most significant pharmaceutical trend in the 1960s. The increased interest in recognition and treatment of depressive disorders helps to explain the rise in antidepressants. Replacement, not necessarily wisely, of barbiturates by Mandrax and Mogadon partly explains their increase, but taken overall the figures show either a marked increase in mental disturbances in the population or an increasing tendency on the part of doctors to prescribe psychotropic drugs as a placebo.

From the viewpoint of the pharmaceutical industry these trends are altogether praiseworthy, and indeed the new drugs have done much undoubted good. One is nonetheless compelled to wonder about the long-term effects on society of an ever-increasing dependence on psychotropic drugs, and this is an area where much research remains to be done.

Notes and references

General There are two classic books on the development of medicine written in preantibiotic days: Paul de Kruif, *Microbe Hunters*, Jonathan Cape, 1927 but reprinted frequently, and H. W. Haggard, *Devils, Drugs and Doctors*, USA 1929. More modern is Brian Inglis, *Revolution in Medicine*, Hutchinson, London, 1958.

Much of this chapter is based on the relevant sections of Kirk-Othmer and on the excellent series of booklets *Studies in Current Health Problems* published by the Office of Health Economics (OHE) 162 Regent Street, London, W1. We are also indebted to this body for several useful discussions.

For general information on pharmacology we have drawn on W. C. Bowman, M. J. Rand, and G. B. West, *Textbook of Pharmacology*, Blackwell, Oxford, 1968, and the obsolescent G. M. Dyson, *May's Chemistry of Synthetic Drugs*, Longman, London, 1958. There is a useful introductory paper-back, T. F. McCombie, *The Fine Organic Chemicals Industry*, Arnold, London, 1971. A standard work on the chemistry of drugs is A. O. Bentley and J. E. Driver, *Text Book of Pharmaceutical Chemistry* 8th edn. Oxford University Press, Oxford, 1969.

The British Pharmacopoeia, General Medical Council, 1968 and supplements, is the most important UK pharmacopoeia. There is also the *British Pharmaceutical Codex*, 1968 and supplements, produced by the Pharmaceutical Society of Great Britain, and the *British National Formulary*, British Medical Association and Pharmaceutical Society, 1966.

The equivalent books in the USA are *Pharmacopoeia of the U.S.A.* US Pharmacopoeial Convention, New York, *The National Formulary*, American Pharmaceutical Association, Washington, D.C. and the *Homeopathic Pharmacopoeia of the U.S.A.*, Boericke and Tafel, Philadelphia, all of which appear in new editions about every five years.

The World Health Organization has launched an *International Pharmacopoeia*, WHO, Geneva. Less official than the above, but crammed with useful information on both chemicals and drugs is the encyclopaedic *Merck Index*, Merck, New Jersey, 1968.

Section 13 Figure 13.1 comes from an OHE information sheet, November, 1970. Data in Figs. 13.2, 13.3 and 13.4 are from the *Registrar General's Decennial Supplement for England and Wales, 1931*, HMSO, London, and *The Registrar General's Statistical Review for England*

364   *Pharmaceuticals*

and Wales, HMSO, London, various years, but the graphs come from OHE publications. The survey of unwanted tablets was reported in *The Times*, 2 September 1969.

Section 13.1   Table 13.1 was compiled by the OHE. In Table 13.2, English data came from the *Department of Health and Social Security, Annual Report for 1969*. HMSO, London, the US data from the *Prescription Drug Industry Fact Book, 1969*, Pharmaceutical Manufacturers Association, Washington. Further details of the drugs mentioned as examples can be found in the various formularies and pharmacopoeias already cited.

Section 13.2   The role of the pharmaceutical industry in the UK economy and the prices which it charges were the subject of much debate in the mid-1960s. The attack was spearheaded by the Sainsbury Report, *Committee of Enquiry into the Relationship of the Pharmaceutical Industry with the National Health Service, 1965–67*, HMSO London, 1967, Cmnd 3410. The industry was defended by M. H. Cooper in a hastily written book, *Prices and Profits in the Pharmaceutical Industry*, Pergamon, 1965, and in two collections of papers edited by G. Teeling Smith, *Science, Industry and the State*, Pergamon, London, 1965 and *Innovation and the Balance of Payments: The Experience in the Pharmaceutical Industry*, OHE, London, 1967.

Table 13.4 is taken mainly from the US Tariff Commission report and from the UK Board of Trade, *Census of Production*, HMSO, London, various years. Export data came from Stirling and Co.

Section 13.3   The 1958 UK *Census of Production*, op. cit., gave numbers of aspirin and aspirin-containing tablets consumed but does not appear to have done so since.

Section 13.4   Sulphonamide data came from 'Molecular Modifications in Drug Design', *Advances in Chemistry*, **45**, ACS, Washington, 1964.

Section 13.5   Suicide statistics are recorded by the *Registrar General's Statistical Review*, op. cit., and cases of poisoning are analysed by poison in the *Pharmaceutical Journal*, June 12th, 1970.

Section 13.6.1   The effects of modifying the side chain in penicillin, along with several other related topics are discussed in 'Molecular Modifications in Drug Design', op. cit. An interesting book spanning scientific and economic approaches is 'Drug Discovery' *Advances in Chemistry*, **108**, ACS, Washington, 1971.

Section 13.6.2   For a less cursory description of the chemistry of tetracyclines see M. L. Burstall, 'The Tetracyclines', *Manufacturing Chemist*, November, 1960, p. 474.

Section 13.7   *Chemistry in Britain*, March, 1972, **8**, was a special issue devoted to 'Drugs', meaning psychotropic drugs. It covers the topic at non-specialist level from almost every viewpoint and provides a useful bibliography. For consumption trends we have drawn on P. A. Parish *The Prescribing of Psychotropic Drugs in General Practice* supplement 4, November 1971, **21**, *J. Royal. Coll. Gen. Pract.*

14 Heavy inorganic chemicals

The heavy inorganic chemical industry is engaged in the production of very large quantities of fairly simple inorganic substances. Most of the processes it uses date back to the First World War or earlier though, of course, their efficiencies have been substantially improved since then. The industry is thus highly traditional. The technologies are well understood and largely in the public domain. It is possible to buy ammonia manufacturing plant of any desired tonnage virtually off the shelf. (In 1963 a 'package' plant producing 60 tons/day cost $1·8 m. in the USA.) Consequently, no one firm has any great advantage in the market and the profit margins of almost all heavy inorganic chemical manufacturers are low.

The main divisions of the industry are shown in Table 14.1 together with the production levels of some of the products. For comparison, the outputs of some other industries are shown in Fig. 14.1. It will be seen that cement, pig iron and crude oil are used on a scale an order of magnitude larger than any of the conventional 'chemicals', while coal

Table 14.1 The heavy inorganic chemical industry

Sector	USA Production m. tons 1970	USA Price[1] ¢/lb.	Western Europe production m. tons 1969	UK Production m. tons 1971	UK Price[1] £/ton
THE NITROGEN INDUSTRY					
Ammonia					
Synthetic anhydrous	11·8	1·30[3]	8·2	~1·0*	12·1
Nitrogenous fertilizer					
(as nitrogen)	6·8*	—	8·1	0·78*	—
Nitric acid	5·85	4·15	—	0·011‡	38·0
Ammonium sulphate					
Total	2·26	1·25	n.a.	n.a.	14·3
from $NH_3 + H_2SO_4$	0·58	—	—	—	—
from coke ovens	0·54	—	—	—	—
by-product[2]	1·14	—	—	—	—
Ammonium nitrate	5·61	2·20	0·97[5]	0·35*	37·5
Urea	2·81	2·65	n.a.	~0·3	41·25
Methyl Alcohol	1·88	3·07	n.a.	n.a.	33·7
SULPHURIC ACID					
Sulphur	12·9[6]*	1·30[7]	6·47	0·05[8]*	33·13
Sulphuric acid					
Total	26·8	1·41[9]	23·5	3·40	15·4[9]
by lead chamber process	0·3	—	—	—	—
by contact process	26·5	—	—	—	—

Table 14.1 The heavy inorganic chemical industry – continued

Sector	USA Production m. tons 1970	USA Price[1] ¢/lb.	Western Europe production m. tons 1969	UK Production m. tons 1971	UK Price[1] £/ton
THE ALKALI INDUSTRY					
Sodium chloride					
Total	40·1*	1·54[10]	30·6	8·5	6·6
for use in chemicals	29·2*	—	20·8	5·6	—
Sodium hydroxide					
solution	8·70†	3·5†	$\sim$3·45**	$\sim$0·95**	—
solid	0·43†	6·8			45·6
Chlorine	8·50†	3·75	$\sim$5**	$\sim$0·85**	37
Sodium carbonate	7·15	1·65	n.a.	n.a.	16·3
synthetic	4·28†	—	—	—	—
natural	2·87†	—	—	—	—
Hydrochloric acid	1·84	1·5[15]	n.a.	0·132‡	16[15]
Total hypochlorites	n.a.	—	n.a.	0·197‡	18·4[17]
THE PHOSPHORUS INDUSTRY (see Table 14.8)					
Elemental phosphorus	0·50[16]†	19	n.a.	n.a.	—
Phosphoric acid (as H_3PO_4)	4·95[16]†	6·95	n.a.	n.a.	151·6
Phosphate fertilizers (as P_2O_5)	4·17†	—	5·9	0·45*	—
THE BORON INDUSTRY (see Table 14.9)					
Borax as $Na_2B_4O_7.10H_2O$	0·57*	25·5	n.a.	n.a.	53
THE POTASH INDUSTRY (see Table 14.10)					
Potassium sulphate (as K_2SO_4)	0·94*	19	0·91[11]	Nil[12]	23
Other potash fertilizers (as K_2O)	2·03*	—	4·22	Nil[12]	—
Potassium hydroxide	0·17†	12[13] 4·2[18]	n.a.	n.a.	105[13] 43[18]
HYDROGEN PEROXIDE					
Hydrogen peroxide	0·06†	16[14]	n.a.	$\sim$0·028	112

[1] Prices are from *European Chemical News*, April 14 1972; *Chemical Marketing Reporter*, 27 March 1972 and *Chemical Age*, 14 April 1972.
[2] Probably entirely from caprolactam manufacture. [3] 30% aqueous solution.
[4] 25% aqueous solution. [5] OECD data said to be for fertilizers but probably for total production.
[6] USA and Canada. [7] Ex terminal Tampa, Florida or Rotterdam, Netherlands.
[8] No natural deposit. Sulphur is mined mainly in Spain, Canada, France, Japan and Mexico.
[9] 66°Bé. [10] Chemical grade. [11] Excluding France.
[12] No UK potash deposits exploited; 485 000 tons net imports of all potash fertilizers (as K_2O) in 1971.
[13] As a solid. [14] 35% solution. [15] 18% Bé
[16] World production of elemental phosphorus in 1969 was about 1 million tons and of phosphoric acid about 20 million tons P_2O_5 content.
[17] Value. [18] 45% solution.
* 1969. ** 1967. † 1971. ‡ Sales.

is greater by almost another order of magnitude. Prices range from about £6/ton for common salt and slaked lime to about £100/ton for phosphate fertilizers; synthetic polymers at about £150/ton are more expensive than any of them. It is interesting to note that beer, an agricultural product, is consumed on the same scale as industrial chemicals and at a similar price (~ £250/ton in pint containers delivered over the bar).

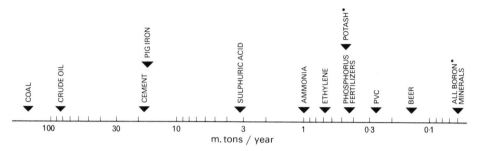

Figure 14.1 UK production of various materials 1968. *Consumption.

The sub-divisions listed in Table 14.1 are part of the chemical industry proper and involve the frequent application of chemical principles. The silicate industry (cement, bricks, glass and ceramics), the iron and steel industry and the non-ferrous metal industry are less interesting from this viewpoint and will not be discussed here.

14.1 The nitrogen industry

Mankind consumes chemically combined nitrogen both in peaceful and warlike activities. Nitrogenous matter in food is obtained directly or indirectly from the soil and the soil either becomes exhausted, or its nitrogen content has to be replenished by the use of nitrogenous fertilizers. Living matter is unable to 'fix' nitrogen from the air at the rates required for intensive agriculture and before the First World War the majority of nitrogenous fertilizer came from the large but not inexhaustible deposits of Chile saltpetre and from gasworks ammonium sulphate. The 'fixed' nitrogen, once it has been used, only partly returns to agricultural soil. Excreted nitrogen is oxidized in sewage works and carried out to sea; human bodies are buried in the earth – but churchyards are not cultivated for food – or they are cremated and the fixed nitrogen converted to elementary nitrogen.

In wartime, the depletion of 'fixed' nitrogen accelerates. Trinitrotoluene and nitroglycerine, both important explosives, require nitric acid for their manufacture but yield only gaseous nitrogen on combustion.

In 1898 Sir William Ramsey, discoverer of the inert gases, pointed out the depletion of world supplies of fixed nitrogen. In view of the rapidly rising population and increasing intensity of agriculture, he predicted world disaster due to a 'fixed' nitrogen famine by the middle of the twentieth century. That this has not occurred, and mankind has gone on eating and blowing itself up, is due to the discovery by F. Haber of a method of 'fixing' atmospheric nitrogen by reacting it with hydrogen to give ammonia.

This formed the basis of the nitrogen industry. Ammonia can react with acids to give ammonium salts, many of which are used as fertilizers, and it can also be oxidized to nitric acid, which in turn can give nitrates. Reaction of ammonia with carbon dioxide gives urea, which is also useful as a fertilizer and a plastics intermediate. In addition, liquid ammonia itself can be applied by a special machine directly to the soil. The method is gaining rapidly in popularity, and over half of the fertilizer nitrogen in the USA in 1969 was applied in this way. Ammonia is sold either as the liquified gas in pressurized tanks or as a concentrated solution containing about 29% ammonia.

Fertilizer is overwhelmingly the most important outlet for ammonia and this presents quite a different market from that for industrial chemicals. Political factors and government policies govern demand in most countries.

World consumption of nitrogen as a fertilizer in 1967–68 was 24·4 m. tons which represented 85% of total fixed nitrogen consumption. This quantity is analysed geographically in Fig. 14.2 in which data are also presented for phosphatic and potash fertilizers. The underdeveloped world accounts for less than a quarter of fertilizer nitrogen consumption although it numbers two-thirds of the world's population and contains over half the arable land. The USA and Western Europe account for half of total consumption while the whole of Africa uses less than 1%.

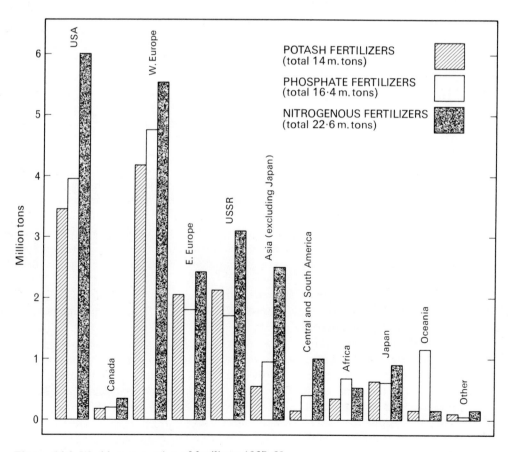

Figure 14.2 World consumption of fertilizers 1967–68.

Nonetheless, the rate of growth of fertilizer consumption in underdeveloped countries is high – over 15% per annum – and this reflects not only the realization by some of their governments of the importance of food production but also the volume of world aid, particularly from the much maligned USA. Technical aid from organizations such as FAO and AID is also important in providing conditions where full benefits from increased fertilizer use can be obtained.

The world industrial market for ammonia takes up the other 15% of world ammonia consumption – about 4·5 m. tons of nitrogen in 1967–68. The USA accounts for 40% of this, Western Europe for over a quarter. The major end-use is in synthetic fibres which, in the UK for example, account for a third of industrial nitrogen consumption. The greater part of this goes into nylon 66 but consumption for acrylonitrile and caprolactam (for nylon 6) is also growing (section 10.7). Acrylic and urea–formaldehyde plastics and ammonium nitrate explosives are also major end-uses.

It is interesting to note that the manufacture of caprolactam generates about four tons of ammonium sulphate for every ton of product. World production from this source in 1968 was about 6 m. tons and by the early 1970s may well overtake 'voluntary' ammonium sulphate production. Indeed, some nylon producers are using phosphoric instead of sulphuric acid in caprolactam production because ammonium phosphate is a better fertilizer and more saleable.

14.1.1 The Haber process

The role of the Haber process in the First World War was described in Section 1.9. The modern version of the process will be described here. The overall reaction is:

$$\tfrac{1}{2}N_2 + \tfrac{3}{2}H_2 \rightleftharpoons NH_3 \quad \begin{array}{l} \Delta H_{25°C} = -11 \text{ kcal/mole } (-46 \text{ kJ/mole}) \\ \Delta H_{400°C} = -13 \text{ kcal/mole } (-54 \text{ kJ/mole}) \\ \Delta S = -23\cdot7 \text{ cal/mole/°C } (-99 \text{ J/mole/°C}) \end{array}$$

It involves a decrease in pressure and is exothermic. The right-hand side of the equilibrium will therefore be favoured, on Le Chatelier's principle, by high pressure and low temperature. The extent to which the temperature can be lowered will, however, be limited by the necessity for a reasonably high rate of reaction.

The equilibrium constant K_p is given in terms of the partial pressure of the reactants* p_{NH_3}, p_{N_2} and p_{H_2} by

$$K_p = \frac{p_{NH_3}}{p_{N_2}^{1/2} \cdot p_{H_2}^{3/2}}$$

and its value at any given temperature and pressure can be calculated via the Van't Hoff isochore and isotherm.

K_p at 25 and 400°C, for example are 820 and 0·026 respectively. If P is the total

---

*To be strictly accurate, fugacities rather than partial pressures should be used.

pressure, and α is the percentage conversion to ammonia, the expression for K_p can be rewritten:

$$K_p = \frac{P\alpha}{\left[\frac{P}{4}(1-\alpha)\right]^{1/2}\left[\frac{3P}{4}(1-\alpha)\right]^{3/2}}$$

$$= \frac{16\alpha}{3\sqrt{3}P(1-\alpha)^2}$$

Suitable algebraic manipulation gives a value for α. Figure 14.3 shows the percentage of ammonia in the equilibrium mixtures at various temperatures and pressures. The equilibrium would be achieved, however, only after a very long tme and a factor called the efficiency can be defined such that

$$\text{Efficiency} = \frac{\%\ NH_3\ \text{formed}}{\%\ NH_3\ \text{at equlibrium}}$$

The Haber process is usually operated in such a way that the efficiency is about 0·5. We face the dilemma that increase of temperature increases the rate of achievement of equilibrium but decreases the proportion of ammonia in the equilibrium mixture.

A compromise must be found. Temperatures below 400°C are never used because no catalyst is known which will operate at such low temperatures. The favoured temperature nowadays is approximately 450°C. Formerly, pressures up to 350 or even 600 atm were used. These were obtained with reciprocating pumps which required a great deal of maintenance. Nowadays, highly efficient centrifugal compressors are used which can achieve pressures of 250 atm at relatively low cost, and it is more economic to operate at this pressure and accept the lower conversions which it implies.

In addition to temperature and pressure, the catalyst is a vitally important variable in the system, since an efficient catalyst permits lower operating temperatures. The usual catalyst in the Haber process is 'doubly promoted iron oxide' that is iron oxide mixed (by addition of the promoters to the fused iron oxide) with 0·35% potassium oxide, and 0·84% aluminium oxide and a trace of cupric oxide. Once on stream, the iron oxide is reduced by the hydrogen to give an $Fe-K_2O-Al_2O_3$ catalyst which is pyrophoric and is poisoned by contact with phosphorus, arsenic and sulphur and, to a lesser extent, carbon monoxide. The effect of the promoters is illustrated by the following figures: Under similar conditions, at 200 atm and 400°C:

Iron oxide catalyst gives 3–5% ammonia in exit gases
Singly promoted iron oxide gives 8–9% ammonia in exit gases
Doubly promoted iron oxide gives 13–14% ammonia in exit gases

The role of the catalyst, indeed the role of all heterogeneous catalysts, is not fully understood. Nitrogen is known to adsorb on most metals so that both nitrogen atoms are bonded to the metal (Fig. 14.4(a)).

The heat evolved is of the order of 10 kcal/mole (42kJ/mole) and nitrogen adsorbed in this way does not react with hydrogen. When nitrogen comes into contact with some transition metals, however, dissociative adsorption occurs (Fig. 14.4(b)). The heat of adsorption is much higher and the nitrogen will react with hydrogen to give ammonia. Indeed, it is found that the rate of ammonia formation is approximately equal to the rate of nitrogen adsorption. The best catalyst will be the metal which leads to the dissociative adsorption but shows the lowest heat of adsorption. Such a metal will allow

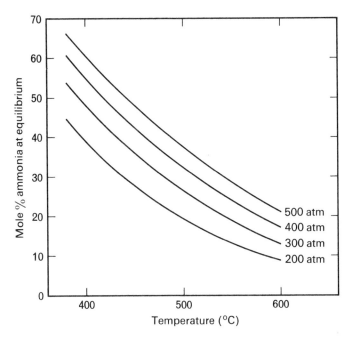

Figure 14.3 Percentage of ammonia at equilibrium in stoichometric N_2/H_2 mixtures.

the ammonia to desorb the most easily. Thus of the metals given in Fig. 14.4, iron will be the best catalyst. Ruthenium and osmium are as good as iron if not a little better, but they are very expensive.

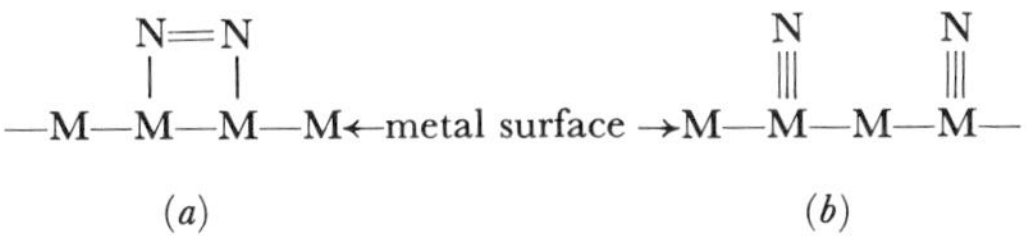

		Metal	$\Delta H_{adsorption}$	
			kcal/mole	kJ/mole
N=N N N		Most metals	~ -10	~ -42
| | ‖ ‖		Ta	−140	−540
—M—M—M—M ← metal surface → M—M—M—M—		W	−95	−400
(a) (b)		Fe	−70	−293

Figure 14.4 Adsorption of nitrogen on metals.

The catalyst, of course, does not alter the position of equilibrium in the reaction; which is determined by ΔG, the overall free energy change. Instead, it increases the rate of attainment of equilibrium, i.e. it accelerates equally the forward and back reactions.

372   *Heavy inorganic chemicals*

A block diagram of the equipment for carrying out the Haber process is shown in Fig. 14.5. Nitrogen and hydrogen ('synthesis' or 'make-up' gas) the preparation of which will be described in the next section, are passed through a guard vessel. This is to cope with traces of carbon monoxide and carbon dioxide, which could poison the catalyst, and which have, due to some accident, not been removed at a previous stage.

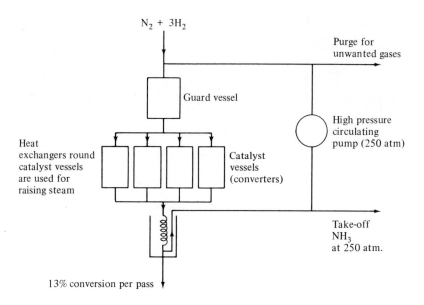

Figure 14.5 The Haber process.

The gases then pass to a number of catalyst vessels (converters) in parallel. This enables one converter to be shut down for catalyst renewal and general maintenance without the plant stopping. The catalyst is slowly poisoned and its efficiency drops to about half over two years, after which it is replaced.

The gases leaving the converters are cooled to -10 or $-20°C$ and some of the ammonia liquefies and can be run off. The rest of the gases is recirculated. Part of the recycled gas is 'purged' to prevent the accumulation of diluents such as argon which would lower the partial pressures of nitrogen and hydrogen in the reaction mixture and thus lower the yield. The purged gases contain about 12% methane which is used as fuel and 10% argon which is recovered and sold for argon arc welding.

The conversion per pass is about 13% and the overall yield is between 85 and 90%. The converters are surrounded by heat exchangers and the reaction exothermicity is used to raise steam.

14.1.2 'Synthesis' or 'make-up' gas

The manufacture of ammonia requires as raw material a mixture of hydrogen and nitrogen in the ratio of 3:1 by volume. This mixture, together with carbon monoxide and dioxide and the rare gases, was traditionally manufactured from coal.

A reactor filled with coke was designed so that it could be fed alternately with air or steam. First of all, air was passed through the bed so that the coke was heated up to and remained at 1 000°C. The two exothermic processes:

$$C + O_2 \longrightarrow CO_2, \quad \Delta H = -97 \text{ kcal/mole } (-407 \text{ kJ/mole})$$

$$C + \tfrac{1}{2}O_2 \longrightarrow CO, \quad \Delta H = -26 \text{ kcal/mole } (-107 \text{ kJ/mole})$$

took place, giving 'producer' gas which consists of carbon monoxide and dioxide, nitrogen and the rare gases, mainly argon. After a time, the air was replaced by steam and the endothermic reaction

$$C + H_2O \longrightarrow CO + H_2, \quad \Delta H = 31 \text{ kcal/mole } (130 \text{ kJ/mole})$$

took place giving a carbon monoxide–hydrogen mixture known as 'water gas'. The times devoted to each half of the cycle were adjusted to give the appropriate ratio of hydrogen to nitrogen in the product. Subsequent stages (q.v.) removed the carbon oxides to leave the 'synthesis gas' (section 8.4.5).

This process is obsolete because of its high operating costs and the high cost of coal. It was replaced by the pressure steam reforming of naphtha. Paraffin hydrocarbons will react with steam between 700 and 900°C to give hydrogen:

$$C_nH_{2n+2} + nH_2O \rightleftharpoons nCO + (2n+1)H_2$$

$$C_nH_{2n+2} + 2nH_2O \rightleftharpoons nCO_2 + (3n+1)H_2$$

High temperatures favour the production of carbon monoxide rather than dioxide. Excess steam pushes both equilibria to the right. The reactions are endothermic and heat must be supplied. ICI, who were one of the originators of the process, used light naphtha (typically C_5H_{12} to C_7H_{14} hydrocarbons) as feedstock. This is cheap but is tainted with sulphur compounds which must be removed as they are potent catalyst poisons. The main ones are thiols (RSH_2), mercaptans (R_2S), thiolane  and sulpholane.

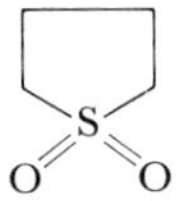

As coal was replaced by naphtha, so naphtha is being replaced by natural gas (section 7.9) the methane in which behaves as a typical paraffin hydrocarbon in the above reactions.

The process described in the following pages is the one used for naphtha but it can be applied to natural gas with only slight modifications. The main one is that desulphurization is not normally necessary. The costs of the various processes are shown in Fig. 14.6.

The sulphur compounds are first removed from the naphtha in a so-called hydrodesulphurizer. The naphtha is mixed with a little hydrogen and passed (as a vapour) through zinc oxide and cobalt molybdate–alumina beds at 400°C. The thiols and mercaptans are reduced to hydrogen sulphide which reacts with the zinc oxide:

$$2H_2 + R_2S \longrightarrow H_2S + 2RH$$

$$H_2S + ZnO \longrightarrow ZnS + H_2O$$

374   *Heavy inorganic chemicals*

The cobalt molybdate–alumina is very effective at catalysing the reduction of the mercaptans but does not retain the hydrogen sulphide. The zinc oxide charge is changed when it is about 20% sulphur by weight. The technique reduces the sulphur level from around 300 ppm to 1 ppm, most of which is sulpholane.

The desulphurized naphtha is preheated to about 250°C, mixed with steam at the same temperature and passed into the primary reformer (Fig. 14.7) at about 25 atm. The theoretical steam:carbon ratio is 3:2 but a ratio of 3:1 is used to minimize deposition of carbon on the nickel oxide catalyst.

The reactors are heated directly by the burning of the 'heavy distillate' petroleum fraction and the flue gases which emerge at 640°C are used for raising high pressure steam. The effluent gases (apart from the unreacted steam) have the approximate composition of town gas (Table 14.2).

Air is added and the gases at 250 atm and 110°C are passed to a secondary reformer also

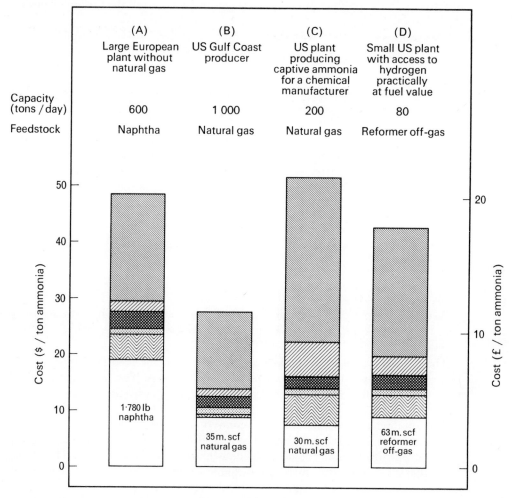

Figure 14.6 The economics of ammonia production vary greatly with circumstances. Four cases are given above all of which include the generation of synthesis gas.

Table 14.2 Composition of reformed naphtha (by volume)

	From primary reformer (%)	From secondary reformer (%)
H_2	60	69
CH_4	4–8	0·2
CO	3–4	2
CO_2	1	6
Steam	30	—
N_2	—	23

containing a supported nickel oxide catalyst. The secondary reformer serves two purposes. It reduces the methane content to less than 0·2%, and the introduction of air provides the nitrogen required in the Haber process. The quantity of air is adjusted to give the correct $H_2:N_2$ ratio, and the oxygen is of course removed by combustion of the methane and carbon dioxide and some hydrogen. The gases from this reactor contain traces of carbon oxides and methane. They emerge at 900°C and the heat is again used to raise steam. In an integrated ammonia plant, the overall thermal efficiency may be as high as 90%. Steam is added to the mixture and the remaining carbon monoxide is converted to carbon dioxide by the water gas shift reaction:

$$CO + H_2O \underset{\substack{Cr_2O_3 \\ 450°C}}{\overset{\text{Iron oxide}}{\rightleftharpoons}} CO_2 + H_2, \quad \Delta H = -9·8 \text{ kcal/mole} \; (-41 \text{ kJ/mole})$$

This reduces the carbon monoxide level to 0·2%. The gases are then washed with water

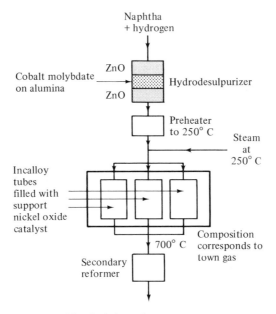

Figure 14.7 Naphtha reformer.

under a pressure of 50 atm. The carbon dioxide dissolves completely and is removed. The pressure on the solution is reduced in two stages. The first stage gives rise to 99% pure CO_2 which can be condensed to solid and liquid carbon dioxide or reacted with ammonia to give urea (section 14.1.6).

The second stage gives 78% CO_2 which can be reacted with ammonia to give ammonium carbonate which in turn reacts with gypsum to give ammonium sulphate and calcium carbonate (section 14.1.4).

The gases, now minus carbon dioxide, are compressed to 250 atm and washed with cuprous chloride. The residual CO reacts to give a complex of structure

$$\begin{array}{c} OC \diagdown \quad Cl \diagdown \quad Cl \\ \quad Cu \quad \quad Cu \\ Cl \diagup \quad Cl \diagup \quad CO \end{array}$$

which is removed and which, on depressurization, regenerates CO which is used for carbonylation.

The nitrogen–hydrogen mixture is now pure enough for use in the Haber process, the only impurities being traces of methane and some argon and other rare gases.

14.1.3 Nitric acid

Nitric acid is one step further than ammonia towards the production of fertilizers such as ammonium nitrate (which consume about three-quarters of nitric acid production) and explosives (which consume about one-seventh). The remainder is used for other nitrations (section 16.8) and oxidations (section 16.1).

Ammonia is converted to nitric acid in three stages; it is first catalytically oxidized to nitric oxide, this is then oxidized with air to nitrogen dioxide which is dissolved in water to give nitric acid:

$$4NH_3 + 5O_2 \longrightarrow 4NO + 6H_2O \quad \Delta H = -226 \text{ kcal/mole } (-950 \text{ kJ/mole}) \quad (a)$$

$$2NO + O_2 \rightleftharpoons 2NO_2 \quad \Delta H \approx 0 \quad (b)$$

$$3NO_2 + H_2O \rightleftharpoons 2HNO_3 + NO \quad \Delta H = -28 \text{ kcal/mole } (-117 \text{ kJ/mole}) \quad (c)$$

The first stage is carried out by the passage of anhydrous ammonia and air at 7 atm over a platinum gauze catalyst at 900°C. A fast flow rate is normally used, with a catalyst contact time of about 10^{-4} s. If the temperature is too low or the contact time too long, the side reactions

$$4NH_3 + 3O_2 \longrightarrow 2N_2 + 6H_2O \quad \Delta H = -303 \text{ kcal/mole } (1\,270 \text{ kJ/mole})$$

$$2NH_3 \rightleftharpoons N_2 + 3H_2 \quad \Delta H_{900°C} = 31.4 \text{ kcal/mole } (131 \text{ kJ/mole})$$

become important. Furthermore if the gas flow is too fast, the unchanged ammonia makes the platinum crystalline and brittle and also reacts with the nitric oxide:

$$4NH_3 + 6NO \longrightarrow 5N_2 + 6H_2O, \quad \Delta H = -476 \text{ kcal/mole } (-2\,000 \text{ kJ/mole}).$$

In practice 90% air and 10% ammonia enter the burner at 300°C and emerge at 900°C giving a 93% yield of nitric oxide. This is cooled to 25°C and more air added so that reaction (b) takes place, to give the reddish brown nitrogen dioxide. The reaction is remarkable both in being one of the few known termolecular reactions and one of the even rarer reactions whose rate decreases with increasing temperature, i.e. it appears to have a negative activation energy. Thus the reduction in temperature is a positive advantage.

The nitrogen dioxide is then dissolved in water (reaction (c)) and gives a 61–65% solution quite near the 68% HNO_3 constant boiling point (122°C) nitric acid. For nitration purposes the acid is required in a more concentrated form, which may be prepared by treating the 65% acid with concentrated sulphuric acid in a stone tower. The nitric acid is fed in at the top of the tower, the sulphuric acid some way below it. The vapour emerging from the top of the tower is concentrated nitric acid while the bottoms consist of relatively dilute sulphuric acid which is then reconcentrated for further use. Dilute nitric acid produced in the process is converted to ammonium nitrate (section 14.1.5).

14.1.4 Ammonium sulphate

Ammonium sulphate was once the most widely used nitrogenous fertilizer, though its use is decreasing. It contains 21% nitrogen and is made by a variety of processes. The oldest involves the scrubbing of ammonia-containing streams from coking plant with sulphuric acid. Alternatively, synthetic ammonia can be neutralized with sulphuric acid. A third process, which uses gypsum (calcium sulphate) instead of sulphuric acid, involves the intermediate production of ammonium carbonate:

$$NH_4OH + CO_2 \longrightarrow (NH_4)_2CO_3$$

$$CaSO_4 + (NH_4)_2CO_3 \longrightarrow (NH_4)_2SO_4 + CaCO_3$$

The by-product calcium carbonate is used for cement manufacture.

14.1.5 Ammonium nitrate

Ammonium nitrate is an important nitrogenous fertilizer (34.5% nitrogen) and about 10% of output is still used as an ingredient in blasting explosives. Small tonnages are also used as a source of the anaesthetic nitrous oxide, which it gives on heating:

$$NH_4NO_3 \longrightarrow N_2O + 2H_2O$$

It is prepared by neutralization of weak (40–60%) nitric acid with gaseous ammonia. It is deliquescent and difficult to crystallize. The heat of reaction results in some evaporation of water and an 85% solution of ammonium nitrate is obtained. Vacuum evaporation concentrates this to about 95% and the solution is then sprayed from the top of a 240 ft (75 m) 'prilling' tower countercurrent to a flow of dry air. The ammonium nitrate solidifies into small pellets which are coated with an anti-caking agent.

Alternatively, the solution is mixed with chalk and spray dried to give the so-called 'nitro chalk' fertilizer which, although its nitrogen content is half that of pure ammonium nitrate, is not deliquescent and is a convenient way of adding lime to the soil.

14.1.6 Urea

Urea is another nitrogenous fertilizer and contains 46% nitrogen. It competes directly with ammonium nitrate, its higher price being balanced by a higher nitrogen content and greater ease of handling. Urea–ammonium nitrate mixtures are also sold. In addition, urea is a plastics intermediate and is used in U/F resins (section 10.4.2.1).

It is manufactured by reaction of ammonia with carbon dioxide to give ammonium carbamate which loses water to give urea:

$$CO_2 + 2NH_3 \rightleftharpoons O{=}C\begin{array}{l}{\diagup ONH_4}\\{\diagdown NH_2}\end{array} \rightleftharpoons O{=}C\begin{array}{l}{\diagup NH_2}\\{\diagdown NH_2}\end{array} + H_2O$$

Ammonium carbamate   Urea

The two gases are compressed to about 350 atm at which pressure they are both liquid. A high pressure reactor with a silver liner is used but corrosion is still a major problem. The heat of reaction raises the temperature to about 200°C and it is maintained at this value. After equilibrium has been reached, the pressure is released and the unreacted ammonia and carbon dioxide recycled. The urea solution is evaporated and the urea crystallized.

14.1.7 Methanol and formaldehyde

Methanol manufacture is part of the nitrogen industry although its molecule contains no nitrogen. It is made, however, by a process similar in many ways to the Haber process and uses raw materials which are available on an ammonia complex. Thus the manufacture of ammonia and methanol go together and this allows flexibility in plant use.

Carbon monoxide and hydrogen from the ammonia process are reacted together at 200–300 atm and 350°C:

$$CO + 2H_2 \rightleftharpoons CH_3OH, \quad \Delta H = -22 \text{ kcal/mole} (-91 \text{ kJ/mole})$$

An H_2:CO ratio of 10:1 is used and the catalyst is zinc oxide promoted with chromium sesquioxide. The formation of methanol is favoured by low temperature and high pressure.

The methanol is condensed out and small amounts of gas purged, as in the Haber process, to prevent build-up of inert gases such as argon and methane. Unreacted hydrogen and carbon monoxide are recirculated.

About 40% of methyl alcohol production goes into methyl esters, amines, etc., and 10% is used as a solvent. The remaining 50% is mixed with air and passed over a silver catalyst for about 0·01 s at 450–600°C when either dehydrogenation or oxidation takes place to give formaldehyde:

$$CH_3OH \longrightarrow HCHO + H_2, \quad \Delta H = -20 \text{ kcal/mole} (-84 \text{ kJ/mole})$$

$$CH_3OH + \tfrac{1}{2}O_2 \longrightarrow HCHO + H_2O, \quad \Delta H = -36·8 \text{ kcal/mole} (-154 \text{ kJ/mole})$$

Formaldehyde is used with phenol, urea or melamine in thermosetting resins. Other uses, in pentaerythritol, hexamethylene tetramine and polyacetals, are minor. Formaldehyde is usually sold as a 37% aqueous solution stabilized with methyl alcohol to inhibit polymerization. Alternatively, formaldehyde polymers may be purchased and depolymerized for use. The two principal ones are paraformaldehyde, a mixture of linear polyoxymethylenes, $HO(CH_2O)_nH$ where $n = 8$–50, which is a white powder, and trioxane, a cyclic trimer $(CH_2O)_3$ which forms colourless crystals. Market data are given in Table 14.3.

Table 14.3 Methyl alcohol, formaldehyde and pentaerythritol

	USA 1969 Production (m. lb)	Sales (m. lb)	Value (¢/lb)	USA Price 4/72	UK 1971 Production ('000 tons)	Price 4/72 (£/ton)
Methyl alcohol	4 206	1 621	3	3·07	—	33·7
Formaldehyde	1 630	545	2*	3·5*	112	32·6*
Pentaerythritol	92	74	23	24	—	200

* Value per lb of 37% aqueous solution.

14.2 Sulphuric acid

Sulphuric acid is one of the most widely used of all chemicals. UK consumption is nearing 4 m. tons/year and the uses are so numerous that they can only be conveniently listed by groups (Table 14.4). The ubiquitous uses of sulphuric acid mean that its level

Table 14.4 Sulphuric acid

	'000 tons	%	Consumption of raw material '000 tons
UK production (1971)			
From elementary sulphur	2 389·9	67·9	789·8
pyrites	259·7	7·6	201·6
spent oxide	66·5	2·0	50·0
anhydrite	539·0	15·8	1033·6
zinc concentrates	211·9	6·2	296·8
Total	3 404·6	99·5	
UK consumption (1971)		'000 tons	%
Fertilizers/agriculture (including phosphate fertilizers 973·8; ammonium sulphate 226·7)		1 215·3	32·4
Paints/pigments (mainly in titania production)		627·6	16·7
Natural and man-made fibres, transparent cellulose film		517·4	13·8
Chemicals: including plastics 125·3; aluminium, barium, copper sulphates, etc. 98·3; hydrofluoric acid 77·0; hydrochloric acid 46·9		449·9	12·0
Detergents and soap		344·9	9·2
Metallurgy/steel pickling		121·3	3·2
Dyestuffs/intermediates		102·1	2·7
Oil/petrol		51·5	1·4
Other uses		324·4	8·7
Total		3 754·4	

Table 14.5 Commercial grades of sulphuric acid

Grade	Degree Bé 60°F or 15·6°C	Specific gravity 60°F or 15·6°C	% H_2SO_4
Battery acid	29·0	1·250	33·4
Chamber acid, fertilizer acid	50	1·526	62·18
Glover or tower acid	60	1·706	77·67
Oil of vitriol, concentrated acid	66	1·835	93·19
98% Acid	66·2	1·841	98·0
Monohydrate, H_2SO_4	65·98	1·835	100
20% Oleum or fuming acid	—	1·927	104·5 (20% free SO_3)
40% Oleum or fuming acid	—	1·965	109·0 (40% free SO_3)
65% Oleum or fuming acid	—	1·990	114·6 (65% free SO_3)
70% Oleum or fuming acid	—	—	115·8 (70% free SO_3)

of consumption can be taken as measure of a country's technical development (Chapter 6). Furthermore, problems of bulk storage mean that production responds rapidly to changes in consumption, hence the level of economic activity in a country can be gauged from its sulphuric acid production. The close correlation between the latter and the gross national product explains why sulphuric acid is known as 'the grandfather of business indicators'.

Sulphuric acid is sold in the form of various solutions of sulphuric acid in water, or sulphur trioxide in sulphuric acid. The former are sold, by custom, according to their specific gravity or the related degree Bé (Baumé), measured at the rather low temperature of 60°F (15·6°C) which testifies to the age of the tradition. Above 100% acid, specific gravity ceases to be a useful measure of strength (Table 14.5) and the solutions of sulphur trioxide in sulphuric acid, called oleum, are sold according to their free SO_3 content, estimated by electric conductivity or titration.

Difficulties of storing such a corrosive material and the age of the technology involved mean that most large users manufacture their acid *in situ*. All grades of acid can be transported in glass carboys, dilute acid can be stored in rubber or lead lined containers and iron or steel can be used to store concentrated (above 65°Bé) acids.

Disposal of unwanted oleum is an even greater problem than storage since dilution generates much heat. Many picturesque stories are told of carboys of oleum dumped at sea and shelled from a distance, a direct hit resulting in a gratifying explosion. Recent concern about the environment may well result in even this method being forbidden.

14.2.1 The lead chamber process

The lead chamber process involves the oxidation of sulphur dioxide to trioxide by air in the presence of nitric oxide as a catalyst. It is possible to write a very simple mechanism:

$$2NO + O_2 \rightleftharpoons 2NO_2$$
$$NO_2 + SO_2 \rightleftharpoons SO_3 + NO$$

but this is belied by the dependence of the reaction on the presence of water and also by the appearance during the reaction of a variety of coloured crystals of intermediate products. The actual mechanism is very complicated but may be approximated by the following:

$$2NO + O_2 \rightleftharpoons 2NO_2$$
$$SO_2 + H_2O \longrightarrow H_2SO_3$$
$$H_2SO_3 + NO_2 \longrightarrow H_2SO_3.NO_2 \text{ (violet acid or nitrosulphuric acid)}$$
$$2H_2SO_3.NO_2 + \tfrac{1}{2}O_2 \longrightarrow H_2O + 2HSO_3.NO_2 \text{ (nitrosyl sulphuric acid or chamber crystals)}$$
$$2HSO_3.NO_2 + SO_2 + 2H_2O \rightleftharpoons 2H_2SO_3.NO_2 + H_2SO_4$$
$$H_2SO_3.NO_2 \rightleftharpoons H_2SO_4 + NO$$
$$2HSO_3.NO_2 + H_2O \rightleftharpoons 2H_2SO_4 + NO + NO_2$$
$$HSO_3.NO_2 + HNO_3 \rightleftharpoons 2NO_2 + H_2SO_4$$
$$3NO_2 + H_2O \rightleftharpoons 2HNO_3 + NO$$

The lead chamber process was discovered in 1746 by a Dr Roebuck of Birmingham who operated it as a batch rather than a continuous process. The Gay–Lussac tower for recovery of nitrogen oxides was introduced in 1827 and the Glover tower in 1859. The process became obsolete before the Second World War but has taken a long time to die and there is still a handful of lead chamber plants in operation both in the UK and the USA.

A sulphur dioxide and oxygen feedstock (section 14.2.3) is passed up a Glover tower down which trickles a stream of nitrous vitriol (72·8% H_2SO_4) obtained from the bottom of the Gay–Lussac towers diluted with 50°Bé acid. Oxides of nitrogen from an ammonia oxidation unit are also introduced. Partial oxidation of the sulphur dioxide to trioxide occurs at the top of the Glover tower and 60°Bé acid emerges at the bottom. The gaseous mixture is passed to the lead chambers where oxidation is completed and the sulphur trioxide absorbed by water sprayed into the chambers. Finally the remaining gases pass to the Gay–Lussac tower where the oxides of nitrogen are absorbed in the 60°Bé acid from the Glover tower to give the nitrous vitriol mentioned previously.

A simplified diagram of the process is given in Fig. 14.8.

The function of the Gay–Lussac tower is thus to permit the recovery and recycling of the oxides of nitrogen, while the Glover tower concentrates the chamber acid, removes the oxides of nitrogen from the Gay–Lussac acid, and cools the gases from the burners which produce sulphur dioxide to the point at which they can enter the chamber.

Lead chamber plants rarely produce acid above 60°Bé and while subsequent concentration of the acid is possible, it also adds to the expense. The 60° acid is suitable for making superphosphate and sulphate fertilizers. The lead chamber process has the advantage of operating more readily than the contact process with feedstocks containing impurities

382 *Heavy inorganic chemicals*

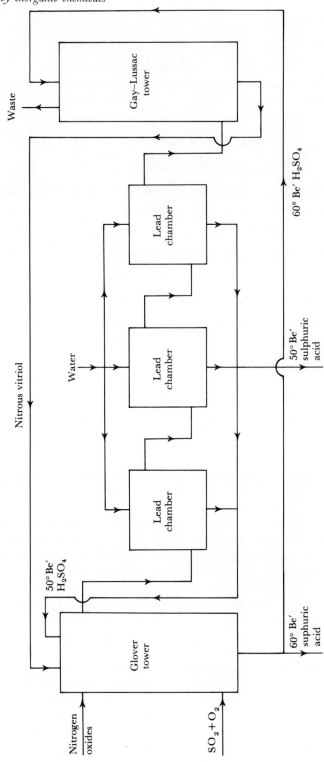

Figure 14.8 Lead chamber process for sulphuric acid.

which would poison the catalysts used in the contact process. Nonetheless, few lead chamber plants have been built since the Second World War, and whereas 53% of US sulphuric acid production in 1937 came from the lead chamber process, by 1963 the figure had dropped to 8% and it is now negligible.

14.2.2 The contact process

Like the lead chamber process, the contact process involves the oxidation of sulphur dioxide by air to sulphur trioxide which is dissolved in water (or, more accurately, in concentrated 98·5% sulphuric acid). It is a catalytic process and involves the simple equilibrium

$$2SO_2 + O_2 \rightleftharpoons 2SO_3 \quad \Delta H = -23 \text{ kcal/mole } (-96\cdot 5 \text{ kJ/mole})$$

The right-hand side of the equation is favoured by low temperatures but below 400°C the rate of attainment of equilibrium is negligible. Figure 14.9 shows conversion and rate at various temperatures. The equilibrium favours SO_3 production up to about 550°C and the usual practice is to flow sulphur dioxide and oxygen over catalyst at 575°C when the rate is high and about 80% of the sulphur dioxide is rapidly oxidized. The gases are then cooled and blown over another portion of catalyst at about 450°C which increases the yield to about 97%.

Increase of pressure above atmospheric would increase the yield of sulphur trioxide but the effect is only small and the expense is high. Similarly, a large excess of oxygen would lead to the conversion of a greater percentage of the sulphur dioxide, but again the effect is small and uneconomical, and a gaseous feedstock from the sulphur dioxide

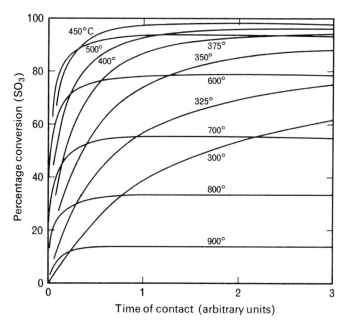

Figure 14.9 Relation between conversion and time of reaction at various temperatures, and under typical conditions of pressure and feedstock composition.

plant which contains 7–10% SO_2 and 11–14% O_2 is regarded as containing a satisfactory excess of oxygen.

The contact process was in use in Europe before the First World War since oleum was required by the dyestuffs industry but was uncommon in the USA. The catalyst was platinized asbestos. A typical platinum catalyst is prepared by the washing, drying and carding of long fibre asbestos which is then dipped in chloroplatinic acid H_2PtCl_6 or ammonium chloroplatinate. It is dried and fluffed first by machine and then by hand and contains 5–10% platinum by weight. When put on stream, the chloroplatinate is rapidly reduced to platinum black.

Between the wars, vanadium catalysts began to replace platinum. Soluble vanadium compounds (reduced in use to V_2O_5) promoted with potash are deposited on very finely divided pumice and the resulting mass pelleted. The advantages of platinum catalysts are that they are effective at lower temperatures than vanadium (heat costs money!) and the original metal may be recovered after use. On the other hand they are not active for as long as vanadium catalysts, they are more easily poisoned (although this is of little importance nowadays), they are more difficult to handle and have a higher initial cost. These drawbacks are now thought to outweigh the advantages and since 1945 few platinum catalyst plants have been built.

Typical sulphuric acid plants produce 100–5 000 tons/day. Above about 1 600 tons/day, economies of scale level out and when a greater capacity is required, it is usual to build a number of 1 600 tons/day units. This has the added advantage that a breakdown usually only affects one of the units. About 99% of the sulphur input is converted to sulphuric acid. The remaining 1% gives rise to air pollution. As a proportion this is small, but a 1 600 tons/day plant will produce 32 tons/day of pollutant (measured as SO_2) so the problem is by no means insignificant.

14.2.3 Sulphur dioxide production

The raw material for both the chamber and contact processes is sulphur dioxide. The various sources of this material are shown in Table 14.6. About 60% of sulphuric acid in the UK is made from sulphur dioxide obtained by the burning of elementary sulphur. The resulting SO_2 is generally filtered and dried before being passed to contact process plant. The chamber process would accept unpurified gases.

Another source of sulphur dioxide is iron pyrites, which contains 42–45% sulphur. It is roasted in a fluidized bed to give sulphur dioxide and iron oxide but the former contains such impurities as dust, moisture, arsenic and halogens which have to be removed by filtration, scrubbing and drying before the material can be used in the contact process. As a step in the preparation of many metals is the roasting of their sulphides, waste gases containing sulphur dioxide are common in metallurgical plants, especially those making zinc, and such gases are also used as feedstock for sulphuric acid.

Sulphur dioxide is also obtained by oxidation of the hydrogen sulphide occurring in waste refinery gases, by decomposition of spent sulphuric acid from petroleum refining, and alkylation and nitration plant, and from the anhydrite ($CaSO_4$) used in cement production. Portland Cement is a mixture of tri- and dicalcium silicates, tricalcium aluminate and tetracalcium aluminoferrite. Its approximate analysis is CaO 64%,

Table 14.6 Sources of sulphur and sulphuric acid 1969 ('000 tons)

	Total	Production Native	Recovered	From pyrites	Consumption
SULPHUR					
European OECD countries	6·47	0·079	2·11	3·41	n.a.
USA	10·37	7·26	1·45	0·37	9·50
Japan	2·75	0·26	0·14	1·45	2·68

	Production	Consumption Total	Into fertilizers
SULPHURIC ACID			
European OECD countries	23·50	22·59	11·81
USA	26·06	26·83	13·15
Japan	6·76	6·83	2·79

SiO_2 21%, Al_2O_3 5·4%, Fe_2O_3 4·8%, MgO 2·2%, SO_3 1·7%. If anhydrite, silica (SiO_2) and limestone ($CaCO_3$) are mixed in the right proportions, cement is produced and sulphur dioxide, which may be collected and used, is given off.

14.3 The alkali industry

The traditional alkali industry of the nineteenth and early twentieth centuries was based on limestone, ammonia and common salt (see Chapter 1), (Fig. 14.10). The alkali manufacturing complex was based on the Solvay process which produced soda ash (a powdery, almost anhydrous, form of sodium carbonate) which was treated with slaked lime from limestone to give caustic soda:

$$Na_2CO_3 + Ca(OH)_2 \longrightarrow 2NaOH + CaCO_3$$

In the 1880s, however, various methods were developed for making sodium hydroxide by electrolysis of brine:

$$2NaCl + 2H_2O \xrightarrow{electrolysis} 2NaOH + H_2 + Cl_2$$

The inevitable by-products of this process are hydrogen and chlorine, and the ratio of $NaOH:Cl_2:H_2$ of 80:71:2 is fixed by the molecular weights, and cannot be varied. Hydrogen finds ready outlets (e.g for ammonia or methanol manufacture) and the economics of the process thus hinge on the markets for sodium hydroxide and chlorine. Ideally, sufficient chlorine to satisfy the market is produced by brine electrolysis and any deficit of sodium hydroxide is made up by use of the Solvay process. This system has proved adequate in the past. In the modern chemical industry, however, large quantities of chlorine are required not only for bleaching and disinfecting as formerly but also for manufacture of polyvinyl chloride, DDT, alkyl halides, chloro-hydrocarbons, chloro-ethylenes, and other 'sophisticated' organic chemicals (Fig. 14.11). The consumption of more sodium hydroxide than chlorine is now the mark of an underdeveloped country, and the developed countries are faced by a growing surplus of electrolytic caustic soda.

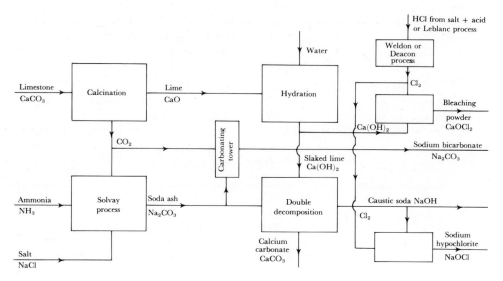

Figure 14.10 The traditional alkali industry.

Rather than sodium hydroxide being the product and chlorine the by-product as in the past, chlorine is now the product and sodium hydroxide the by-product and the alkali industry could more accurately be called the chlorine industry. In consequence, the number of Solvay plants is falling rapidly and the present prices of sodium hydroxide and chlorine may be expected to fall and rise respectively. Figure 14.12 shows US pro-

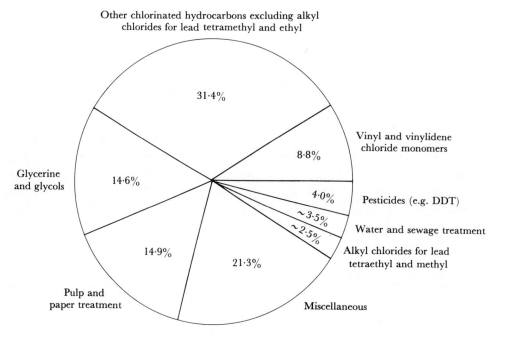

Figure 14.11 Chlorine end-use pattern – USA 1964.

duction of chlorine, total sodium hydroxide and electrolytic sodium hydroxide over the past 70 years, together with their prices for the past 50. It illustrates the growth of the organic chemical industry from 1920 onwards, with its increased demand for chlorine, and the consequent pushing-out of the Solvay process.

The lines for electrolytic sodium hydroxide and chlorine production almost coincide in spite of the fact that electrolysis should give an 11% excess of sodium hydroxide. This is because about 6% of chlorine arises from electrolytic production of potassium hydroxide and sodium metal.

The fact that the prices of chlorine and sodium hydroxide solution are rising more or less in step suggests that producers' complaints in the past about a glut of caustic soda were largely tendentious, though relative prices have changed to some extent since the forties.

14.3.1 Electrolysis of brine

In principle, sodium hydroxide, chlorine and hydrogen could be made simply by immersion of graphite electrodes in a solution of sodium chloride, passage of electricity through the solution and collection of the gases evolved. The efficiency of such a process, however, is reduced by two factors: first, hydroxyl ions from the sodium hydroxide product migrate to the anode and give oxygen which oxidizes the graphite of the electrode reducing its life:

$$2OH^- \longrightarrow H_2O + \tfrac{1}{2}O_2 + 2e^-$$

and second, mixing of the chlorine and sodium hydroxide products leads to the formation of hypochlorites:

$$NaOH + Cl_2 \longrightarrow NaOCl + HCl$$

It is important, therefore, to keep apart the anode and cathode products of electrolysis and this is achieved by the use of either a diaphragm or a mercury cathode cell.

In the latter, the mercury acts as the cathode and forms an amalgam with the metallic sodium released as the primary product of electrolysis. The amalgam is then allowed to react with water in another compartment of the cell to give very pure sodium hydroxide solution (20–70% concentration) and hydrogen.

The diaphragm cell contains a porous asbestos diaphragm to separate the anode and cathode, which are made of graphite and steel respectively. The asbestos allows ions to pass through but reduces diffusion of the anode products (sodium hydroxide and hydrogen) into the cathode compartment where chlorine is produced. An approximately 12% solution of caustic soda is obtained, but it is contaminated with both brine and sodium chlorate.

The mercury cell is more expensive to build than the diaphragm cell because of the capital cost of the mercury, and to operate because it requires a higher voltage. Mercury is lost in the brine leaving the cell and in the sodium hydroxide liquor; some is carried away in the hydrogen, presumably as fine droplets, and pilfering is always a problem. Mercury losses contribute about £0·50/ton to the cost of sodium hydroxide. On the other hand, the mercury cell produces a purer caustic soda, because the latter is completely separated from the brine when it is formed, and a more concentrated solution which can be used directly by rayon manufacturers, while the output from diaphragm cells must

Heavy inorganic chemicals

Table 14.7 Sodium hydroxide

1967	USA	EEC	UK
Production '000 tons	6 900	3 450	~950
Price (1972, solid, $/ton)	150	90–121	109
End-uses			
Textiles	700	650	200
Paper	1 150	250	30
Cellophane	280	60	35
Soap and detergents	300	140	70
Organic and inorganic chemicals	3 000	918	~250
Aluminium	530	300	30
Other uses	940	482	~155
Exports	500	650	~180

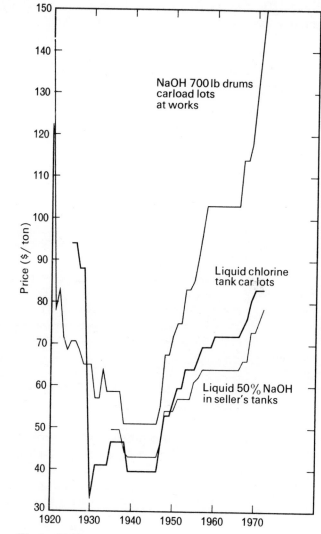

Figure 14.12

be further purified. Almost 80% of US caustic soda is currently made in diaphragm cells, 15% in mercury cells and the other 5% by miscellaneous methods. The trend is towards mercury cells.

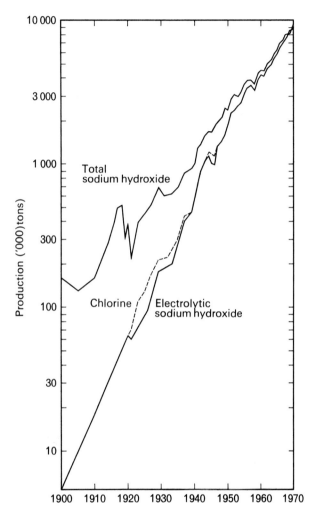

Figure 14.12 US Production of chlorine, total sodium hydroxide and electrolytic sodium hydroxide 1900–70 and prices of chlorine, sodium hydroxide and sodium hydroxide solution 1920–1970.

Pure caustic soda consists of white deliquescent lumps or sticks which absorb both carbon dioxide and water from the air. Shipment clearly presents problems and many consumers prefer to take delivery of a 70% solution. End-uses are indicated in Table 14.7.

14.3.2 Bleaching powder and calcium hypochlorite

Bleaching powder is made by treatment of slaked lime with chlorine:

$$Ca(OH)_2 + Cl_2 \xrightarrow{50°C} Ca\begin{smallmatrix}Cl\\OCl\end{smallmatrix}\cdot H_2O$$

Unfortunately, bleaching powder decomposes on standing to give calcium chloride and oxygen, and also can absorb carbon dioxide from moist air giving calcium carbonate and hypochlorous acid. It has therefore been largely replaced by calcium hypochlorite which is more stable, not hygroscopic, and twice as strong (i.e. it contains two hypochlorite groups). It is made by chlorination of mixed sodium and calcium hydroxides under refrigeration to give the salt $Ca(OCl)_2 \cdot NaOCl \cdot NaCl \cdot 12H_2O$ from which the calcium hypochlorite is separated by treatment with chlorinated lime slurry.

14.3.3 Sodium hypochlorite

Sodium hypochlorite is another important oxidizing agent and is used as a bleach in laundries, as a disinfectant and for sewage treatment. It was made in the past by 'inefficient' electrolysis of brine in which mixing of the anode and cathode products was encouraged, but nowadays caustic soda solution is simply treated with gaseous chlorine:

$$2NaOH + Cl_2 \longrightarrow NaOCl + NaCl + H_2O$$

14.3.4 Sodium carbonate, bicarbonate and sesquicarbonate

Sodium carbonate is still made by the heating of the crude sodium bicarbonate product from the Solvay process in which carbon dioxide is passed into ammoniacal brine. Sodium bicarbonate, surprisingly enough, is not generally made by refining the crude material formed in the above reactions, but by treatment of sodium carbonate with carbon dioxide. The reason for this is that it is very much easier to purify the carbonate than the bicarbonate.

Manufacture of sodium carbonate provides a use for the Solvay plants which are not required nowadays for caustic soda manufacture; few if any have been built since the Second World War. About half of all soda production is used by the glass industry and other uses are varied and relatively small.

Sodium bicarbonate is made on a scale of about 2% of sodium carbonate. As baking powder, its main end-use is in foods.

Sodium sesquicarbonate is a stable mixed carbonate–bicarbonate

$$Na_2CO_3 \cdot NaHCO_3 \cdot 2H_2O$$

and is used in detergents and water treatment.

14.3.5 Chlorine – the Deacon process

The main source of chlorine is the electrolysis of brine as described in section 14.3.1. Recent shortages of chlorine have resulted in the development of oxychlorination processes where hydrogen chloride is oxidized by air, and have also caused a revival of interest in the Deacon Process (section 16.10.3).

14.4 The phosphorus industry

Like fixed nitrogen, phosphorus is essential to plant growth. Bird dung (guano) was used as a fertilizer in Carthage in 200 BC and bone meal has also been used since ancient times.

In 1841 John Bennett Lawes patented a process for making 'superphosphate' fertilizers from bone meal and sulphuric acid and a year later extended the process to include mineral phosphates. By 1862, UK production had soared to 200 000 tons.

Meanwhile, Arthur Albright (who gave his name to Albright and Wilson) and Edward Sturge started in 1844 to make yellow phosphorus for matches by charcoal reduction of phosphoric acid derived from bone ash. Contact of the workers in match factories with this material gave rise to phosphorus necrosis ('phossy jaw') and it was later replaced by red phosphorus and phosphorus sesquisulphide. Conditions of work in the industry were still appalling and the match girls' strike in the Bryant and May factories in 1888 was a testimony to this and also a turning point in the history of social reform in Britain.

Table 14.8 The phosphorus industry

PHOSPHORIC ACID PRODUCTION ('000 tons)

	Total	'Wet' process	Electrothermal process
USA 1971	4 950	4 020	930
Western Europe 1965	1 954*	1 542	77†

Phosphate fertilizers ('000 tons P_2O_5 content)

	Single Super-phopshate	Concentrated super-phosphate	Basic Slag	Other simple phosphates e.g. ammonium	Complex fertilizers	Total
United Kingdom 1969	59	47	92	10	244	452
USA 1971	607	1 340	~11	1 900	327‡	4 596
Western Europe 1969	1 231	683	1 214	334	2 314	5 889

US production of miscellaneous phosphates ('000 tons) in 1971

			Prices (¢/lb)
Tribasic sodium phosphate	Na_3PO_4	47·7	12·1
Sodium metaphosphate	$NaPO_3$	69·0	11·9
Sodium pyrophosphate	$Na_4P_2O_7$	39·8	9·1
Sodium tripolyphosphate	$Na_5P_3O_{10}$	94·0	7·6
Dibasic calcium phosphate	$CaHPO_4$	37·7§	3·7‖
Potassium pyrophosphate	$H_4P_2O_7$	41·0	16·7

* Excludes UK and Greece.
† Many figures not disclosed, so real value is much higher.
‡ This figure is described as 'other phosphatics' and we may be wrong in identifying it as the P_2O_5 content of complex fertilizers.
§ 1969 figure.
‖ Per ton $CaHPO_4 \cdot 2H_2O$.

392 *Heavy inorganic chemicals*

The most important phosphorus compounds marketed at present are superphosphate fertilizers (Table 14.8), basic slag (a phosphorus-containing by-product of the steel industry), phosphoric acid, sodium tripolyphosphate and related salts and a number of organic derivatives such as triphenyl and tricresyl phosphate plasticizers (section 10.3.4.3). The market for phosphorus compounds is growing much faster than that for heavy inorganic chemicals in general.

14.4.1 Superphosphate fertilizers

'Superphosphate' is the monocalcium salt of phosphoric acid and is made by digestion of naturally-occurring phosphate rock in sulphuric acid:

$$[Ca_3(PO_4)_2]_3CaF_2 + 7H_2SO_4 + H_2O \longrightarrow 3CaH_4(PO_4)_2 + 7CaSO_4 + 2HF$$
$$\text{Phosphate rock} \hspace{4cm} \text{Superphosphate}$$

The product contains about 30% monocalcium phosphate monohydrate, 10% dicalcium phosphate, 45% calcium sulphate and 15% impurities amounting to 18–21% available P_2O_5, which is the way in which the phosphorus content of fertilizers is expressed.

If phosphate rock is dissolved in phosphoric acid rather than sulphuric, 'triple superphosphate' is formed which contains a higher proportion of monocalcium phosphate and 43–50% available P_2O_5:

$$[Ca_3(PO_4)_2]_3CaF_2 + 14H_3PO_4 \longrightarrow 10CaH_4(PO_4)_2 + 2HF$$
$$\text{Phosphate rock} \hspace{3.5cm} \text{Triple}$$
$$\hspace{5cm} \text{superphosphate}$$

14.4.2 Phosphoric acid and phosphates

Phosphoric acid is made by two main methods, the electrothermal process in which elemental phosphorus is burned in air to give phosphorus pentoxide which is then hydrated to phosphoric acid:

$$2P + 2\tfrac{1}{2}O_2 \longrightarrow P_2O_5$$

and the 'wet' process which involves the treatment of phosphate rock with excess sulphuric acid, the by-product calcium sulphate being precipitated:

$$[Ca_3(PO_4)_2]_3CaF_2 + 10H_2SO_4 + 20H_2O \longrightarrow 10CaSO_4 \cdot 2H_2O\downarrow + 2HF + 6H_3PO_4$$
$$\text{Phosphate rock} \hspace{4cm} \text{Calcium sulphate} \hspace{2cm} \text{Phosphoric acid}$$

Phosphoric acid gives rise to a wide range of double salts, partly dehydrated salts, etc., and the phase diagrams are very complicated.

Treatment of phosphoric acid with milk of lime gives dicalcium phosphate ($CaHPO_4 \cdot 2H_2O$) which is widely used as a source of phosphorus in animal feeds.

Sodium tripolyphosphate ($Na_5P_3O_{10}$) is a double salt of sodium metaphosphate ($NaPO_3$) and sodium pyrophosphate ($Na_4P_2O_7$). It is obtained when appropriate quantities of these salts are heated at 300–500°C and then allowed to cool slowly. The product is used in large quantities as a 'builder' in detergents (section 11.3.3). Potassium pyrophosphate, sodium metaphosphate and other sodium phosphates are also used in

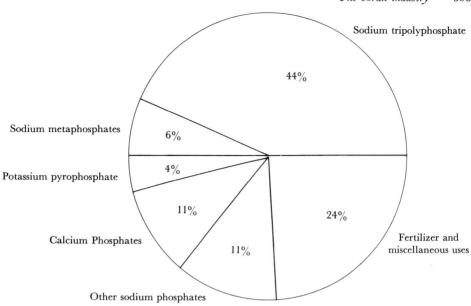

Figure 14.13 End-uses of electrothermal phosphoric acid – USA 1969.

detergent and water conditioning outlets, the first going into liquid heavy duty detergents.

World production of elemental phosphorus in 1969 was about 1 m. metric tons and 85–90% of this was converted to phosphoric acid by the electrothermal route giving about 2 m. tons of the latter on the basis of P_2O_5 content. About nine times as much acid was produced by the 'wet' route. The relative economics of the processes hinge on the cost and supply of sulphuric acid, which in turn depend on sulphur prices and supply. Any easing of these factors favours the 'wet' process while most technical factors favour the electrothermal route.

Approximately 93% of world 'wet' acid production is used in fertilizers. The end-use pattern for electrothermal acid in the USA, where 60% of all elemental phosphorus is made, is shown in Fig. 14.13. In Western Europe the pattern is different in that sodium tripolyphosphate tends to be made from 'wet' acid.

14.5 The borax industry

The basis of the borax industry is borax itself, sodium tetraborate $Na_2B_4O_7 \cdot 10H_2O$ which is obtained in the United States by purification of a crude naturally-occurring sodium borate ore and in Turkey from colemanite, a naturally-occurring calcium borate. It is also obtained as a by-product in the production of potash from brine from Searles Lake, California (section 14.6).

Borax itself is the most important boron compound both in tonnage and value. Annual world production is over a million tons. In particular borax is used to reduce the coefficient of expansion and hence provide thermal shock resistance in the manufacture of borosilicate (Pyrex) glass which contains about 14% B_2O_3. Sodium borate is similarly used in glass fibre manufacture, though in Europe it has largely given way to cleaned colemanite (calcium borate) for this purpose.

Treatment of borax with sulphuric acid gives boric acid H_3BO_3 which together with borax is a constituent of glazes and enamels. It is also used in nylon manufacture (section 10.7.2) to modify the course and selectivity of the liquid phase oxidation of cyclohexane to cyclohexanol and cyclohexanone. The boric acid is not merely a catalyst; it dehydrates to metaboric acid above 160°C and esterifies the cyclohexanol which helps the separation of the latter from the cyclohexanone. The boric acid is partially recovered in the subsequent hydrolysis of the ester. Boric acid and borax also serve as precursors for a wide variety of borates of which the most familiar is sodium perborate $NaBO_3.4H_2O$ which is made from borax, sodium hydroxide and hydrogen peroxide:

Table 14.9 The Borax Industry.

PRODUCTION	'000 tons	
USA		
Total Boron minerals and compounds 1968	900	
Above measured as B_2O_3	471	
Boric acid 1969 as H_3BO_4 (unit value $121/ton)	126	
Technical grade borax $Na_2B_4O_7.10H_2O$ in 1968 excluding anhydrous and pentahydrate forms (unit value $65/ton)	565	
Turkey		
Colemanite 1968	250	

PRICES	USA ($/ton)	UK (£/ton)
Borax	56	53
Boric acid	138	77
Sodium perborate	304	126

CONSUMPTION

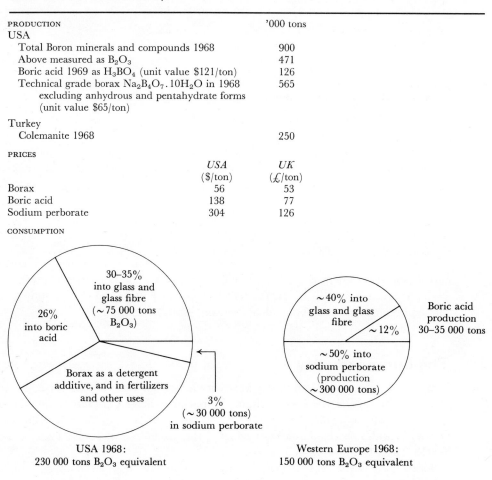

USA 1968:
230 000 tons B_2O_3 equivalent

Western Europe 1968:
150 000 tons B_2O_3 equivalent

Borax has small fertilizer and miscellaneous uses in Western Europe; other consumption estimates give total B_2O_3 consumption of up to 200 000 tons which presumably allows for these.

$$Na_2B_4O_7 + 2NaOH \longrightarrow 4NaBO_2 + H_2O$$
$$NaBO_2 + H_2O_2 + 3H_2O \longrightarrow NaBO_3 \cdot 4H_2O$$

and which is added at a level of about 20% to detergent powders as a bleach. At room temperature it is stable but above 50°C in solution it decomposes to give oxygen:

$$2NaBO_3 \longrightarrow 2NaBO_2 + O_2$$

It is thus more stable than chlorine bleaches and is without their unpleasant smell.

Boron carbide B_4C is the hardest known artificial abrasive and is made by fusion of boric oxide with carbon in an electric furnace. Its largest use currently is in ceramic armour for military aircraft and personnel, and US forces in Vietnam were equipped with bullet-proof boron carbide vests after 1968.

Market estimates for boron compounds are given in Table 14.9. An important difference between the European and North American markets is that, in the latter, it is borax not sodium perborate which is used as a detergent additive. It has excellent water-softening properties but is not a bleach and its use in laundering is to some extent traditional.

14.6 Potash

Potassium hydroxide is the oldest alkali. Potassium carbonate used to be leached from wood ashes at Pompeii, 'strengthened' with lime (i.e. converted to the hydroxide) and used for soap-making. As western civilization developed, more and more wood was consumed for this purpose, so that the forests of Europe were threatened, and it was only the advent of the Leblanc process at the time of the French revolution which saved the day. The switch to sodium rather than potassium based alkalis means that the major outlet for potassium salts is now in fertilizers.

Potassium is essential in the soil, in addition to phosphorus and nitrogen compounds, if plants are to grow normally. Potassium chloride (muriate of potash) is the form in which it is normally applied, but the fertilizer content is usually expressed in terms of potash content (K_2O). Market data are given in table 14.10.

Table 14.10 The potash industry (figures in thousand tons of K_2O in 1969)

	Production*					Prices†	
	K_2SO_4	Other potash fertilizers	Total	Imports	Exports	K_2SO_4 (50% K_2O)	KCl (60% K_2O)
UK	nil	nil	nil	483	5	£23·13/ton	£23/ton
USA	512	2 032	2 544	2 123	701	$76/ton†	$35/ton†
Western Europe	485‡	4 227	4 712	1 245	2 293	—	—

Principal producers are USA, USSR, Germany, France, Canada, Spain and Israel. See also Fig. 14.2.
* Production from imported materials excluded.
† f.o.b. Carlsbad or Saskatchewan.
‡ Excluding France.

396 *Heavy inorganic chemicals*

Potassium chloride has been extracted from 'rubbish salt' at the historic Strassfurt salt mines since 1860, and until 1914 Germany was the sole world supplier of potash. Together with dyestuffs this trade established Germany's early lead as chemical manufacturers. The Strassfurt and most other European deposits take the form of carnallite ($KCl.MgCl_2.6H_2O$).

In America potash is extracted from Searles Lake, California, brines which are also a source of borax. Other American sources include the Carlsbad mines in New Mexico and deposits in Saskatchewan only recently exploited. American deposits are usually in the form of sylvinite (42·7% KCl, 56·6% NaCl).

A few per cent of world potash comes from the brines of the Dead Sea and is extracted at Sodom, the somewhat dull successor to the biblical town. This extraction will also be described since it involves the carnallite separation which is applicable to other carnallite deposits, and also because it provides a classical, or perhaps biblical, example of the application of the phase rule.

14.6.1 Extraction from sylvinite by hot leaching

The traditional method of extracting potassium chloride from sylvinite, still widely used especially in the USA (Carlsbad, New Mexico), depends on the fact that whereas potassium chloride is very much more soluble in hot water than in cold, sodium chloride is only slightly more soluble at 100°C than at 20°C (Fig. 14.14). Indeed in solutions saturated with potassium chloride, it is actually less soluble at the higher temperature.

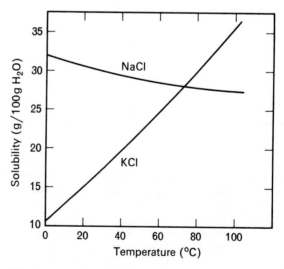

Figure 14.14 Solubilities of KCl and NaCl in solutions saturated with respect to both.

The sylvinite ore is first crushed and then mixed with brine which has been recycled from the previous extraction. It is heated almost to boiling so that the potassium chloride dissolves; the recycled brine is already saturated with sodium chloride so none of that dissolves. When the solution is saturated with potassium chloride, it is cooled by vacuum evaporation. The potassium chloride crystallizes out and is filtered off, washed and dried.

14.6.2 Extraction from sylvinite by flotation

In the flotation process, the finely ground sylvinite ore is treated with a 'collector', that is a hydrophobic material such as an aliphatic amine, which selectively coats the potassium chloride crystals. Air is then bubbled in to give a froth. The bubbles attach themselves to the coated particles of potassium chloride and carry them to the top of the flotation cell where they are removed and filtered off. The uncoated sodium chloride particles sink and are removed from the bottom.

In some older plants, different collectors are used and the sodium chloride is floated while the potassium chloride sinks.

It is noteworthy that the flotation method only works because sylvinite is a physical mixture of sodium and potassium chlorides and not a true double salt.

14.6.3 Potash from the Dead Sea (via carnallite)

Brine from the Dead Sea contains about 25% dissolved solids. These are mainly chlorides of sodium, potassium and magnesium but some bromides are also present and the Dead Sea is one of the world's major sources of bromine.

The extraction of potash should be illustrated by the phase diagram for the NaCl–MgCl$_2$–KCl–H$_2$O system, but this is highly complicated and the three-component triangular phase diagram (Fig. 14.15) is normally used, it being understood that the system is saturated with sodium chloride at all times.

The diagram is complicated by the existence of a double salt, the mineral carnallite, whose composition corresponds to the point D, and also by the existence of a hydrated

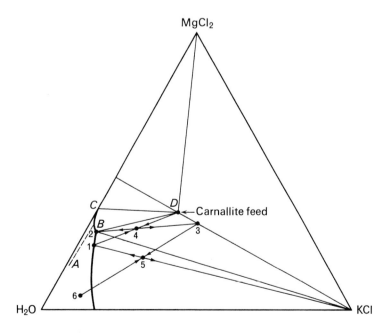

Figure 14.15 MgCl$_2$–KCl–H$_2$O phase diagram for a system saturated with NaCl.

salt, $MgCl_2 \cdot 6H_2O$, corresponding to point C, and this is the form taken by many European deposits.

Dead Sea brine has a composition corresponding approximately to point A. It is evaporated by the sun so that it loses water and the composition moves along the line AB. As the solution at A is saturated with NaCl, some of this crystallizes out.

When the composition reaches B, carnallite begins to separate and the brine is pumped to a fresh set of ponds. The solid carnallite depletes the solution of potassium and its composition moves along the line BC. When it reaches C, all the potassium has separated as carnallite and further evaporation would result in the crystallization of $MgCl_2 \cdot 6H_2O$ as a second solid phase. By application of the phase rule (three phases, three components, temperature and pressure fixed) it can be seen that C is an invariant point and the composition will not change further until the liquid phase has disappeared.

In practice, as soon as all the carnallite has precipitated, it is dredged out together with the sodium chloride which was precipitated simultaneously. The carnallite is then mixed with an interstage liquid of composition 1 (which is later regenerated) to give a mixture of average composition 4. Liquid 1 can exist in equilibrium with solid KCl but not with carnallite and it leaches magnesium chloride out of the carnallite so that its composition moves to point 2 and the carnallite splits into a mixture of two solid phases (KCl and carnallite) of average composition 3. Point 2 is also an invariant point in that it is the only composition of liquid that can exist in equilibrium with both solid KCl and carnallite. Liquid 2 is, however, discarded and the solids mixture 3 is treated with waste liquid from a crystallizer of composition 6 (water would be perfectly satisfactory for this purpose but is less readily available) to give a mixture of composition 5. There is no question of the unsaturated solution 6 being in equilibrium with the two solid phases at 3, and hence one solid phase, the carnallite, disappears leaving solid KCl which is removed – regenerating the interstage liquid 1 which is used for a subsequent extraction.

Sodium chloride is precipitated along with potassium chloride, so a sylvinite separation process of the kind already described is used to give the final yield of potash.

One of the problems associated with the extraction is that the bottoms of the evaporation ponds are always covered with white crystals. Sunlight is not absorbed by water and tends to be reflected by the crystals. Evaporation can be very slow and a blue dye is usually added to the water to increase absorption of solar energy.

The brine contains only 1% KCl and the amount of water which must be evaporated to obtain it would render the process uneconomic were it not for the extreme heat and sunshine at Sodom. Even so, the construction of the ponds and the handling of large quantities of liquids and solids is very expensive.

14.7 Hydrogen peroxide

Hydrogen peroxide was made in the nineteenth century by the action of dilute acid on barium peroxide, itself prepared by the roasting of barium oxide in dry air. Other methods were subsequently used including the liquid-phase oxidation of isopropyl alcohol:

$$CH_3 \cdot CHOH \cdot CH_3 + O_2 \longrightarrow \underset{\substack{\text{Acetone} \\ \text{by-product}}}{CH_3 \cdot CO \cdot CH_3} + H_2O_2$$

and the hydrolysis of a solution containing the persulphate ion ($S_2O_8^=$). The persulphate is prepared by anodic oxidation of the corresponding sulphate in an electrolytic cell:

$$2NH_4HSO_4 \xrightarrow{electrolysis} (NH_4)_2S_2O_8 + H_2$$
$$(NH_4)_2S_2O_8 + 2H_2O \longrightarrow 2NH_4HSO_4 + H_2O_2.$$

The most popular route, however, is the auto-oxidation method in which an anthraquinone derivative is reduced by hydrogen on palladium to the hydroquinone, and then reoxidized with air to give hydrogen peroxide and the original anthraquinone:

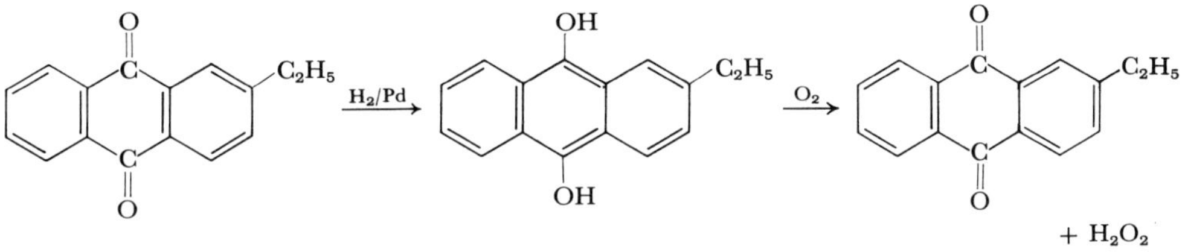

The end-use patterns of hydrogen peroxide differ in the UK and USA because the Americans do not use sodium perborate in their detergents. UK production of hydrogen peroxide in 1970 was about 28 000 tons of which 40–50% was used in perborate manufacture, the rest being used for textile bleaching, paper and pulp bleaching, and the production of organic and inorganic peroxides such as benzoyl peroxide and epoxidized oil plasticizers.

Notes and references

General We are indebted to our colleague Dr. R. J. Irving (formerly an employee of ICI, Billingham) for much of the information in this chapter. General books on the chemical industry such as are cited in our general bibliography are strong on heavy inorganic chemicals. In addition D. M. Samuel, *Industrial Chemistry – Inorganic, ordinary and advanced levels*, R.I.C. teachers monographs no. 10, RIC London, 1966, provides a useful introductory text. There is much marketing information in the ECMRA Conference, *Heavy Chemicals and their Raw materials in Europe*, London, 1969. USA production data are to be found in US Department of Commerce – Bureau of the Census, *Current Industrial Reports – Inorganic Chemicals*, series M28A, Washington D.C.

Section 14 Data in Fig. 14.1 and Table 14.1 come mainly from the *UN Statistical year book*, *Inorganic Chemicals M28A*, op. cit., and OECD, *The Chemical Industry*.

Section 14.1.1 C. A. Vancini, *Synthesis of Ammonia*, Macmillan, London, 1972, provides a massive technical write-up of the Haber process. Catalysts for ammonia and hydrogen manufacture are discussed at length in the ICI, *Catalyst Handbook*, Wolfe, London, 1970. Ammonia markets are considered by J. P. Johnson, ECMRA London, 1969, op. cit.

Sections 14.1.3 and 14.1.6 Nitric acid and urea were the subject of ECN process surveys on 30th January, 1970 and 17th January 1969 respectively.

Section 14.1.7 Data from US Tariff commission and *Business Monitor*.

Section 14.2 Consumption and production data come from the UK National Sulphuric Acid Association and OECD, *The Chemical Industry*. Technical data are mainly from Shreve. See also M. N. J. Horseman and D. L. Mermikides, *The Place of Europe in*

400   *Heavy organic chemicals*

the World Sulphur/Sulphuric Acid Scene, ECMRA London, 1969, which is the source of Table 14.6. Fig. 14.9 is taken from Kirk-Othmer.

Section 14.3   Economic data have been partly drawn from C. Borromee, *The Market for Caustic Soda* and G. J. Lewis, *A Forecast of the Chlorine Requirements of Western Europe*, ECMRA London, 1969. An updating of Fig. 14.12 would show smaller proportions of chlorine going into glycols and PVC because of the abandonment of the chlorhydrin route to ethylene oxide, and the development of oxychlorination.

Figure 14.11 involves production data from a variety of sources but all derive ultimately from the US Bureau of the Census. Prices until the early sixties came from Kirk-Othmer; subsequent prices came from journals. Historically interesting discussions of chlorine, and electrolytic versus lime–soda are to be found in *Chemical and Metallurgical Engineering*, August 1944, p. 115, and *Industrial and Engineering Chemistry*, October 1949.

Section 14.4   The figure of 200 000 comes from Shreve. Hardie and Davidson Pratt (see bibliography to chapter 2) give a figure of 40 000 tons for 1862 phosphate production but this is presumably P_2O_5 content. An excellent review of technology and market trends for phosphoric acid is given by F. M. Cussons and W. M. Karn, ECMRA, London, 1969.

Section 14.5   We wish to thank Borax Consolidated Research Centre, Chessington, Surrey for discussion and advice. We have also drawn on *Boron, Industrial Minerals*, July, 1969; R. Thompson *The Boron Products Industry*. 9th Commonwealth Mining and Metallurgical Congress, 1969; R. Thompson *Boron Chemistry in Industrial Perspective*, Chemistry in Britain, 1971, **7**, 140, and Stirling and Co's report on Laporte. We have written sodium perborate as $NaBO_3.4H_2O$. This is sometimes written as $NaBO_2.H_2O_2.3H_2O$ but there is now ample evidence for the peroxyborate ion and very little for hydrogen peroxide of crystallization.

Section 14.6   Table 14.10 is based on OECD *The Chemical Industry*. Figures 14.14 and 14.15 are taken from Kirk-Othmer.

Section 14.7   Stirling and Co's report on Laporte contains useful information on the UK hydrogen peroxide market.

Part IV Large scale chemistry

15 The scaling-up of laboratory experiments

Chemical plants making ammonia or sulphuric acid often have capacities of about 1 000 tons per day, polyethylene, acetic acid and phenol plants operate at about 100 or 200 tons per day. Continuous esterification plant for making dialkyl phthalates might produce 20–30 tons per day, and even an aspirin manufacturer would be unlikely to install plant with capacity of less than 2 tons per day. If a typical laboratory sample is 10 gm, then a major industrial chemical is made on a scale some 10^8 times as large. This is the ratio of the size of the world to that of a cricket ball, or of the size of a marble to that of an atom. It is hardly surprising, therefore, that the quantitative differences between chemical synthesis in the laboratory and in industry are sometimes so large as to have become qualitative differences; indeed it is possibly more of a surprise to see how close industrial practice is in certain aspects to what the chemist is used to. Industrial distillation columns, batch reactors and stirrers for example are recognizably distillation columns, batch reactors and stirrers.

The differences between industry and the laboratory are nonetheless basic and arise from a number of constraints which are partly economic and partly technical, with a degree of overlap between the two.

15.1 Economic factors in scaling-up

It is important for industry to make as large a profit as possible. The reagents used in industrial chemistry tend to be very cheap; for example, reductions are carried out using hydrogen and not lithium aluminium hydride. Insofar as it is possible, reagents should be easy to handle (i.e. not solids) and unit costs must be kept low. A corollary of this is that chemical plant is built with a large capacity because of the economies of scale and consequent reduction in unit cost.

15.1.1 Yields

The size of the yield in an industrial process is also of great importance. Apart from the expense of purifying seriously contaminated products, there are two reasons for this. First, an increase in yield increases profits out of proportion to the size of the increase. Suppose a firm manufactures a chemical selling at £100/ton and the plant costs and raw material costs for sufficient raw materials to make 100 tons of product if the plant were 100% efficient are £8 000. Then, if the plant operates at 90% efficiency, the profits will be £1 000 whereas if the efficiency were increased by 5% to 95% the profits would increase by 50% to £1 500. Second, a low yield might produce a serious effluent problem.

Plant operating on a scale of 100 tons/day produces by-products very rapidly. At 95% efficiency these accumulate at 5 tons/day and if they are unsaleable they have to be disposed of in some other way. Sometimes they can be burned and thus have a certain value (about £10/ton is typical) as fuel. Sometimes it is possible to recycle them so that they are reprocessed and ultimately contribute to a high overall yield after several passes. Sometimes, however, the material has simply to be dumped in a tip or possibly in sewers or in the sea. The cost of tipping is high and the more slowly a firm fills its tip, the better. Overall yields of over 90% are the rule rather than the exception in the chemical industry and figures of over 95% are not uncommon.

15.1.2 Planning

The scale of operation of modern chemical plant is so huge that the industrial chemist cannot afford to 'suck it and see'. Anything which is built must have been planned beforehand in great detail; the process must have been worked out on the laboratory scale, and on one or even two pilot plants before there can be any question of the capital being found for a commercial unit. Possibly pilot plant production will actually be sold in order to assess the market, but in any case the technical side of the process must be thoroughly understood. Modern accounting techniques such as discounted cash flow (section 5.7) highlight the losses to a company if its plant fails to come on stream at the planned time. This is not to say that chemical plant never has teething troubles; indeed there are many occasions where a process has worked perfectly on a smaller scale but has failed completely for unforeseen reasons on being scaled up. It must nevertheless be realized that this represents disaster for the company concerned and that a delay of even a few months in a plant coming onstream may make all the difference between an economically viable process and one which never recovers its development costs and capital outlay. The problems of Laporte with its new titanium dioxide plant using the chloride process (section 10.5.6) illustrate this point only too well.

15.2 Technical factors in scaling-up

The technical problems of scaling-up may be illustrated by consideration of a specific example. Suppose we are interested in manufacturing vinyl acetate by the Wacker Process (section 10.5.1). We can set up a bench-top laboratory rig to do this without very much trouble. We take a 500 ml 3-necked flask and equip it with an inlet for reactant gases, a thermometer, a stirrer and a condenser (Fig. 15.1). We remember to pass the cooling water from the bottom to the top of the condenser so that air bubbles get carried away instead of accumulating in the cooling jacket and reducing the area available for cooling.

We half-fill the flask with glacial acetic acid plus the platinum chloride/cupric chloride catalyst system, heat the bath till the reacting system boils gently and pass in ethylene and oxygen:

$$C_2H_4 + CH_3COOH + \tfrac{1}{2}O_2 \xrightarrow[\text{in HCl}]{PdCl_2\text{-}CuCl_2} CH_3COOC_2H_3 + H_2O$$

$$\Delta H = -41 \text{ cal/mole } (-172 \text{ kJ/mole})$$

Technical factors in scaling-up 405

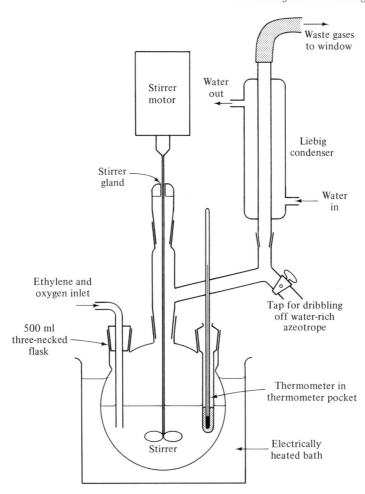

Figure 15.1 Laboratory apparatus for making vinyl acetate.

The water which is formed is a nuisance because it promotes a side reaction to give acetaldehyde (section 9.1.4) but we find we can remove most of it by draining off a water-rich azeotrope at the tap at the bottom of the condenser while allowing the remainder of the condensate to flow back into the reaction vessel. We throw the azeotrope down the sink.

After we have run the experiment for an hour or two, we switch off the apparatus, allow it to cool and analyse the mixture in the flask. It contains vinyl acetate, so the experiment has been a success. It also contains assorted glycols and aldehydes, especially acetaldehyde, dissolved carbon dioxide, and the catalyst system plus unchanged acetic acid. We distil a few times, discarding by-products, and eventually obtain an adequately pure vinyl acetate.

It turns out that our yield on ethylene was very low because most of it bubbled straight through the reaction mixture and vented to the window. Indeed, at one point when repeating the experiment we had a small explosion due to a man outside lighting a cigarette and igniting the ethylene/air mixture. Fortunately no one was hurt and only a few pounds' worth of ground glass equipment was destroyed. We are concerned, therefore, to get better absorption of gases in the liquid and this is described as a *mass transfer* problem.

Much encouraged, nonetheless, we build our Mark II apparatus which is the same as Mark I but stands on the floor, is bigger all round and has a 5-litre reactor. This time, the heat produced in the exothermic reaction causes the mixture to boil vigorously; the condenser cannot carry it away fast enough to keep the system refluxing and it froths up and explodes. Some expensive glassware is fragmented, a technician is slightly injured by boiling glacial acetic acid and a nasty yellow mark is left on the ceiling. Our *heat transfer* arrangements have proved inadequate and we realize that we have no real temperature control in that we can only stabilize at the boiling point of the mixture under atmospheric pressure.

We try to cope with these problems in our 50 litre, Mark III apparatus which we build in stainless steel, a material for which medium-scale chemists have great fondness. We arrange to operate under pressure because that will give us better absorption and also enable us to control temperature more easily because it will stabilize at whatever is the boiling point of the reaction mixture at the pressure under which we are working. We replace our mercury-in-glass thermometer by built-in thermocouples and pressure gauges which feed to chart recorders and we redesign and enlarge our condenser to be sure it is of sufficient capacity.

Mark III operates quite well, and we get a lot of information about the kinetics of the reaction and optimum reaction conditions. After some time, however, we notice some corrosion at the bottom of our condenser. This presumably indicates local reducing conditions which destroy the oxide layer which is the reason for stainless steel's resistance to corrosion. We resolve to replace the crucial section in titanium. [This principle was less well-established around 1960 when The Distillers Company built their 40 000 tons/year PFD plant for acetic acid at Hull (section 9.4). The bottom twelve plates of a giant stainless steel distillation column simply dropped out.] In addition to our corrosion problem, various colleagues start to complain about smelly drains and an odour of rotten apples, probably due to traces of acetaldehyde, which, now we have scaled up our plant from 500 ml to 50 litres, are beginning to be noticeable.

These problems and many others will have to be tackled before full scale plant can be constructed. Meanwhile our parable has thrown up a number of problems. We require better absorption of gases and better control of the exotherm. We need to design an efficient reactor, preferably one which operates continuously, and we have to determine optimum temperatures and pressures. We have to separate our mixture of products and build our plant of materials which will not corrode seriously. We have to pump our materials around the plant and try not to waste heat. In addition we must cope with fire and explosion hazards and dangers from toxic chemicals and we must arrange for monitoring and control of temperatures, pressures and flow rates around the plant. The technology required to solve these problems is complex but fairly well understood. It will be touched on in the following chapter and more detailed treatments will be found in standard texts particularly those on chemical engineering. Finally, a description of the scaled-up plant will be given. Discussion of effluent disposal is postponed till chapter 17.

15.3 Reactor design and optimization of reaction conditions

The laboratory chemist can afford to be sanguine about the feasibility of the preparations he attempts to carry out. If a reaction goes at 1% of the expected rate, he can take a month's holiday; if it fails altogether no one is much worse off. The industrial chemist on the other hand demands to know of any reaction if it is feasible (i.e. what is the percentage conversion if the reaction goes to equilibrium), how fast it goes towards equilibrium, and how far he can optimize yields, maximize throughput, discriminate against side reactions, prolong catalyst life etc. Equilibrium constants can be calculated by classical thermodynamics (Le Chatelier's principle, the Gibbs–Helmholtz equation, the Van't Hoff isochore and isotherm) and tables of thermochemical data. Reaction rates cannot be calculated *ab initio* but semi-empirical methods are of value. Other factors are usually determined empirically on the basis of laboratory or pilot plant experience. Reactions themselves can be sub-divided into homogeneous (taking place in a single phase) and heterogeneous (taking place at a phase boundary, usually the surface of a catalyst). The amount of heat emitted or absorbed during reaction is important when a reactor is designed, and the order of reaction is also significant.

The reaction vessel or reactor is the heart of any chemical plant. Every process is unique and requires a special design which takes account of the temperatures, pressures, quantities and chemical substances involved. The resemblances between the pot used for an esterification at 140°C and the electric arc furnace used to make calcium carbide at 2 100°C are tenuous or non-existent, so that only very broad principles governing reactor design can be enunciated.

The purpose of a reactor is to bring about reaction in the most economical way possible. It must have facilities by which it can be charged with reactant and discharged of product. As reaction rates and side reactions depend critically on temperature, it must provide means for this to be controlled and it must also bring the reactants into contact with the catalyst if one is to be used.

15.3.1 Batch and continuous reactors

Reactors may be subdivided into two main classes – continuous reactors and batch reactors or autoclaves. A batch reactor is charged with material which is allowed to react and the product subsequently discharged for purification. The reactor is then recharged with fresh materials. Operation is thus intermittent and labour charges tend to be higher than with continuous reactors where materials are supplied and withdrawn continuously during operation. In general, batch reactors tend to have higher variable and lower fixed costs than continuous ones.

In addition to their high labour costs, batch reactors are unsuitable for fast exothermic reactions since the initial reaction might get out of control. On the other hand, they are useful for slow reactions and are flexible in use since the conditions can be changed for each run. They are widely used for low-tonnage preparations (e.g. cosmetics, pharmaceuticals and speciality plasticizers) and preparations where there is an explosion hazard or contamination risk (e.g. explosives, some polymerizations, and biological fermentations).

Continuous reactors have lower labour costs and are relatively easily automated. In general they produce a more uniform product than batch reactors and are indispensable for the very large tonnage processes and the many gas phase reactions in modern industrial chemistry. A variety of continuous reactors exists, the two predominant forms being the tubular flow and tank flow reactors. In the former, the reactants travel down a tube reacting as they go and the products emerge from the far end (Fig. 15.2a). Ideally, in a

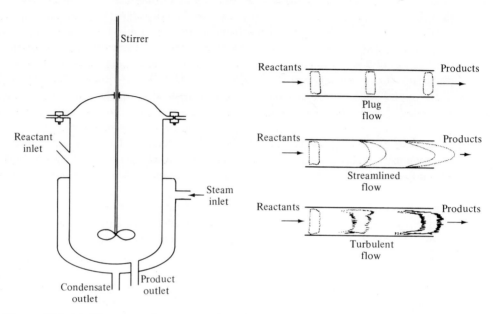

Figure 15.2 (a) Diagrammatic representation of a tank reactor. If reactants do not flow through continuously during reaction, it is equivalent to an autoclave. (b) Diagrammatic representation of a tubular flow reactor, showing various flow patterns.

case referred to as 'plug flow', no mixing of fluids entering the reactor at different times takes place. In a tank flow reactor, stirring is very efficient and the contents of the tank (Fig. 15.2b) have a constant composition which, since product is withdrawn continuously, must be identical with that of the product mixture. Tank reactors are valuable in situations where a reaction might get out of control in the early stages, since the reactant concentration is never high. Similarly they may be used when particular compositions have to be avoided, perhaps because they are explosive.

A major difference between these types of reactor is illustrated by the direct chlorination of benzene to chlorobenzenes:

$$C_6H_6 \xrightarrow{Cl_2} C_6H_5Cl \xrightarrow{Cl_2} C_6H_4Cl_2 \xrightarrow{Cl_2} C_6H_3Cl_3$$

In a tubular flow reactor, very little chlorine would come into contact with monochlorobenzene and this would be the major product; in a tank flow reactor the reverse would be true and a higher proportion of polychlorobenzenes would be formed.

In the absence of such situations, however, tubular flow reactors are generally preferred. The reaction rate in a tank flow reactor is low and a large amount of product is effectively 'stored' in it. Thus for a given throughput a tank flow reactor must be very much larger than a tubular reactor and other things being equal this size factor is critical.

15.3.2 Reactors for heterogeneous reactions

The above descriptions of reactions have assumed a homogeneous reaction in which mixing and temperature control are relatively easy, and any catalyst is distributed uniformly throughout the reaction mixture. In heterogeneous reactions, by contrast, there are problems of mass transfer (getting the reactants into contact with the catalyst and the products away from it) and heat transfer (preventing localized heating at the catalyst surface). There is also a possible problem with catalyst regeneration.

Most heterogeneous reactions are carried out in tubular flow reactors either 'fixed bed' or 'fluidized bed' under conditions as close as possible to 'plug flow'. In a typical fixed bed reactor a number of long parallel tubes are filled with catalyst pellets about 5 mm in diameter but with a high internal pore area. The reactants are passed over them. In a fluidized bed reactor, the catalyst is in the form of a fine powder. Reactant gases are passed through it at such a speed that the particles are suspended in the gas stream and the gas/solid system takes on many of the properties of a fluid.

The advantages of the fixed bed reactor are its simplicity and its lack of problems if the rate of gas flow varies. Its drawbacks are the difficulty of preventing temperature gradients in the catalyst, the low rate of diffusion of reactants into and products out of the pellets, and the expense of catalyst regeneration.

Fluidized bed reactors on the other hand give excellent heat transfer. Diffusion to and from the smaller catalyst particles is quicker, and the quasi-fluid can easily be induced to flow into a separate vessel for regeneration. Their main problems are concerned with the maintenance of the quasi-fluid which is subject to such ills as slugging (masses of catalyst particles hanging together in lumps instead of being distributed evenly throughout the gas), channelling (the gas managing to find a catalyst-free channel up the side of the reactor) and attrition (the breaking down of catalyst particles to give a fine dust which is carried over with the products).

These problems can be minimized by use of tubular flow reactors with a low height: diameter ratio. Plug flow is then more difficult to achieve and one of the main advantages of the tubular flow reactor may be lost unless a suitable compromise can be found.

The choice between fixed and fluidized bed reactors depends on a careful analysis of each individual case. Suffice to say that in the UK there is a slight preference for fluidized beds while the USA has a penchant for fixed beds.

Table 15.1 Unit operations

Fluid Dynamics	Pumping, stirring and other aspects of liquid and gas handling.
Solids Handling	Mixing, sedimentation, filtration, screening, crystallization, centrifuging, size reduction, fluidization, polymer and rubber processing.
Heat Transfer	Heat exchange, evaporation, liquefaction, heating of furnaces, ovens etc.
Mass Transfer	Humidification, gas absorption, solvent extraction, distillation, sublimation, drying.

15.4 Unit operations

The chemical industry, in common with other modern industries, uses the concepts of mass production. The complicated overall processes by which materials are manufactured are simplified by being broken down into simple individual steps. For example, a number of procedures leading to physical and mechanical changes, such as distillation, filtration and grinding, are used in the chemical industry in otherwise widely different processes and these are described as unit operations. The principal unit operations are shown in Table 15.1. Laboratory preparations, of course, involve similar operations and this is recognized implicitly by the use, for example, of interchangeable ground glass equipment. The chemical industry, however, lays great stress on these unitary concepts because problems arising from them will be solved using similar theories and similar and often interchangeable equipment. We shall confine ourselves here mainly to the problems arising out of our vinyl acetate plant and refer more avid readers to the bibliography.

15.4.1 Fluid dynamics

In our original vinyl acetate unit, inadequate mixing led to poor heat transfer and inadequate absorption of gas. Our Mark III apparatus required more efficient stirring and a compressor to improve gas absorption and both of these will need to be scaled-up.

These, however, are only two of the problems concerned with fluid dynamics. The ease with which a laboratory chemist can pour a beaker of liquid into a filter funnel cannot be matched in industry. A large part of any chemical industrial complex is devoted to pipes, pumps and other equipment for transferring solids, liquids and gases from one place to another. The Esso petroleum refinery at Fawley, England contained about 500 miles of piping in the late 1960s and this figure had probably doubled by 1972. Transfer of gases and liquids is relatively easy and even live steam can be carried in lagged pipes with fairly low heat loss. Solids, on the other hand, are usually expensive to handle and it is thought worthwhile, for example, to ship and handle phthalic anhydride in bulk at 140°C – at which temperature it is liquid – to avoid the problems of solids handling. Similarly carbon dioxide is transported under both pressure and refrigeration so that it remains liquid. In spite of the ease of handling fluids, the pumping and piping of them still add to the cost of a process. Every opportunity is taken to keep the number of pumps and length of piping to a minimum and to let gravity (which is provided free) do the work. All too often, the major problem of scaling up is to persuade all the liquids to run through the pipes in the right direction.

15.4.2 Solids handling

A major aim of the chemical industry is to avoid handling solids and such problems did not arise in our vinyl acetate plant. Techniques are available but will not be discussed here. A specialized branch of solids handling is the fabrication of plastics and this is summarized in section 16.12.2.

15.4.3 Heat transfer

Heat costs money and in the same way that industry tries to make efficient use of its supplies of chemicals, it also tries to use efficiently its supplies of energy. Piping etc. is often lagged to minimize heat losses, and the hot products of a reaction are often cooled by their being used to heat up the incoming cold reactants. This is achieved by the use of heat exchangers and related equipment (section 15.4.3.4).

Many large chemical manufacturing complexes use fuel oil etc. to generate high pressure steam at 130–160 atmospheres. This is used to drive electricity-generating turbines and the waste steam, usually at about 6 atmospheres pressure, is used for evaporation, distillation and general heating around the plant. The high latent heat of evaporation of steam is an advantage. Methods of heating apparatus other than steam (direct or indirect firing, direct gas heating, electrical heating etc.) are considerably less economic and plant is designed wherever possible to operate on steam.

Almost every process and operation used in the chemical industry involves the production or absorption of heat. At the reactor stage, heat may have to be added to bring about a reaction or absorbed to prevent its getting out of control; at the separation stage heat may have to be added to operate a distillation column or removed to condense the products.

When two objects at different temperatures are brought together, heat flows from the hotter to the colder object. Spontaneous flow in the opposite direction would infringe the second law of thermodynamics. There are three mechanisms by which heat moves: conduction, convection and radiation.

15.4.3.1 Conduction Conduction involves the passage of heat through a body unaccompanied by any flow of material. It is the only form of heat transfer in opaque solids. Its rate depends on the size of the body, on the temperature difference across which heat flow takes place and on the coefficient of thermal conductivity of the substance of which the body is made. The latter is defined as the heat flow in unit time through a body of unit length and unit cross-sectional area with unit temperature difference between its ends. Some typical values are given in Table 15.2. Copper is easily the best conductor; water has an unusually high conductivity for a liquid, the value for acetone being more typical. Hydrogen has a high conductivity for a gas because of its high molecular velocity. Air is a poor conductor and many heat insulating materials (e.g. foamed polystyrene) act by trapping air and thus minimizing convection as well.

Table 15.2 Thermal conductivities of some common materials at ordinary temperatures

	$Joules\ m^{-1}\ °C^{-1}\ sec^{-1}$	$Btu\ ft^{-1}\ °F^{-1}\ hr^{-1}$
Copper	385	223
Aluminium	238	138
Mild steel	46–50	26–29
Stainless steel	25	14·5
Glass	0·85–0·105	0·5–0·6
Brick	0·6	0·34
Water	0·58	0·34
Acetone	0·18	0·10
Hydrogen	0·168	0·097
Wood	0·12–0·17	0·07–0·1
Asbestos	0·125	0·073
Air	0·024	0·014

(Units: $1\ cal\ cm^{-1}\ °C^{-1}\ sec^{-1} = 418\ Joules\ m^{-1}\ °C^{-1}\ sec^{-1} = 242\ Btu\ ft^{-1}\ °F^{-1}\ hr^{-1}$)

15.4.3.2 Convection When a fluid travels from one place to another it carries with it a certain amount of heat. This heat transfer, which is associated with mass transfer, is called convection. If it occurs spontaneously as a result of differences in buoyancy between hot and cold materials (e.g. the rising column of warm air in front of a domestic electric radiator) it is called natural convection. If the fluid flow is caused deliberately by a fan or agitator (e.g. the warm air impelled out of a domestic fan heater or cool air into an automobile cooling system) it is called forced convection. The two can, of course, occur together in the same fluid.

15.4.3.3 Radiation Any body above the absolute zero of temperature emits electromagnetic radiation and absorbs part of the radiation from other bodies falling on it. The 'thermal' radiation which results in heat transfer lies in the infra-red and visible region of the spectrum between 0·8 and 400 microns, and for most purposes nothing above 25 microns is important. The emission of visible radiation, of course, is shown by the object becoming red or white hot. A black body is defined as one which absorbs all wavelengths of radiation falling upon it. No body is ever completely black in practice, but many approach acceptably close to the ideal. The total amount of radiation Q emitted by a body of area A and absolute temperature T is given by a modification of the Stefan–Boltzmann Law

$$Q = \epsilon b A T^4$$

where b is the Stefan-Boltzmann constant (0.174×10^{-8} Btu/hour/sq. ft/°F^4 or 5.7×10^{-8} Joules. m^{-2} sec^{-1} °K^{-4} and a factor ϵ called the emissivity has been added to account for deviations from black body behaviour. The emissivities of polished metals are as low as 0·03 to 0·08: those of oxidized metals, bricks and painted surfaces range from 0·60 to 0·95.

Radiation is the only kind of heat transfer possible across a vacuum so that, apart from conduction through the stopper and the point where the walls are joined, a vacuum flask cools entirely by radiation. In order to decrease heat losses from this source, the walls of the flask are silvered and their emissivity thus reduced.

Calculations of radiative heat loss must take account of heat radiated back to a body from its surroundings, though the occurrence of T^4 in the Stefan–Boltzmann law shows that for a reasonable temperature difference, the radiation back to a body from its surroundings is insignificant.

Radiation heat losses are comparable with those due to conduction and convection when a hot surface borders on a material transparent to electromagnetic radiation. For example, a steam radiator or hot pipeline in a room loses heat more or less equally by radiation and conduction/convection. On the other hand, a lagged pipe will lose scarcely any heat by radiation since its surface is only slightly above ambient temperature.

15.4.3.4 Heat exchangers The simplest form of heat exchanger is the Liebig condenser (Fig. 15.3a). The hot fluid flows through the central tube and is cooled by water flowing through the surrounding jacket. In a counter-current heat exchanger, the coolant enters at the end of the tube where the fluid is coolest and emerges where it is hottest. Fig. 15.3b shows a temperature profile along the heat exchanger.

This is not always the optimum system, and sometimes it is useful to reverse the direction of flow of the water so that it travels in the same direction as the fluid to be cooled. The system is then called a parallel current heat exchanger. The temperature difference between fluid and water is very large in the early stages and decreases continuously. Figure 15.3c shows the sort of temperature profiles obtained.

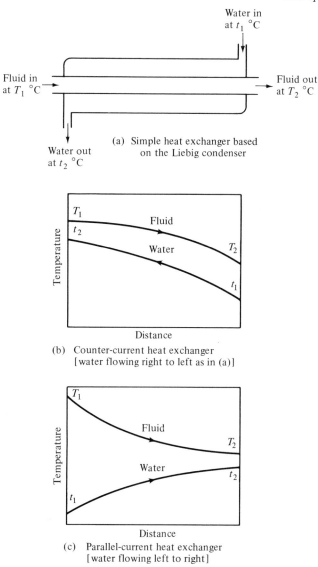

Figure 15.3 Counter and parallel-current heat exchange.

In the case of counter-current flow, the temperature difference between fluid and water is more nearly constant along the length of the tube and all the surfaces are therefore transferring similar amounts of heat. Also, the final temperature of the fluid can be lower than the exit temperature of the water, so that a larger proportion of the heat content of the hot fluid can be removed for a given initial temperature of water. In the case of parallel current flow, the exit temperature of fluid cannot be greater than that of the water, and the effectiveness of unit area of surface at the fluid entrance is much greater than a similar area near its exit.

Counter-current flow is therefore to be preferred in most applications, the main exception being the case where a reaction has to be quenched. The initial rapid cooling provided by parallel-current flow is then a great advantage.

The problem now arises as to how this simple heat exchanger is to be scaled up to an industrial level. It will be assumed, as it has been up till now, that the heat exchanger is required to cool a fluid and that water is the coolant, though neither of these assumptions has any particular significance. If a giant Liebig condenser were built, it would not work very efficiently because whereas the fluid throughput, water consumption etc. increase with the cube of the linear dimensions, the cooling area only increases with their square. To put this in another way, the enlarged condenser would be inefficient because a central core of fluid would travel through the tube without being cooled and the cooling water near the edges of the jacket would travel through without doing any cooling. The efficiency of such a heat exchanger could be increased by increase in the temperature difference, the area available for heat transfer or the heat transfer coefficient. The first of these is usually decided by external circumstances, e.g. it would be difficult to have the cooling water at less than ambient temperature. Increase of cooling surface area leads to increased capital outlay but this may be the best solution. Increase in heat transfer coefficient may be brought about by use of more water (which causes turbulence and will also increase the temperature difference at the exit end of the heat exchanger) and by employment of materials of construction with high thermal conductivities. The choice is the inevitable one between solutions involving high capital outlay (large surface area for heat exchange, cooling surfaces made of copper) and high running costs (increase in the flow of water).

Figure 15.4 shows the simplest heat exchanger, a pipe still, with a flame playing on a pipe. Figure 15.5 shows a typical single pass tubular heat exchanger, designed in this case for heating a liquid with steam. It consists of a bundle of tubes, heated by steam around them, and through which the liquid flows. The whole is contained in a cylindrical case. The arrangement of the tubes depends on how often they will need to be cleaned (Fig. 15.6). The system permits very large heat exchange areas in a relatively small volume. Its drawback is that the tubes have to be long and fairly robust. This means that their wall thickness is large and a lot of space is wasted. One possible solution is the multi-pass heat exchanger in which there are many more tubes which are shorter than those in the single pass exchanger. Systems of baffles ensure that liquid flows up one set of tubes,

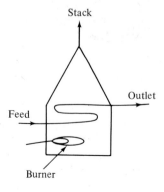

Figure 15.4 Pipe still.

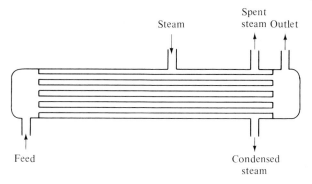

Figure 15.5 Single-pass tubular heat exchanger.

back down another, up a third, down a fourth and so on. The multipass exchanger is more complicated than the single pass device and pumping costs are higher. For equivalent performance it is smaller so that pumping costs have to be set against construction costs.

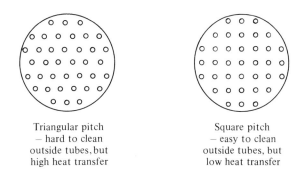

Triangular pitch
– hard to clean
outside tubes, but
high heat transfer

Square pitch
– easy to clean
outside tubes, but
low heat transfer

Dirty fluid is usually passed tube side as tubes are easier to clean than the shell.

Figure 15.6 Arrangement of tubes.

If a liquid is to be used as the cooling medium, it is important that its flow be turbulent (to increase the heat transfer coefficient) and its path length as long as is consistent with reasonable pumping costs. Figure 15.7 shows a multipass heat exchanger with baffles to ensure mixing of the cooling liquid.

Frequently it is economically advantageous to re-use heat to warm up reactants etc. If a reaction proceeds at 200°C, it is wasteful to have to warm up the reactants from ambient temperature and then cool the products, and a system as shown in Fig. 15.8 can cut heating costs dramatically. An arrangement of this kind would be used on our eventual vinyl acetate reactor and, indeed, each distillation column would be rigged up so that heat given up by liquids condensing was used to warm other liquids which it was desired to vaporize.

416   *The scaling-up of laboratory experiments*

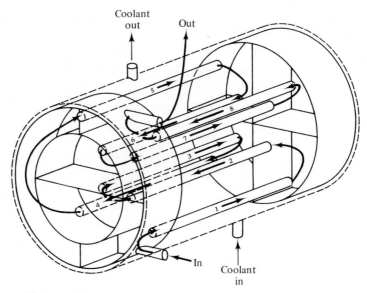

Figure 15.7 Multipass heat exchanger with baffles. Two distributing heads, divided by baffles into five (near end) and four (far end) compartments are linked by numerous short tubes. For simplicity only one tube per pass is shown. The numbers and arrows show the sequence of the passes and direction of fluid flow. The whole is immersed in a cooling jacket (heavy dotted lines).

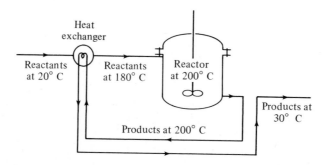

Figure 15.8 Use of heat in products to warm reactants.

15.4.4 Mass transfer and distillation

Most mass transfer and a number of solids handling operations in the chemical industry are concerned with the separation and purification of the products of a reaction. The importance of separation processes must be emphasized. A miracle catalyst which simultaneously produced every important industrial chemical directly from a single petroleum feedstock would be valueless because the problem of separating the products would be intractable.

In a given situation, such as the purification of vinyl acetate from our reactor, the various methods of separation such as solvent extraction, gas absorption and distillation are considered from the technical and economic standpoints. In the majority of cases distillation does the job better and more cheaply than the other methods and is thus the most widely used industrial separation process.

15.4.4.1 Application of distillation theory Distillation is the separation by vaporization of a homogeneous liquid mixture of volatile substances into individual components or groups of components. The design of equipment is dictated by the degree of purity of products which is required. Table 15.3 (page 420) gives three examples. The more stringent the specification the more expensive it becomes to achieve it.

In the laboratory the commonest form of distillation is where a mixture is placed in a flask and heated, and the vapour (which is richer in the more volatile component) condensed and collected (Fig. 15.9a). The degree of separation achieved is low and the amount of material which can be handled is small.

In 'equilibrium' or 'flash' distillation, the flask is fed continuously with mixture, distillate and residue are withdrawn continuously and a coarse separation is achieved (Fig. 15.9b).

Figure 15.10 is a conventional boiling-point diagram for a mixture of benzene and toluene and it can be seen that equilibrium distillation of a mixture containing 50% benzene cannot give a distillate containing more than 72% benzene. The distillate from a single flask could, however, be led to another flask and the distillate from that to another flask and so on (Fig. 15.9c). Each successive distillation would give a purer product. Similarly the bottom products would descend the chain of flasks and ultimately, fairly pure benzene and toluene could be recovered respectively from the top and bottom of the chain.

This method is not used in practice as it is expensive in terms of heat and equipment. Instead, a fractionating column is used in which reflux (or rectification) achieves the above idea more cheaply. A fractionating column may be visualized as a number of equilibrium distillations arranged in series. The liquid which has condensed at the top of the column flows back down again (the reflux) and is met by vapour of different composition coming up. The column may be thought of as divided into zones (labelled 0, 1, 2, ... M ... N in Fig. 15.9d). Each zone corresponds to a single equilibrium distillation, i.e. the liquid emerging from the bottom of it is in thermodynamic equilibrium with the vapour emerging from the top. Such zones are called theoretical plates and one theoretical plate is thus equivalent to one equilibrium distillation. The bottom stage (zone 0) is heated, and each stage has two feeds, a liquid and a vapour stream. The material to be separated is inserted as a third feed to zone M. In each zone, the upward flowing vapour is brought into intimate contact with the downward flowing liquid so that the streams leaving each zone are in thermodynamic equilibrium. All the zones above the feed zone are called the rectifying section and all those below, including the feed zone itself are called the stripping section.

The theory of design of a fractionating column is relatively complicated but the following example illustrates how it is done in practice in a straightforward case.

Suppose we wish to separate a mixture of equal mole fractions of benzene and toluene (mole fraction of benzene in feed, $x_f = 0.5$) to give a benzene product 95% pure ($x_D = 0.95$) and a toluene product 90.5% pure $x_w = 0.095$). We must first construct a boiling point diagram or equilibrium curve on the basis of laboratory measurements in which

418   *The scaling-up of laboratory experiments*

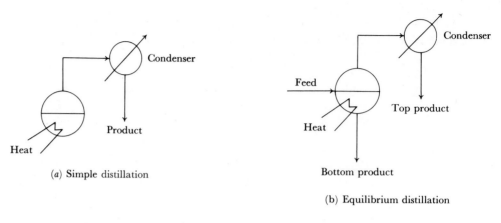

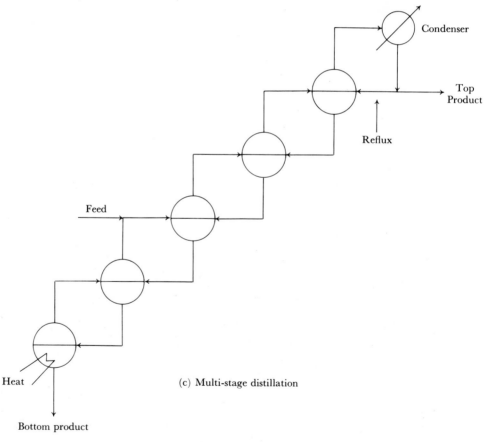

Figure 15.9 Simple, equilibrium, multi-stage and fractional distillation.

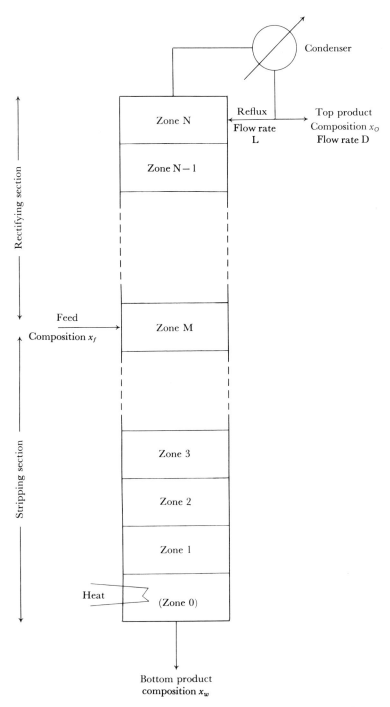

(d) Each of the sections marked on the column corresponds to a single equilibrium distillation i.e. a single theoretical plate.

420   The scaling-up of laboratory experiments

Table 15.3   Application of distillation

(a) Both products required fairly pure

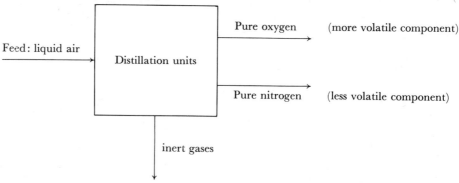

(b) Only one product required pure

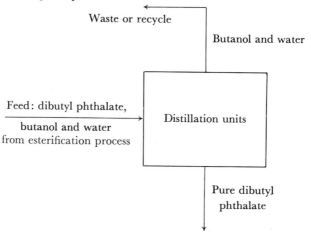

(c) Mixtures defined by boiling point range obtained rather than pure products

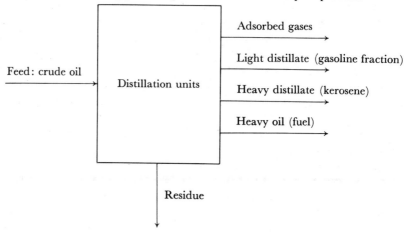

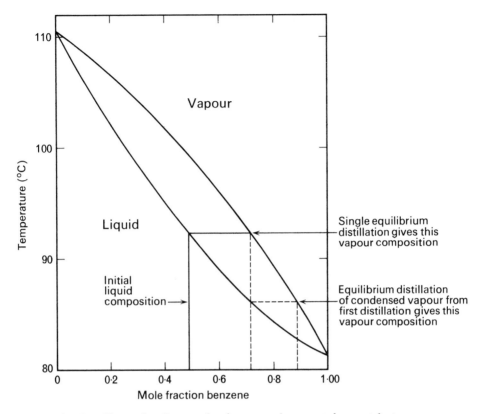

Figure 15.10 Boiling point diagram for the system benzene-toluene at 1 atm.

the mole fraction of benzene in the liquid (x) is plotted against the mole fraction of benzene in the vapour in equilibrium with it (y) (Fig. 15.11a).

We then choose a reflux ratio. This is the ratio of the amount of distillate returned down the column to the amount retained as product, and at this stage any reasonable value, say $R = 9:1$ is satisfactory. We calculate the value of $x_D/(R + 1)$ and insert a point this far up on the y axis. We draw in the point (x_D, x_D) and join these two points by a line called the rectifying line. We also construct the feed line, $x = x_f$ (Fig. 15.11b).

The point (x_w, x_w) is located and a line called the stripping line drawn between this point and the point of intersection of the feed line and the rectifying line (Fig. 15.11c).

Starting at the top right corner of the diagram, the constructions are carried out which give the compositions of the liquids and vapours on the top (Nth), one from the top ($N - 1$th) plates etc. (Fig. 15.11d).

The construction is continued until x_w is reached with the proviso that once values of x which are less than x_f are reached, the stripping rather than the rectifying line is used (Fig. 15.11e). This condition gives optimum operation for the column.

The number of theoretical plates required for the separation can now be counted, each step which touches the equilibrium line counting as one. In the example given,

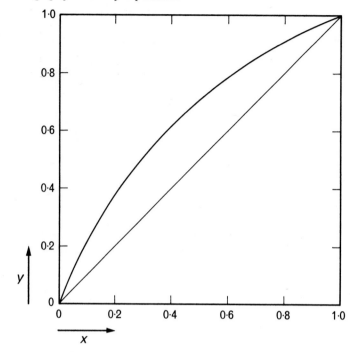

Figure 15.11(a) Construction of equilibrium curve.

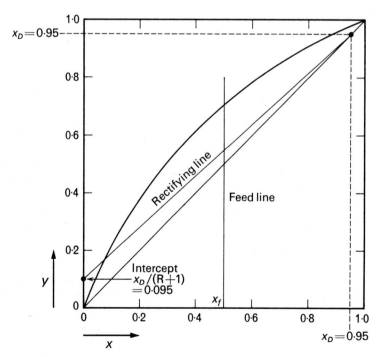

Figure 15.11(b) Insertion of feed line and rectifying line.

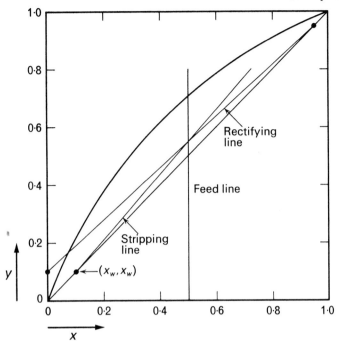

Figure 15.11(c) Insertion of stripping line.

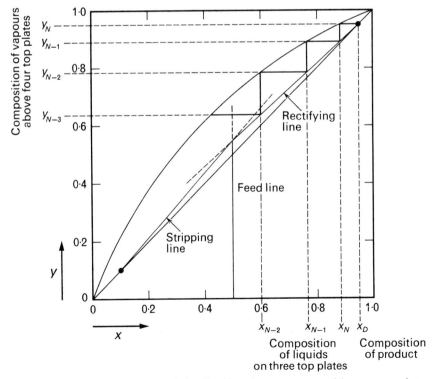

Figure 15.11(d) Construction giving liquid and vapour compositions on top plates.

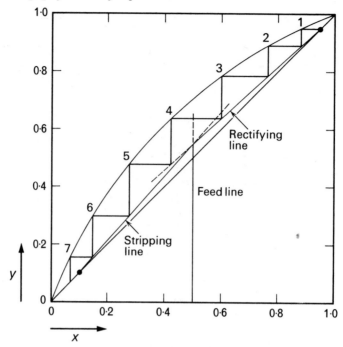

Figure 15.11(e) Transfer to stripping line and completion of construction to the point at which the bottom product is sufficiently pure.

seven theoretical plates are required in addition to the boiler. It can also be seen that the feed should enter on the fourth plate.

15.4.4.2 Variation of reflux ratio If the reflux ratio R is increased, the intercept of the rectifying line on the y-axis $(x_D/(R + 1))$ becomes smaller and in the limiting case of total reflux $(R = \infty)$ the intercept is zero. The operating lines become co-linear and coincide with the diagonal. Under these conditions, the number of plates required to affect separation is a minimum but the allowable offtake is zero (Fig. 15.12a). This case, involving an infinite reflux ratio and a minimum number of plates, is one extreme in the operation of fractionating columns. Its opposite is that of a minimum reflux ratio R_{min} which gives a maximum possible offtake, but separation requires an infinite number of plates. This occurs when either or both of the operating lines touches the equilibrium curve. The situation is represented in Fig. 15.12b and it can be shown that $R_{min} = (x_D - y_f)/(y_f - x_f)$. These quantities can be obtained by measurement of Fig. 15.12b and the value of R eventually chosen must be between R_{min} and total reflux.

15.4.4.3 Optimization of reflux ratio If the fractionation column is operated at just less than total reflux, a column of very large cross-section will be required to obtain a given throughput of material, but the number of plates required will be close to the minimum. If on the other hand the reflux ratio is just greater than the minimum value, the throughput will be high and the column can therefore be relatively narrow. The number of plates required and therefore the height of the column will, however, be very large. The precise variation with reflux rate of number of plates required for separation, and cross-section of column required for a given offtake D depends on the shape of the

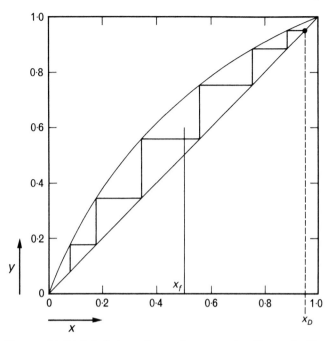

Figure 15.12(a) Infinite reflux ratio – operating lines coincide with diagonal – minimum number of plates required for separation (cf. Fig. 15.11).

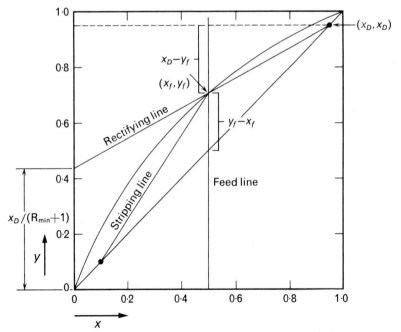

Figure 15.12(b) Minimum reflux ratio – infinite number of plates required for separation.

equilibrium curve, but to a first approximation the cross-section required (A) will be proportional to the liquid flow L down the column, i.e.

$$A = kL = kDR \quad (k \text{ is a constant of proportionality}; R = L/D \text{ by definition})$$

therefore for a given offtake, A is directly proportional to R (Fig. 15.13).

The height of column required will be roughly proportional to the number of plates. At infinite reflux, the number of plates approaches a minimum value, P_{min}, and at R_{min} the number of plates required is infinite. The graph of P vs R will therefore resemble the curve shown in Fig. 15.13 and will approach P_{min} and R_{min} asymptotically.

The capital cost of building a column can be obtained by asking contractors for estimates; suppose that it is proportional to the volume of the column, i.e. the product of the cross-sectional area and the number of plates. Figure 15.13 shows the product of the curves of A and P, and in Fig. 15.14 the same curve represents the fixed cost of the column.

In addition to the capital cost of building the column, there are the variable costs of operating it. The column requires labour and maintenance which are independent of reflux ratio, plus heat and cooling water both of which are used in smaller quantities as the reflux ratio decreases, but which never drop to zero. Suppose they are represented by the variable cost line in Fig. 15.14. The total cost will be the sum of the fixed and variable costs and from it the optimum reflux ratio can be found. It will always be slightly less than the optimum reflux ratio if fixed cost alone is considered. A convenient rule of thumb is that it tends to be about $1 \cdot 5 \, R_{min}$ but, of course, this is only an order of magnitude figure.

Once the optimum reflux ratio is known and the offtake required has been specified, all the other variables can be calculated and the column designed and built.

The problem of designing a distillation column has been considered in some detail as it is typical of the optimization problems which face the chemist in industry. In every case the problem is to design equipment which will just succeed in carrying out the required operation (with a slight safety margin to be on the safe side) at minimum cost.

15.4.4.4 Types of fractionating column   Figure 15.15 shows some designs of fractionating columns used in the laboratory and in industry. The design of the laboratory columns and the packed column (which is used on all scales from laboratory, through pilot plant up to industrial) hinders the flow of liquid back down the column and provides a large surface area, so that the ascending vapour comes into contact with as large an area of liquid as possible. Both of these factors promote the attainment of equilibrium between liquid and vapour.

Packed columns are usually filled with Raschig rings. These are small cylinders open at both ends and with a height equal to their diameter. They vary from about 5 to 75 mm in size, and are made from ceramics, metals or plastics.

The bubble cap and sieve plate columns operate differently. The columns themselves are divided by actual plates on each of which a depth of liquid rests. Vapour from the preceding plate is blown through the liquid and comes into equilibrium with it. Liquid flows over a weir and down a pipe to reach the plate beneath. In a typical column, the plates are set about 50 cm apart. The distance is governed by the amount of turbulent liquid, foam and mist above each plate, and space must be allowed for the de-entrainment of droplets from the ascending vapour phase.

In the bubble cap column the vapour ascends through 'chimneys' which are covered by bubble caps which force the vapour into the liquid. In the sieve plate column, the

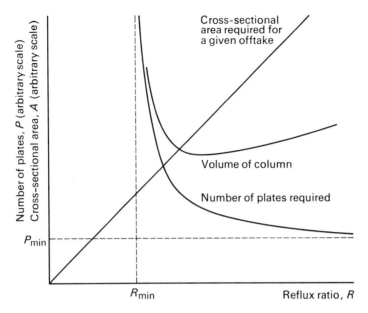

Figure 15.13 Variation of column dimensions with reflux ratio.

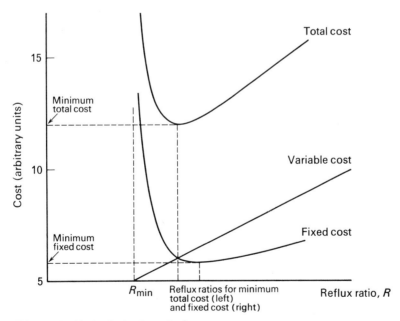

Figure 15.14 Optimization of column costs.

vapour ascends through 5 to 15 mm holes in the plate forming small bubbles. Except at exceptionally low flow rates the velocity of the vapour is sufficient to prevent the liquid from leaking back through the 'sieve' holes.

428 *The scaling-up of laboratory experiments*

Sieve plate columns are at present favoured by most large chemical companies for large-scale distillation, and it appears that they are normally the most economic. They have to be carefully designed and used, however, otherwise the advantage is lost. Bubble cap columns have the advantage that they are easy to design and to operate in that even if the vapour flow is interrupted, the liquid remains on the various plates instead of flowing down the column. Design information on sieve plate columns is scarce and expensive and a manufacturer who wants a margin for error may well instal bubble cap columns with the money saved on a design consultant.

Packed columns are of use in highly corrosive environments in that it is much cheaper to replace Raschig rings than sieve plates, and in any case ceramic packing will be attacked less easily than steel plates. The chief disadvantages of packed columns are that

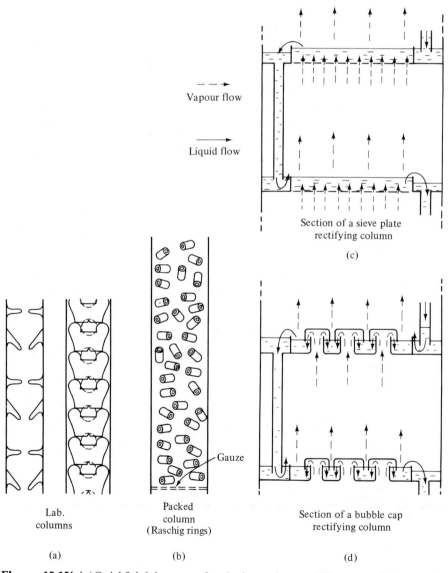

Figure 15.15(a) 'Quickfit' laboratory fractioning columns. (Vigreux and Pear bulb columns). **(b)** Packed column (Raschig rings). **(c)** Section of a sieve plate rectifying column. **(d)** Section of a bubble cap rectifying column.

the descending liquid tends to flow down the walls of the column instead of over the packing, and the weight of the packing limits the size of the column. Very few packed columns of more than 3 ft in diameter are used. Use of plastic packings reduces the weight problem but most plastics tend to soften above 100°C so their scope is limited.

It is fair to conclude that packed columns are for medium scale distillations, particularly under corrosive conditions, that sieve plates provide optimum efficiency and that bubble caps are for those who prefer to play safe. Our scaled-up vinyl acetate plant will involve ten columns each requiring separate calculations and designs.

15.5 Fire and explosion hazards

In our example, we need to know what concentrations of ethylene in air are explosive. In fact, explosive mixtures contain 3–28% by volume of ethylene; these limits are wider if pure oxygen is used instead of air. Furthermore, rust can catalyse combustion. We must either avoid explosive concentrations of gases or sources of ignition or both. In either case maximum safeguards must be provided for operatives. At all chemical plant, employees and visitors are required to surrender lighters and matches at the main gate. Fire hazards of materials can be gauged by knowledge of their flash points, that is the temperature above which they will burn if ignited. Such values are tabulated in technical literature. Related dangers include explosive polymerization of monomers, and handling of materials such as explosives which are dangerous in themselves.

15.6 Toxicity and related hazards

Many chemicals burn the skin or are poisonous or even carcinogenic. Plant operatives are exposed to materials in larger quantities and for longer periods than laboratory workers and it is even more important that they should take precautions. Concentrated acids and alkalies can cause burns, dusty materials can irritate the lungs, dermatitis arises mainly among operatives handling epoxy resins, chromium compounds and rubber additives, and so on. The toxicities of most industrial chemicals have been thoroughly investigated and are listed in Sax's *Dangerous Properties of Industrial Materials*.

15.7 Corrosion and materials of construction

Laboratory processes are carried out in glass equipment. The use of glassware on a substantially larger scale as a primary material of construction would be both expensive and impractical. Even if glass equipment of the appropriate size could be built, it would be liable to break from thermal or mechanical shock. (The use of glass as a secondary material in the chemical industry is, however, increasing. It is used as piping or as a coating on cast iron or steel or as glass fibre in insulation etc.). One seeks the cheapest material – cheapest over the lifetime of the plant – that will do the job properly, taking any scrap value into account. Mild steel is cheap, strong, easily fabricated and has a high thermal conductivity and is used wherever possible. It is, however, liable to corrosion and other materials frequently have to be used instead. A list of the more important ones is given in Table 15.4. Often, it is not possible to avoid corrosion completely (as it usually is with glass) and the industrial chemist may well have to plan to replace various of his reactors and columns after a certain time. This may be acceptable so long as the corrosion products do not appear in the final product at an objectionable level.

Table 15.4 Materials of construction

Material	Use	Corroded by	Resistant to
Mild steel	General purpose construction. Cheap. Used wherever feasible. Alloy steels give improved high and low temperature properties	Dilute acids, moisture in the air (rusting!)	Cool conc H_2SO_4, alkali if not concentrated, neutral organic liquids.
Stainless steel	General purpose construction. Three times as expensive as mild steel	Acids under reducing conditions, chloride ion (stress cracking)	Organic acids and nitric acid. Good for highly oxidizing conditions.
Copper	Easily fabricated but very expensive and tends to colour products. Alloys obsolescent	Ammonia Amines Mercury	Alkali (better than steel), organic acids. Good for reducing conditions and low temperatures.
Aluminium	Light, easily fabricated	Halogen acids or alkalis	Strong organic acids, nitric acid and nitrates.
Nickel	Especially used in food industry (Monel metal is 67% Ni 30% Cu)		
Lead	Poor mechanical properties. Used in Pb chamber process. Obsolescent		Dilute sulphuric acid.
Titanium	Usually coated on to mild steel to give a CLAD STEEL. Other clad steels have other desirable properties without the expense of using the pure metal		All strengths of acetic acid under reducing conditions.
Paint	Very cheap; affords protection (e.g. against rusting) in some circumstances		
Wood	Cheap; has some uses e.g. barrels	Biological activity – dry rot, woodworm etc.	
Glass	Piping, linings for other materials and, as fibre, for insulation etc	Hydrofluoric acid	Almost everything.
Ceramics (stoneware porcelain)	For strong acids		Acids.
Concrete	Storage tanks	Acids	
Rubber and plastics.	Hoses and gaskets, linings for drums etc		Hydrochloric acid, hydrofluoric acid and non-aqueous solvents.
Earth	Frozen holes are used for methane storage		

15.8 Instrumentation and control

Another important difference between industrial and laboratory practice is the area of instrumentation and control. A laboratory chemist is able to watch his various thermometers at frequent intervals, observe his reaction mixture in its glass apparatus and make continual adjustments to his rate of heating to take account of what he sees. These adjustments are all the more effective in that his apparatus will have a fairly small thermal capacity and changes in heating rate will have an immediate effect. In large scale plant, on the other hand, the reactors, columns etc. may be widely separated, the reaction mixture and products will be out of sight and the system will have a high thermal capacity so that quick adjustments are out of the question. In addition, the labour force as a whole will be less well trained than a laboratory chemist and the scale of the operation will mean much greater danger should any process get out of hand. The way in which these problems are tackled has changed over the years in a way which reflects both the increasing cost of labour and the increasing sophistication of measuring and feedback devices.

Traditionally, measuring instruments – thermometers, flowmeters etc. – were attached to the pieces of apparatus with which they were concerned. A number of workmen would walk around examining the various meters and making any corrections necessary on nearby valves or other controls. Particularly important valves would have a workman to themselves. A well-trained group of workmen could operate plant quite efficiently in this way, but it was difficult for one man to know everything that was happening at a given moment and therefore coordination was difficult.

The first modification of this was the establishment of a control room. Plant was so designed that the various pipes all passed through the control room. Thermometers, flowmeters etc. could be attached to them and a single operator could see at a glance many of the variables that affected the working of the plant. Many of the valves were also in the control room so he could also exercise a measure of control over the process. The method suffered from the disadvantage that additional piping was required to enable all equipment to be monitored from the control room. Greater lengths of piping meant higher pressure drops along their lengths and hence greater pumping costs. The method was not applicable to really large plant because the piping involved would not fit into a control room of acceptable size.

The most modern method still involves a control room, but temperatures and flows are measured by instruments attached to the process equipment and the results transmitted to the control room either pneumatically or electrically. Corrections to these variables are usually made entirely automatically by a feedback system. The operator is able to see a continuous record of every important variable and to know that this will be held automatically at any specified value. If the automatic system is unable to cope with a particular situation, a system of klaxons or flashing lights will inform the operator who can take appropriate action. As a rule, the system is fail-safe, that is in a real emergency the plant is automatically closed down.

15.9 The scaled-up plant

Figure 15.16 shows a proposed flow-sheet for making vinyl acetate. It consists of a titanium-lined pressure reactor and ten distillation columns. The most numerous items

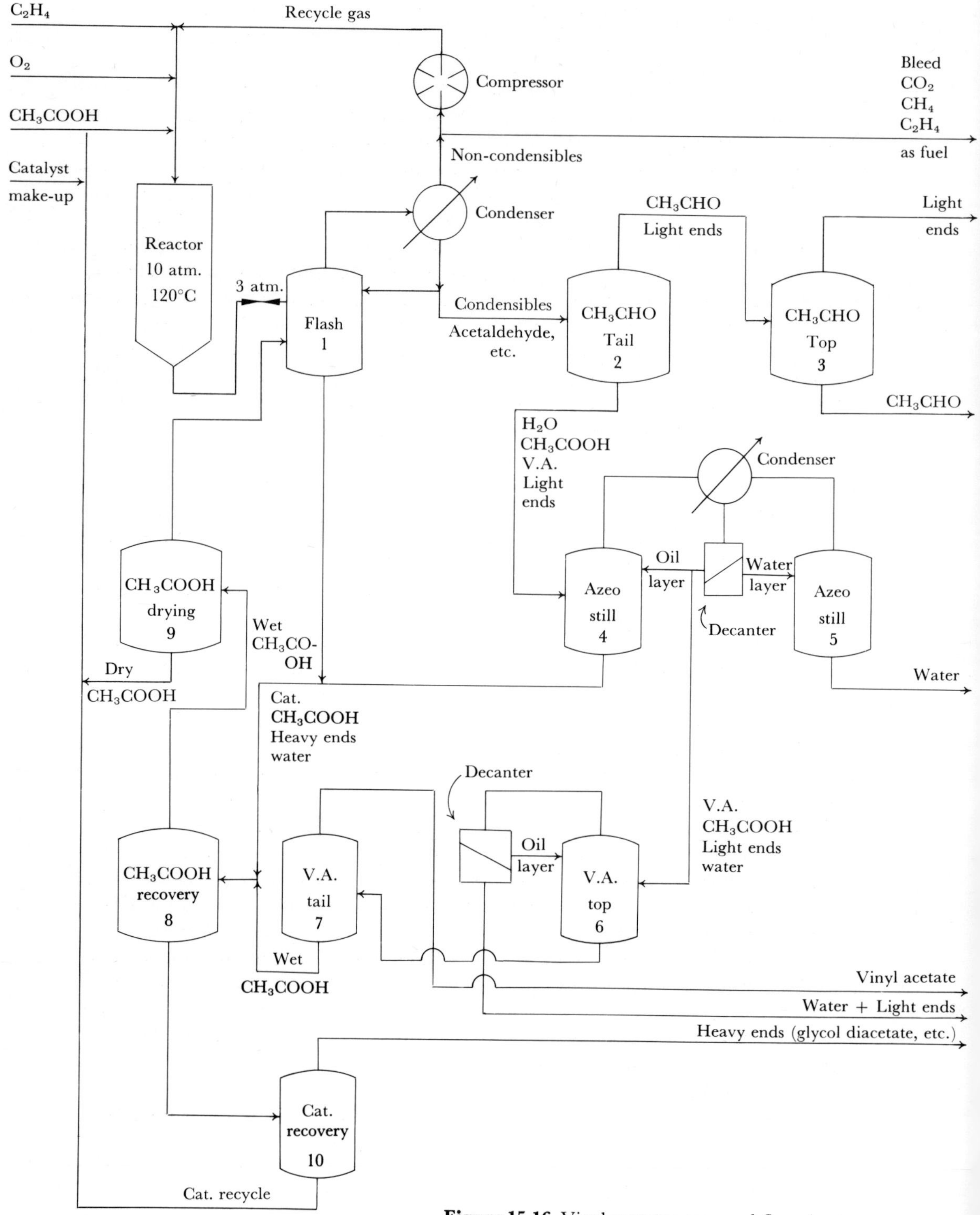

Figure 15.16 Vinyl acetate: proposed flow sheet.

on the plant, pumps and heat exchangers, have been omitted from the diagram to avoid over-complicating it. Almost all the columns come into contact with acetic acid and must be titanium lined.

Ethylene, oxygen and glacial acetic acid are mixed and fed into the reactor at 120°C and 10 atmospheres. About ten times as much ethylene as oxygen is used and this sweeps the volatile products from the reaction zone. Reaction takes place while the materials pass through, and the effluent stream contains vinyl acetate (boiling point at 1 atmosphere 72·7°C), acetaldehyde (20·9°C), acetic acid (118·1°C), ethylidene diacetate $CH_3CH(O \cdot OC \cdot CH_3)_2$ (168°C), ethylene glycol diacetate (190·5°C), ethylene, water, methane, butane, carbon dioxide and minor glycol, glycol acetate and aldehyde impurities. The pressure is reduced to 3 atmospheres by flash distillation which also brings about a crude separation of the product mixture. The pressure is not let down to atmospheric immediately because if we run the acetaldehyde 'topping' and 'tailing' stills, 2 and 3, at 3 atmospheres, the boiling point of acetaldehyde is raised sufficiently for the condensers to operate on ordinary cold water instead of requiring refrigerated water. Subsequent columns can be operated at atmospheric pressure. Returning to the flash distillation column, 1, its bottom product contains the catalyst system, untreated acetic acid, high boiling glycol acetates etc. ('heavy ends') and water. At the condenser on top of the column, the vinyl acetate, acetaldehyde, some acetic acid and water and the more volatile impurities ('light ends') condense, while the gases (unreacted ethylene, carbon dioxide, methane and oxygen) pass straight on and may be alkali-scrubbed to remove carbon dioxide before being compressed and recycled. To prevent build-up of gases such as methane in the recycle stream, a proportion is bled off and burnt as fuel.

The products which condensed at the top of the column pass to column 2, an acetaldehyde 'tailing' column which separates acetaldehyde plus some light ends from vinyl acetate, water, acetic acid and the rest of the light ends. The distillate passes to column 3 where the light ends are separated from the acetaldehyde and both are pumped to storage. The bottom product from column 2, meanwhile, flows to two azeotroping stills, 4 and 5. Column 4 gives much of the acetic acid and some water at the bottom and a 92·7% vinyl acetate/7·3% water azeotrope (b.p. 66°C) is taken overhead where it condenses to give two layers, the oil layer containing 98·7% vinyl acetate/1·03% water and the aqueous layer containing 2·0% vinyl acetate/98·0% water. The oil layer is decanted off and the aqueous layer flows to column 5 where it is separated into water, which is discarded, and the 7·3% water azeotrope which is returned to the decanter. Most of the water can be removed from the mixture in this way and it is a useful general technique for azeotropes.

Some of the oil layer from the decanter provides reflux down column 4 while the rest, which has lost most but not all of the water, acetic acid and light ends formerly present, flows to column 6, the vinyl acetate 'topping' column. Here the last traces of water are carried overhead as azeotrope and when it condenses, the oil layer is decanted and refluxed while the small amount of aqueous layer which contains 2·0% vinyl acetate plus the remaining light ends is discarded.

The bottom product from column 6 is vinyl acetate from which all the more volatile impurities have been removed, that is it has been 'topped'. It now flows to column 7 where the last traces of water and acetic acid are removed as tails, and the pure vinyl acetate distils over and is pumped to storage.

Meanwhile, the bottom products from columns 1, 4 and 9 remain untreated and contain sizeable quantities of acetic acid, catalyst and heavy ends. The three streams are united and flow to column 8 where wet acetic acid distils over and is dried on column 9. The dry acetic acid is recycled to the reactor while the wet distillate goes back to the bottom of column 1. The bottom product from column 8 goes through a further separation process which divides the catalyst from the heavy ends. Some catalyst is inevitably lost during the process so some 'make-up' catalyst is added to the recovered material before it is recycled to the reactor. Some hydrogen chloride is also added at this stage because experience has shown that chloride is lost somewhere though no chlorinated products have been identified.

The overall yield on plant of this type is very high and the major by-product, acetaldehyde, is saleable, as is ethylene glycol diacetate. Any unsaleable light or heavy ends can be burned, so the only true effluents are the virtually pure water emerging from column 3 and the small amount of 2% aqueous vinyl acetate from column 6.

Notes and references

General The design of chemicals manufacturing plant falls in the province of chemical engineering but is not customarily discussed in terms of scaling up. Chemical engineering texts dealing with individual aspects of plant design will be cited below. The best way to get an overall view of chemical plant is from the excellent series of process surveys in *Chemical and Process Engineering*.

Details of commercial equipment for large-scale chemistry, of process development and of factory organization are given by J. Manning *An Introduction to Chemical Industry*, Pergamon, London, 1965. J. Davidson Pratt and T. F. West *Services for the Chemical Industry* Pergamon, London, 1968, present a series of articles on neglected topics such as water supplies, steam supplies, corrosion, plant maintenance, patents etc.

Section 15.1 J. Heppel, *Chemical Process Economics* Wiley, New York, 1958 contains many useful rules of thumb for costing chemical plant, which can fairly readily be brought up-to-date.

Section 15.3 The following are convenient sources of thermodynamic data:

Thermochemistry of Organic and Organometallic Compounds, J. D. Cox and G. Pilcher, Academic Press, London, 1970.

Selected Values of Chemical Thermodynamic Properties, F. D. Rossini, D. D. Wagman, W. H. Evans, S. Levine, I. Jaffe, N.B.S. Circular 500, U.S. Government Printing Office, Washington 1952 and subsequent N.B.S. circulars in series 270 which will eventually replace it.

J. H. Perry (Ed.), *Chemical Engineer's Handbook*, 4th Ed. McGraw Hill, New York, 1963.
D. R. Stull and H. Prophet (Eds) *JANAF Thermochemical Tables*, 2nd ed., U.S. Dept. of Commerce, Washington, 1971.

E. S. Domalski, 'Selected values of heats of combustion and heats of formation of organic compounds containing the elements C, H, N, O, P and S, *J. Phys. Chem. Ref. Data* **1**, 222 (1972).

Reactor design is neither a unit process nor a unit operation and so is omitted from many chemical engineering textbooks. Three useful sources of information are S. M. Walas, *Reaction Kinetics for Chemical Engineers*, McGraw Hill, New York, 1959, H. Kramer and K. R. Westerterp, *Elements of Chemical Reactor Design and Operation*, Chapman and Hall, London, 1963 and the new, third volume of Coulson and Richardson (see below).

There is an excellent elementary treatment of reactor theory by D. A. Blackadder, J. Bridgewater and R. M. Nedderman, *Education in Chemistry*, **8**, 219 (1971).

Section 15.3.2 We have omitted all discussion of catalysts on the grounds that they are a form of applied witchcraft. A rational exposition is given by C. C. Bond, *Catalysis by Metals*, Academic Press, London, 1962, and in two multi-volume series *Catalysis* (ed. P. H. Emmett) Reinhold, New York, 1954 onwards and *Advances in Catalysis* Academic Press, New York, 1948 onwards. A wealth of practical detail is provided by C. L. Thomas, *Catalytic Processes and Proven Catalysts*, Academic Press, New York, 1970 and ICI's *Catalyst Handbook*, Wolfe, London, 1970, the latter coming from the Agricultural Division and dealing mainly with ammonia and hydrogen production. Delightfully written though somewhat academic is A. J. B. Robertson *Catalysis of Gas Reactions by Metals*, Logos, London, 1970, while more elaborate and mathematical treatments are given by J. M. Thomas and W. S. Thomas *Introduction to the Principles of Heterogeneous Catalysis*, Academic Press, New York, 1967, and C. N. Satterfield and T. K. Sherwood *Role of Diffusion in Catalysis*, Addison-Wesley, Massachusetts, 1963.

Section 15.4 Unit operations are the heart of chemical engineering, and distillation and heat transfer are bread-and-butter topics. The most popular undergraduate text appears to be W. L. Badger and J. T. Banchero, *Introduction to Chemical Engineering*, 2nd ed. McGraw-Hill, 1963, and we have drawn on it in our text. More comprehensive is J. M. Coulson and J. F. Richardson, *Chemical Engineering*, Pergamon, London. Volume 1 (1966) deals with fluid flow, heat transfer and mass transfer, volume 2 (2nd ed., 1968) with unit operations and volume 3 (1971) with reactor design, process control, computers and biochemical engineering. A concise, modern book is D. A. Blackadder and R. M. Nedderman, *A Handbook of Unit Operations*, Academic Press, London, 1971, and a practical approach is adopted in E. J. Henley and E. M. Rosen, *Material and Energy Balance Computations*, Wiley, New York, 1969.

Section 15.4.3 For a detailed discussion see D. Q. Kern, *Process Heat Transfer*, McGraw-Hill, New York, 1950 and W. H. McAdams, *Heat Transmission*, 3rd ed. McGraw-Hill, New York, 1954.

Section 15.4.4 Further references on distillation include R. E. Treybel, *Mass Transfer Operations*, 2nd ed., McGraw-Hill, New York, 1968, and R. J. Hengstebeck, *Distillation*, Reinhold, New York, 1961. Sources of vapour-liquid equilibrium data are listed by E. Hala, J. Pick, V. Fried and O. Vilim, *Vapour-Liquid Equilibrium* 2nd ed. Pergamon, London, 1967.

Section 15.6 The toxicity 'bible' is N. I. Sax, *Dangerous Properties of Industrial Materials*, 3rd ed. Reinhold, New York, 1968.

Section 15.9 This vinyl acetate process was described in *Chemical and Process Engineering*, March 1967 p. 71. I.C.I. closed their plant which was not titanium-lined because of corrosion problems and it is uncertain if the exact process is anywhere in operation.

16 Unit processes

The term 'unit operations' describes the unitary physical and mechanical changes brought about in the chemical industry; the unitary *chemical* changes are referred to as unit processes. They are of widely differing importance. Some, like oxidation, are used in a large number of ways. Others, like causticization, find application only in a single industry. Furthermore, the concept of unit processes is of uncertain value, for while all polymerizations or halogenations have something in common, there is only a tenuous relationship between the hydrogenation of natural oils to margarine and of nitrogen to ammonia. In this chapter, many reactions already discussed will be re-classified under unit process headings and others which have so far been overlooked will be discussed in more detail.

16.1 Oxidation

Oxidation is probably the most widely used process in the chemical industry. The availability of hydrocarbon feedstocks gives scope for many oxidation reactions, and the avail-

Table 16.1 Industrial consumption of oxygen, USA 1966

Chemical uses	*'000 tons/year*
Acetylene by partial combustion of methane	1 606
Ethylene oxide	389
Methyl alcohol	99
Acrolein and hydrogen peroxide	73
Titanium dioxide	45
Chlorine by HCl oxidation	11
Oxidation of liquefied petroleum gas	128
Other uses	388
Total chemical uses	2 639
Non chemical uses	
Steel production	6 420
Metal-working	576
Aerospace	422
Miscellaneous uses (including medical)	1 002
Total consumption	11 100
Unit value	$23.3/ton

ability of air means that a free oxidizing agent is available to carry them out. Few large tonnage processes use oxidizing agents other than air or oxygen though a wide variety of compounds (e.g. ozone, nitric acid, hypochlorous acid, potassium permanganate) find occasional application. Industrial consumption of oxygen is shown in Table 16.1; data for air do not exist.

Most oxidations are exothermic and it is usually important for heat to be removed efficiently otherwise the whole system might go to carbon dioxide and water. Catalysts tend to be metals or metal oxides. Examples of industrial oxidations brought about by air or oxygen are alcohols to aldehydes and ketones (sections 9.1.3, 9.2.1, 9.3). Ethylene to ethylene oxide (section 9.1), primary flash distillate to acetic acid (section 9.4), cumene to cumene hydroperoxide (section 9.5.1.1), naphthalene and benzene respectively to phthalic and maleic anhydrides (sections 9.5.4 and 9.5.1) and the Wacker process (sections 9.1.4 and 10.5.1).

Processes using oxidizing agents other than air or oxygen include the oxidation of propylene by hypochlorous acid (section 9.2.2), the oxidation of cyclohexanol/cyclohexanone with nitric acid (section 10.7.2) and the oxidation of allyl alcohol to glycerol by hydrogen peroxide (section 16.5).

Ozone, produced by a silent electrical discharge in air, is used to split the double bond in oleic acid (ozonolysis) to give a mixture of azelaic and pelargonic acids:

$$CH_3(CH_2)_7CH\!=\!CH(CH_2)_7COOH \xrightarrow{O_3}$$
Oleic acid

$$CH_3(CH_2)_7COOH + HOOC(CH_2)_7COOH$$
Pelargonic acid   Azelaic acid

Hypochlorous acid and hydrogen peroxide are used extensively for the bleaching of textiles and sodium perborate is incorporated in many detergents for a similar purpose.

16.2 Hydrogenation

Hydrogenation is not nearly so widely used a process as oxidation. Feedstocks susceptible to hydrogenation are not common, and hydrogenation usually requires high pressure so capital costs are high.

Hydrogen is an expensive material, but because it is used where it is made there are difficulties in attaching a realistic price to it. This is particularly true when it is formed as a by-product of petroleum cracking processes.

Hydrogen made from natural gas on the Gulf Coast in plant of 10–60 m. ft^3/day in 1969 was estimated to cost \$0.25–0.35/1 000 ft^3 (\$100–140/ton) which represents a lower limit. Purified hydrogen would cost three or four times this and pure electrolytic hydrogen twice as much again. This final figure is made up largely of electricity costs and would be lower in countries with cheap electricity. A notional value for hydrogen of about \$150/ton in the US and about £70/ton in the UK is probably a reasonable estimate for most purposes.

Until the Second World War, 80% of world hydrogen came from coal and coke (section 14.1.2). By 1961 this proportion had dropped to 40%, and another 40% was coming from refinery gases and natural gas and the remaining 20% from fuel oil and by

438   Unit processes

electrolysis of water. In the USA 75% of hydrogen came from natural gas and the world trend since 1961 has undoubtedly been away from the coal based route.

United States hydrogen production in 1968 is given in official statistics as 201 752 m. ft^3 (490 000 tons) but this figure excludes hydrogen used for ammonia, methyl alcohol, fuel, and in petroleum refineries. The actual production is probably of the order of 6 m. tons.

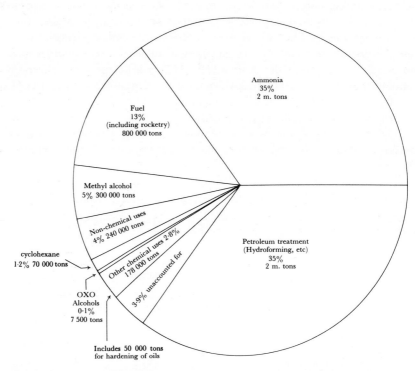

Figure 16.1 Hydrogen end-uses – USA 1969. Total production ~6 m. tons.

UK production is of the order of a million tons, but a mere 2 400 tons were actually sold in 1970, and, because of the expense of shipping, the price was in the region of £1 000/ton.

Hydrogenation catalysts are usually metals or metal oxides. Nickel, cobalt and iron catalysts together with molybdenum and tungsten oxides or sulphides tend to lead to vigorous hydrogenation, e.g. aldehydes are reduced beyond the alcohol to the hydrocarbon. Copper, zinc oxide, chromium oxide, manganese oxide, platinum, palladium and vanadium pentoxide are components of milder hydrogenation catalysts. The economic feasibility of hydrogenation depends on the amount of hydrogen required and the value of the final product. Margarine, for example, requires small amounts, synthetic gasoline very large amounts.

Industrially important hydrogenations (Fig. 16.1) include:
The Haber process (section 14.1.1), methyl alcohol from carbon monoxide (section 14.1.7), primary alcohols from aldehydes (section 16.3) and cyclohexane from benzene (section 9.5.1). The largest outlet for hydrogen excluding ammonia is the upgrading of

petroleum refinery products. Olefinic compounds are treated with hydrogen at high pressure to give a more fully saturated product. Good yields of gasoline and lubricating oils with a high viscosity index, low carbon residue on combustion and good resistance to oxidation are obtained. The process was used during the Second World War for the production of aviation fuel. It was insufficiently cheap for peace time use until the nineteen fifties when the development of catalytic reforming processes made large quantities of by-product hydrogen available. The necessity of using this on site made the improvement of petroleum products by hydrogenation an attractive possibility and it is now used not only for removing olefinic double bonds but also for removing sulphur compounds.

Hydrogenation is also used as a method of 'hardening' natural oils such as cocoanut oil to give margarine or other cooking fats. By this means, the triglycerides of unsaturated fatty acids are converted into triglycerides of saturated fatty acids, e.g. the triglyceride of oleic acid is converted into the triglyceride of stearic acid:

$$\begin{array}{l} H_2C-O.OC(CH_2)_7CH=CH(CH_2)_7CH_3 \\ | \\ H.C-O.OC(CH_2)_7CH=CH(CH_2)_7CH_3 + 3H_2 \\ | \\ H_2C-O.OC(CH_2)_7CH=CH(CH_2)_7CH_3 \end{array} \xrightarrow[\text{Ni Catalyst}]{100-250°C,\ 100\ \text{psi}} \begin{array}{l} H_2C-O.OC(CH_2)_{16}CH_3 \\ | \\ H.C-O.OC(CH_2)_{16}CH_3 \\ | \\ H_2C-O.OC(CH_2)_{16}CH_3 \end{array}$$

Double bonds which are not hydrogenated tend to isomerize. The natural, low-melting *cis* form of oleic acid gives high melting *trans* forms, known as isooleic acids. The process is economically attractive because complete hydrogenation of a ton of oleic acid requires only 7 kg hydrogen.

Hydrogenation of bituminous or brown coal to give gasoline is a process of considerable historical interest. It was discovered by Bergius in 1913 and the first commercial plant was opened in Germany in 1927. It was further developed by I.G. Farben and used in Germany during the Second World War (section 8.4.6). The process was not economically viable even then if wartime needs were discounted, and the relative prices of coal and oil have changed since 1945 to weight the scales even more heavily against coal-based petrol.

16.3 The OXO process

The Fischer Tropsch process (section 8.4.5) was another way of deriving petrol from coal and involved the reaction between carbon monoxide and hydrogen. In the OXO process, these two materials plus an olefine are passed over a cobalt catalyst to give the aldehyde one step up the homologous series, e.g.

$$C_5H_{11}CH=CH_2 + CO + H_2 \longrightarrow \underset{\underset{\text{Isooctaldehyde}}{\overset{|}{CHO}}}{C_5H_{11}CHCH_3} + \underset{\text{n-octaldehyde}}{C_5H_{11}CH_2CH_2CHO}$$

1-Heptene

Straight chain olefines give similar amounts of the two isomers; the position of the double bond is relatively unimportant.

Aldehydes obtained by the OXO process are usually hydrogenated to primary alcohols, many of which are used in plasticizers (Fig. 16.2).

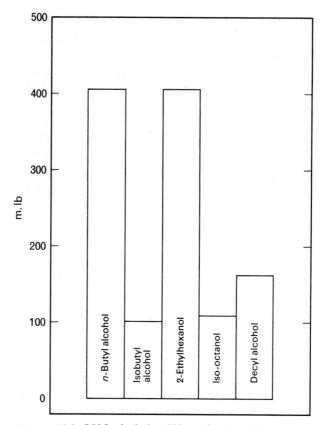

Figure 16.2 OXO alcohols – US production 1969.

16.4 Esterification

Esterification is a relatively simple process. Equipment for a batch preparation is cheap, the temperatures required are easily attainable and no excess pressure is necessary. In general, the acid is mixed in the reactor with excess alcohol and a sulphuric acid catalyst and the mixture heated:

$$ROH + R'COOH \longrightarrow R'COOR + H_2O$$

The reaction is reversible and the equilibrium is pushed over to the right by removal of the water which distils from the reaction vessel along with the excess alcohol which is recycled. Sometimes an additional entraining agent such as benzene is added to help removal of the water.

Once esterification is complete, the ester must be purified and decolorized. Volatile esters such as butyl acetate can be distilled, but many are involatile, especially plasticizer esters which are useful for this very reason, and these are known as residue products. They are treated variously by washing, by nitrogen stripping, in which a current of nitrogen is blown through the material to remove traces of excess alcohol, by potassium permanganate which improves the colour, and by vacuum distillation where a premium product is required. Careful control of mixing and temperature in the esterification kettle can often make most of these unnecessary.

Acetic anhydride is sometimes used in esterifications instead of acetic acid. Maleic and phthalic anhydrides are always used instead of their acids since they are more readily available and permit milder reaction conditions. Cellulose acetate, vinyl acetate and methyl methacrylate (sections 10.7.1, 10.5.1 and 10.2.4) are made by atypical methods.

Over 400 esters are commercially available, the most important conventional ones being shown in Fig. 16.3. Most esters are made on a medium scale and batch processes are widely used. It becomes economic to build a continuous unit if an output of more than about 1 000 tons/year is required.

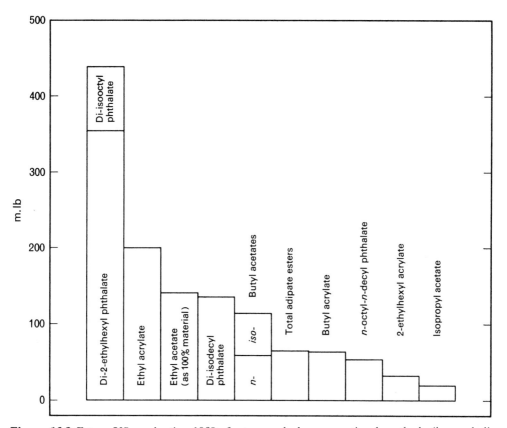

Figure 16.3 Esters. US production 1969 of esters made by conventional methods (i.e. excluding vinyl acetate, cellulose acetate and methyl methacrylate).

16.5 Hydrolysis and hydration

In its narrowest sense, hydrolysis is the opposite of esterification: in the presence of excess water and an esterification catalyst esters give acids and alcohols. The manufacture of soap from natural fats and oils (olive oil, cocoanut oil, inedible beef tallow from the stockyards) is an example of this, the raw materials being mainly the triglycerides of stearic, palmitic and oleic acids, and the products glycerol plus the fatty acids or their sodium salts. The crude glycerine, incidentally, is either used for the manufacture of

442 *Unit processes*

dynamite (section 16.8) or further purified. The economics of the soap trade hinge on the price obtainable for glycerol and in addition to the above method, it can be obtained by hydrolysis of epichlorhydrin (section 10.4.6) and from propylene by the routes shown in Fig. 16.4. Market data are given in Table 16.2.

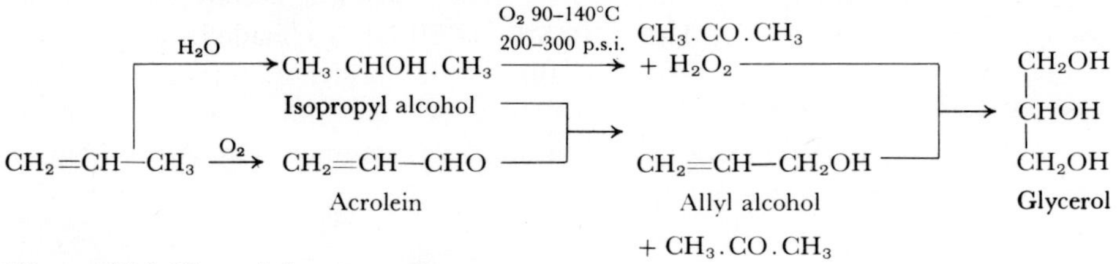

$$CH_2=CH-CH_3 \xrightarrow{O_2} CH_2=CH-CHO \text{ (Acrolein)}$$

$$CH_2=CH-CH_3 \xrightarrow{H_2O} CH_3.CHOH.CH_3 \text{ (Isopropyl alcohol)} \xrightarrow[200-300 \text{ p.s.i.}]{O_2 \; 90-140°C} CH_3.CO.CH_3 + H_2O_2$$

$$CH_2=CH-CHO \longrightarrow CH_2=CH-CH_2OH \text{ (Allyl alcohol)} + CH_3.CO.CH_3$$

$$CH_2=CH-CH_2OH + H_2O_2 \longrightarrow \begin{array}{c} CH_2OH \\ | \\ CHOH \\ | \\ CH_2OH \end{array} \text{ Glycerol}$$

Figure 16.4 Glycerol from propylene.

Hydrolysis is usually interpreted more broadly, however, so that it is any process of the type

$$AB + H_2O \longrightarrow HA + BOH$$

and the hydrolysis of chlorobenzene to phenol (section 9.5.1.1) comes into this category as does the hydrolysis of starch and of pentosans.

Table 16.2 Glycerol production ('000 tons) 1969

	Natural	*Distilled*	*Price (4/72)*
USA	166·8*	70·3	22·75 ¢/lb
UK	25·5†	32·1	£230/ton

* Includes 96 580 tons synthetic glycerol.
† Sales.

The former, in the presence of hydrochloric or sulphuric acids at 150°C, gives dextrose or corn syrup which, though containing some disaccharides, is mainly D-glucose. Starch consists of long chains of anhydroglucose units joined by glucoside linkages which is where hydrolysis takes place:

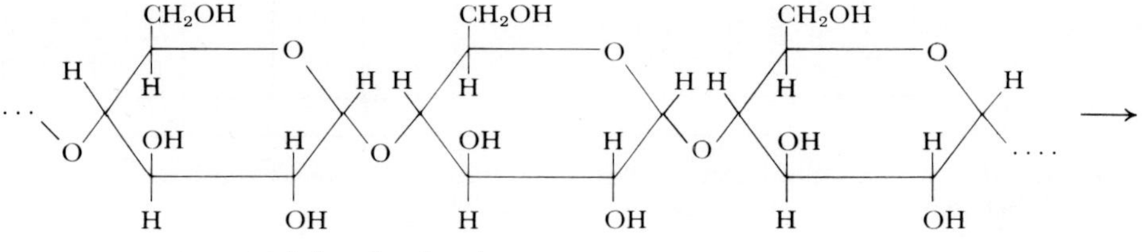

starch (amylose form)

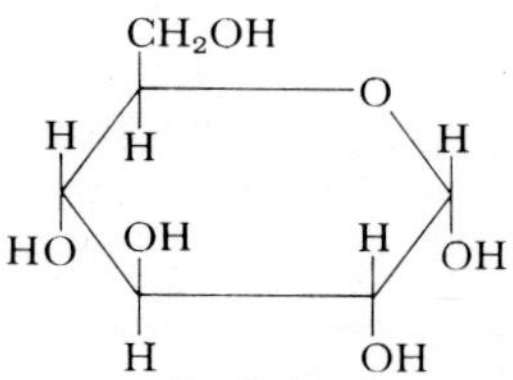

D-glucose

Hydrolysis of oat hulls, corncobs, rice hulls and cottonseed bran with dilute sulphuric acid and 3–6 atmospheres steam gives furfural. The reactants are polymeric pentoses which hydrolyse to pentoses, mainly xylose, which then cyclize and lose water again.

$$(C_5H_8O_4)_n \xrightarrow{H_2O} HOCH_2(CHOH)_3CHO \longrightarrow \underset{\text{Furfural}}{\begin{array}{c} HC-CH \\ \| \quad \| \\ HC \quad C-CHO \\ \diagdown \; \diagup \\ O \end{array}} + 3H_2O$$

Pentosans   Xylose

Hydration is taken to mean processes of the type

$$AB + H_2O \longrightarrow HO-AB-H$$

and may be exemplified by the overall processes by which aliphatic alcohols are produced from olefines (sections 9.1.3, 9.2.1, 9.3) and by the manufacture of glycols from olefine oxides (sections 9.1.1.1, 9.2.2.1).

16.6 Alkylation

Alkylation is the replacement by an alkyl radical of a hydrogen atom in an organic compound. In the majority of industrial alkylations, the hydrogen is attached to a carbon atom but there are cases where it is attached to oxygen, nitrogen, or sulphur atoms or in which metal alkyls are produced. In a high proportion of these latter processes, alkylation is brought about by treatment with ethyl chloride. The following are examples of the various types of alkylation.

1. Treatment of an aromatic compound with an olefine gives an alkaryl compound. This is the Friedel Crafts reaction and the catalyst is usually aluminium chloride and hydrogen chloride though boron trifluoride has been used in certain processes and sulphuric acid and hydrofluoric acid in others. The process is used to make ethylbenzene, cumene, dodecyl benzene and octyl phenols (sections 10.2.3, 9.5.1.1, 11.4.2 and 11.6 respectively). With $AlCl_3$, the reaction has been shown by isotope tracer techniques to go via the $AlCl_4^-$ cation and a similar protonation mechanism applies in the other cases.

$$AlCl_3 + HCl + R.CH{=}CH_2 \rightleftarrows R.\overset{+}{C}H{-}CH_3 + AlCl_4^-$$

$$\underset{}{\bigcirc} + R.\overset{+}{C}H{-}CH_3 \longrightarrow \underset{}{\bigcirc}{-}\underset{CH_3}{\overset{R}{\diagdown\;\diagup}}{CH} + H^+$$

2. Alkylation of paraffins by olefines gives a material called 'alkylate' which is a constituent of high quality aircraft and automobile fuels. The highest quality product is made by alkylation of isobutane with isobutene to give 2,2,4-trimethylpentane

('isooctane'). The alkylation can be carried out thermally (now obsolete) or catalysed by sulphuric or hydrofluoric acid:

$$\begin{array}{c} H_3C \\ C=CH_2 \\ H_3C \end{array} + \begin{array}{c} H_3C \\ CH-CH_3 \\ H_3C \end{array} \xrightarrow{27\text{-}46°C} \begin{array}{c} H_3C \\ CH-CH_2-\overset{\overset{\displaystyle CH_3}{|}}{\underset{\underset{\displaystyle CH_3}{|}}{C}}-CH_3 \\ H_3C \end{array}$$

The 2,3,4 and 2,3,3 trimethylpentanes are also produced in substantial quantities along with small amounts of other branched chain hydrocarbons. Low temperatures favour the formation of organic fluorides while higher ones reduce the octane number of the fuel. Care must be taken to avoid side reactions such as the polymerization of the olefine, and a large excess of paraffin is normally used.

3. The substitution by an alkyl group of a hydrogen atom attached to an oxygen leads to an ether linkage. Diethyl ether itself is made from ethanol and sulphuric acid, but two important solvents – Cellosolve and Carbitol – are made by alkylation of ethanol with ethylene oxide:

$$C_2H_5OH + H_2C\overset{\displaystyle\diagdown\diagup}{\underset{\displaystyle O}{-}}CH_2 \longrightarrow C_2H_5.O.CH_2CH_2OH$$
Ethylene glycol ethyl ether (Cellosolve)

$$\Big\downarrow H_2C\overset{\displaystyle\diagdown\diagup}{\underset{\displaystyle O}{-}}CH_2$$

$$C_2H_5.O.CH_2CH_2.O.CH_2CH_2OH$$
Diethylene glycol monoethyl ether (Carbitol)

Ethylene and propylene oxides undergo many similar reactions (sections 9.1.1 and 9.2.2).

Sodium cellulose (cellulose in sodium hydroxide solution) is alkylated by treatment with ethyl chloride to give ethyl cellulose, a plastic material which is more stable than the cellulose esters. With sodium chloroacetate, sodium cellulose gives sodium carboxymethyl cellulose:

$$[C_5H_7O_4.CH_2ONa]_x + xCl.CH_2.COONa \longrightarrow [C_5H_7O_4.CH_2OCH_2.COONa]_x + xNaCl$$

Sodium cellulose Sodium carboxymethylcellulose (CMC)

CMC is used as a soil suspending agent in detergents (section 11.3.3) and as a viscosity improver in paints, foods (ice cream), inks, latexes etc. Market data are given in Table 16.3.

4. The alkylation of aniline to dimethylaniline is an example of the alkylation of a hydrogen attached to nitrogen. It is made by heating aniline and methanol under pressure:

$$\text{C}_6\text{H}_5\text{NH}_2 + 2\text{CH}_3\text{OH} \xrightarrow{H_2SO_4} \text{C}_6\text{H}_5\text{N}(\text{CH}_3)_2 + 2\text{H}_2\text{O}$$

Table 16.3 Amines, cellulose ethers, lead alkyls, resorcinol, nitro-compounds and halogenated hydrocarbons

	Production m. lb	USA 1969 Sales m. lb	Value ¢/lb
Aniline	334·0	205·0	11
Diphenylamine	34·7	34·6	21
Methylamine[5]	22·2	14·8	9
Dimethylamine[5]	74·0	35·1	10
Trimethylamine[5]	21·2	14·9	10
Ethylamine	15·5	14·8	15
Total propylamines	14·1	12·5	26
Total butylamines	26·2	20·4	25
Dimethylaniline	n.a.	12·8	19
Sodium carboxymethyl cellulose	61·6	62·3	42
All other cellulose ethers	60·8	53·5	72
Resorcinol[3]	n.a.	n.a.	55
Tetraethyl lead[6]	371	365	53
Tetramethyl lead[1, 6]	76	n.a.	n.a.
Tetra (methyl–ethyl) leads[6]	431	424	56
Nitrobenzene	484·0	9·9	8
Pentaerythritol tetranitrate	5·4	3·6	60
2,4(&2,6)dinitrotoluene	259·0	n.a.	n.a.
Methyl chloride	403·0	166·0	5
Methylene chloride	366·0	338·0	8
Chloroform	216·0	172·0	6
Carbon tetrachloride	883·0	786·0	5
Chlorodifluoromethane	n.a.	77·9	56
Dichlorodifluoromethane	368·0	335·0	25
Trichlorofluoromethane	239·0	204·0	18
Ethyl chloride	679·0	268·0	6
Ethylene dichloride	6 037·0	1 227·0	3
Ethylene dibromide	310·0	n.a.	n.a.
Trichloroethylene	597·0	562·0	7
1,1,1-trichloroethane	324·0	299·0	11
Tetrachloroethylene[2]	635·0	612·0	7
Vinyl chloride, monomer	3 736·0	434·0	33
Chlorinated paraffins	61·9	59·1	13
Chlorobenzene	602·0	93·1	6
o-dichlorobenzene	70·3	53·0	10
p-dichlorobenzene	52·0	53·9	9
Chlorotoluene (benzyl chloride)	74·7	19·3	14
Chloral	62·4	n.a.	n.a.
DDT	123	80	14[4]

[1] Includes production used in synthesis of tetra (methyl–ethyl) leads
[2] Perchloroethylene.
[3] The sole UK plant was shut down for a large part of 1971 but is expected to reopen early in 1972. UK production (1970), nil. Consumption 1 200 tons. Price $\sim$ £1 000/ton.
[4] 1972 price, 22¢/lb.
[5] UK production of the three methylamines (1970) $\sim$ 7 000 tons. Price $\sim$ £115/ton.
[6] UK consumption of lead alkyls (1970), $\sim$ 9 000 tons.

It is used in an intermediate in the production of the explosive 'Tetryl' (section 16.8).

Similarly, a wide range of alkylamines, including almost every mono-, di- and tri-alkylamine from methyl to amyl plus cyclohexylamine, is made on a commercial scale by reaction of ammonia with the relevant alcohol or alkyl chloride. Separation of the mixed mono-, di- and tri-alkylamines which result from a single preparation may be troublesome as the boiling points tend to be close, but distillation techniques are used. There is a related problem of trying to arrange the reaction conditions so that the relative amounts of amines produced fit in with the markets available. Data are given in Table 16.3.

5. Lead tetraethyl is made by treating a lead-sodium alloy in an autoclave with ethyl chloride:

$$4PbNa + 4C_2H_5Cl \longrightarrow Pb(C_2H_5)_4 + 4NaCl$$

Mixed with ethylene dibromide, it is added to petrol at a level of about 0·5 g/litre as an octane improver and anti-knock agent. The bromide converts the lead and lead oxide formed in the cylinder of the engine into the relatively volatile lead bromide. The possible pollution problem is discussed in section 17.4.3. Other lead alkyls are made similarly. Market data are given in Table 16.3.

16.7 Amination

The industrial preparation of compounds containing amine groups bears some resemblance to laboratory methods. As noted in section 16.6 a wide range of alkylamines may be made by reactions of ammonia or derivatives with alcohols or alkyl chlorides. Such processes are said to give amines by ammonolysis, though they can also be regarded as alkylations of ammonia. Since ammonia is cheap, as many amines as possible are made by this method. The most important aliphatic amines are the methylamines. Monomethylamine is used in agrochemicals and in the USA in rocket fuels; dimethylamine is the starting material for dimethylformamide, and trimethylamine is the starting material for choline chloride, an animal food additive.

Amines are also made by reduction of nitro compounds. Aniline is the most important and comes from reduction of nitrobenzene with either iron borings and hydrochloric acid in the liquid phase, or hydrogen and a copper catalyst in the gas phase:

$$C_6H_5NO_2 + 4H \longrightarrow C_6H_5NH_2 + 2H_2O$$

The vapour phase route operates continuously on a 10 000–30 000 ton/year scale; the liquid phase route is a batch process and problems of obtaining and handling iron borings confine it to below 10 000 tons/year. An ammonolysis route using chlorobenzene plus ammonia has been operated but is uneconomic unless surplus chlorobenzene, perhaps from a chlorobenzene/phenol plant, is available.

Aniline was of great industrial importance in the nineteenth century as an intermediate for synthetic dyestuffs. This now accounts for about a sixth of its market. Rubber chemicals (antioxidants and vulcanization accelerators) account for over half, and there

16.8 Nitration

Nitration is brought about in the liquid phase by 'mixed acids', i.e. by a mixture of concentrated nitric and sulphuric acids which give rise to the nitronium ion NO_2^+. In the gas phase, it is accomplished by vaporized nitric acid or by nitrogen dioxide. In general, aliphatic nitrations are carried out in the gas phase and aromatic ones in the liquid phase. Examples are:

1. Nitrobenzene is prepared by reaction of benzene with 'mixed acids' at 50–55°C

$$C_6H_6 + HNO_3 \xrightarrow{H_2SO_4} C_6H_5.NO_2$$

The reaction is highly exothermic and stirring and cooling are important to avoid formation of di-nitrobenzene. In spite of the strength of the acids, a cast-iron or steel reactor is satisfactory. The spent sulphuric acid may be recovered, concentrated and recycled. The only important use of nitrobenzene is as a raw material for aniline (section 16.7).

2. Nitromethane, nitroethane and the two isomeric nitropropanes are made by vapour phase nitration of methane, ethane or propane with nitric acid vapour at 350–500°C and 5–10 atmospheres. They are used as propellants, chemical intermediates and solvents but are still not large-tonnage products.

Nitration of natural gas gives about 30% nitromethane and 70% nitroethane. These may be condensed with formaldehyde and the product nitrated to give nitro-nitrates which are useful explosives.

$$CH_3NO_2 + 3H.CHO \longrightarrow O_2N-C{\begin{smallmatrix}CH_2OH\\CH_2OH\\CH_2OH\end{smallmatrix}} \xrightarrow{HNO_3} O_2N-C{\begin{smallmatrix}CH_2ONO_2\\CH_2ONO_2\\CH_2ONO_2\end{smallmatrix}}$$

$$CH_3CH_2NO_2 + 2H.CHO \longrightarrow O_2N-C{\begin{smallmatrix}CH_2OH\\CH_3\\CH_2OH\end{smallmatrix}} \xrightarrow{HNO_3} O_2N-C{\begin{smallmatrix}CH_2ONO_2\\CH_3\\CH_2ONO_2\end{smallmatrix}}$$

3. Almost all high and low explosives are nitrates or nitro compounds. A high explosive is one which is insensitive to normal shocks but explodes with great violence when detonated by an initiating explosion from a detonator. The latter is a very sensitive explosive such as lead azide or mercury fulminate.

A low explosive burns rather than explodes and evolves large quantities of gas. Its combustion is much more controllable and low explosives tend to be used in mining and blasting rather than for military purposes.

An explosive must undergo very rapid chemical reaction and evolve large amounts of gas and heat. The carbon and hydrogen combust using the oxygen in the nitro groups to give carbon dioxide and water. The 'oxygen balance' of an explosive is the

Unit processes

Table 16.4 Explosives based on nitrates or nitro compounds

Explosive	Type	Made by the nitration of	Formula	Maximum detonation rate* (m/sec)
Cellulose nitrate (nitrocellulose or guncotton); smokeless powder and cordite contain additives	High	Cotton linters or specially prepared wood pulp	(cellulose nitrate ring structure with ONO_2, CH_2ONO_2, NO_2 groups)	~ 7 300
Trinitrotoluene (TNT)	High	Toluene	2,4,6-trinitrotoluene (CH_3 benzene ring with three NO_2 groups)	7 400
2.4.6(trinitrophenyl) methyl nitramine (Tetryl)	High	Dimethylaniline	(trinitrophenyl methyl nitramine structure)	7 700
Picric acid	High	Mono and di-phenol sulphonic acids (the related compound, ammonium picrate, is made by neutralizing picric acid with ammonia)	(2,4,6-trinitrophenol structure with OH and three NO_2)	7 500
Pentaerythritol tetranitrate (PETN)	High	Pentaerythritol	O_2NOCH_2 — C — CH_2ONO_2 / O_2NOCH_2 CH_2ONO_2	8 100
Cyclotrimethylene trinitramine (Cyclonite)	High	Hexamethylene tetramine, plus ammonium nitrate and acetic anhydride	(cyclic ring with three N–NO_2 and three CH_2 groups)	8 350
Black powder (Gunpowder)	Low	Potassium nitrate is made by double decomposition of potassium chloride and sodium nitrate and mixed with sulphur and charcoal	KNO_3 (75%) Sulphur (10%) Charcoal (15%)	~ 1 000
Nitroglycerine (glyceryl trinitrate is absorbed on wood pulp to give dynamite)	High	Glycerol	$CH_2.O.NO_2$ $CH.O.NO_2$ $CH_2.O.NO_2$	8 500
Ammonium nitrate	High	Ammonia	NH_4NO_3	2 700

* The speed at which a detonation travels is a crude measure of the power of an explosive. The rate depends also on the state of compaction.

number of grams of oxygen lacking or in excess for the complete combustion of 100 g of the explosive. Many explosives, notably the aromatic nitro compounds, have a negative oxygen balance and combust to give CO instead of CO_2. This could produce a 'fire damp' in mines and lead to a further explosion, so when used in mining the explosives are blended with an oxygen carrier (chlorate, nitrate or perchlorate) which ensures complete oxidation.

The manufacture of explosives requires elaborate and specialized safety precautions. A list of the main explosives produced by nitration is given in Table 16.4.

16.9 Sulphation and sulphonation

Sulphation (the preparation of sulphate esters R—O—SO_2—OH) and sulphonation (leading to sulphonic acid derivatives R—SO_2—OH) are brought about by a variety of reagents but the most important are sulphuric acid in the liquid phase and sulphur trioxide in the vapour phase. These processes are used to a significant extent but are still only minor consumers of sulphuric acid (Table 14.4).

Examples include the preparation of sodium dodecylbenzene-sulphonate and sodium lauroyl glyceryl sulphate (section 11.4.2) which are used in detergents, and the sulphation of propylene and butenes to their sulphates (sections 9.2.1 and 9.3).

Sulphonation of aromatics, often with chlorosulphonic acid, $ClSO_3H$, made from sulphur trioxide and hydrogen chloride, gives sulphonic acids which are useful intermediates in the production of azo dyes, pharmaceuticals and other fine chemicals (sections 12.3 and 13.4).

Alkaline fusion of sulphonic acids gives phenolic compounds (section 9.5.1.1) and preparation and fusion of naphthalene-β-sulphonic acid and benzene metadisulphonic acid are still the only routes to β-naphthol and resorcinol. The former is used in dyes, the latter in premium resorcinol-formaldehyde adhesives (Table 16.3).

16.10 Halogenation

The introduction of halogen atoms into an organic compound is usually brought about by treatment with the elementary halogen or the halogen hydracid. Halogens are very reactive and the conditions required are usually mild. Halogenated organic compounds find wide application as refrigerants, insecticides, anaesthetics, petrol additives and solvents. Market data on the more important compounds are given in Table 16.3.

16.10.1 Fluorination

Elementary fluorine is too reactive to be used directly in industrial synthesis and fluorination is usually carried out with hydrofluoric acid, obtained by the action of concentrated sulphuric acid on fluorspar (CaF_2). Even the handling of hydrofluoric acid is extremely dangerous and the problems of fluorination centre on safety precautions and waste disposal.

The most important industrial fluorine compounds are the low boiling so-called fluorocarbons, which are in fact chlorofluorohydrocarbons in which most or all of the hydrogens of hydrocarbons from C_1 to C_4 have been replaced by chlorine and fluorine. The group may be exemplified by fluorocarbon 12, dichlorodifluoromethane, b.p. $-29.8°C$, m.p. $-158°C$.

It is prepared by the action of hydrofluoric acid on carbon tetrachloride in the presence of antimony pentachloride:

$$CCl_4 + 2HF \xrightarrow[SbCl_5]{60-110°C} CCl_2F_2 + 2HCl$$

A mixture of chlorofluorocarbons is obtained which are separated by distillation.

An alternative method involves the preparation of vinylidene fluoride by the action of hydrogen fluoride on acetylene followed by high temperature chlorination:

$$CH{\equiv}CH + 2HF \longrightarrow CH_2{=}CF_2 \xrightarrow[775-850°C]{Cl_2} CF_2Cl_2 \; (+CCl_4, CCl_3F, C_2Cl_4 \text{ etc.})$$

The fluorocarbons are odourless, inert, non-inflammable gases of low toxicity. They boil not much below room temperature and so can be maintained as liquids under slight pressure. They are therefore useful as refrigerants for domestic refrigerators and air conditioners, and are indispensable as aerosol propellants. They are also used as blowing agents for urethane foams, and as sources of monomers for fluorocarbon polymers (section 10.2.8).

Hydrofluoric acid is also used to make aluminium fluoride and synthetic cryolite for the aluminium smelting industry.

16.10.2 Chlorination

Chlorination is a process carried out on a scale an order of magnitude larger than fluorination. End uses are shown in Fig. 14.11. Chlorine atoms are introduced into organic molecules in a variety of ways.

1. Elementary chlorine can displace hydrogen atoms from hydrocarbons, e.g. treatment of excess methane with chlorine at 370°C gives a mixture of methyl chloride CH_3Cl, methylene dichloride CH_2Cl_2 and chloroform $CHCl_3$. Traces of carbon tetrachloride CCl_4 are also formed and this can be made the main product by the recycling and rechlorination of the other products. The by-product is hydrogen chloride, so half the chlorine is liable to be wasted. This is a problem with most chlorine substitution processes of this type

$$CH_4 \xrightarrow{Cl_2} CH_3Cl + HCl \xrightarrow{Cl_2} CH_2Cl_2 + HCl \xrightarrow{Cl_2} CHCl_3 + HCl \xrightarrow{Cl_2} CCl_4 + HCl$$

The products may be separated by distillation. Methyl chloride is mainly used for the manufacture of silicones, tetramethyl lead (to some extent replacing tetraethyl lead as a petrol additive) and butyl rubber. Methylene dichloride is useful as a paint remover. Chloroform, once an important anaesthetic, and carbon tetrachloride, once an important dry-cleaning and degreasing solvent, are now used primarily as intermediates for fluorocarbons and fluorocarbon polymers. Carbon tetrachloride fumes appeared to cause cirrhosis of the liver (more enjoyably obtained by sustained ingestion of alcoholic drinks) and it has been replaced by trichloroethylene in degreasing and perchloroethylene in dry-cleaning. These are also prepared by halogenation processes:

Halogenation 451

$$C_2H_2 + 2Cl_2 \xrightarrow{SbCl_3} \underset{\underset{\text{Tetrachloroethane}}{CHCl_2}}{CHCl_2} \xrightarrow{BaCl_2} \underset{\underset{\text{Trichloroethylene}}{CCl_2}}{CHCl} + HCl \xrightarrow{Cl_2} \underset{\underset{\text{Pentachloroethane}}{CCl_3}}{CHCl_2} \xrightarrow{Ca(OH)_2} \underset{\underset{\text{Perchloroethylene}}{CCl_2}}{CCl_2}$$

Acetylene

This four-step route to perchloroethylene is being displaced by one based on the simultaneous chlorination and pyrolysis of hydrocarbons

$$\text{e.g. } C_3H_8 + 8Cl_2 \longrightarrow CCl_2{=}CCl_2 + CCl_4 + 8HCl$$

$$2\,CCl_4 \longrightarrow CCl_2{=}CCl_2 + Cl_2$$

2. Chlorine can add on to double bonds to give dichlorocompounds such as ethylene dichloride (section 10.3.1).
3. Hydrogen chloride can be used as a chlorinating agent by formation of chloride 'esters', e.g. ethyl chloride which is used mainly for tetraethyl lead production.

$$C_2H_5OH + HCl \xrightarrow[145°C]{ZnCl_2} C_2H_5Cl + H_2O$$

4. It can also add across double bonds to give monochlorocompounds, e.g. acetylene plus hydrogen chloride provided the second step in the 'balanced chlorine economy' for vinyl chloride production (section 10.3.1).
5. Hypochlorous acid is sometimes added across double bonds to give chlorhydrins (section 9.2.2).
6. A range of highly chlorinated aromatic compounds, DDT, BHC, aldrin, chlordane, dieldrin, endrin, heptachlor, methoxychlor, toxaphene and lindane, are used as insecticides and are manufactured by fairly complicated halogenations and other reactions. They may be exemplified by dichloro-diphenyl-trichloroethane (DDT) which is the most widely used. It is prepared by reaction of chloral with chlorobenzene in the presence of sulphuric acid, the two precursors being made by halogenation processes:

$$\underset{\text{Acetaldehyde}}{CH_3CHO} \xrightarrow{Cl_2} \underset{\text{Chloral}}{CCl_3CHO} + 3HCl$$

$$\underset{}{C_6H_6} \xrightarrow{Cl_2} \underset{\text{Chlorobenzene}}{C_6H_4Cl} + HCl$$

$$\xrightarrow{H_2SO_4} \text{DDT}$$

Chloral may also be made by chlorination of ethanol, and chlorobenzene may be made by an oxychlorination process – see below. Worries about the accumulation of DDT residues in the fatty tissues of mammals and birds have led to a drop in world consumption from 400 000 tons in 1963 to between 200 and 250 000 tons in 1971.

16.10.3 Oxychlorination

Chlorine is a fairly expensive chemical because its production by electrolysis of brine involves high and unavoidable electricity costs. In many of the reactions in which it is involved, hydrogen chloride is evolved as a product so that half the chlorine in each molecule is wasted. This was an unsatisfactory situation and the first response to it was the development of 'balanced chlorine economies' (section 10.3.1). Vinyl chloride, for example, was produced by two routes, acetylene plus hydrogen chloride and ethylene plus chlorine so that all the chlorine was used. Unfortunately, acetylene is now scarcely economic as a raw material and there is increasing interest in other methods of using by-product hydrogen chloride. The most hopeful involves oxychlorination which is based on the Deacon process. The latter has been known since 1868 and was the first process to use heterogeneous catalysis. Air and hydrogen chloride were passed over bricks soaked in cupric chloride:

$$4HCl + O_2 \xrightarrow[500-500°C]{CuCl_2} Cl_2 + 2H_2O$$

and this process has itself been improved and modified, so that it involves a lower temperature, a fluidized bed and a didymium-promoted catalyst. If another compound is added to the gas stream, however, it may be chlorinated in situ, e.g. benzene gives chlorobenzene (section 9.5.1.1) and ethylene gives ethylene dichloride obviating the need for the acetylene based route to vinyl chloride.

16.10.4 Bromination

Bromine is obtained from some concentrated brines, in which it exists as bromide ions, by the action of gaseous chlorine which liberates free bromine which is blown out of solution with air. Extraction from sea water is similar but more complicated:

$$2Br^- + Cl_2 \longrightarrow Br_2 + 2Cl^-$$

Sea water contains only about 65–70 ppm bromine but natural brines in Michigan contain up to 1 300 ppm and Dead Sea water contains 4–5 000 ppm. In 1970, the Dead Sea works produced about 12 000 tons of bromine, representing some 5–6% of total world supply.

About 70–80% of world bromine production is used to make ethylene dibromide as a scavenger for lead in petrol. Ethylene is usually brominated directly at the site of bromine production.

$$C_2H_4 + Br_2 \longrightarrow CH_2Br—CH_2Br$$

No other brominations are carried out on a significant scale.

16.10.5 Iodination

Iodine occurs as sodium iodate in the mother liquors left after the recovery of Chile saltpetre and is released by reduction of this with sodium bisulphite:

$$2NaIO_3 + 5NaHSO_3 \longrightarrow 3NaHSO_4 + 2Na_2SO_4 + H_2O + I_2$$

It is also extracted from certain brines and from seaweed. Because it occurs in such low concentrations, its recovery is expensive and its high equivalent weight means that it forms a high percentage by weight of its compounds (e.g. methyl iodide is 89·5% iodine) so it is uneconomic for it to be used in any but the most specialized applications. World production is only about 1 000 tons/year and this is used mainly as alkali iodides in photography and pharmaceuticals.

16.11 Polymerization

The term polymerization covers the general class of processes by which monomers are converted to polymers. Traditionally, this class was sub-divided into addition polymerizations, in which no by-product molecule was eliminated as the monomer polymerized, and condensation polymerizations in which simple by-product molecules (water, ammonia, etc.) were produced. The production of Nylon 6 from caprolactam (section 10.7.2) however does not involve the elimination of a by-product molecule and yet is clearly similar to the production of the other nylons by unequivocal condensation processes. Analogous considerations apply to the production of polyethers from ethylene oxide (sections 9.1.1, 10.4.6, 11.6). It is probably more helpful, therefore, to define condensation polymerizations as those in which the intermediates as the chains grow are stable chemical entities which can be isolated and purified, while in addition polymerizations, the intermediates are transient species such as free radicals, carbonium ions, and carbanions.

The importance of polymerization processes and the resultant polymers in the modern chemical industry has already been stressed. The properties and ease of processing of the polymers is dependent on such things as their chain length and state of physical subdivision and these in turn depend on how polymerization is carried out.

16.11.1 Condensation polymerization

Condensation polymerization is similar to a simple chemical reaction carried out repeatedly. For example, in the production of a polyester, a bifunctional alcohol and bifunctional acid are mixed and an esterification catalyst added:

$$HO.R.OH + HOOC.R'.COOH \xrightarrow{H_2SO_4} HO.R.O.OC.R'.COOH + H_2O$$

$$\downarrow HO.R.OH$$

$$HO.R.O.OC.R'.CO.O.R.OH + H_2O$$

$$\downarrow HOOC.R'.COOH$$

$$HOOC.R'.CO.O.R.O.OC.R'.CO.O.R.OH + H_2O$$

and so on.

As in a conventional esterification, the reactions are reversible, and the equilibrium is pushed over to the right hand side by removal of water as it is formed. The water either boils off because of the high temperature used, or is entrained in benzene or xylene as in a simple esterification.

Examples of polyesterification reactions may be found in sections 10.3.4.4, 10.5.3 and 10.4.4 among others. Other condensation polymerizations are exemplified by phenoplasts and aminoplasts (sections 10.4.1 and 10.4.2), nylons and polyethers.

A major difference between a simple condensation reaction and a polycondensation reaction is that the development of high molecular weight products in the latter case leads to an increase in the viscosity of the reaction mixture, if the polymer is soluble in the monomer. If not, it precipitates. This problem is overcome in one of two ways. Either the reaction is carried out in a solvent, a method which is particularly useful if the finished polymer is to be used in a surface coating, or it is interrupted at a stage where the resin is still fusible and soluble. The final 'in situ' curing process is then relied on to build up the molecular weight.

16.11.2 Addition polymerization

Addition polymerization usually involves the polymerization of substances with the general formula $CH_2=CHR$. The intermediates are transient species – either free radicals or ions. In principle, the purest and best polymers would be obtained by allowing the pure monomer to stand at atmospheric temperature and pressure until it had polymerized. Life is too short, however, for this to be a useful industrial technique and the process is speeded up by the introduction of catalysts which react to produce the initial transient species on which the polymer chains build. The speeding up of polymerization brings further problems in its train, especially the dispersal of the heat generated in the reaction:

$$n\ CH_2=CHR \longrightarrow (CH_2-CHR)n$$

A double bond is weaker than two single bonds so that a typical addition polymerization is exothermic by about 40 kcal (170 kJ) per mole of monomer.

Addition polymerization can be catalysed by free radical, anionic or cationic initiators, and polymerization can be carried out in bulk or, when heat transfer and other problems make it necessary, in solution, emulsion or suspension. A summary of which catalysts and production techniques are suitable for which monomers is given in Table 16.5.

Table 16.5 Techniques of polymerization

Monomer	Bulk	Solution	Suspension	Emulsion	Anionic Conventional	Anionic Ziegler	Cationic	Free radical
Ethylene	A	B	B	B		✓		✓
Propylene		A				✓		
Vinyl chloride	B	B	A	B				✓
Styrene	A	B	A	B	✓	✓	✓	✓
Vinyl acetate	B	B	B	A				✓
Acrylonitrile	B	B	A		✓			✓
Acrylates and methacrylates	A	A	B	B*	✓	✓		✓
Butadiene		A		B†	✓	✓		✓
Isoprene		A			✓	✓		
Isobutene		A					✓	
Chloroprene				A				✓
Ethylene-propylene terpolymer		A				✓		

A = commonly used method.
B = occasionally used method.
* Emulsion polymerization is important for acrylates.
† A redox system is used for SBR rubbers.

16.11.3 Free radical polymerization

A free radical initiator (or catalyst) reacts to give free radicals ($X^{\cdot}$). A free radical can react with an olefinic molecule of monomer to give another free radical and this reacts with another molecule of monomer, and so on. In general, the odd electron present in the free radical tends to locate itself on the carbon atom of the —CHR end of the growing polymer and this then attacks the —CH$_2$ end of the next molecule of monomer, so that the —R groups are arranged regularly down the polymer chain.

$$X^{\cdot} + CH_2{=}CHR \longrightarrow XCH_2{-}\dot{C}HR \xrightarrow{CH_2=CHR} XCH_2{-}CHR{-}CH_2{-}\dot{C}HR$$

$$\downarrow CH_2{=}CHR$$

$$XCH_2{-}CHR{-}CH_2{-}CHR{-}CH_2{-}\dot{C}HR$$

$$\downarrow (n-2)\ CH_2{=}CHR$$

$$X(CH_2{-}CHR)_n CH_2{-}\dot{C}HR$$

The chains go on growing until two of them meet when they can either couple or disproportionate to give stable polymers:

$$\ldots\ldots CH_2{-}\dot{C}HR + RH\dot{C}{-}CH_2 \ldots \longrightarrow$$

$$\ldots\ldots CH_2{-}CHR{-}RHC{-}CH_2\ldots \quad \text{(coupling)}$$

$$\ldots CH_2{-}CH_2R + RHC{=}CH\ldots \quad \text{(disproportionation)}$$

Chains can also be terminated by reaction with materials which react with free radicals to give inert products. Such materials are called inhibitors and examples are quinones and methylene blue. In addition to the above reactions, chain transfer processes may take place:

$$\ldots\ldots CH_2{-}\dot{C}HR + M \longrightarrow \ldots CH{=}CHR + \dot{M}H$$

M can be a molecule of monomer or of polymer chain which has stopped growing or any other molecule. Having become a free radical, it then starts 'growing' and a new chain builds up.

Free radical initiators are usually either organic peroxides such as benzoyl peroxide or azo compounds such as 2,2'-azobisisobutyronitrile. These decompose readily to give free radicals:

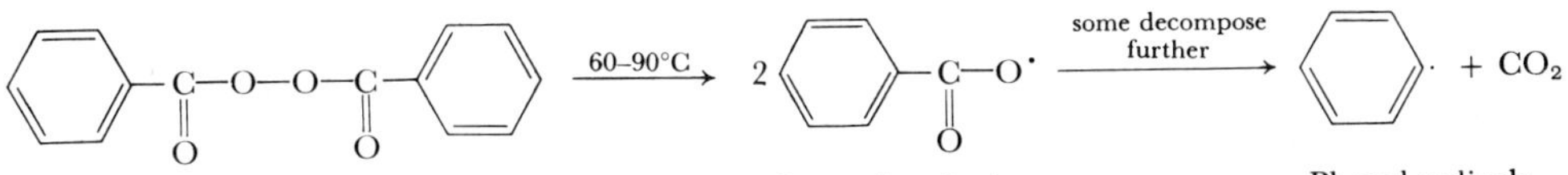

Ph–C(=O)–O–O–C(=O)–Ph $\xrightarrow{60-90°C}$ 2 Ph–C(=O)–O$^{\cdot}$ $\xrightarrow{\text{some decompose further}}$ Ph$^{\cdot}$ + CO_2

Benzoyl peroxide → Benzoyl radicals → Phenyl radicals

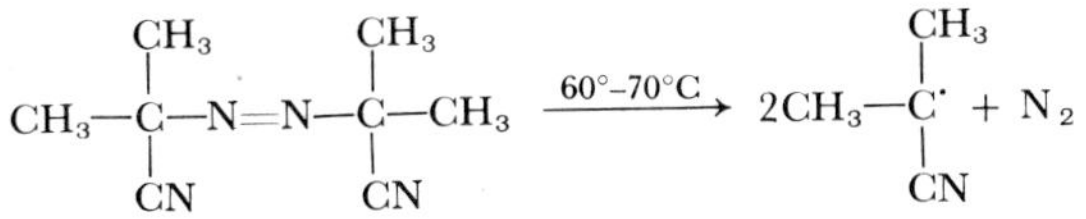

$$CH_3{-}\underset{CN}{\underset{|}{\overset{CH_3}{\overset{|}{C}}}}{-}N{=}N{-}\underset{CN}{\underset{|}{\overset{CH_3}{\overset{|}{C}}}}{-}CH_3 \xrightarrow{60°-70°C} 2CH_3{-}\underset{CN}{\underset{|}{\overset{CH_3}{\overset{|}{\dot{C}}}}} + N_2$$

2,2'-azobisisobutyronitrile

Benzoyl peroxide is easily the most widely used initiator. US sales in 1969 were 2 900 tons at 94¢ per lb.

Organic peroxides and azo compounds tend to be soluble in non-aqueous solvents but not in water. Therefore, a group of composite catalysts called 'redox' initiators is sometimes used. They consist of a mixture of a *re*ducing agent and an *ox*idizing agent such as ferrous ammonium sulphate and hydrogen peroxide. In the absence of monomer, the former would reduce the latter in a two-stage process.

$$Fe^{++} + H_2O_2 \longrightarrow Fe^{+++} + OH^\cdot + OH^-$$

$$Fe^{++} + OH^\cdot \longrightarrow Fe^{+++} + OH^-$$

If monomer is present, however, the hydroxyl free radical can initiate polymerization. Other redox systems include potassium peroxydisulphate/sodium bisulphite, benzoyl peroxide/ferrous ammonium sulphate and hydrogen peroxide/dodecyl mercaptan.

16.11.4 Cationic polymerization

Cationic polymerization uses acidic catalysts such as boron trifluoride or aluminium chloride together with a proton releasing substance such as water. It is effective with electron-donating monomers and will operate at temperatures round about $-100°C$. Chain initiation involves production of hydrogen ions:

$$BF_3 + H_2O \rightleftharpoons F_3BOH^- + H^+$$

followed by

$$H^+ + CH_2=CHR \longrightarrow CH_3-\overset{+}{C}HR \xrightarrow{CH_2=CHR} CH_3-CHR-CH_2-\overset{+}{C}HR$$

$$\downarrow (n-1) CH_2=CHR$$

$$H(CH_2-CHR)_n CH_2-\overset{+}{C}HR$$

Chains do not terminate by coupling or disproportionation because their positive charges prevent them from coming together. They can dissociate to regenerate hydrogen ions or can react with an anion:

$$\ldots\ldots CH_2-\overset{+}{C}HR \longrightarrow \ldots CH=CHR + H^+$$

$$\searrow^{F_3BOH^-}$$

$$\ldots\ldots CH_2-CHR-OH + BF_3$$

16.11.5 Anionic polymerization

Anionic polymerization uses basic catalysts such as sodium in liquid ammonia, or sodium cyanide in dimethylformamide. The mechanism is presumably analogous to cationic polymerization. Anionic catalysts were of relatively little importance till the development of stereospecific catalysts, of which Ziegler catalysts are the main examples, in the mid nineteen fifties. Their use results in the highly crystalline isotactic or syndiatactic polymers described in section 10.2.1. The principal examples are aluminium alkyls mixed with titanium tetrachloride, or lead tetra-alkyls plus titanium compounds.

They are normally used as suspensions in inert solvents but can also be dissolved in organic solvents if converted to their cyclopentadienyl complexes.

Ziegler catalysts need not be used only for macromolecules, and are also the basis for the Alfol process which gives mixtures of straight chain primary alcohols with even numbers of carbon atoms between C_2 and C_{20}. Ethylene is polymerized onto aluminium triethyl and the resultant aluminium trialkyls are oxidized with air and hydrolysed with sulphuric acid:

$$Al(C_2H_5)_3 \xrightarrow{C_4H_2} Al\begin{matrix} \diagup C_xH_{2x+1} \\ -C_yH_{2y+1} \\ \diagdown C_zH_{2z+1} \end{matrix} \xrightarrow{air} Al\begin{matrix} \diagup OC_xH_{2x+1} \\ -OC_yH_{2y+1} \\ \diagdown OC_zH_{2z+1} \end{matrix} \xrightarrow{H_2SO_4} \begin{matrix} C_xH_{2x+1}OH \\ C_yH_{2y+1}OH \\ C_zH_{2z+1}OH \end{matrix} + Al_2(SO_4)_3$$

The molecular weight distribution depends on the conditions, but the process is usually run to give mainly C_6 to C_{12} alcohols for detergents and plasticizers.

16.11.6 Bulk polymerization

It is important not only to choose the correct catalyst for a polymerization process, but also the appropriate production method. Bulk polymerization where the monomer is not diluted with solvent or dispersed in any way is the simplest method. Ethylene is usually polymerized in bulk in the gas phase and the molten polymer, being insoluble in the (gaseous) monomer, settles at the bottom of the reactor and is withdrawn. Polymethyl methacrylate is soluble in its monomer, so when it is polymerized in bulk the mixture becomes increasingly more viscous until polymerization is complete and a solid block obtained.

The disadvantages of bulk polymerization are the difficulties of heat removal as the mass becomes more viscous, and control of molecular weight. An advantage is that the polymer contains fewer impurities. The method is widely used for production of small castings. Péchiney St Gobain have developed a method for bulk polymerization of PVC giving a high clarity polymer which is particularly useful for bottles.

16.11.7 Solution polymerization

Monomers may be polymerized in solution in a suitably inert solvent. Heat removal is much easier than in bulk polymerization because the solution is less viscous, and the solvent can be allowed to reflux.

There are, however, certain disadvantages. Polymerization temperature is limited to the boiling point of the solvent, the average molecular weight tends to be very low because of chain transfer to solvent molecules, separation and recovery of solvent may be expensive and may involve a fire hazard and the polymer when formed may be difficult to obtain untainted by residual solvent.

The method is particularly suitable for preparation of polymers required for surface coatings and adhesives where they will be used anyway in solution.

16.11.8 Emulsion polymerization

Instead of the carrying out of polymerization in solution, it is frequently better to make an emulsion of the monomer in water containing a suitable emulsifying agent. The monomer exists as colloidal spherical particles of 1–10μ diameter covered by a molecular layer of surface active agent, in much the same way as dirt is held in 'solution' by a detergent (section 11.2). A water-soluble polymer-insoluble initiator (e.g. a redox system) is used. Chains thus are initiated in the aqueous disperse phase but diffuse into the monomer micelles and initiate polymerization. The polymer is obtained either as a precipitate or as a very stable emulsion called a latex. The technique gives the highest molecular weight product of any of the methods described here.

After polymerization, if the polymer is precipitated, it is filtered off and washed to remove the emulsifying agent but this is difficult to do efficiently and contamination always occurs. Heat transfer problems are negligible and the method is particularly useful for materials such as polyvinyl acetate which can be used as a latex (section 10.5.1).

16.11.9 Suspension polymerization

In suspension polymerization, the monomer is maintained as small globules in an aqueous medium by rapid stirring. Suspension stabilizers such as gelatine, ethyl or methylcellulose or starch are added to help keep the globules separate. A monomer soluble, water insoluble catalyst such as benzoyl peroxide is used (cf. emulsion polymerization) and the product takes the form of tiny beads or pearls, hence the alternative name of 'pearl polymerization'. The stabilizer and water can easily be removed and a polymer of fairly high molecular weight is obtained. Heat transfer is no problem.

Suspension polymerization can be thought of as emulsion polymerization without the emulsifier, but as the catalyst is in the monomer phase it is probably more accurate to think of it as a very large number of tiny bulk polymerizations.

16.12 Processing and fabrication of plastics

Plastics and also rubbers and fibres are solids. They have properties of size and shape which do not exist for liquids and gases. The plastics industry, consequently, must not only manufacture monomers and polymers, it must process them (e.g. by blending them with suitable additives and altering the size of the particles) and then fabricate them into the multitudinous products of that industry. Though formally an aspect of solids handling; processing and fabrication follow on logically from polymerization and will be discussed here.

16.12.1 Processing of plastics

The material which is fabricated in order to make a plastics article may be a monomer, a partially polymerized monomer or a polymer. It will probably be intimately mixed with assorted additives – plasticizers, stabilizers, fillers, pigments etc. The conversion of the products of the polymerization process to a suitable form for fabrication is usually a matter of grinding or pelleting the material and then blending in the additives. Plastics are usually sold to fabricators in this 'ready-to-use' form.

Thermosetting resins (P/F, U/F resins etc.) emerge from the manufacturing process as brittle lumps. They are reduced to powder by mechanical crushing. Thermoplastic resins made by bulk polymerization are less brittle and are reduced by rotatory cutters. Other methods of polymerization produce solutions, powders or granules *ab initio*.

Processing and fabrication of plastics 459

The plastic is then blended with the necessary additives to give 'compound'. In 'dry blending' the ingredients are mixed mechanically at room temperature; in 'hot blending' the mixing is performed above the softening temperature of the polymer, which is simultaneously melted and compounded, the resulting 'compound' then being cooled and diced into approximately 2·5 mm cubes. It is now in an easily handleable form, is the appropriate colour, contains the required stabilizers, plasticizers etc. and is ready for fabrication.

16.12.2. Fabrication of plastics

Table 16.6 summarizes the commonest fabrication processes for plastics.

1 Compression moulding
Compression moulding is practically the oldest method of fabricating polymers and is still widely used. The polymer is placed in one half (the 'male' half) of a mould and the second or 'female' half compresses it to a pressure of round about one ton/sq. inch (Fig. 16.5). The powder is simultaneously heated, which causes the resin to cross-link. Transfer moulding is a cross between this technique and injection moulding.

2 Casting
Casting was also used before the Second World War. In the sheet casting of polymethyl methacrylate, monomer is partly polymerized and the viscous liquid then poured into a cell made up of sheets of glass separated by a flexible gasket which allows the cell to contract as the casting shrinks.

3 Injection moulding
In injection moulding, polymer is softened in a heated volume and then forced under high pressure into a cooled mould where it is allowed to harden (Fig. 16.6). Pressure is released, the mould opened, the moulding expelled and the cycle repeated. Injection moulding is a versatile technique and can be used for bottle manufacture by a method identical with blow moulding except that the initial 'bubble' is injected rather than extruded.

4 Extrusion
Extrusion is a method of producing lengths of plastics materials of uniform cross section. The extruder is similar to a domestic mincing machine with the added facility that it can

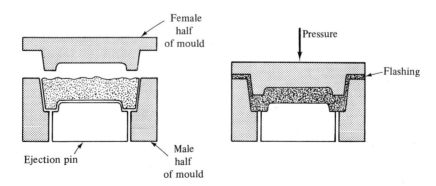

Figure 16.5 Compression moulding.

460 *Unit processes*

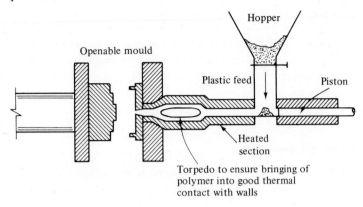

Figure 16.6 Diagrammatic representation of an injection moulding machine.

Table 16.6 Summary of the commonest fabrication processes for plastics

Fabrication process	Used with the following polymers	Examples of products
Compression or transfer moulding	P/F, U/F and M/F resins and some rubbers	Electrical fittings, toilet seats, ashtrays
Casting	Polymethyl methacrylate, epoxy resins	'Perspex' sheets, 'potted' electrical components
Injection moulding	Most thermoplastics	'Vintage' car bits, combs, washing-up bowls
Extrusion	Most thermoplastics particularly polyethylene and PVC	Hosepipe, curtain rails, insulated electrical wiring
Blow moulding	Most thermoplastics particularly polyethylene and PVC	Bottles, thin film
Vacuum forming (thermo-forming)	Polystyrene, cellulose acetate, cellulose acetate/butryrate, ABS, polymethyl methacrylate. Polyolefines not really suitable	Refrigerator interiors, 'contour' maps, chocolate box interiors, aircraft cockpit covers, 'perspex' advertising signs
Calendering	Several thermoplastics, particularly PVC but not polyolefines	Imitation crocodile-skin, floor tiles
Slush moulding, dipping, rotational casting, paste spreading	PVC (very occasionally, low MW polyethylene)	PVC gloves, PVC coated washing up racks, children's toys, PVC leathercloth
Laminating	P/F, U/F, M/F, unsaturated polyesters, epoxy resins	Decorative laminates, plywood, tubes for conveying oil, structural components for industry
Foaming	Polyurethanes, polystyrene, PVC	Cushioning materials, heat insulation, foam backed carpets

be heated and cooled (Fig. 16.7). The pellets enter the screw section via the hopper, are melted and then pass through the breaker plate into the die. The plastic material is forced out of the die with its cross section determined by the shape of the die, but not identical with it because of stresses induced by the extrusion process.

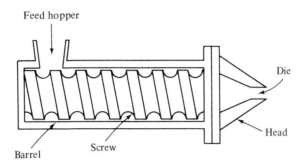

Figure 16.7 Extrusion.

Extrusion can be used for coating electrical wiring by means of a cross-head die of the kind shown in Fig. 16.8 and for filament and sheet by use of pinhole and slot dies.

5 *Blow moulding and vacuum forming*

Blow extrusion, in which the initial lump of polymer is formed by an extrusion process, is

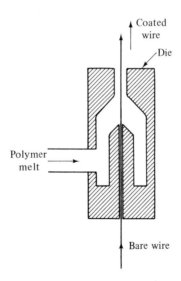

Figure 16.8 Wire coating.

the commonest form of blow moulding and is shown diagrammatically in Fig. 16.9. A short length of plastic tubing is extruded through a crossed die and the end is sealed by the closing of the mould. Compressed air is passed into the tube and the 'bubble' is blown out to fill the mould.

462 Unit processes

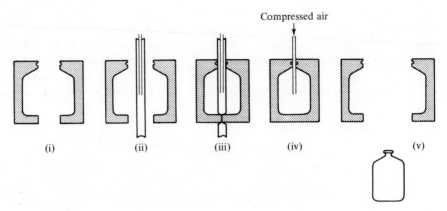

Figure 16.9 Blow extrusion.

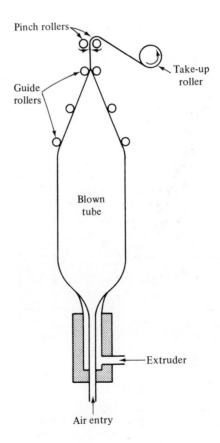

Figure 16.10 Production of film by extrusion and blowing.

Processing and fabrication of plastics 463

A variation of this process is used for the manufacture of thin film. A tube of plastic is continuously extruded and expanded by being blown to a large volume and consequently a small wall thickness (Fig. 16.10). The enormous bubble of plastic is cooled by air jets, and is continuously taken up on rollers. This material can either be slit down the side to give plastic film, or turned into plastic bags by the making of a single seal across the bottom of the tube. Blown diameters of up to 7 feet have been achieved leading to flat film widths of 24 feet.

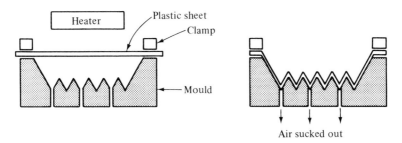

Figure 16.11 Vacuum forming.

Vacuum forming, in a sense, is the opposite of blow moulding. A sheet of heat-softened plastic is placed over a mould and the air sucked from the mould (Fig. 16.11). The plastic is drawn down and its surface conforms to the shape of the mould. It is then allowed to cool and removed from the mould.

6 Calendering

In the calendering process, a preheated polymer mix is turned into a continuous sheet by being passed between two or more heated rolls which squeeze it to the appropriate thickness. If fabric or paper is fed through the final rolls, the plastic can be pressed on to it and a plastic coated material results.

Table 16.7 Foamed plastics markets, 1970 (m. lb) USA

	Urethanes		Styrene	Vinyl	Olefin	Others	Total
	Flexible	Rigid					
Furniture	195	10	5	110	—	—	320
Packaging	9	7	243	—	5	—	264
Transportation, i.e. seats in vehicles	174	50	10	20	5	—	259
Insulation (excludes vehicles)	—	106	80	—	—	—	186
Miscellaneous (including epoxies)	205	37	48	120	15	35	460
Total	583	210	386	250	25	35	1 489
(%)	39	14	26	17	2	2	—

USA 1966: Cellular rubber—75 m. lb.; latex foam rubber—110 m. lb.

7 Slush moulding, dipping, rotational casting, paste spreading

The mixing of PVC powder with relatively large quantities of plasticizers leads to PVC pastes or plastisols, that is dispersions of PVC which will flow in a liquid or quasi-liquid fashion.

In slush moulding, a mould is filled with plastisol and placed briefly in an oven where the plastic gels to a thickness of a few millimetres. The excess plastisol is poured out and reused and the mould is placed in another oven where curing is completed.

In 'dipping', the article to be coated is preheated and dipped into a plastisol. Some of it gels, the surplus is allowed to drain off and the coated object is cured in an oven.

In rotational casting, a small amount of plastisol is placed in a heated mould which is then closed, and rotated in all directions in an oven so that its inside becomes uniformly coated.

In paste spreading, PVC paste runs on to a moving belt of fabric. This is spread into a uniform layer by a doctor knife and is then cured by passage through an oven.

8 Laminating

The best known laminates are the 'formica' type decorative laminates for household use. Brown paper is impregnated with an alcoholic solution of a resol (section 10.4.1.2) and is then cut up and arranged in piles with a suitably printed melamine-formaldehyde-impregnated decorative sheet on top. The whole is then pressed at about 150–180°C and 3 tons/sq. inch to give the finished laminate.

9 Foaming

Foamed or expanded plastics have a cellular structure and are subdivided into flexible foams with an open, interconnecting, cell structure and rigid foams with closed cells. Market data are given in Table 16.7.

The foaming may be produced in many different ways. Blowing agents such as volatile aliphatic hydrocarbons or their chloro or fluoro derivatives may be incorporated into the polymer mix and caused to evaporate by heating or decompression. Sometimes a chemical blowing agent is employed which, instead of vaporizing, decomposes on heating to give carbon dioxide (e.g. sodium bicarbonate) or nitrogen [e.g. 4,4'-oxybis (benzenesulphonylhydrazide), dinitrosopentamethylene tetramine, N,N'-dimethyldinitrosoterephthalamide, azobisformamide and urea-biuret mixture].

Some polyurethanes give off carbon dioxide on cross-linking and act as their own blowing agents. They are used for flexible foams. Finally, latex rubber is produced by a mechanical frothing process similar to the making of meringues.

16.13 Other unit processes

Table 16.8

Process	Reagents	Examples
Combustion (completed oxidation)	Fuel and air	Heating of furnaces and boilers, disposal of waste materials
Neutralization	Acid or alkali added to alkali or acid	Sodium nitrate from soda and nitric acid. Ammonium nitrate from ammonia and nitric acid
Silicate formation	Clay, feldspar and quartz	Manufacture of ceramics, cement and glass
Causticization	Calcium hydroxide	Production of caustic soda from sodium carbonate
Electrolysis	Electricity	Production of caustic soda and chlorine from brine, H_2 and O_2 from water; Al, Mg and Na from fused salts
Double decomposition		Phosphoric acid from calcium phosphate and H_2SO_4, many Na, K, Mg and Ca salts
Calcination	Carbon dioxide	Sodium carbonate from sodium bicarbonate
Condensation	Heat is often used, sometimes in the presence of acids	Diethyl ether from ethanol
Diazotization and coupling	Sodium nitrite and hydrochloric acid	Azo dyes and pigments, e.g. Hansa Yellow G from 2-nitro-p-toluidine and acetoacetanilide
Fermentation	Micro-organisms	Antibiotics, alcoholic drinks, lactic acid from hexoses, citric acid from sucrose.
Pyrolysis	Heat	Hydrocarbon cracking, destructive distillation of coal, carbon black from natural gas
Aromatization or hydroforming	Catalyst e.g. Al_2O_3 + 0.25% Pt in the 'platforming process'	Production of aromatics (benzene, toluene, xylenes) in natural or thermally cracked gasolines
Isomerization	Catalysts	p-xylene (for 'Terylene') from mixed xylenes
Ion exchange	Ion exchange resins	Water treatment

Notes and references

General The classic book on unit processes is P. H. Groggins, *Unit Processes in Organic Synthesis*, 5th ed. McGraw-Hill, New York, 1958. It gives a greater insight into chemical engineering aspects of the subject than has been attempted here but is inevitably somewhat dated. Figures 16.2 and 16.3 and Table 16.3 are based mainly on US Tariff Commission figures.

466 *Unit processes*

Section 16.1 Units are a problem with much of the literature. 1 million standard cubic feet of hydrogen weigh 2·42 tons, and of oxygen 40 tons. Table 16.1 is derived from Kirk-Othmer. US oxygen production in 1969 was 12·3 m. tons. UK sales in 1971 were 400 000 tons at an average of £3·75/ton.

Section 16.2 Data on hydrogen price and end-use come from *Reports on Progress in Applied Chemistry* **17**, 1969, R. A. Johnson and M. Litwak, *Chem. Eng. Prog* **65** (3) 21, 1969, and *Hydrocarb. Proc.* **48** (5) 15, 1969. The figures do not agree, in that the end-use pattern allocates 35% to ammonia production. US Dept of Commerce gives ammonia production in 1969 as 11·76 m. tons implying total hydrogen production of at least 6 m. tons. Of this 13%, i.e. 780 000 tons, should go into chemicals but only 550 000–600 000 tons appear to do so, leaving about 4% of production unallocated.

Section 16.5 Table 16.2 is based on the **OECD** *Annual Report*.

Section 16.8 Table 16.4 is based on J. N. Shreve, *Chemical Process Industries*, 3rd ed. McGraw-Hill, New York, 1967. See also C. S. Robinson, *Explosions, their anatomy and Destructiveness*, McGraw-Hill, New York, M. A. Cook, *the Science of High Explosives*, Reinhold, New York, 1958 and T. Urbanski, *Chemistry and Technology of Explosives*, Pergamon, Oxford, 1964.

Section 16.10.2 Technical and commercial aspects of perchloroethylene and trichloroethylene are discussed by G. Harris *ECMRA, Vienna, 1971*. The drop in DDT consumption was announced in *Chem. Age*, Oct. 15, 1971.

Section 16.10.3 Royal Dutch Shell produced a delightful booklet in 1964 to celebrate the 50th anniversary of Koninklijke Shell Laboratorium, Amsterdam. Entitled *Trail into the Unknown*, it described their re-investigation of the Deacon Process.

Section 16.11 We have drawn primarily on D. C. Miles and J. H. Briston, *Polymer Technology*, Temple Press, London, 1965. Table 16.5 is an expansion of one in Interscience's *Encyclopaedia of Polymer Science and Technology*.

Section 16.12 The standard book on this topic is *Polymer Technology*, op. cit. There is a useful chapter in L. K. Arnold, *Introduction to Plastics*, see notes to Chapter 10.

Section 16.12.2 The US plastic foam market is surveyed by D. A. Patriarca, *ECMRA, Amsterdam, 1970* and data in Table 16.7 are taken from this source. Blowing agents are particularly well discussed in the *Encyclopaedia of Polymer Sci. and Tech.*, op. cit.

17 The effluent society

After the Lord Mayor's show comes the Corporation dustcart. After every pageant of mankind there is the problem of clearing up the mess. As the standard of living in a country and its population both rise, the amounts increase of waste products of life which must in some way be disposed of. In this chapter the problems of waste disposal and of pollution will be discussed.

17.1 Historical background

The most primitive method of scavenging is when a tribe throws all its rubbish into the open where it is eaten by pigs or dogs or allowed to putrify. If the smell makes life unbearable, the tribe moves on and repeats the process elsewhere.

The next stage, which in London was reached in the Middle Ages, was the collection of rubbish and excreta by carts and its removal to a tip. As long as a town used little water and was fairly small this was just tolerable. Inhabitants threw refuse out of their windows and it rotted in the roadway or was collected in carts or washed away by the rain.

The influx of population into the towns in the nineteenth century put an intolerable strain on this system of refuse disposal. There were already a few sewers in the richer areas of town but the new urban proletariat had to rely on the mediaeval method.

'In one part of Leeds in 1839 it was discovered that "the streets had become so full of ashes, filth and refuse of every description that their surfaces were far above their original level". The Leeds Corporation admitted that "the greater part of the town is in a most filthy condition". Of the 586 streets in the town, 231 were classified as "bad" or "very bad" and some were actually impassable for the rubbish heaped up in them. Much of this rubbish consisted of human excrement. In three adjoining streets containing 452 persons, there were only two conveniences, neither of them fit to use. In Boot and Shoe yard with 340 people there were only three privies one of which – not surprisingly perhaps since the nearest water supply was a quarter of a mile away – had not been cleaned for seven years. When the dirt was at last cleared away from them it filled seventy cartloads.'

When a town was fortunate enough to have a stream flowing through it, it was rapidly turned into a sewer. London was intersected by the West and East Bourne, the Tyburn, the Old Bourne and the Fleet. These all flowed into the Thames, as indeed they still do though via underground conduits, and they provided natural sewers for much of the population.

The system was not effective for two reasons. First, at high tide, water came up the drains forcing the sewage back into the houses. Second, the Thames was not only the destination of the sewage; it was also a source of drinking water and this encouraged the spread of water-borne diseases such as dysentery, typhoid fever and cholera. Beer was easily the safest drink in nineteenth century London, since the brewing process killed the microorganisms in the water. In the cholera epidemic in Soho in 1854, the workers at the Broad Street brewery escaped unharmed because they drank only beer.

The next stage in effluent disposal was the building of sewers and the construction of cesspools. Cesspools are a relatively primitive technique, but there are still many houses not very far out in the British countryside which have cesspools rather than being connected to the main sewage system, and this is even more true in the United States.

This stage was reached in London about 1860. Epidemics of cholera in 1831, 1848, 1854 and 1866 and an understanding of their cause resulted in the construction, between 1860 and 1870, of an elaborate system of sewers and a tremendous effort was made both to carry away sewage and to provide a supply of filtered, clean water.

Even so, no effort was made to purify the sewage; it was merely led into the larger watercourses. As the population grew, so did the pollution of rivers. London still got its water from the Thames and thus Londoners were drinking much-diluted sewage from Oxford and High Wycombe.

It is difficult to determine when sewage began to be treated on a considerable scale. In the USA which is relatively thinly populated and has huge waterways, sewage treatment began between 1920 and 1930. In Britain, with a smaller population, very small waterways and a longer tradition of government and civic intervention it got under way between 1890 and 1900.

There are nonetheless still many places in Britain, particularly seaside towns, which do not treat their sewage. Most famous is Edinburgh where the city fathers have repeatedly preferred concert halls, sports stadia, etc., to sewage plant. In the USA in 1953 there were 16 661 incorporated townships which contained 100 m. of the 160 m. population. Of these, 8 743 had sewers and 6 816 treated the sewage. Of this group, 41% of the total, about one-third had plant achieving 90% purification, another third had plant achieving 60% while the remainder merely had mesh screens, chlorination or other limited devices. In hygienic Switzerland, there was an outbreak of typhoid in Zermatt due to inadequate sewers, and most of the famous lakes almost 'died' from discharge of untreated effluent; only strenuous effort over the past few years has saved them.

Sewage treatment is thus a relatively new idea and one which in many ways has followed the growth of the chemical industry as a whole. Unlike the chemical industry, however, it has not offered the prospects of profits to the people who pioneered it. In spite of the picture painted earlier in this chapter of nineteenth century apathy, Britain had an acceptable sewage treatment system before most other countries, and the fact that this was so must reflect some credit on the public-spiritedness of a substantial body of men late in the nineteenth and early in the twentieth centuries.

17.2 Domestic sewage treatment

Domestic sewage is a very dilute liquid. The average water consumption in the USA is about 100 gallons per person per day; it is about half that in Britain. Once used, whether for baths, washing up or flushing the toilet, it goes down the drains mixed with only a small amount of impurity. The average adult production of faeces for example is only about 1·5 ft^3/year and of urine, which is mainly water anyway, about 18 ft^3. Strong sewage contains only about 500 ppm (0·05%) of impurities. Industrial effluents may be more concentrated but even so the problem of sewage treatment is one of dealing with huge quantities of almost pure water. Enormous tank capacities, large pumps and extensive piping systems account for a high proportion of the cost.

The reasons why sewage should be treated are obvious. Untreated sewage leads to stench, spread of disease, pollution of watercourses and a general deterioration of the environment. The processes of sewage decay involve the reduction of sulphate ions by organic material in the sewage. If the organic material is written CHX where X covers all elements except carbon and hydrogen then the decay can be written:

$$CHX + SO_4^= \longrightarrow CO_2 + OX + H_2S \text{ or RHS}$$

In other words, the organic matter in sewage reduces sulphates to give foul smelling hydrogen sulphide and worse-smelling mercaptans. The important thing about sewage from a chemist's point of view is that it is a powerful reducing agent. In addition to the above, the break-up of proteins and other nitrogenous matter in the sewage leads to amines which mix with unpleasant nitrogen compounds already present such as skatole (which gives the smell to faeces) to accentuate the odour from the sulphur compounds.

Organic sewage can also act as a reducing agent by absorbing from river water the oxygen on which fish and underwater plants live and the river can thus 'die'.

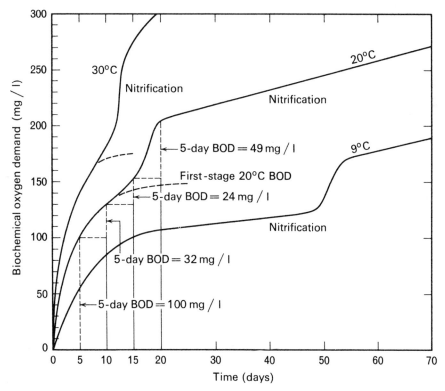

Figure 17.1 Progress of biochemical oxygen demand at 9, 20 and 30°C (after Theriault).

Regarded in these terms it can be seen that the treatment of sewage must involve the oxidation of organic material to the point at which it is unable to reduce sulphates or absorb oxygen from river water.

There would be a bonus if it were possible to oxidize nitrogenous matter to nitrites or nitrates. Once this had occurred, there would be no food for putrefying bacteria. All sewage treatment thus involves the oxidation of sewage and the various methods can be classified as either aerobic:

$$CHX + O_2 \longrightarrow CO_2 + H_2O + OX$$

or anaerobic:

$$CHX + H_2O \longrightarrow CO_2 + CH_4 + X$$

and these will be discussed later in this chapter (sections 17.2.3 and 17.2.4).

In order to be able to treat sewage successfully, methods are needed by which its strength can be measured. A variety of parameters can be followed (e.g. a modified Kjeldahl experiment gives the total nitrogen content) but the two most useful and widely used estimates of strength are the 'suspended solids' and 'biochemical oxygen demand' (abbreviated to BOD).

Suspended solids are measured by filtration of a known volume of sewage through a Gooch crucible. BOD is measured by the mixing of a sample of the sewage with aerated water. The mixture is maintained at 20°C for five days and then the amount of oxygen removed from the water is estimated by an iodine–thiosulphate titration.

Typical raw sewage contains about 500 ppm suspended solids and has a BOD of about 200 ppm. The standards for effluents (i.e. treated sewage) recommended by the

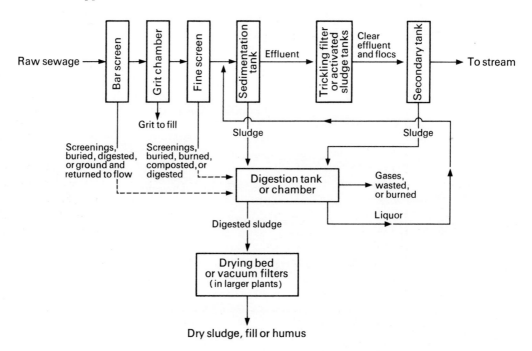

Figure 17.2 Complete treatment of raw sewage (trickling filter process).

Royal Commission and which came into force in 1969 were 30 ppm suspended solids and 20 ppm BOD.

The BOD test is a purely arbitrary one and is not entirely satisfactory, for by the time it is discovered that an effluent is outside the statutory limits it will already have been in the river for five days. It is a widely accepted test, however, and shows up any long term defects in the operating of the plant. Figure 17.1 shows the variations of BOD with time. Before nitrifying bacteria get to work, the oxidation of organic matter follows a first-order rate law with a half-life of about 0·4 days. The variation of rate with temperature may be represented by the Arrhenius equation $k = A \exp(-E/RT)$ with an activation energy of approximately 13 kcal/mole (55 kJ/mole).

A block diagram of a typical domestic sewage system is shown in Fig. 17.2. The raw sewage goes through screening and sedimentation stages which split it into a sludge and a watery effluent and these two fractions are both oxidized.

17.2.1 Screening

The raw sewage first goes through a series of screens which remove nominally 'inorganic' matter – old prams, grit, rags, etc. Sometimes the grit is washed and used for road mending. Otherwise, the detritus has to be burned, composted or buried. At this stage there is a weir which is designed to take storm water. A sewage works is normally designed to take up to three times the 'dry weather flow' of sewage. If the flow is between three and six times the dry weather value, the surplus runs over the weir into settlement tanks and then to the river. Above six times the dry weather flow, the surplus runs over a second weir and reaches the river having only been screened. This is not as bad as it sounds in that rainwater is pure anyway. A relatively new idea involves the sending of the first rush of liquid through the sewers from a rainstorm (which consists of the material already in the sewers before the rain) to a storage tank. The subsequent flow which is mainly rainwater is then directed over the weirs. The first rush of liquid can then be sent through the main sewage system at any convenient time.

17.2.2 Sedimentation

The sewage after screening contains only fine solids, emulsions and colloids. It flows to sedimentation tanks where a sludge settles out at the bottom. Rectangular settling tanks with sloping bottoms were at one time used for sedimentation, and the sludge settled at the deep end. When the tank was satisfactorily full, the supernatant liquid was drained away and the sludge shovelled out by hardy labourers. Labour costs were high and labourers not enthusiastic about the job. This type of sedimentation tank has therefore been largely replaced by the design shown in Fig. 17.3, which permits continuous operation and cuts down labour costs although, of course, capital costs are higher.

The settling out of the sludge is sometimes encouraged by the addition of a flocculating agent. The theory of colloids suggests that ions with a high charge should be most effective in precipitating colloids as may be observed in the use of alum as a styptic pencil. Iron alum is indeed found to be a very efficient flocculating agent for sewage but its use tends to be expensive. The addition of 2 ppm of iron alum to 21 m. gallons of sewage a day – the input to the sewage works at Bradford, a typical industrial town of

472 *The effluent society*

300 000 inhabitants – would involve 420 lb of alum per day which adds significantly to the running costs. A fairly new idea is the use of polyacrylamide, a cross-linkable polymer, to entrain the colloid solids.

The sludge left in the sedimentation tank is still about 95% water. The liquid which is run off the top after sedimentation is called the clear effluent.

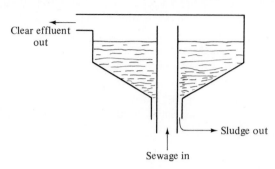

Figure 17.3 Modern sedimentation tank.

17.2.3 Treatment of clear effluent

The clear effluent from the sedimentation tanks contains the really fine colloids and has lost over half of its suspended solids. Its BOD, however, is still high – about 70% of the original value – and must be reduced further. Two methods are used both involving aerobic biological oxidation – the trickling filter and the activated sludge method.

The trickling filter is not a filter at all but a bed 3–6 ft in depth filled with pieces of broken rock 2–4 in in diameter. Furnace slag is usually used, but sandstone is a possibility and even coal was used when it was cheaper. The clear effluent is sprayed on this bed (Fig. 17.4). Biological growths develop on the rock surface and these live on the sewage constituents and oxidize them. To some extent they also convert nitrogenous matter to nitrites and nitrates. Some of the micro-organisms involved in this process are shown in Fig. 17.5.

The effluent from the trickling filters is odourless and colourless and has a BOD less than 10% of the original value. It contains lumps of biological growth, however, which have broken away from the rocks in the trickling filter. This humus is allowed to settle out in a secondary sedimentation tank and the remaining effluent is discharged into the river.

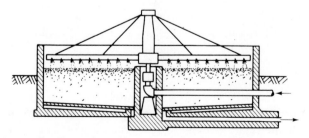

Figure 17.4 Trickling filter with rotary distributor (after Fair and Geyer).

Domestic sewage treatment 473

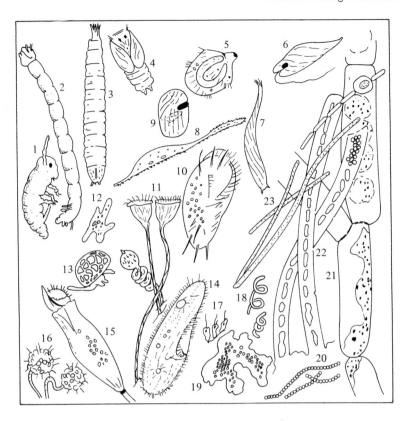

Organisms associated with the biological treatment of sewage.

Nos. 1 to 4 – Insects × 5

1. Water Springtail, *Podura;* the genus found on trickling filters is *Achorutes*.
2. Larva of Bloodworm, *Chironomus*.
3. Larva of Filter Fly, *Psychoda*.
4. Pupa of Filter Fly, *Psychoda*.

Nos. 5 to 17 – Protozoa × 150 *

5. *Didinium*.	9. *Colpidium*.	13. *Arcella*.
6. *Euglena*.	10. *Stylonichia*.	14. *Paramecium*.
7. *Choenia*.	11. *Vorticella*.	15. *Opercularia*.
8. *Lionotus*.	12. *Amoeba*.	16. *Anthophysa*.

17. *Oikomonas* × 1,500

Nos. 18 to 23 - Bacteria and Fungi × 1,500

18. *Thiospirillum*.	20. *Streptococcus*.	22. *Sphaerotilus*.
19. *Zooglea ramigera*.	21. *Leptomitus*.	23. *Beggiatoa*

* Excepting No. 17, *Oikomonas*

Figures 17.5

474 *The effluent society*

In the activated sludge process, the clear effluent is pumped to a tank which has on top of it a spinning cone which throws the sewage out in a spray and aerates it thoroughly (Fig. 17.6). In certain designs of equipment, the effluent is aerated by the bubbling of compressed air through the tank. The liquid is inoculated with activated sludge which is

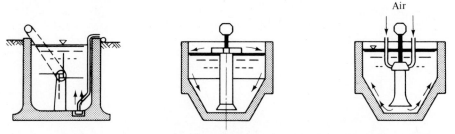

Figure 17.6 Common forms of mechanical aeration units. Left to right: paddle mechanism with diffused air, spray mechanism and aspirator mechanism.

a collection of micro-organisms similar to those found in trickling filters. They flourish in the aerated sewage and oxidize it. The effluent is pumped to secondary sedimentation tanks and the humus is allowed to settle as described previously.

The trickling filter uses more land than the activated sludge method but requires little supervision and no motive power – the distributor jets keep themselves moving by the force of the liquid which emerges from them. The activated sludge method requires less land but electricity costs are high, and more maintenance is required. It is claimed that the activated sludge method gives better purification, but this is doubtful. As land is cheaper around small towns, it tends to be the smaller local authorities who use trickling filters but this is by no means a general rule.

17.2.4 Sludge treatment

What to do with the sludge from the sedimentation tanks is the main problem of modern sewage treatment. Some local authorities put it into drained beds and hope that the sun will dry it. The British climate being what it is, this savours of gross optimism. Even under the best possible conditions, the hydrophilic groups in the sludge (—OH, —COOH, etc.) make it impossible to reduce the water content of the sludge below 55%. It is nasty to handle, and farmers have to be paid to cart it away and use it as fertilizer.

A less haphazard method involves anaerobic oxidation. The sludge is pumped to a closed tank where it putrefies to give methane and carbon dioxide. The methane is taken off and burned, usually to provide electricity for the activated sludge process. About 75% of the calorific value of the sludge can be recovered in this way. The process removes the hydrophilic groups from the sewage and thus the digested sludge has no odour and dries easily. It has little or no value as a fertilizer having been too thoroughly oxidized, and it has to be disposed of in a tip. This method of sludge digestion is obviously most suitable for use in conjunction with the activated sludge method of treating clear effluent.

A third possibility for sludge treatment is filter pressing. It gives a sludge containing 30% water which, if left in heaps, will self-heat to give a powdery product containing about 10% water and this can be sold as a fertilizer. In spite of this, the filter pressing method appears to be uneconomic in most cases.

Table 17.1 shows treatment costs of the Greater London Council averaged over their plant of all types and ages.

Table 17.1 Cost of domestic sewage treatment
(Average over all Greater London Council plant, 1970/71)

	Cost per '000 gal flow (£)
Reception of sewage into Council sewers and conveyance to plant	0·01483
Primary treatment (screening and sedimentation)	0·00920
Biological treatment (mainly by the activated sludge method)	0·01230
Sludge treatment (mainly by anaerobic digestion) and disposal	0·01540
Total	0·05173

17.3 Industrial effluents

Domestic wastes from whichever town they originate have a similar composition. The same is far from true of industrial wastes – each industry has its own special problems. Sometimes industrial wastes can be treated in sewage plant by modifications of the standard method but this is not always possible and special techniques must be devised. We mention here examples of each of the above cases.

17.3.1 Effluents from electroplating

The electroplating industry is to some extent tied to the automobile industry and is concentrated in the Midlands. A typical works may plate with zinc or copper from alkaline solutions containing sodium cyanide and with nickel and chromium from acid solutions. The dilute rinse liquors and waste liquors from these processes contain copper and zinc salts and up to 500 ppm cyanide in the former case, and similar amounts of nickel and chromium salts in the latter. A large works produces about 3 500 000 gallons per week of these effluents which are toxic both to fish and to the microbiological species in sewage works which are entirely unable to cope with them. The firms themselves are held responsible for rendering them harmless.

The acid and alkaline effluents must be rigorously segregated into separate drains, otherwise the acid would liberate poisonous gaseous hydrogen cyanide from the cyanide;

$$NaCN + HCl \longrightarrow NaCl + HCN$$

The alkaline cyanide-containing effluent is then treated with chlorine or sodium hypochlorite which destroys the cyanide and leaves a solution of zinc and copper salts:

$$NaCN + NaOCl + H_2O \xrightarrow{pH\ 11} CNCl + 2NaOH$$

$$2CNCl + 3NaOCl + 2NaOH \longrightarrow N_2 + 2CO_2 + 5NaCl + H_2O$$

The chromium in the acid effluent, which exists as chromate, is reduced to a chromium salt with sodium bisulphite or sulphur dioxide:

$$2H_2CrO_4 + 3SO_2 \longrightarrow Cr_2(SO_4)_3 + 2H_2O$$

The two effluent streams are mixed and treated with slaked lime $Ca(OH)_2$ which is available very cheaply as a by-product of acetylene production. The nickel, chromium, zinc and copper are precipitated as their hydroxides. Because slaked lime is only sparingly soluble, precipitation is slow and has to be carried out in lagoons which take up a lot of land. The optimum pH for precipitation is different for different metals (e.g. zinc can give soluble zincates in highly alkaline solution) so the efficiency of precipitation is not high. Furthermore, the precipitate is gelatinous and after settling still contains 99% water. It is pumped out and taken away in tank cars by contractors who presumably dump it in pits somewhere, though they are reluctant to divulge if and where this happens.

The supernatant liquid in the lagoons contains no chromium, nickel, zinc or copper but it is very 'hard' due to dissolved calcium, and it also contains a high concentration of chloride ions. It sometimes has to be diluted before it can be pumped into a waterway, and the electroplaters are engaged in a constant battle with the river authorities on this account.

17.3.2 Woollen industry wastes

The treatment of wastes from the wool textile industry is an example of the modification of standard sewage treatment. The following technique is used at Esholt Sewage Works, Bradford, England. Bradford is still the largest woollen town in the world and every week some 200 tons of wool grease extracted from wool during the scouring process empty into its sewers. Wool grease consists of fatty acids, hydrocarbon waxes and 15% cholesterol and is the only natural grease which is not a triglyceride. It comes down the sewers in a semi-colloidal form mixed with soap and if untreated would choke the biological filters in a matter of days.

The normal method of sewage treatment has therefore been modified so that, after it has been screened, the sewage is treated with sulphuric acid till it reaches pH 3. In theory this is done with the aid of an electrical pH meter; in practice a few drops of methyl orange are sprinkled on the effluent and the colour estimated by an experienced workman. The sulphuric acid breaks the grease-soap colloid and the grease is precipitated in the sedimentation tanks.

Sewage works like to be self-sufficient, as is illustrated by the burning of methane from anaerobic oxidation to give electricity to operate other plant. Bradford therefore likes to manufacture its own sulphuric acid, and this was done in a lead chamber plant – one of the last in the country – until 1962 when the building on top of a Gay–Lussac tower caught fire. It was replaced by a small plant operating the contact process, and surplus production is sold.

The clear effluent from the sedimentation tanks is oxidized in trickling filters in the conventional manner. It is interesting to note that the flow to the filters is at about pH 5 and yet the micro-organisms on the filters appear to flourish even under these acid conditions.

The mixture of grease and sludge from the sedimentation tanks is heated with live steam and filter-pressed through nylon bags. The grease and water come through and a sludge is left which dries easily and can be sold as a fertilizer. The grease solidifies on cooling and can be sold as a lubricant and has certain other end-uses.

As Bradford does not digest its sludge anaerobically there is no methane produced to generate electricity for aerators. Consequently it is more sensible for them to use trickling filters for oxidizing their clear effluent.

17.3.3 Petrochemical effluents

Because of the scale on which it operates, the petrochemical industry produces large quantities of water contaminated with small amounts of hydrocarbons, phenol and their derivatives. These materials cannot be dealt with by the micro-organisms in a conventional sewage works but it has proved possible to develop strains of micro-organism which can cope with them. Petrochemical complexes therefore tend to have 'bug pits' in which heavy organic chemical-eating 'bugs' digest the petrochemical effluents. Adequate aeration is essential.

17.4 Air pollution

Pollution is very much in the air at present – and in the rivers and on the fields. It is presented by the media as a consequence of the grasping materialism of twentieth century life in marked contrast to the happy innocent pastoral nature of the world we have lost.

This being so, it is interesting to find that the first laws in England to limit the smoke from coal-burning were passed in 1273. In 1578 Queen Elizabeth I stayed away from London because of the 'noysomme smells of coal smoke' and in 1661 John Evelyn, the diarist, addressed a pamphlet on the subject to King Charles II (Fig. 17.7). He describes the evil as 'epidemicall: endangering as well the health of your subjects as it sullies the glory of this Your Imperial Seat'. Evelyn suggested that factories using coal should be moved further down the Thames valley and that a green belt of trees should be put around the centre of the city 'to rid London of the columns and clouds of smoke which are belched forth from the sooty throats of (those shops) rendering (the city) in a few moments like the picture of Troy sacked by the Greeks or the approaches of Mount Hecla'.

With the increase in the use of coal during the industrial revolution, the situation worsened. In addition to coal fumes, the Leblanc process (section 1.3) poured gaseous hydrogen chloride into the atmosphere and gave a solid waste of mixed sulphides and sulphate. Rain picked up acid from the air and liberated hydrogen sulphide from the sulphides and these pollutants account for the 'blasted heath' appearance of the countryside around Widnes in Lancashire to this very day.

In 1863 Victorian reformers passed the Alkali Act which forbade the emission of hydrogen chloride and also set up the Alkali Inspectorate which still exists and is one of the major controllers of pollution in the UK.

Legislation against smoke from fuel burning had to wait another 93 years until the Clean Air act of 1956 which came about as a direct consequence of the great fog of London in 1952 which was responsible for the deaths of 3 500–4 000 people. Subsequent pollution arising from the petrol engine (e.g. Los Angeles smog) has brought pollution to the front of the political stage.

> # FUMIFUGIUM:
> ## OR,
> ### The Inconvenience of the AER,
> ### AND
> ### SMOAKE of LONDON
> ### DISSIPATED
> #### TOGETHER
> ### With some REMEDIES humbly proposed
> ### By John Evelyn Esq;
> ### To His Sacred MAJESTIE,
> ### AND
> ### To the PARLIAMENT now Assembled.
>
> *Published by His Majesty's Command.*
>
> ---
>
> Lucret. l. 5.
> Carbonumque gravis vis, auque odor insinuatur.
> Quam facile in Cerebrum?——
>
> ---

Figure 17.7 An early protest against air pollution.

Air pollution affects the population in four ways:

1. Health – e.g. pollution may lead to lung cancer or bronchitis.
2. Amenity – e.g. before the Clean Air Act Central London received considerably less sunshine than its suburbs.
3. Danger – e.g. a Diesel lorry emitting thick black smoke may obscure the vision of someone trying to overtake.
4. Annoyance – e.g. clothes need to be washed and cleaned more frequently in areas of high dirt precipitation.

Some of the problems of air pollution have already been tackled. Most of them are tractable but solutions are often expensive and the question arises as to what a community

Air pollution 479

is prepared to pay to have cleaner air. This is difficult to judge since the issues are so complex. A few of them will be outlined here.

17.4.1 Cigarette smoke

Cigarette smoke is the most virulent air pollutant of which we know. Smoke particles have to be below 20 microns in size to be harmful otherwise they are removed by the cilia (hairs) of the nose; cigarette smoke consists of sub-micron particles. In addition it contains quantities of carbon monoxide. Car exhaust fumes are a major source of carbon monoxide yet the concentration of the latter in Fleet Street in Central London at rush hour is only 20 ppm while in a Southern Region railway compartment for smokers it rises as high as 40 ppm together with 5 mg/m^3 solids. The ill effects of carbon monoxide are due to the fact that it combines much more readily with the blood than does oxygen to give carboxyhaemoglobin. The level of this in the blood can easily be measured and the value found in people exposed to petrol engine exhaust fumes rarely reaches the value which is commonplace in smokers.

In addition to this simple role as a pollutant, cigarette smoking is a major factor causing both lung cancer and bronchitis. Lung cancer is a disease characteristic of the middle of the twentieth century. In the 60 years 1900–1959 there were were 286 000 deaths from lung cancer, while in the ten years 1960–1969 there were 250 000, a figure only slightly smaller. In 1968 there were 24 048 male deaths from lung cancer, this being 8·2% of all male deaths. Among women there were only 4 997 fatalities (1·8%) representing the lower rate and shorter history of smoking among women. Furthermore, lung cancer has changed from being a disease of extreme old age, as it was in the mid-nineteenth century, to being one of middle age.

The data linking smoking and lung cancer are now extremely good. Particularly convincing are the results of a prospective survey where a group of British doctors was selected and their smoking habits and their death rate from lung cancer were followed over a number of years (Fig. 17.8).

Various objections have been put forward to the evidence linking smoking and lung cancer. It has been suggested that the cancer increase could equally well correlate with the increased use of coal, of petrol or of Diesel fuel. None of these would explain the difference in male and female deaths. Furthermore, while coal usage has been declining for some time, male lung cancer deaths have only just reached a maximum and female deaths are still increasing because female smoking is still on the increase. Diesel fuel was not widely used till the 1930s by which time lung cancer deaths were already climbing, and petrol cannot be the factor because death rates from lung cancer among garage hands, traffic policemen, etc., are the same as those of the rest of the population.

The mechanism by which smoking causes lung cancer is still uncertain but one should appreciate that the lungs present something like 90 m^2 of surface to attack by smoke. The surface tension of the fluid lining the lungs is almost zero – this must be so or it would be impossible to breathe – and there is some evidence that smoking affects this surface tension. There is also evidence that soot particles promote the growth of certain bacilli. For example, *Haemophylus influenzae* thrives on a smoke-particles-and-blood culture but will not grow on blood alone.

In spite of the mechanism not being fully understood, it is beyond reasonable doubt that there is no safe level of smoking, and that the chance of dying of lung cancer is

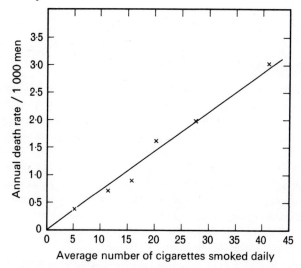

Figure 17.8 Death rate from lung cancer, standardized for age, among men smoking different daily numbers of cigarettes at the start of the inquiry (men smoking pipes or cigars as well as cigarettes excluded).

linearly proportional to the number of cigarettes smoked. The effect is to some extent reversible and if one stops smoking one's life expectancy will have returned to normal after ten years (Fig. 17.9).

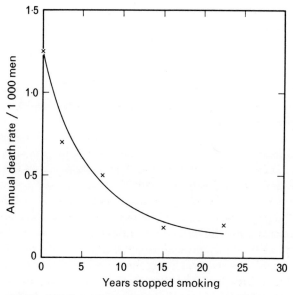

Figure 17.9 Death rate from lung cancer, standardized for age and amount smoked, among men continuing to smoke cigarettes and men who had given up smoking for different periods (men who had regularly smoked pipes or cigars as well as cigarettes excluded). The corresponding rate for non-smokers was 0.07 per 1 000.

17.4.2 Domestic and industrial fuel burning

The inhaling of cigarette smoke, at any rate in its higher concentrations, occurs as part of a conscious decision by the victim. Inhalation of pollutants in the urban atmosphere is far less a matter of choice in that it is not possible for everyone to go and live in the country. Most of these pollutants, in Britain at any rate, arise from the burning of carbonaceous fuels by industry and especially by domestic users. Consumption of coal, for centuries the main source of heat and energy, reached a peak of 214 m. tons in 1956 and has since declined. Much of it is now burned in installations such as power stations where smoke emission can be controlled. The role played by electricity, gas and fuel oil has correspondingly increased.

The main pollutants in urban air arising from fuel burning are particulate matter, sulphur dioxide and minute amounts of polycyclic hydrocarbons. The burning of soft coal in domestic grates, in particular, is less a matter of combustion than of destructive distillation. Minute tarry droplets are produced together with black smoke. In larger installations, although the fuel is burned smokelessly, ash particles sometimes find their way into the atmosphere. In a typical city the air contains 100 $\mu g/m^3$ of particles but figures of twenty times this have been recorded.

Both coal and the heavier fractions of oil contain small amounts of sulphur and virtually all of it goes up the chimney, as sulphur dioxide, during combustion. Under humid conditions this may react with other pollutants to give sulphuric acid and sulphates. A typical level of pollution is 200 $\mu g/m^3$ and again an increase of twenty times would represent severe pollution and has been recorded.

Some of the polycyclic hydrocarbons occurring in smoke, such as benzpyrene, are known to be carcinogenic (cancer-producing) when painted on the skins of experimental animals. There is, however, no real evidence linking their presence in the air with cancer in humans.

The Clean Air Act of 1956 was promulgated as a result of the great fog of London which occurred between December 5th and 9th, 1952. About 3 500–4 000 died as a result of the fog and, in the week ending December 13th, the 2 484 deaths in the administrative County of London alone were about three times the number that might have been expected at that time of year. Sulphur dioxide reached a peak of 0·7 ppm and smoke rose to 1 600 $\mu g/m^3$. The Clean Air Act controlled the emission of smoke from industrial premises all over the country and encouraged the establishment of smokeless zones where the emission of smoke by the burning of domestic fuel was also forbidden.

The effect of the act has been dramatic, especially in London. In 1950, 2·3 m. tons of smoke were emitted in the UK (165 000 tons in London) and by 1968 this had dropped to 0·9 m. tons (25 000 tons in London). More sunshine is being recorded in Central London than at any time in the past hundred years and there is no longer a difference between London and the suburbs. The nationwide drop in industrial smoke has been the most marked. Between 1950 and 1968 it dropped from just under 1 m. tons to 150 000 tons, a factor of six, while domestic smoke dropped from just over 1·3 m. tons to 0·75 m. tons, a factor of less than two (Fig. 17.10). The level of sulphur dioxide has dropped only slightly but the present level, though occasionally unpleasant, does not appear to be dangerous.

The level of smoke pollution varies widely in different areas of the country and is linked with domestic coal burning as is shown in Table 17.2. The differences in coal

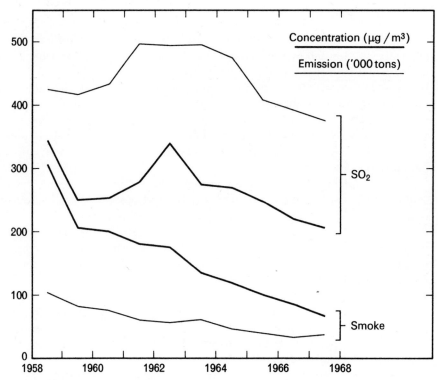

Figure 17.10 Smoke and SO_2 concentrations – London 1958–68

Table 17.2 Coal consumption and dirt precipitation UK 1968/69

Area	Coal consumption (tons/head/year)	Dirt precipitation (mg/m²/year)
North-west (Lancashire/Cheshire)	0·6	109
North	0·7	108
West	0·45	63
South-east	0·17	39

consumption are much too large to be due merely to the north of England being colder than the south. Neither can they be explained as a consequence of concessionary coal for miners in mining areas, since the coal consumption in Belfast, where there is no concessionary coal, is similar to that in the north of England. It appears to be related to cultural factors and the desire of many people, particularly in the north, to have 'a nice coal fire' instead of central heating. Indeed, the much publicized 'dirtiness' of many northern industrial towns is not nowadays due to emissions from the industry located in them but to a series of, at least partly voluntary, decisions by their inhabitants to go on burning coal in open fires.

One of the consequences of air pollution in England appears to be the high incidence of bronchitis, chronic bronchitis and emphysema.* Bronchitis is the British disease *par excellence*. Figure 17.11 shows the death rate in a number of countries. In 1967, in the 45–64 age group, 6·8% of all male and 2·6% of all female deaths were attributed to this cause.

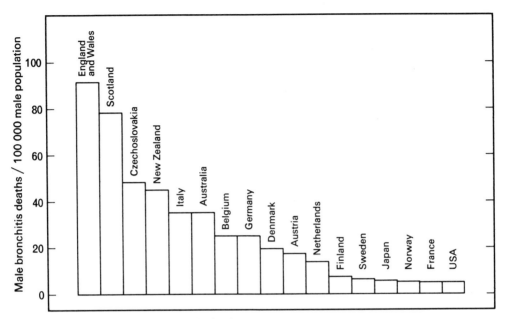

Figure 17.11 International bronchitis death rates.

The link between air pollution and bronchitis is brought out by a comparison of death rates from it in conurbations and rural areas, the former being twice as great. There is also, not surprisingly, a link between bronchitis and smoking illustrated above by the discrepancy between male and female death rates (Fig. 17.12). On the other hand, the British are not the only people to smoke and breathe polluted air and it may well be that there is an additional climatic factor which predisposes them to bronchitis. This argument is less strong when it is realized that the Dutch have a climate similar to the British but have a much lower bronchitis death rate, and that the British and some East Europeans are virtually the only people in the world who still burn coal in open fires.

These and other data show that it is impossible to attribute the incidence of bronchitis to a single cause. A Royal College of Physicians report concludes that:

'There is some evidence of a combined effect of air pollution and cigarette smoking on chronic bronchitis. By irritating the bronchial tubes cigarette smoke brings on cough and expectoration. High levels of air pollution in British cities maintain and aggravate the bronchitis thus initiated, with the risk of further disease of the lung and premature death. In the absence of cigarette smoking, these effects of air pollution are much less.'

While air pollution from fuel burning at its present level appears to be a secondary rather than a primary cause of disease, one should not forget the improved environment

* A condition in which the air spaces in the lung enlarge and break down. This interferes with the even flow of air and diminishes the effective gas-exchanging area of the lung.

(increased sunlight, better vegetation in city parks, etc.) which has resulted from the reduction in smoke since 1956 and which could undoubtedly be further improved by reductions in smoke and sulphur dioxide levels.

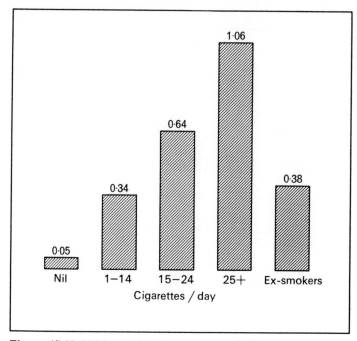

Figure 17.12 UK bronchitis death rates per 1 000 persons per year.

17.4.3 Automobile exhaust fumes

The increasing density of motor traffic in almost all urban areas of the world has given rise to questions about pollution by motor vehicle exhaust gases. The composition of a typical internal combustion engine exhaust is shown in Table 17.3. When the engine is idling a great deal more unburned petrol is discharged and, of course the proportions of the various gases can be altered by changes in carburation, but the figures are broadly true. Carbon dioxide, oxygen, nitrogen, hydrogen, water and ammonia are certainly harmless at the levels at which they are emitted. Carbon monoxide combines with the blood to give carboxyhaemoglobin and to sit in a car in an enclosed space with the engine running might well be fatal, but the level of carbon monoxide in drivers' blood is less than its level in the blood of smokers. It is possible that the carbon monoxide inhaled by a driver in a prolonged traffic jam in an inadequately ventilated street might impair his efficiency but even this has not been proved. The contribution of car exhausts in the UK to the atmospheric levels of sulphur dioxide and carcinogenic hydrocarbons is much smaller than that of coal burning and there is no suggestion that the small amounts of lower hydrocarbons (produced by cracking of the petrol) are in any way toxic although they play an important role in the production of Los Angeles smog. The aldehydes plus smaller quantities of ketones and organic acids are apparently harmless though they are probably responsible for the characteristic smell of petrol and diesel exhausts.

Lead occurs in petrol exhausts because it is commonly added to petrol in the form of lead tetraethyl as an octane improver, and the addition of 0·1% by weight will raise the octane rating by approximately five points. Ethylene dibromide is also added to petrol to scavenge the lead and the latter is emitted in the exhaust in inorganic form as the oxide, the bromide or metallic lead. It is readily absorbed into the body by inhalation and if it is inhaled faster than it is excreted the blood and urine level rises and 'lead intoxication' sets in. This occurs above 80 mg/100 ml blood. US surveys revealed levels of over 60 mg/100 ml in only 11 out of 2 300 city dwellers and the UK medical research council has claimed that only 8% of the lead taken up by the average Briton each year comes from motor vehicle fumes; indeed New Guinea natives far from any motor car show similar lead levels in their bodies. D. Bryce Smith dissents strongly from this view and claims to have found lead levels in the bodies of UK schoolchildren near those believed to result

Table 17.3 Petrol engine exhaust

	%		ppm
CO_2	9	CH_4	178
O_2	4	C_2H_4	231
H_2	2	C_2H_2	104
CO	4–10	Propylene and isobutane	89
Hydrocarbons	<0·20	Isobutylene	
		Isobutane	40
Aldehydes	0·004	1:3 Butadiene	
Nitrogen oxides	0·06–0·4	Benzene	26
SO_2	0·006	3:3 Dimethyl pentane	
		Toluene	42
NH_3	0·0006	2:3:4 Trimethyl pentane	
Particulate matter:			
Pb	4 mg/m³	Xylenes	21
Other	6 mg/m³		

in brain damage. The result of the debate is in doubt; meanwhile the lead levels in petrol in many countries are being limited.

The production of oxides of nitrogen is one of the more curious aspects of the internal combustion engine. The $N \equiv N$ bond in nitrogen has a strength of 225 kcal/mole (940 kJ/mole) and it is extremely hard to write feasible free radical or molecular reactions that would result in its fission. The initial product is nitric oxide and it appears that the latter must be formed at the very high temperatures accompanying the explosion shock front in the cylinder, but the mechanism is still uncertain.

When the nitric oxide comes into contact with molecular oxygen, it can undergo the reaction:

$$2NO + O_2 \longrightarrow 2NO_2$$

The reaction is thermodynamically favourable and one would expect all nitric oxide to be converted rapidly to the dioxide. In fact, polluted air contains about ten times as much NO as NO_2. The explanation depends on the fact that the above reaction is one of the few termolecular gas reactions and its rate is proportional to $[NO]^2[O_2]$. If the partial pressure of nitric oxide is small, as it is in a typical street, then it will oxidize only very

486   *The effluent society*

slowly. By the time it has oxidized, it may well have been blown away and there will be fresh nitric oxide being formed to maintain the high [NO]/[NO$_2$] ratio.

Nitrogen dioxide in high concentrations is poisonous and causes pulmonary oedema but at low concentrations it tends to dissolve in water in the mouth, nose and throat to give very dilute nitric acid which is harmless. Levels of 5 ppm for 8 hr exposures are considered safe in industry though there is a powerful lachrymatory effect. It is claimed that nitric oxide is more toxic than nitrogen dioxide because it is insoluble in water and can therefore be inhaled deep into the lungs where it can oxidize, albeit slowly, to nitrogen dioxide. The highest level of nitrogen oxides recorded in London is about 0·2 ppm which seems acceptable.

Dispassionate examination would suggest therefore that legislators in the UK would do better to concern themselves with cigarette smoking and coal fires than to worry about car exhausts. Legislation about car exhausts did not, in fact, originate in the UK but in the USA and it came about as a consequence of the problems of a single city – Los Angeles – whose experience is very much the exception. Nonetheless, Los Angeles smog has had such an effect on car exhaust specifications in the USA that it is impossible to ignore it.

Table 17.4 Composition of Los Angeles smog

	'Smoggy' day (*ppm*)	*'Non-smoggy' day* (*ppm*)
Aldehydes	0·05–0·6	0·05–0·6
CO	8–60	5–50
Hydrocarbons	0·2–2·0	0·1–2·0
Nitrogen oxides	0·25–2·0	0·05–1·3
Oxidant*	0·2–0·65	0·1–0·35
Ozone	0·2–0·65	0·05–0·3
SO$_2$	0·15–0·7	0·15–0·7

plus minute traces of peroxyacetyl nitrate $CH_3C(=O)O-O-NO_2$ and its propionyl and butanyl homologues.

* This is a measure of the amount of oxidizing agents in the atmosphere. UK fogs contain mainly SO$_2$ and have reducing properties.

Los Angeles smog is a greenish-brown haze which covers the city at certain times. It makes breathing difficult, it is powerfully lachrymatory, and it kills vegetation. The concentrations of the various atmospheric pollutants on a smoggy and a non-smoggy day are shown in Table 17.4. Los Angeles smog differs from a traditional British smog in three ways:

1. It is associated with warm and dry rather than cold and damp weather.
2. It has oxidizing rather than reducing properties (SO$_2$ is a reducing agent).
3. It is associated with burning of petrol rather than coal.

The damaging materials in the smog appear to be the oxides of nitrogen which are lachrymatory and the ozone and peroxyacetyl nitrate which are vegetation killers. The ozone is frequently present at levels approaching those known to cause lung damage.

There are 7 m. people in Los Angeles using 4 m. cars and 30 m. litres of petrol per day

Air pollution 487

plus 0·5 m. litres diesel fuel and 2·5 m. litres aviation fuel. The Los Angeles basin is 3 200 km² in area and the pollution is often confined by meteorological inversions which are usually stable layers of cool air trapped by an upper layer of warmer air. Into this space are emitted each day some 2 000 tons unburned or partly burned petrol, 1 000 tons nitrogen oxides and 10 000 tons carbon monoxide.

The exact mechanism by which smog is produced is still obscure, but the broad outline is clear. Nitric oxide is emitted by vehicle exhausts. This is oxidized by atmospheric oxygen to nitrogen dioxide (Fig. 17.13, reaction 1). Nitrogen dioxide absorbs ultra-

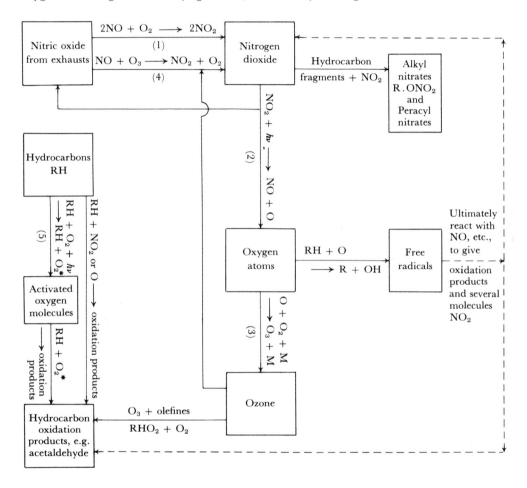

Figure 17.13 Mechanism of Los Angeles smog production.

violet light intensely in the spectral region 3–4×10^{-7} m and this wavelength is amply present in bright Los Angeles sunlight. The products of the photolysis (reaction 2) are nitric oxide, which is thus regenerated, and oxygen atoms which can go on to react with molecular oxygen giving ozone (reaction 3). The ozone can react with nitric oxide to regenerate nitrogen dioxide and oxygen (reaction 4). There is thus a complete cycle involving nitrogen- and oxygen-containing molecules which would lead to a steady but low concentration of ozone, and a much lower nitrogen dioxide concentration than is actually found.

It is the presence of hydrocarbons, particularly olefines, which makes the difference. They absorb energy from sunlight and transfer it to oxygen molecules producing an activated form (section 5) and they are oxidized in one way or another by active oxygen molecules, oxygen atoms, nitrogen dioxide and ozone. Even this would not lead to exceptionally high concentration of ozone and nitrogen dioxide, and it appears that oxygen atom attack on some hydrocarbon must result in the oxidation of more than one molecule of nitric oxide to nitrogen dioxide, that is to say there is a chain branching step in the reaction. It is this part of the mechanism which is not fully understood. Side reactions between nitrogen dioxide and partially oxidized hydrocarbon fragments lead to alkyl and peroxyacyl nitrates.

The Californian authorities have passed legislation designed to reduce this problem. Fuel oil containing more than 0.25% sulphur has been banned (petrol contains almost no sulphur) and the controls shown in Table 17.5 are already coming into effect.

Table 17.5 Californian emission requirements (g/mile of 'average' driving cycle)

	Prior to control	1970	1975
Hydrocarbons	11	2.2	0.5
CO	80	23	12
Nitrogen oxides	4	4 (1971)	1

At one stage it was hoped that careful carburation could cut down emissions to an acceptable level. Conventionally, petrol engines operate with excess petrol and the excess appears in the exhaust gases. If excess oxygen is used instead, the emitted hydrocarbons drop sharply but oxides of nitrogen increase and unfortunately, even using stoichiometric mixtures, the combined level of the two pollutants is still unacceptable. The motor industry therefore decided that it would have to remove pollutants after the petrol had burned and recent engine designs involve a reactor to remove nitrogen oxides and an afterburner to remove carbon monoxide and hydrocarbons. Because so much pollution arises when the engine is idling, crankcase emission must also be controlled.

The afterburners which have so far been designed depend on catalyst systems some of which are poisoned by lead, and it may be that lead-free petrol will have to be marketed for this reason and not because of health hazards of lead. The motor industry claims that the 1972 Californian requirements add little to the cost of a car but that the 1975 standards might add £100–£200.

It is of interest to consider whether this is worthwhile outside Los Angeles and Tokyo which has recently had its first outbreak of photochemical smog. All men of goodwill are in favour of reducing pollution, but as the Royal College of Physicians report on *Air Pollution and Health* puts it, 'The burning of fuel for heat and power is the prime cause of pollution of air in British towns . . . most of the pollution likely to affect health comes from the burning of coal fires in homes'. It might perhaps be better to concentrate resources on the implementation of the Clean Air Act especially in areas in the North of England where soft coal is still widely used as a domestic fuel. It is also of importance for the sulphur dioxide level in London to be watched and every encouragement given to the

17.5 Long-term pollution of the biosphere

The effects of pollution so far discussed have been immediate and local involving deterioration of health or 'killing' of rivers. The long-term and less direct effects of industrial activity are much more difficult to assess.

17.5.1 Medium-scale effects

Examples of medium-scale effects are: Tees-side mist, 'heat haze' in Ireland, acid rain in Sweden and the Laporte anomaly.

White Tees-side mist occurs near the chemical complex in Billingham-on-Tees when there is an escape of ammonia from some of the plant. This combines with atmospheric sulphur trioxide, from the burning of fuel, to give a thick white mist of ammonium sulphate.

Laporte is a small town east of Chicago. Its rainfall has increased steadily over past years while that in South Bend, 25 miles away, has remained constant. The rain appears to correlate with days of smoke haze over Chicago but the reason is unknown.

Similarly the reasons are not understood for the occurrence of acid rain of pH 3 in Sweden or of 'heat haze' in Ireland when there is a westerly wind but no especial heat. Again, a link-up with industrial activity is assumed.

17.5.2 The 'greenhouse' effect

Increase in human population and of the amount of fuel consumed by each one has resulted in an increase in the world-wide level of carbon dioxide in the atmosphere. From 290 ppm in the nineteenth century, it has now risen to 318 ppm. Carbon dioxide is transparent in the visible and ultra-violet regions of the spectrum and therefore does not prevent light of these frequencies from reaching the earth. Having reached the earth, however, the energy is dissipated as heat which the earth then tries to re-radiate as infra-red radiation. Carbon dioxide absorbs strongly at three frequencies in the infra-red and so absorbs some of this radiation and gains in temperature. The more carbon dioxide there is in the atmosphere the more heat will be prevented from leaving the earth and the warmer it will become. This is called the 'greenhouse' effect and it was calculated that a 10% change in carbon dioxide level would raise the average world temperature by 0·3°C with possible effects on cloud cover, the polar ice cap, etc.

In fact, the world temperature rose as predicted until 1940 when it went through a maximum and then started to fall – this effect was over and above the well-known long-term secular variation in world temperature. It transpired that the 'greenhouse' effect was being overshadowed by a sharp rise in the turbidity of the earth's atmosphere which was preventing a proportion of the sun's radiation which would normally have reached the earth from doing so.

The world temperature trend has therefore been exactly the opposite of what was predicted by the best scientific opinion and, if nothing else, this illustrates the dangers of making any sort of forecasts about the biosphere. The further north one goes, the worse the temperature drop has been. Franz Joseph Land has suffered most and Britain has also suffered. The warm water drift across the Atlantic, which results in the UK having a warmer climate than most countries in its latitude, is wind driven and the number of days of westerly wind has dropped catastrophically.

It is tempting to attribute the increased turbidity to industrial activity, and satellite photographs indeed show a haze over Western Europe and the USA. It is much worse however over Africa and Asia where it may be a silicate haze from deserts. Over tropical forests there is a blue haze which may be due to the photochemical oxidation of naturally occurring terpenes and would thus be a 'vegetable smog'. It is increasing at the high rate of 30% per decade and shows strong seasonal variations being densest in spring and summer. For this reason, it is probably not industrial in origin. It may be related to the dust raised by mechanized agriculture but it is difficult to see how. There is a presumption that it is related to man's activities but even that is not certain, and many meteorologists feel that the earth's climate is determined by sunspot activity and similar factors and is totally unaffected by anything man might do.

17.5.3 Insecticides

One of the major causes for optimism in the 1940s and 1950s was the possibility of making a better life for people, especially those living in tropical areas, by improvement of agriculture and elimination of insect-borne disease, such as malaria, by use of insecticides. These hopes were not unfounded. The spraying of the breeding grounds of mosquitoes has eliminated malaria from large areas of the world, and crop yields and quality have been dramatically increased by insect control.

The insecticide which has been responsible for this miracle is, of course, DDT (dichlorodiphenyltrichloroethane). Fortunately from the point of view of its insecticide action, but unfortunately from the ecological point of view, DDT is fat-soluble but not water-soluble. DDT which is ingested by living beings is not excreted but retained in the fatty tissue. Above a certain level it is a health hazard and can lead to sterility. Insects have or can develop a tolerance to small doses of DDT and the bodies of many live insects therefore contain low, non-lethal concentrations of DDT. Some of these are eaten by birds or frogs who concentrate the DDT in their own fatty tissue. Subsequent predators eat the birds and frogs and concentrate the DDT further. Eventually, the DDT may build up to toxic levels in the tissues of animals or birds a long way down the life cycle. Human beings are most likely to absorb DDT by drinking milk from cows who have ingested it through its having been sprayed on the grasses, etc., which make up their diet.

The rapidity with which DDT contamination spreads was established by the discovery of relatively high concentrations in penguins' eggs in the antarctic, and its effect on animal populations is illustrated by the virtual disappearance of frogs from Britain.

The US Environmental Protection Agency banned the use of DDT in the USA, with trivial exceptions, from the end of 1972. Many Americans, plagued by termites, mosquitoes etc. nonetheless feel that they cannot do without it. The discovery of a substitute is clearly a matter of priority, but the solution may well be to use less effective insecticides for general purposes and for the government to monopolize the use of DDT and reserve it for situations where no other material is suitable.

17.5.4 High level jet aircraft flights

Supersonic jet aircraft such as the Concorde are expected to cruise at a height of about ten miles in a rarefied zone known as the stratosphere. Engine exhausts from four hundred such aircraft would increase its natural uptake of water by a third and it has been suggested that this could form additional clouds and reflect a proportion of the sun's heat away from the earth. In addition, the exhaust gases are likely to contain about 1% of nitrogen oxides and these might interfere with the photochemically maintained ozone layer which protects the earth from damaging ultra-violet radiation. Various mechanisms for nitrogen oxide catalysed destruction of ozone have been proposed and are being studied. The ozone layer might also be endangered.

17.6 Conservation and regeneration

The previous sections have dealt with some of the problems posed by the waste materials produced in an industrial society. It is not a complete list. No mention has been made for example of the problems of pollution by oil spillage in accidents involving tankers, organo-mercury poisoning, or disposal of waste plastics. The list of actual or potential dangers to the environment is seemingly endless. As will be explained in Chapter 18, we do not consider that the end of the world is at hand, and we believe that many predicted disasters are either illusory or can be prevented. On the other hand it is vital for the scientifically trained members of the community to be aware of the sort of problems which can arise so that they can assess their magnitude and if necessary press for solutions.

This book has laid great stress on costs. When an individual or an industry puts waste products down the drain or into the air, he is passing the cost of processing these materials on to the community (or, if the materials are allowed to pollute the environment, the community is bearing a cost in terms of health and amenity). Every community has to decide how far it will pay for communal waste disposal out of general taxation, e.g. by garbage collection from houses, how far it will insist on citizens processing their own wastes, e.g. by fitting afterburners to their cars, and to what extent it will permit deterioration of the environment, e.g. by permitting the emission of smoke from coal fires. There are no fixed answers to these questions and each community will find a different one based on its size, density of population, wealth and level of culture. The scientist in this situation has a special responsibility towards the community to see that the correct decisions are taken.

With reference to the long term dangers already discussed, it is reassuring that of the various pollutants which can be detected in the earth's atmosphere as a whole, only one, carbon monoxide, is produced in greater quantities by man than by biology in general, and even then it is produced on almost the same level by a marine organism which uses it in order to float. All other air pollutants – carbon dioxide, methane, hydrogen sulphide, sulphur dioxide, nitrogen dioxide, nitrous oxide and ammonia – when considered on a global scale are produced in greater quantities by nature.

The solution to environmental problems is not a simple one. Every decision to preserve amenities requires diversion of strictly limited resources from other activities. Insecticide control may mean less food for starving people. Strict effluent control may mean that a chemical firm goes out of business because it can no longer compete economically with a similar firm in a more permissive country. It is no use living in a prosperous country if we are all being poisoned by our own effluvia but it is equally no use living in a pristine and clean country if we are all unemployed and starving. Environment must compete with other desirable aims for communal resources, the main difference being that pollution charges are paid by the community rather than the individual.

This last proviso leads to the political wrangles that affect any legislation or communal expenditure. In Britain, it is interesting to note that the Labour Party is enthusiastic about cleaning up waterways but less keen on enforcing the Clean Air Act because the major air pollution comes from domestic fires in mining areas which traditionally vote Labour. The Conservative party via Sir Gerald Nabarro, on the other hand, was responsible for the Clean Air Act but is less enthusiastic about cleaning waterways because this would mean added costs for the industrialists who are the principal contributors to Conservative party funds. In many towns with a large chemical industry, labour and management seem united in their opposition to effluent control, the latter fearing for their profits, the former for their jobs. In a democratic country one can only hope that an enlightened legislature backed by an informed public opinion will succeed in taking and enforcing the most beneficial decisions.

Notes and references

General The flood of publications on the environment in recent years has become a source of pollution in itself. The following are books we have actually used but there are undoubtedly many excellent ones we have omitted. B. A. Southgate, *Water Pollution and Conservation*, Thunderbird, London, 1969, is a useful book with an extensive bibliography. Royal College of Physicians, *Air Pollution and Health*, Pitman, London, 1970, is an authoritative report, and the *Daily Telegraph Magazine*, May 28th, 1971, produced some excellent photographs and maps.
ECRMA Conference, Vienna, 1971, *Chemical Markets Affected by the Environment*, London, 1971, shows the situation from the point of view of the chemical industry.
Cleaning our Environment – The Chemical Basis for Action, ACS Washington, 1969, is a report by the ACS sub-committee on environmental pollution, which is a sub-committee of the Committee on Chemistry and Public affairs.
An Introduction to Sewage Treatment, Institute of Sewage, London, 1960.
W. Strauss (ed.), *Air Pollution Control*, Wiley, New York, 1971.
A. C. Stern (ed.), *Air Pollution*, 2nd edn, 3 vols., Academic Press, New York, 1968.
Section 17.1 The description of Leeds is quoted in N. Longmate's, *Alive and Well*, Penguin, Harmondsworth, 1970, a popular and well written study of public health since 1830. Data on US sewage systems come from Kirk-Othmer, op. cit.
Section 17.2 This section draws heavily on K. Imhof and G. M. Fair, *Sewage Treatment*, Wiley, New York, 1956, and Figs. 17.1, 17.4, 17.5 and 17.6 are taken from this source. K. Imhof, W. J. Müller and D. K. B. Thistlethwayte: *Disposal of Sewage and Other Waterborne Wastes*, Butterworth, London, 1971, is an updated version of this text. Analytical methods are to be found in *Methods of Chemical Analysis as applied to Sewage and Sewage Effluents*, HMSO, London, 1956, which is used by local authorities and *Recommended methods for the Analysis of Trade Effluents*, ABCM, London, which is used by the trade. Treatment costs in Table 17.1 were quoted by the Greater London Council.

Section 17.4 John Evelyn, *Fumifugium*, 1661, reprinted by National Society for Clean Air, 1961.

Section 17.4.1 In addition to the RCP report on *Air Pollution*, op. cit., there is a new report by the same body on *Smoking and Health Now*, Pitman, London, 1971. US data is summarized in US Dept. of Health, Education and Welfare Chart Book on *Smoking, Tobacco and Health*, PHS Publication no. 1937, obtainable from US Govt. Trinity Office. Data on cancer deaths came from the Registrar General, *Statistical Review of England and Wales* HMSO, London, various years. Figures 17.8 and 17.9 came from R. Doll and A. B. Hill, 'Mortality in relation to smoking: Ten years observations of British Doctors', *Brit. Med. J.*, 1964, **1**, 1399 and 1460.

Section 17.4.2 Much of the information on smoke in the UK came from the Warren Spring Laboratories and the RCP report, op. cit. Figure 17.11 is from J. Crofton and A. Douglas, *Respiratory Diseases*, Blackwell, Oxford, 1969, who plotted WHO data.

Section 17.4.3 Tables 17.3 and 17.4 are from A. C. Stern, op. cit. The problem of lead poisoning has been widely ventilated. See, for example, P. J. Placito and M. J. Bennet, *ECMRA Vienna, 1971*, op. cit.; B. Silcock, *Sunday Times*, 25th April, 1971, D. Bryce Smith, *Chem. in Brit.*, 1971, **7**, 54 and 284, A. L. Mills, *Chem. in Brit.*, 1971, **7**, 160, and W. E. Broghton, *Petroleum Review*, October 1971. An excellent article by J. R. Goldsmith on 'Los Angeles Smog' with two splendid colour photographs appeared in *Science Journal*, 1969, **5**, 44. See also *Chemistry in Britain*, June 1972.

Section 17.5.4 See R. E. Newell, *Nature*, 1970, **226**, 71. The whole problem of supersonic flights in the stratosphere is being considered in a three year, $20 m. programme supervised by the Climatic Impact Committee of the US National Academy of Science and financed by the US Department of Transportation. Oxygen atoms are present in the ozone layer and reaction cycles such as:

$$NO_3 + O_3 \to NO_2 + O_2$$
$$NO_2 + O \to NO + O_2$$

could remove ozone. The situation is much more complex than this, however, and the rate constants of at least 100 important reactions are being determined as part of the above programme which also involves studies of engine emission and biological effects of ultra-violet radiation, monitoring of the stratosphere and modelling of gas transport in it.

Part V **The future**

18 And will the trees grow up to the sky?

18.1 Yesterday's tomorrows

The first commercial air services appeared in Europe immediately after the First World War. By the middle 1920s, a sparse network of highly unreliable services enabled the adventurous traveller to proceed between the major European capitals at speeds not much greater than those of the best express trains. The central problem was that of aircraft reliability. This was solved by the general introduction from 1930 onwards of the large rigid airship which made it possible to proceed in near-total comfort not merely from London to Paris or Rome but to Bombay, to Singapore, to Sydney and, from 1938 onwards, to New York. The airship had, however, one major inherent defect. Its speed is limited even today to no more than 100 knots. Forward-looking engineers continued the development of the potentially faster heavier-than-air craft with increasing success. In 1970 the first transatlantic aeroplane service was introduced. It carried 20 passengers at the hitherto unparalleled speed of 150 m.p.h.

The book from which this remarkable glimpse of the future is taken was published in 1929. The writer was Neville Shute Norway, now remembered as the popular novelist Neville Shute, but then a rising young aeronautical engineer. As this example shows, futurology is a particularly unforgiving science. Futurologists tend to be remembered for their errors; their correct predictions are seen as obvious truisms. Nevertheless it is essential for us to have some view of the future and even the greatest opponents of forecasting show by their everyday lives that they are making unspoken assumptions about what the morrow will bring. It is not the purpose of this book to encourage the view that it is better to avoid error than to risk being wrong. It is rather our aim to encourage the reader to think, to examine his assumptions, and to make his mistakes on paper. We therefore make no apology for concluding with a free examination of what the next fifty years holds for the chemical industry.

18.2 Five versions of the future

From what has been said in previous chapters, it is obvious that the chemical industry is inextricably linked with the world economy as a whole. Its future depends not only on endogenous factors such as advances in technology but also, and probably to a greater extent, on exogenous factors such as wars and political upheavals. Here are five popular versions of what the next two generations might experience.

18.2.1 The trees *will* grow up to the sky

The future will be very much like the present only more so. The advanced nations will become rapidly richer. They will be joined by certain countries which are in an intermediate stage of development such as South Korea, Taiwan and perhaps India. The differences in prosperity between the advanced and less advanced countries will remain large but will tend to diminish since countries starting from a low base can achieve a high growth rate by using freely available technology. Differences in prosperity between advanced nations will mirror present trends. By the year 2000 Japan will be the largest single industrial nation in the world. In contrast, Britain will have sunk to the bottom of the pecking order for industrial nations. In the chemical industry, too, present trends will continue unbroken. The production of plastics will become ever more increasingly its central activity. By 1985, plastics production will equal steel production; by 1990 the total world production of plastics will have reached approximately 1 000 m. tons per year. No major changes will take place in political organization or in value systems except that backward countries will become more like advanced countries. There are no major problems which cannot be relatively easily overcome by man's all-embracing ingenuity.

This somewhat Panglossian view probably represents the unspoken set of assumptions upon which most senior managers, politicians and economists of the present day work. It is most forcibly presented in the writings of Herman Kahn and at a more popular level in the *Economist* and other journals which live by telling their readers what they like to hear. The level of scholarship of this school is not high. Dr. Kahn's well-known writings resemble nothing so much as the pre-suppositions of a Middle-Western business man turned into computer print-out. His extrapolations can be criticized on intellectual grounds (section 18.3.1), and on aesthetic ones: it seems improbable that the future could be so banal, so obvious. The history of the last hundred years suggests, in fact, that life is a great deal more unpredictable than straight (or even slightly curved) line extrapolation predicts. When one comes, moreover, to questions of individual industries, one notices that the history of the science-based industries has been one of sharp changes rather than continuous unmodified growth for long periods. It may well be that the future holds great things for the chemical industry but it seems less likely that the industry in the year 2000 or 2050 will simply be a larger version of what we have today.

18.2.2 Doom is just around the corner

Population, standards of living, pollution and the consumption of irreplaceable natural resources are all linked in a fairly simple way. All are increasing exponentially. World population will rise rapidly in the next 50–100 years and reach a point at which available stocks of raw materials are consumed. A major demographic crisis will result and by the year 2100 the people of the world will be sadder, wiser, poorer and considerably smaller in number than they were a short time before. If raw material resources prove to be larger than is now supposed then the increase in pollution will produce the same effect a little later. Any limited problem which one can solve is part of a larger problem which one cannot solve.

In many ways this view, widely popular among the more extreme environmentalists, is simply the Kahn model stood on its head. Present trends will continue indefinitely and there is nothing much that can be done about it. The forecasts of this school are more categorical if less felicitously phrased than those of the prophet Jeremiah.

The models on which these gloomy predictions are based, however, are distinctly suspect. It is far from clear that the variables considered are relevant. No allowance is made for market forces which send up the price of a disappearing resource so that it lasts longer, nor for the substitution of a disappearing resource by an alternative one. Moreover, it seems likely that the relationships assumed between the variables are in error. Thus the model of Professor Forrester postulates that investment of capital is inevitably linked to an increase in pollution, an assumption which would certainly be contested by the Tennessee Valley Authority among others. Furthermore, when the model of Professor Forrester is run backwards, and it should after all be able to predict the past as well as the future, it emerges that the late nineteenth century was a period of catastrophic population decline, a conclusion strikingly at variance with that recorded by more conventional historical methods. When the coefficients in the model are adjusted to give a correct version of past population history, the disaster predicted in the future also disappears. Again, it appears that the model is very sensitive to small changes in the values of some of the variables. For example, a 1% change in the rate at which new resources were discovered or pollution control technology was improved would lead to very much more optimistic conclusions.

Above all, it is difficult to ignore the element of mixed *schadenfreude* and selfishness which pervades so much of the thinking of the extreme environmentalist movement, many of whose activities appear motivated by a desire both to keep scientists in their place and the underdeveloped countries underdeveloped. When Brazil invites any firm plagued by pollution problems at home to set up in business in Brazil, she reflects in an extreme way the fact that the environmentalist movement is essentially middle class and confined to countries whose populations live considerably above subsistence level.

Yet it would be unfair to dismiss the work of Professors Forrester and Meadows as a fraud. Undoubtedly there are interrelationships which involve all countries, even though we do not know what they are. The world economy is more than a figure of speech. Moreover, one must agree that the world's resources are not limitless and that, at the very least, a large scale process of substitution is bound to take place during the next century, as indeed it has during this one. The danger in the Forrester–Meadows approach is that its predictions may paralyse action rather than stimulate it.

18.2.3 Rich world, poor world

The advanced countries continue to prosper and become steadily more closely connected. Crippled by high birth rates and value systems which emphasize *machismo*, the underdeveloped countries remain poor. With the end of American–Russian hegemony, the great powers – the USA, USSR, China, Japan and the Common Market – manipulate the nations of the Third World to maintain their access to raw materials, to keep the poor countries in their place and to score points off each other. Immigration from underdeveloped countries is stopped. Diplomatic incidents and *coups d'etats* abound, and the great powers fight proxy wars by sponsoring movements of 'national liberation' in each other's client states. Urban guerrilla movements, neo-anarchist in philosophy, indiscriminately violent in practice, dedicated to the chimera of world revolution and financed by rival great powers, provide outlets for both the mentally unhinged and the genuinely oppressed. The situation resembles a large scale replay of the European diplomatic game of 1890–1910, with disputes about drilling rights off the Maldives or the supply of fighter aircraft to Burundi taking the place of squabbles about the control of the Moroccan police or the Sandjak of Novibazar.

In many respects this is a realistic scenario. The majority of the underdeveloped nations have shown little sign that they will 'take off' into sustained growth in the foreseeable future. This is especially true of the Arab countries. Iraq is a case in point: possessed of vast mineral resources and land which was once exceptionally fertile and which could be made fertile again, it remains poor, backward and thinly populated. Traditionally the site of the Garden of Eden, it might still be flourishing were it inhabited by Chinese rather than Iraqis. The countries of the Arab Middle East are more fortunate than many in that they are thinly populated and control the bulk of the world's readily accessible oil reserves. Yet their system of values is so unsuited to a modern industrial country that they seem unlikely to be able to exploit their temporary riches for permanent ends. Many other underdeveloped countries are less fortunate in that they possess no obvious sources of wealth. The widely prevalent traditions of violence and political instability are further obstacles to economic growth. Moderate political leaders, offering a modest prosperity in return for hard work and family planning, tend to be pre-empted by demagogues offering the Kingdom of God on Earth as a result of violent revolution. The latter, even if successful, is scarcely likely to result in stable or benevolent government.

The interest of the great powers in 'taking underdeveloped countries in hand' is likely to be small. Economic and political suzerainty is cheaper than straightforward colonialism, requires fewer troops and leaves a country less open to charges of imperialism. The situation with regard to small scale violence – terrorism, hi-jacking, etc. – depends crucially on the extent to which terrorists can command the necessary fire-power, recruit new members and fade into the surrounding urban population just as Butch Cassidy and the Sundance Kid disappeared into the Nevada Desert. Conversely, the success of the authorities depends on the degree of ruthlessness they are able to show without driving the population into the terrorists' camp and on their success in winning the hearts and minds of their people. How far insurrectionary or indiscriminate violence can be controlled within the framework of a relatively free society will govern the sort of community in which our children will live.

Whichever way the balance tips, in this sad but credible scenario, the rich get richer and the poor remain backward and serve as client states and as suppliers of raw materials to their prosperous brethren. They are rewarded for their good behaviour by the occasional gift of a second-hand Mig fighter or a couple of thousand tons of substandard napalm to use on their neighbours.

18.2.4 Like, there's been a change of values, man

The influence of the Protestant ethic, already declining in the USA, vanishes altogether. The idea of work as a good in itself disappears and the necessity for work diminishes because of automation. A general consensus in favour of self-realization and 'doing one's own thing' emerges. Economic growth ceases and further striving is left to those countries not fortunate enough to have 'made it' yet. The advanced countries settle down to a life which is totally permissive, markedly fun-orientated and rather violent. The more sombre parts of the chemical industry are transferred to the developing countries but those parts of the pharmaceutical industry concerned with psychotropic and hallucinogenic drugs grow with enormous speed.

In many ways this is quite an attractive picture although its economic and social consequences are far from clear. On a small scale, a life of this kind has already emerged in parts of the USA – notably California and Florida – although, as recent sensational episodes have shown, it is not without its disadvantages. It is uncertain whether a totally permissive society on a large scale could organize itself to live above subsistence level. The upholders of the Protestant ethic may be correct when they say that the value-free society is not value free at all but merely values 'doing one's own thing' above 'loving one's neighbour'.

It does appear, however, that as a country becomes affluent, the drive to become more affluent is fairly rapidly attenuated, and other values come to the fore. A country as rich as the USA is in an enviable position to experiment with many life styles, some of which are radically at variance with the prevailing orthodoxy. Whether European countries are in such a position is an open question; whether they should be is a matter for individual taste.

18.2.5 A generation of difficulties

Widely diffused economic growth continues but the problems of pollution and of dwindling energy sources become acute. The former is relatively easily cured by the application of existing technology and the diversion of comparatively small amounts of national incomes. The second is more fundamental. From a technical point of view, oil is by far the most attractive fuel, but the world's reserves of oil are limited and consumption rises faster than the discovery of new reserves. The Arab nations exploit to the full their position as owners of the world's cheapest and most accessible reserves, and oil rapidly becomes more expensive. Roundabout the turn of the century, in spite of extensive exploration of the sea-bed, oil becomes scarce. By this time, however, commercial thermonuclear power is becoming available albeit at a relatively high price. From this point on, world industry is based entirely on this new source of energy which uses an infinitely available raw material and produces no poisonous effluents whatsoever. The chemical industry grows rapidly after 1970 but slows down sharply in the eighties and nineties as oil becomes more expensive. Once cheap electricity is available, it starts to grow again though on the basis of a completely different technology.

In certain respects, this seems the most likely of all the scenarios which have been considered. Although the oil companies are professional optimists, it seems that the age of cheap oil has gone for ever, and that not only will oil increase in price relative to other sources of energy and of carbon but that the petrochemical industry will suffer accordingly. The price of oil is to some extent limited by the possibility of its substitution by coal or its extraction from shales or tar sands, but swingeing increases could occur nonetheless. Some consequences of this possibility are discussed in section 18.3.4 below.

18.3 But what's in it for chemistry?

No one of these scenarios is entirely exclusive. Elements from all might conceivably be combined. Other scenarios are also possible, although that favourite of the 1950s, the all-out nuclear war, has become much less fashionable. As far as the world economy is concerned, however, it will be seen that the chemical industry might flourish in both developed and developing countries, as in section 18.2.1, different sections might flourish in both, as in section 18.2.4, and it might flourish in developed but not underdeveloped countries, as in section 18.2.3. It might suffer difficulties, as in section 18.2.5 or flourish and then collapse utterly, as in section 18.2.2.

It is, however, notorious that birds-eye views of this kind are all too often strictly for the birds. An alternative to this highly aggregated approach is to consider the future of the chemical industry in terms of the factors which have affected the past and present of its several parts. To this end we now examine the future of the British chemical industry – typical of that of an advanced nation – on the assumption that no major changes in its technological bases are to be expected (section 18.3.2), and, more briefly, that of the world chemical industry (section 18.3.3). We conclude with an attempt to identify the major innovations of the next thirty years and to estimate their impact (section 18.3.4).

Before we proceed, however, we must consider a major conceptual problem in futurology: if the trees do not grow up to the sky, when do they stop growing?

18.3.1 The logistic curve and problems of forecasting

One of the more interesting world obsessions since the Second World War is with the idea of sustained economic growth. A government is expected to manage its economy in such a way that the gross national product increases each year at some arbitrary figure – 3%, 15% or intermediate figures depending on the country and the government. These figures are high even by the standards of nineteenth century industrial expansion, and yet countries growing at, say, 2% per year are seen both by themselves and others as being in a relative decline.

The origins of the obsession are obscure, but it is undoubtedly true that if many measures of a country's industrial capacity or wealth are plotted against time, an exponentially rising curve is obtained. That is to say that the population, gross national product, sulphuric acid production or whatever other measure one takes, frequently turns out to have grown at a constant rate of compound interest (so many per cent per year) over a sustained period. Such is the case for caustic soda and chlorine production, as illustrated in Fig. 14.12, and indeed for polystyrene production as may be seen if the data in Fig. 2.3 are plotted on a semi-logarithmic scale. The law of exponential growth has many applications and the naive futurologist is frequently tempted to predict the future by plotting his data on a semi-logarithmic scale, drawing the best straight line through his points, and extrapolating to the year 2000 or whatever other period interests him.

Cases where this procedure fails are legion. The trees do not grow to the sky. The depth of horse manure in British streets, extrapolated from mid-nineteenth century data, should by now engulf her citizens; instead the horse is virtually an extinct animal. Coal consumption has not soared as it should have done and the number of pawnbrokers long ago passed its maximum. Figure 18.1 shows the USA total population and the number of Ph.Ds in science and engineering plotted on a semi-logarithmic scale. Extrapolation shows that in another 150 years, there will be more Ph.Ds than people in the USA. Whilst the award of research degrees to chimpanzees or dolphins cannot be entirely discounted, the absurdity of the extrapolation technique is clear.

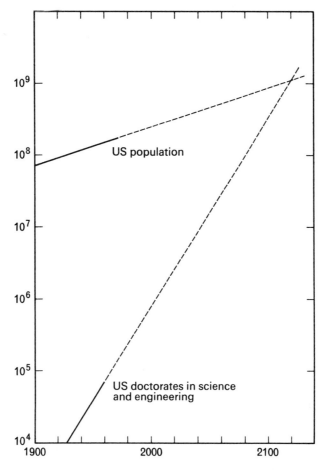

Figure 18.1 Total population and numbers of doctorates in science and engineering in the USA.

In the real world things do not grow and grow until they reach infinity. Rather, exponential growth slackens and approaches some sort of saturation level. A function more realistic than steady exponential growth is the S-shaped logistic curve shown in Fig. 18.2. This curve is also found in nature: for example it is the way in which the length of a beanstalk varies with age. It also represents fairly well such things as the ownership of television sets or the annual world production of zinc. It starts at zero growth, goes through a period of exponential growth, then linear growth and then bends over to approach a saturation limit.

As this function is so much more realistic than the exponential one, why is it not used instead? The answer is that it is virtually impossible to construct it unless one knows the saturation limit or, which amounts to the same thing, the mid-point about which the curve is symmetrical. Indeed, over a large portion of its length the curve is effectively exponential. Figure 18.3 shows the curve in Fig. 18.2 plotted on a semi-logarithmic scale. Production rises virtually exponentially from 1 ton to 100 000 tons and it would be

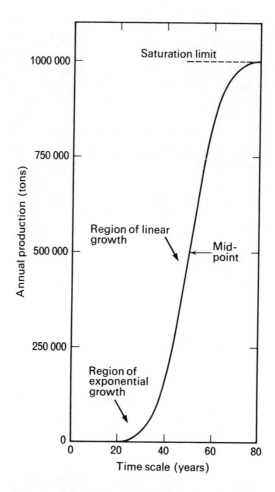

Figure 18.2 Logistic curve plotted on a linear scale. (The figures on the axes are arbitary: it is the shape of the curve which is significant.)

a bold forecaster who would discount the inevitable scatter in his data and predict that a saturation limit of a million tons was not far away.

A major problem of forecasting, therefore, is the estimation of saturation limits. In the case of television sets or refrigerators, this is not too hard; in the case of chemicals and plastics it is exceptionally difficult. It is possible to split plastics consumption into the various end-uses and make forecasts of each end-use, but though a better technique, this still involves a high degree of guess-work in the estimation of various saturation limits.

In industrial life, furthermore, the form of the logistic curve is rarely ideal. When a product reaches its saturation limit, it does not usually cease to grow, but grows instead at the same rate as the economy as a whole. Certain products show cyclical behaviour. For example, if television were suddenly introduced into a country, annual sales of TV sets would rise very rapidly till almost every household had one. They would then drop

But what's in it for chemistry? 505

to a level determined by the number of new households being established and would subsequently rise as the earlier sets needed replacement. Nonetheless, the idea of the logistic curve should be borne in mind when forecasts are being made.

If one moves down the production chain further away from the consumer, it sometimes happens that the demand for a resource – graduates, machine tools or chemical process plant equipment for example – varies not with the level of industrial activity but with its differential. This idea is less familiar than that of the logistic curve and we shall therefore illustrate it by a simplified case.

Consider a firm which started up in 1950 with forty graduates, one of every age between 25 and 64. The retiring age is 65. Suppose that the technology is constant and that one graduate is required for every 2 500 tons of output. The output in 1950 was thus

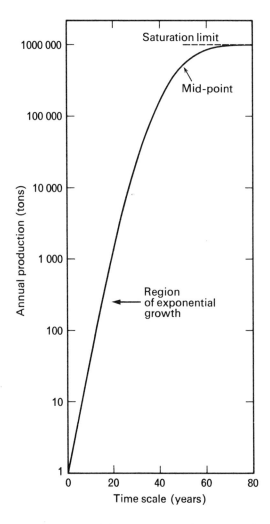

Figure 18.3 Logistic curve plotted on a semi-logarithmic scale.

506   *And will the trees grow up to the sky?*

100 000 tons. Between 1950 and 1970 the firm expanded at a steady 10% per year, i.e. it remained on the exponentially rising part of the logistic curve. Each year it engaged one new graduate to replace the one who was retiring, plus a number of new ones to cater for the increase in output. In 1951, this was 10 000 tons, so four additional new graduates were taken on making a total of five. 1960 output is 260 000 tons. The firm employs 104 graduates of whom 30, now aged between 35 and 64, have been with them since the start while the rest are younger. The 1961 intake will be $10 + 1 = 11$.

By 1970, output is 673 000 tons and there are 269 graduates of whom only 20, now all senior managers aged between 45 and 64, are from the original intake. Note the way the age distribution has become skewed. This is a characteristic of almost all science-based industries.

If the firm continues to grow at 10% per year then the graduate intake in 1971 will be $27 + 1 = 28$. If, however, the growth rate drops to 4% then the firm will recruit only

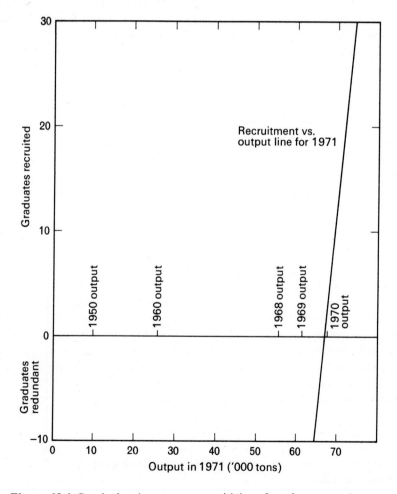

Figure 18.4 Graph showing extreme sensitivity of graduate recruitment to small changes in output.

11 + 1 = 12 graduates and at zero growth only one graduate will be required. Recruitment is thus related to the *increase* in output each year not to its absolute magnitude. In this specific case, until the first intake of graduates have all retired,

$$\text{Recruitment} = (\text{increase in output per year in tons}/2\,500) + 1$$

Figure 18.4 shows recruitment plotted against output. Even with production at a very high level, higher than it was, say, in 1969 when 25 or 26 graduates were recruited, there may be zero recruitment or even redundancy. A similar argument applies among other things to machine tools and chemical process plant. Because of the sensitivity of the demand for these items to small changes in economic climate, such industries are said to be cyclical. Furthermore, recruitment of graduates by the machine tool industry will be doubly sensitive to small changes, that is it will depend on the second differential of industrial activity.

To sum up, short-term forecasting can be done crudely on the assumption of exponential growth. In the long-term, the logistic curve is likely to be a better approximation but there are formidable difficulties in defining saturation limits. Certain industries are cyclical and depend on the first or second differentials of other growth curves. Finally, it must be remembered that logistic curves can come down as well as going up – Fig. 10.13 (water-based paints) shows an excellent example.

18.3.2 The British chemical industry

18.3.2.1 Britain in the 1970s Whichever of the scenarios discussed in section 18.2 comes to pass it is difficult to envisage the British economy growing at more than 2–3% per annum, the rate which has characterized it for many years. The decline of the Protestant ethic, the extent to which it is not respectable to make high profits, the intransigence of the trade unions, the apathy of the population and the general lack of entrepreneurial and managerial talent combine to prevent a take-off into higher growth rates. Equally the fact that expectations in Britain are rising and that she is to a considerable extent linked to the world economy suggest that the growth rate is unlikely to drop below 2% for any length of time, if only because of the political drawbacks which would result from the consequent high unemployment.

In a country where both don and docker prefer a quiet life, successive Labour and Conservative governments who have attempted to discipline the trade unions have been both defeated and humiliated. It would therefore be incautious to predict for Britain other than continued wage inflation matched by small gains in productivity and punctuated by occasional devaluations. On the other hand the efforts of all governments since the Second World War have resulted – perhaps unintentionally – in a steady decline in relative wage rates so that Britain is now an area of fairly cheap labour of low productivity.

During the decade we expect a continued swing to the labour-intensive service industries which are already growing rapidly. This will present certain cultural problems in that work in many such industries is not seen as 'a man's job'. At the time of writing there are a million unemployed and 90 000 vacancies in the hotel and catering trades which are in any case largely staffed by imported foreign labour. It is a sombre thought that the mining of unwanted coal under unpleasant and dangerous conditions is protected and subsidized while the hotel industry, which earns so much in foreign exchange, was until 1971 discouraged by a selective employment tax.

18.3.2.2 The chemical industry in the 1970s Assuming a constant technology, the British chemical industry has a number of things running in its favour. Compared with other British industries it is modern and well-managed. Labour relations are good. Entry into the Common Market is likely on balance to favour the industry and the North Sea gas and oil bonanza should stabilize raw material costs for the time being. Working against the industry is another set of factors. The petrochemical and polymer sectors are now approaching maturity; most of the technology is in the public domain and more and more relatively backward countries are likely to make their own chemicals rather than to import them. As the new plant will contain a prestige element it is likely to perpetuate existing over-capacity and to keep world prices and profits low. The costs of dealing with pollution are a further burden. It is perhaps significant that hardly any chemical companies have floated an equity issue on the London Stock Exchange in the past five years, and although profits may well rise from their current depressed level they are unlikely to hit the heights again.

On balance, therefore, we expect the British chemical industry to grow more slowly in the 70s than in the 50s and 60s but still to grow faster than manufacturing industry in general. Growth will, however, be markedly uneven and it is instructive to examine the prospects for the particular sectors.

18.3.2.3 The future of polymers Plastics should continue to grow faster than the economy but less fast than in the past. Saturation limits for plastics are difficult to predict because they are able to replace so many traditional materials. Figure 18.5 reproduces UK production data on low-density polyethylene. Exponential growth at a rate of 13% per year in the early years was later superseded by linear growth. In 1968 the logistic curve showed every sign of bending over. The price (Fig. 2.4) had dropped to a point where further price-cutting to boost demand would have made profitability unbearably low. Production in 1969–71, however, was well above the implied saturation limit and current figures lead to no obvious predictions.

It seems probable that the main growth areas for plastics will be in existing light-duty applications. Plastics will continue to displace traditional materials such as paper, wood, leather, copper and iron. More goods will be packed in polyethylene film, more gutters and waste pipes will be of rigid PVC, more shoes will be made from porous PVC. Plastics will be used in contiguous fields such as the fabrication of car body panels and party walls. There is ample scope for continued growth in these areas. Paper, for instance, is still far more widely used in packaging than plastics: to take a simple example, in 1969 120 000 tons of paper bags were produced and only 45 000 tons of polyethylene bags. More speculatively paper might be displaced in the printing trades by plastic sheet or synthetic pulp. Yet a note of caution is appropriate. The rapid growth of plastics consumption during the past 20 years has been due in no small part to the continuous fall in the real price of plastics (section 2.9.2). If, as seems likely, the increasing cost of crude oil reverses this trend, the effect on the polymer industry could be serious. A doubling of price of crude oil would admittedly increase the price of polyethylene by only 20–25%, but in a highly competitive market this would be a severe blow to its competitive advantage.

The widely canvassed use of plastics for heavy-duty structural purposes appears to be a more remote prospect. No large tonnage polymer is yet able to compete economically

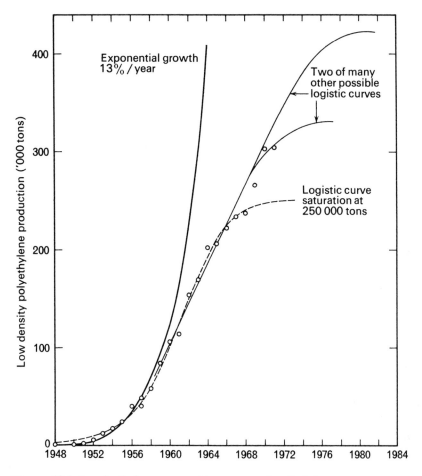

Figure 18.5 Is polyethylene approaching a saturation limit?

with steel and concrete in such uses and none is likely to do so for some time to come. Although carbon fibres have truly remarkable mechanical properties, their cost is astronomical and they are used only when lightness is at an absolute premium. Barring an unforeseeable breakthrough in manufacturing technology they are unlikely to assume major importance in the 1970s. For the next decade the major markets for plastics will be those already developed.

The future prospects of paints, fibres and elastomers are more doubtful. In Western Europe as a whole synthetic fibres – excluding cellulosics – grew from 6 to 30% of total fibre consumption between 1960 and 1970. It is difficult to see them taking more than perhaps 50% of the market by 1980. The position of synthetic rubber is similar. The consumption of rubbers of all kinds is growing steadily rather than dramatically (cf. Fig. 2.1); synthetic rubber already has 60% of the UK market and further growth is bound to be relatively slow. The consumption of paint is intimately related to the state of the national economy and may be expected to grow at much the same unexciting rate. The replacement of traditional ingredients by synthetic materials is largely complete and for the past decade change has meant the replacement of one synthetic resin by another.

18.3.2.4 The future of pharmaceuticals While the future of the polymer industry is subject to doubt and qualification, that of the pharmaceutical industry is extremely bright. Here the trees might indeed grow up to the sky: as yet the consumption of drugs is far below what it might be. A man whose life is saved by drugs lives to take more drugs; as pharmacology advances he survives more illnesses and his lifetime consumption of drugs increases. Many of the major diseases are now curable but others await the biochemical discoveries which will make cures possible. Drugs effective against cancer, arteriosclerosis or schizophrenia will both benefit mankind and make a fortune for those who introduce them. Equally rich rewards await those who cure conditions which are annoying rather than deadly, such as acne or migraine, or whose treatment now requires the exercise of will-power, such as obesity. Moreover, as affluence increases so does hypochondria. The healthy can be readily induced to take harmless placebos for vague or non-existent ills (Snibbo lowers your poculity index); in the USA otherwise sensible citizens take vitamin pills containing vitamin E (α-tocopherol), a factor known to be of great help to pregnant rats which appears to have no effect on humans. Such rituals do not help but neither do they harm, and, to the extent they are believed to be beneficial, they may comfort.

A somewhat more speculative possibility is that synthetic organic compounds may replace traditional mood-enhancing drugs. Totally synthetic cigarettes, formulated to be non-carcinogenic, are very likely to be marketed in the immediate future. As traditional morality continues to decline, it is probable that the laws against such imaginary crimes as the use of non-habit-forming drugs will be repealed. If so, such time-honoured psychotropics as tea, coffee and alcohol may well be replaced by new and superior alternatives. Within a generation the party-giver may be able to buy a whole range of suitably packaged materials, each guaranteed to elevate without affecting the motor centres, damaging the liver or producing a hangover.

18.3.2.5 The future of the old chemical industry For these sectors the future will be no more exciting than the past. Heavy inorganic chemical production will continue to grow at the same rate as the economy. Detergents, a more or less static market, may well grow more slowly. Fertilizer consumption in Britain is likely to rise as the Common Agricultural Policy of the EEC will make cereal production markedly more profitable; on the other hand, the benefits to the fertilizer manufacturers will be reduced by the abolition of the fertilizer subsidy and by increased competition from abroad. Pollution problems may also hold back growth, which is in any event unlikely to be very profitable if the experience of the 60s is typical. The prosperity of the dyestuff and pigment sectors is tied to that of the textile and paint industries; they will continue to grow, but only slowly.

18.3.2.6 The environmental problems of the industry Like other manufacturing industries the chemical industry of the 70s will have to work under much stricter environmental controls. Its past record in this respect is quite good, but the received wisdom of the day favours harsher measures more rigorously enforced. The major problems are financial rather than technical. Practicable methods have been found to avoid virtually all the direct and indirect pollution caused by the industry. The plastic litter problem, much canvassed by environmentalists and others, may well have been solved by Professor W. L. Scott's discovery of a cheap additive which promotes the photochemical oxidation of polyolefins. The investment required, however, is considerable and must inevitably increase UK chemical prices. Other advanced nations are in a similar position but the more ruthless and anti-social semi-developed countries – such as Brazil – may well attempt to enlarge their share of the world trade in chemicals (section 18.2.2). A binding international agreement on effluent standards is much to be desired, if only for reasons of self-preservation. All clouds, however, have a silver lining: if environmental legislation is a burden to most of the industry, it is an opportunity to those who design, build and install pollution control equipment. Fertilizer manufacturers may tremble but sewage engineers rejoice.

18.3.2.7 Is the golden age over? Whichever way one looks at the data the conclusion is inescapable: the growth of the industry will slow down. Indeed the first two years of the decade have already seen large cuts in research expenditure and in plant investment. The planned 1973 investment of £350 m. was cut at the beginning of 1972 to £240 m. and many plans for expansion have been deferred or cancelled. While over-optimism about continued rapid growth has led to world over-capacity in chemicals, the drop in growth-rate has had the effect on graduate recruitment predicted by the model developed in section 18.3.1. There is widespread graduate unemployment in both Britain and the USA and much doubt as to whether the slack will ever be taken up. It may well be that the year 2 000 will see a chemistry degree course esteemed not as a vocational training but, like the study of the classics, as a training for the mind: having graduated, one then takes a course in, say, hotel and catering management to enable one to earn a living.

18.3.3 The world chemical industry

The chemical industries of the mature developed countries are likely to fare only marginally better than that of the UK. The predictions for Western Europe shown in Table 18.1 were made by the President of the Industrial Division of the Chemical Society; they represent the general consensus of informed opinion within the industry. The pharmaceuticals figure is perhaps pessimistic and the 4·8% overall growth figure for the OECD area may be optimistic. Growth in the USA will probably be somewhat slower than that of Western Europe and growth in Japan somewhat faster but by the eighties or nineties all will be mature economies growing at similar rates.

Table 18.1 Economic growth in Western Europe 1970–80

	Growth rate % per year	
	Actual growth 1960–70	Predicted growth 1970–80
Gross national product (OECD figure)	?	4·8
Heavy organic chemicals	13·7	9
Inorganic chemicals	6·1	<6
Plastics	16·8	11 to 12
Synthetic fibres	19·4	8
Pharmaceuticals (value growth)	9·5	<9

In the developed countries we see the growth areas as the service industries – tourism, restaurants, hairdressing and so on – leisure activities, electronics and telecommunications – cable television, telephone access to libraries and computers – environmental control systems and pharmaceuticals. As in Britain, the relative importance of manufacturing industry will decline; as the surviving peasants die off, agriculture will become a smaller and much more efficient part of the economy. The chemical industry will develop along the lines discussed in previous sections, essential but unglamorous, and with much the same image as the soda industry of seventy years ago.

Predictions for the UK, a stable, culturally homogenous country, are likely to be more reliable than those for most other nations. The rich countries will, however, probably remain rich. The underdeveloped countries might develop chemical industries during the next 20 years. If so they will probably be based on fertilisers, other heavy chemicals, and large-tonnage polymers. Their foundation, however, would depend less on commercial considerations than on questions of national prestige, availability of capital, political stability and a wealth of imponderable cultural factors. What chemical industry does develop is unfortunately likely to contribute to the prevailing overcapacity and may even lead to the developed nations edging out of the high pollution, low profitability heavy chemical sectors.

18.3.4 New technologies and their impact

This somewhat gloomy analysis has assumed a constant or slowly changing technology, a convenient but not necessarily correct presupposition. Economists are much inclined to forecasts of the 'other things being equal' type; past history suggests that unforeseen technical change is the main reason why other things are not equal. On the other hand it would be foolish to assume that what is in principle conceivable will automatically be practicable, let alone economic. A successful innovation fills a need in a technically and commercially satisfactory way: to predict the pattern of technical change during the next 30 years one must identify both the likely needs and the likely scientific advances of this period.

At the present time the chemical industry stands at the end of a process of technical development which began some 50–60 years ago (cf. Chapter 2). This period has been dominated by the rise of the petrochemicals and polymers and during the past 20 years these fields have been the subject of massive research expenditure. Yet it is clear that the law of diminishing returns has set in: few new products or processes have appeared since the mid-50s and it is widely believed within the industry that few remain to be discovered. The slump in graduate recruitment is due in part to the severe recession of 1970–72 but in part to a general feeling that the basic technology of the industry is unlikely to change radically in the foreseeable future. This seems to be a short-sighted view: the outside world is not going to arrange matters to support the industry in its present form, nor will enterprising inventors refrain from exploiting the opportunities created by the advance of science.

What changes are likely? A major preoccupation of the industry for the next half-century at least will be the increasing cost of raw materials. As suggested in section 18.2.5 the price of crude oil is likely to rise steadily. Since coal and agricultural products are unlikely to fall in price the industry faces a future of much higher raw material and fuel bills. It is difficult to predict which will be the favoured source of fixed carbon in the year 2000 but further research into coal and fermentation-based processes would be a wise defensive move. Changes in the technology of energy generation may also have a powerful impact on the industry. In many places conventional nuclear power plants are broadly competitive with fossil-fuel installations; in the 80s commercial breeder-reactor units might become available and in the 90s thermonuclear plants. By the end of the century electric power may well be the cheapest source of energy. Electrothermal and electrochemical routes to chemical products would become correspondingly attractive, especially if cheap off-peak power were available. The chemical industry of the early twenty-first century may be one which uses electricity to transform a wide range of rather expensive raw materials into saleable products.

Further possibilities arise from likely developments in chemistry and the related sciences. The work now in progress on ion–molecule and molecule–molecule reactions in the gas phase promises to lead to a quantitative understanding of the factors which control reaction rates. In turn, this might assist the optimization of reaction conditions and the development of selective processes. Even more interesting prospects are offered by heterogenous catalysts. Up to the present time all commercial catalysts have been developed by empirical means and, indeed, something like a century of academic work in this field has produced remarkably few results. Current studies of the mechanism of enzyme action by chemists, biochemists and molecular biologists may well result in the elucidation of the electronic and steric factors involved; conceivably this might lead to the development of industrial catalysts of the same efficiency and specificity but capable of functioning at much higher rates. Such materials would have revolutionary consequences. Many relatively complex compounds would become readily available; thus, for example, it would be practicable to market a large range of tailor-made polymers rather than a handful of general-purpose materials. The use of highly specific reactions would largely eliminate the necessity to purify the product. Plants would be smaller, cheaper and more flexible, and economies of scale less compelling. In the extreme case the industry might become decentralized as the need for the large integrated plant disappeared. Such an outcome would make the chemical industry as a whole much more like the pharmaceutical sector – research rather than capital intensive, flexible, opportunistic and profitable. From the viewpoint of both the shareholder and the working chemist this prospect is extremely attractive.

Thus external factors seem almost certain to force changes in the technology of the industry while the probable developments in chemistry make it clear that such changes could be very radical indeed. Far from being a time of somewhat stagnant calm the next 30 years could equally be characterized by convulsive change. If so, it will be interesting to see how the established giants of the present day cope with the situation. They will to some extent be demoralized by the slowing of their growth rate and their managements will be aging, factors which will increase the pessimism brought about by overcapacity and low profits. Innovation might well be pioneered by small energetic companies rather than the existing monsters who will in any case move more slowly because of their enormous investment in existing technology.

We predict that the chemical industry will survive and prosper but within the lifetime of most readers it will look quite different.

> The old order changeth, yielding place to new,
> And God fulfils himself in many ways,
> Lest one good custom should corrupt the world.

Notes and references

Section 18.1 Shute's views on the future of heavier-than-air craft are in C. D. Burney, *The Air, The Empire, and The Future*, Knopf, London, 1929. The best account of why forecasts fail is in A. C. Clark, *Profiles of the Future*, Pan Books, London, 1960; the record of science-fiction writers is generally better than that of the experts in particular fields.

Section 18.2.1 For typical scenarios of this type see H. Kahn and A. J. Wiener, *The Year 2000*, Collier–Macmillan, London, 1967.

Section 18.2.2 The best-known works are J. W. Forrester, *World Dynamics*, Wright–Allen Press, New York, 1971, and D. H. Meadows, *et al.*, *The Limits to Growth*, Potomac Associates for the Club of Rome, Washington DC, 1972. The 'running backwards' of the Forrester model is described in *Futures*, 1973, **5**, 1. The Brazilian invitation to 'dirty' industries was reported in a number of newspapers to have been given at the UN Conference on the Environment, Stockholm, June 1972. John Maddox, *The Doomsday Syndrome*, Macmillan, London, 1972, argues cogently against the extreme environmentalists.

Section 18.2.3 For a gripping description of the problems which arise in a system containing half-a-dozen great powers and many minor ones see A. J. P. Taylor, *The Struggle for Mastery in Europe*, OUP, 1954.

Section 18.3.1 The uses of the logistic curve are discussed in more detail in D. de Solla Price, *Little Science, Big Science*, Columbia University Press, New York, 1963. Now available as a slim paperback, it is a classic in its field, witty, well-written and penetrating. Figure 18.1 is based on Price's data.

Section 18.3.2 The future of the British chemical industry has been discussed by J. W. Barrett, *Nature*, **234**, 515, 1971. Forecasts for Western European and North American industry as a whole are to be found in *The Growth of Output 1960–1980*, OECD, Paris, 1970. For a highly optimistic view of the future of the plastics industry see NEDO, *The Plastics Industry and Its Prospects*, HMSO, London, 1972.

Section 18.3.4 The quotation is from Alfred, Lord Tennyson, *Morte d'Arthur*.

Compound index

Notes:
1. The compound index contains materials to which a definite molecular formula can be assigned. Other materials (e.g. linseed oil, liquid petroleum gas, polyvinyl chloride) appear in the subject index.
2. Compounds are indexed under their IUPAC names; common trivial names are also included. Pharmaceuticals are indexed under their pharmaceutical names e.g. amphetamine *not* 1-phenylpropyl-1-amine. Dyes are listed under the main chromophoric system, e.g. anthraquinone, monoazo compounds etc. Trade names are not included.
3. Bold figures indicate particularly important sections. 4. n.e.c. = not elsewhere classified.

Acetaldehyde
   manufacture, 202–3, 431–4
   market data, 199
   uses, 202–4, 213, 451
Acetaldehyde cyanhydrin, 301
Acetanilide, 353
Acetic acid
   manufacture, 202–3, 209, **213–14**
   market data, 214, 290
   uses, 216
Acetic anhydride
   manufacture, 203, **215–16**
   market data, 214, 290
   uses, 216, 294–5, 351
Acetoacetanilide, derivatives, 215, 336
Acetone
   manufacture
     by fermentation, 18, 192
     from i-propanol, 207
     via cumene–phenol process, 221–2
   market data, 208–9
   uses, 18, 207–9, 242–3, 269, 294
Acetone cyanhydrin, 208, 242–3
Acetonitrile, 206, 300
Acetylene
   acrylic acid from, 277
   acrylonitrile from, 277–8, 301
   manufacture, 173–5, 189
   market data, 227
   trichloroethylene from, 451
   vinyl acetate from, 274–5
   vinyl acetylene from, 287
   vinyl chloride from, 247–8
Acetylsalicylic acid, *see* Aspirin
Acrolein, 277, 300
Acrylamide, 279
Acrylic acid
   manufacture, 277, 279
   esters of, *see under individual names*

Acrylonitrile
   adiponitrile from, 297
   copolymers of, with
     acrylamide and styrene, 279
     butadiene, 286–7
     butadiene and styrene, 235, 240–2
     vinyl chloride, 301
   manufacture, 115, 277, **300–1**
   market data, 206
   polymerization of, 301, 454
Adipamide, 297
Adipic acid
   manufacture, 297–8
   market data, 290
   uses, 256–7, 297–8
Adiponitrile, 297
Allyl alcohol, 442
Allyl chloride, 206, 269
Aluminium triethyl, 238, 456–7
p-Aminobenzoic acid, 353
ω-Aminoundecanoic acid, 295, 298
Amitryptiline, 360–1
6-aminopenicillanic acid, 356–7
Ammonia
   manufacture, by
     coal carbonization, 187–8
     Haber process,
       history, 17–18, 20–1
       labour requirements, 140
       plant costs, 99, 374
       technology and economics, **369–76**
   market data, 364
   nitric acid from, 376–7
   uses for, 367–9
Ammonium bicarbonate, 15
Ammonium nitrate, 364, **377–8**
Ammonium phosphate, 369
Ammonium picrate, 448

Ammonium sulphate
  history, 13, 17–18
  manufacture, 298, 369, 377
  market data, 11, 22, 364
Amphetamine, 359
Ampicillin, 354
Amylobarbitone sodium, 354
Aniline
  manufacture, 446
  market data, 218, 445
  uses, 353, 444, 446–7
Aniline black, 336
Anthanthrone, 326
Anthracene, 188
Anthraquinone
  acid dyes from, 329
  chromophoric effect of ring system, 327–8
  disperse dyes from, 334
  production of dyes from, 337–8
  vat dyes from, 333
Aspirin, 349–52
Azelaic acid, 256–7
Azo-*bis*-butyronitrile, 247, 455
Azo compounds, *see* Monoazo; Disazo: Trisazo compounds

Benzene
  chlorination of, 220–1, 408
  cumene from, 221–2
  cyclohexane from, 218, 297–8, 438
  dodecylbenzene from, 317
  manufacture, **175–7**, 187–8
  market data, 196–7, 216–19
  nitration of, 447
  phenol from, 220–3
  sulphonation of, 220, 449
  uses n.e.c., 217–19
Benzene sulphonic acid, 220
Benzoic acid, 222, 224, 296
Benzoyl peroxide, 247, 267, 455–6
Benzyl chloride, 354, 445
Bisphenol A, *see* Dimethyldi-*p*-hydroxyphenyl-methane
Bleaching powder, 11, 12, 390
Borax (sodium tetraborate), 366, 393–5
Boric acid, 394
Boron carbide, 395
Boron trifluoride, 456
Bromine, 452
Butabarbitone, 354
Butadiene
  adipic acid from, 297
  copolymers of, with
    acrylonitrile, 286–7
    acrylonitrile and styrene, 235, 240–2
    i-butene, 286
    styrene, 285
  manufacture, 166, 170, 172, 212
  market data, 213
  polymerization, 285–6, 454
n-Butane, 166, 212
i-Butane, 166
Butanol-1 (*n*-butyl alcohol)
  manufacture, 18, 192, 204, 211
  market data, 208
  uses, 211
Butanol-2 (*sec.* butyl alcohol), 211–13
i-Butanol (*iso*butyl alcohol), 208, 211
t-Butanol (*t*-butyl alcohol), 212
Butene-1
  manufacture, 166, 169, 212
  market data, 213
  uses, 212, 287
Butene-2
  manufacture, 166, 169, 212
  market data, 213
  uses, 212, 287
i-Butene
  copolymers with butadiene, 285–6
  manufacture, 166, 212
  polymerization, 454
i-Butyl acetate, 211, 214
n-Butyl acetate, 211, 214
n-Butyl acrylate, 208, 273, 279
Butyl alcohols, *see* Butanols
Butylamines, 445–6
Butyl benzyl phthalate, 254–6
Butyl decyl phthalate, 254
Butyl octyl phthalate, 254
i-Butyraldehyde, 211
n-Butyraldehyde, 211, 276

Calcium bisulphite, 12
Calcium borate, 394
Calcium carbide, 174, 189, 264
Calcium cyanamide, 18, 264–5
Calcium hydrochlorite, 390
Calcium dihydrogen phosphate (superphosphate)
  as fertilizer, 12, 391
  manufacture, 392
  market data, 11, 17, 22, **392**
Calcium propionate, 214
Calcium sulphate, 384
Caledon jade green, 333
Caprolactam, 290, 295–8, 369
Caprolactone, 296
Carbenicillin, 357
Carbitol, *see* Diethylene glycol monomethyl-ether
Carbon black, 336
Carbon dioxide
  greenhouse effect and, 489
  manufacture, 178, 190, **372–6**
Carbon disulphide, 28, 290, 293–4
Carbon monoxide
  in car exhaust, 484–5
  manufacture, 178, 190, **372–6**
  use in Oxo synthesis, 211, 439–40
Carbon tetrachloride, 445, 450
Cellosolve, *see* Ethylene glycol ethyl ether
Cephalosporin N, 356
Cetyltrimethylammonium bromide (CETAB), 318
Chloral
  manufacture, 204, 451

market data, 203, 445
 use in DDT manufacture, 451
Chlordiazepoxide, 359, 361
Chlorine
 manufacture
  by brine electrolysis, 15–16, 385, 387–9
  from HCl, 14, 23, 390, 451–2
 market data, 366, 388–9
 uses, 282, 385–6
Chlorobenzene, 218, 220–1, 408
2-Chloro-1,3-butadiene, 287, 454
Chlorodifluoromethane, 245, 445
Chloroform, 245, 445, 450
Chlorotetracycline, 358
Chlorothenoxazine, 351
Chromic acid, 475
Chlorpromazine, 359
Citric acid, 193–4
Cobalt naphthenate, 272–3, 280
Cortisone, 194, 348
Cupric chloride, 203, 275, 404–5
Cresols, 187–9
Cresyl diphenyl phosphate, 256
Cumene
 manufacture, 221, 443
 market data, 206, 218
 phenol from, 221–2
Cumene hydroperoxide, 221–2
Cyanuric chloride (2,4,6-trichloro-1,3,5-triazine), 335
Cyanamide, 265
Cyclohexane
 manufacture, 216
 market data, 218
 oxidation to cyclohexanol and cyclohexanone, 295–7
 phenol from, 222
Cyclohexanol, 290, 296–7
Cyclohexanone, 290, 296–7
Cyclohexanone oxime, 296–7, 369
Cyclotrimethylene trinitramine (cyclonite), 448

Decanol, 213
Diacetone alcohol, 207–8
Diallyl phthalate, 267
4,4-Diaminostilbene-2,2-disulphonic acid, derivatives, 332, 336
2,4- and 2,6-Diaminotoluenes, 224, 268
Diammonium adipate, 297
Diazepam, 359
Disazo compounds
 acid dyes, 329
 direct dyes, 330
 market data, 337
Dibenzanthrone, 326
1,2-Dibromoethane, 445–6, 452
2,6-D-1-*t*-butyl-*p*-cresol, 239
2-(3′,5′-Di-*t*-butylphenyl)-5-chlorobenztriazole, 239
Dibutyl maleate, 276
Di-*n*-butyl phthalate (DBP), 208, 253–5, 276
Dichlorobenzenes, 218, 445

1,4-Dichlorobutene-2, 297
Dichlorodifluoromethane, 450
1,2-Dichloroethane, 247, 288, 445
1,2-Dichloroethylene (vinylidene chloride), 244
1,1-Di-*p*-chlorophenyl-2,2,2-trichloroethane (DDT), 445, 451, 490
Dicloxacillin, 357
Dicumyl peroxide, 287
1,4-Dicyanobutene-2, 297
Dicyandiamide, 265
Dicyclopentadiene, 287
Di-*i*-decyl phthalate, 253–4
Diethanolamine (DEA), 165, 200
Diethanolamine dodecylbenzenesulphonate, 315
Diethylbarbituric acid, 354
Diethyl ether, 203
Diethylene glycol, 200–1
Diethylene glycol dibenzoate, 255–6
Diethylene glycol monomethyl ether (carbitol), 200, 444
Diethylene triamine, 270
Diethyl phthalate, 203, 253, 255
Di-2-ethylhexyl adipate, 253, 256
Di-2-ethylhexyl maleate, 276
Di-2-ethylhexyl phthalate (DOP), 253–5
Diethylpropion, 359, 361
Di-*n*-heptyl phthalate, 254–5
Di-*i*-hexyl phthalate, 254
Dihydrostreptomycin, 358
Diketene, 215
Dimethylamine, 445–6
Dimethylaniline, 444–5
2,3-dimethylbutadiene, 283
Dimethyldi-*p*-hydroxyphenylmethane (Bisphenol A), 208–9, 269–70
nn-Dimethyldinitrosoterephthalamide, 464
Dimethylformamide (DMF), 174, 301
Di-2-methylheptyl phthalate (di-*i*-octyl phthalate (DIOP)), 251–5
Dimethylolurea, 162
Dimethyl terephthalate, 225, 299
Dinitrosopentamethylene tetramine, 464
2,4- and 2,6-Dinitrotoluenes, 268, 445
Dinonyl phthalate, 254
Dioctyl phthalate (DOP) *see* Di-2-ethyl hexyl phthalate
Di-*i*-octyl phthalate (DIOP), *see* Di-2-methylheptyl phthalate
Diphenylamine, 445
Diphenyl oxide, 220
Di-*i*-propyl sulphate, 206
Dipropylene glycol, 210
Di-tridecyl phthalate, 253–4
Dodecylbenzene, 317, 443

Epichlorhydrin, 269–70
Ethane
 ethylene from, 166, 172–3
 from natural gas, 183–4
 vinylidene chloride from, 244
Ethanol (ethyl alcohol)

Ethanol (ethyl alcohol)–(contd.)
  manufacture, 191–3, **202**
  market data, 199
  units, 228
  uses, 202–3, 213
Ethanolamines, see Monoethanolamine, diethanolamine etc.
Ethyl acetate, 203, 214
Ethyl acrylate, 203, 243, 279
Ethyl alcohol, see Ethanol
Ethylamine, 445–6
Ethylbenzene, 209, 239–40, 443
Ethylchloride, 201–2, 445, 451
Ethylene
  acetaldehyde from, 203
  copolymers of
    with propylene, 287–8
    with vinyl acetate, 276
  ethanol from, 202
  ethyl chloride from, 201
  ethylene oxide from, 199–200
  manufacture, **166–73**
  market data, 196–7
  polymerization, **236–7**, 454
  uses n.e.c., 198
  vinyl acetate from, 275–6, 404–6
  vinyl chloride from, 247–8
Ethylene dibromide, see 1,2-Dibromoethane
Ethylene dichloride, see 1,2-Dichloroethane
Ethylene glycol, 200–1, 298–9
Ethylene glycol diacetate, 433–4
Ethylene glycol ethyl ether (cellosolve), 200, 444
Ethylene glycol monobutyl ether, 200
Ethylene cyanhydrin, see Lactonitrile
Ethylene oxide
  manufacture, 199–200
  market data, 199
  uses, 200, 277, 300, 318–19, 444
2-Ethylhexanol (2-ethylhexyl alcohol)
  esters of, 257
  manufacture, 211
  market data, 208, 213
2-Ethylhexyl acrylate, 257, 273, 279
Ethylidene norbornene, 288

Fenfluramine, 359, 361
Ferric aluminium sulphate, 471–2
Fluorine, 449
Formaldehyde
  manufacture, 378
  market data, 379
  melamine-formaldehyde resins, 264–5
  phenol-formaldehyde resins, 260–2
  polymerization, 30, 244–5
  urea-formaldehyde resins, 262–4
  uses n.e.c., 204, 378
Formic acid, 214
Fumaric acid, 218
Furfural, 165, 443

Gentisic acid, 351
Glucose, 442

Glycerol
  manufacture, 191–3, 315–16, 442
  uses, 278–80, 448
Glycerol dichlorhydrin, 269
Glyceryl triacetate, 295
Glyceryl trinitrate (nitroglycerin), 448

n-Heptanol (n-heptyl alcohol), 255, 298
Heptene-1, 439
n-Hexadecane, 166
Hexafluoropropylene, 245
Hexamethylene diamine, 290, 295–8
Hexamethylene tetramine, 261, 378
Hexylene glycol, 208
Hydrofluoric acid, 449–50
Hydrogen
  manufacture
    from hydrocarbons, 178–9, 373–6
    from coke, 190, 373
  uses, 437–9
Hydrogen chloride
  chlorine from, 14, 23, 247, **451**
  disposal of, 478
  market data, 366
Hydrogen peroxide, 366, **398–9**, 456
Hydrogen sulphide, 165, 384
o-Hydroxydiphenyl, 334
Hydroxylamine, 296, 298
2-Hydroxy-3-naphthoic acid, 334, 336

Ilmenite, 281
Imipramine, 360–1
Indanthrene blue, 333
Indigo, 16, 333
Iodine, 452
Isophorone, 208
Isoprene
  manufacture, 170, 284–5
  polymerization, 285, 454

Ketene, 214–15, 277

Lactic acid, 193–4
Lactonitrile, 277, 300
Lauroyl ethanolamide, 313, 315
Lead carbonate, 251
Lead stearate, 251
Lead tetraethyl, 445–6, 485
Lead tetramethyl, 445
Lysergic acid diethylamide, 360, 362

Maleic anhydride, 218, 276
Melamine, 263–6
Meprobarate, 359
Mercury, 387, 389
Mesityl oxide, 207
Methane
  acetylene from, 173
  carbon monoxide and hydrogen from, 178, 373
  from natural gas, 182–4
Methanol (methyl alcohol), 364, **378–9**
Methaqualone, 359–60

Methicillin, 357
Methyl acrylate, 279
Methyl alcohol, see Methanol
Methylamine, 445–6
Methyl i-butyl carbinol, 207
Methyl i-butyl ketone, 207, 208
Methyl chloride, 445, 450
Methylene chloride, 445, 450
Methyl ethyl ketone (MEK), 212, 213
Methyl methacrylate, 208, **242–3**, 457
α-Methylstyrene, 222
Monoazo compounds
　acid dyes, 329
　basic dyes, 330
　disperse dyes, 334
　market data, 337
　mordant dyes, 329
Monoethanolamine (MEA), 165, 183, 200
Monomethylolurea, 262

Naphthalene, 187–9, 217, **226**
2-Naphthol, 449
Nitrazepam, 359
Nitric acid, 14, 18, 364, **376–7**
Nitric oxide, 376, 380–1, 485–8
Nitrobenzene, 218, 445, 447
Nitroethane, 447
Nitrogen dioxide, 376–7, 380–1, 485–8
Nitromethane, 447
1-Nitropropane, 296
1-Nitroso-2-naphthol, 336
Nonylphenols, 319
NTA, see Sodium nitrilotriacetate

i-Octaldehyde, 439
n-Octaldehyde, 439
i-Octane, see 2,3,4-Trimethylpentane
i-Octanol, 213
n-Octyl-n-decyl phthalate, 253
Oxacillin, 357
4,4'-Oxybis(benzenesulphonyl hydrazide), 464
Oxygen, 178, 199, **436–7**
Oxytetracycline, 358
Ozone, 437, 486–8

Palladous chloride, 203, 275, 404–6
Paracetamol, 351–2
Penicillin K, 356
Penicillin G, 191–3, 355–7
Penicillin F and Dihydro F, 356
Pentaerythritol
　manufacture, 204
　market data, 203
　uses, 268, 273, 278–80, 448
Pentaerythritol tetranitrate, 445, 448
Pentachlorethane, 451
Peroxyacetyl nitrate, 486–7
Phenbenicillin, 357
Phenethicillin, 357
Phenobarbitone, 354–5
Pentabarbitone sodium, 354
Phenol
　manufacture, 218, 220–3
　market data, 218, 223
　polymers from, 259–63
Phenylacetic acid, 355
m-Phenylenediamine, 270
Phosphoric acid, 366, 391, **392–3**
Phosphorus, 366, **390–1**
Phthalic anhydride
　manufacture, 226–7
　market data, 225
　uses, 268, 270, 273, 278–80
Phthalocyanine, 327, 336
α-Picoline, 204
γ-Picoline, 204
Picric acid, 448
Picric acid, see 2,4,6-Trinitrophenol
1,1-Pifluoroethylene, 450
Potassium chloride, 395, 396–8
Potassium hydroxide, 366
Potassium pyrophosphate, 391, 393
Potassium sulphate, 366, 395
i-Propanol (i-propyl alcohol), 206–8
n-Propanol (n-propyl alcohol), 206, 440
Propicillin, 357
β-Propiolactone, 277
Propionic acid, 214
i-Propyl acetate, 214
i-Propyl alcohol, see i-Propanol
n-Propyl alcohol, see n-Propanol
Propylamines, 445–6
Propylene
　acrylic acid from, 277, 279
　acrylonitrile from, 277, 301
　allyl chloride from, 269
　copolymer with ethylene, 287
　cumene from, **221**, 443
　manufacture, 166, **170–3**
　market data, 196–7, 205–6
　polymerization, **238–9**, 454
　i-propanol from, 206
　propylene chlorohydrin from, 209
　propylene oxide from, 209
Propylene chlorohydrin, 209
Propylene glycol, 210, 256
Propylene oxide, 206, **209–10**, 268
Propylene tetramer, see Tetrapropylene
Propylene trimer, 213
Propyl hydrogen sulphate, 206
Pyranthrone, 326
Pyridine, 187–9
Pyromellitic dianhydride, 276

Quinalbarbitone sodium, 354

Resorcinol, 445, 449
Rutile, 282

Salicylamide, 351
Salicylic acid, 350–1
Sebacic acid, 256, 268
Sodium acetylsalicylate, 351
Sodium benzene sulphonate, 220
Sodium bicarbonate
　history, 11, 12, 14–16

Sodium bicarbonate–(contd.)
    manufacture, 14–16, 386, 390
    uses, 390, 464
Sodium carboxymethyl cellulose (CMC),
    273, 314–15, 444–5
CMC, see Sodium carboxymethyl cellulose
Sodium carbonate
    history, 11, 12, **13–16**
    manufacture, **13–16**, 385–6, 390
    market data, 366
    uses, 12, 390
Sodium chloride, 366, 385, 387–9
Sodium chloroacetate, 444
Sodium citrate, 319
Sodium cyanide, 475
Sodium dodecylbenzenesulphonate, 218, 314,
    315, 317–19
Sodium hydroxide
    history, 11, 12–14, 26
    manufacture, 385, 387–9
    market data, 366
    uses, 12, 388
Sodium hypochlorite, 315, 390
Sodium iodate, 452
Sodium lauroyl glyceryl sulphate, 318
Sodium metaphosphate, 391–3
Sodium nitrate, 12
Sodium nitrilotriacetate (NTA), 313, 319
Sodium perborate, 313–15, 394–5
Sodium phenoxide, 220–1, 350
Sodium polysulphides, 288
Sodium pyrophosphate, 314–15, 391–3
Sodium sesquicarbonate, 390
Sodium salicylate, 350–1
Sodium silicate, 314–15
Sodium sulphate, 314–15
Sodium tetrapropylenebenzenesulphonate,
    317–18
Sodium *p*-toluenesulphonate, 314–15
Sodium tripolyphosphate, 313–14, 392–3
Stilbene, derivatives, 332, 336–7
Streptomycin, 349, 358
Styrene
    copolymers of
        with acrylamide and acrylonitrile, 279
        with acrylonitrile and butadiene, 235,
            240–2
        with butadiene, 285
        with unsaturated polyesters, 267
    manufacture, 239–41
    market data, 218
    polymerization, 240–1
Succinic acid, 214, 268
Sulphadimethoxine, 352
Sulphamerazine, 353
Sulphamethizole, 352
Sulphanilamide, 352–3
Sulphisoxazole, 352
Sulphur, 283–4, 287, 365, 385
Sulphur dioxide, 383–4, 480–4
Sulphur trioxide
    air pollution by, 489
    manufacture, 380–1, 383–4

sulphonation with, 317, 449
Sulphuric acid
    history, 11–14, 17, 22, 26
    manufacture, 380–4
    market data, 102, 365, 385
    sulphation and sulphonation with, 449
    units of strength, 380
    uses, 379
Superphosphate, see calcium dihydrogen
    phosphate

TCCA, see Trichloro*iso*cyanuric acid
Terephthalic acid, 225, 299
Tetrachlorethane, 451
Tetrachlorethylene, 445, 451
Tetracycline, 349, 358
Tetrafluorethylene, 245
Tetrahydrocannabinol, 360
Tetrapropylene, 206, 212, 317, 443
Tetryl, see 2,4,6-Trinitrophenylmethyl nitra-
    mine
Thalidomide, 354
Thioindigo, 333
Titanium dioxide, 273, **280–1**, 336, 338
Titanium tetrachloride, 280–1, 456–7
TNT, see 2,4,6-Trinitrotoluene
α-Tocopherol, 510
Toluene
    benzene from, 177, 217, 225
    manufacture, **175–7**, 187–9
    market data, 217, 224
    phenol from, 222
    separation from benzene by distillation,
        417–26
    tolylene di-isocyanates from, 268
    trinitrotoluene from, 448
    uses n.e.c., 224–5
Tolylene 2,4- and 2,6-di*iso*cyanates, 224,
    268–9
Triallyl cyanurate, 267
Trichloroethylene, 445, 451
Trichlorofluoromethane, 445
Trichloro*iso*cyanuric acid (TCCA), 314, 315
Tricresyl phosphate, 256, 276
Triethanolamine, 200
Triethanolamine dodecylbenzene sulphonate,
    314–15
Triethylene glycol, 200–1
Triethylene tetramine, 270
Trimethylamine, 445–6
2,2,4-Trimethylpentane (*i*-octane), 443–4
2,3,3-Trimethylpentane, 444
2,3,4-Trimethylpentane, 444
2,4,6-Trinitrophenol (picric acid), 220, 448
2,4,6-Trinitrotoluene (TNT), 225, 448
2,4,6-Trinitrophenylmethyl nitramine
    (tetryl), 448
Triphenylmethane, derivatives, 326, 330
Tripropylene, 213
Trisazo compounds, 337
Trisodium phosphate, 314, 391
Turkey Red Oil, 318

Urea
  manufacture, 378
  market data, 266, 364
  urea-formaldehyde resins, 259, 262–4, 266
  uses n.e.c., 263, 378

Vinyl acetate
  copolymers of, with
    acrylates, 273, 276
    ethylene, 276
    vinyl chloride, 257
  manufacture, 274–6
    engineering problems of, 404–6, 431–4
  market data, 274
  polymerization, 274–6, 454, 458
Vinylacetylene, 287
Vinyl chloride
  copolymer with vinyl acetate, 257
  manufacture, **247–8**, 451, 457
  market data, 445
  polymerization, 247, 454
Vinylidene chloride, *see* 1,2,-Dichloroethylene
Vinylidene fluoride, *see* 1,1-Difluoroethylene

Water
  cost, 7
  desalination, 147–52
  surface properties, 307–10
  treatment, 468–76

m-Xylene, 176, 225
o-Xylene, 175–6, 225–6
p-Xylene, 175–6, 225
Xylenols, 187–9

Zinc oxide, 316, 336, 338

Subject index

NOTES:
1. This subject index lists materials (e.g. linseed oil, liquid petroleum gas, polyvinyl chloride) which do not have a definite molecular formula and have not, therefore, been listed in the compound index.
2. The book as a whole is based on UK, US and West European practice. Of these three areas only the European countries other than the UK are specifically indexed.
3. Individual people and companies have, in general, only been indexed if their names occur more than once. Trade names of chemicals, with occasional exceptions, have not been indexed at all. Subjects listed in the table of contents have not usually been re-indexed.

ABS resins, see Acrylonitrile–butadiene–styrene resins
AC volatility, 253–4, 302
Acrylics
  in surface coatings, 276–9
  polyacrylamide, 279, 472
  polymerization, 454
  polymethyl methacrylate, 242–3
  see also 'Polyacrylonitrile fibres'
Acrylonitrile–butadiene–styrene resins, 198–9, 235, **240–2**, 460
Advertising, 67, 320–1, 348–50
Aerosolvan process, 176
Airships, 497
Albright and Wilson, 114, 117–18, 140–1, 391
Alkali industry, 13–16, 366, **385–90**
Alkyd resins, 31, 204, 218, 259, 271–4, **278–80**
Aldol condensation, 207, 211
ALFOL process, 457
Allied Chemical Co., 128–30, 265
American Cyanamid Co., 128–31
Aminoplasts, see Urea-formaldehyde; melamine-formaldehyde resins
Ammonia–soda process, see Solvay process
Ammoxidation, 117, **300–1**
Antibiotics, 191–4, 343, 346–7, 349, 352–8
Anti-knock fluids, 202, 224, 445–6, 452, 485, 488
Anti-oxidants, 239, 246
Aromatics
  from coal, 187–9
  from petroleum, 175–7
  uses, 216–27
Asphalts, 165, 184
Assets, 89–92
Associated gas, see Natural gas
Aswan high dam, 137, 146
Athabasca tar sands, see Tar sands

Atlas Chemical Industries Inc., 89, 113, 115
Auxochromes, 328

BASF, see Badische Anilin und Soda-Fabrik AG
BP Chemicals (International) Ltd., see British Petroleum Co.
Badische Anilin und Soda-Fabrik AG, 17, 128, 130–2, 173–4
Baekeland and General Bakelite Corporation, 29, 31, 259
Balanced chlorine economy, 247, 451–2
Bayer, Farbenfabriken AG, 17, 128, 130–2, 259
Beecham Group, 114, 118–19
Belgium
  chemical industry, 120–4, 134
  imports and exports, 70–1
  journals, 4
  pharmaceutical industry, 345
  production of sulphuric acid and plastics, 102
  sources of statistics, 6
Bergius process, 439
Biochemical oxygen demand (BOD), see sewage
Bleaching compounds, 12, 14, 313–15, 317, 320, 386, 390, 394–5, 399, 437
Blowing agents, 268, 450, **464**
Brazil, 102, 499, 511
Britain, Great, see United Kingdom
British Dyestuffs Corporation, 19, 21
British Oxygen Co., 114, 116, 129
British Petroleum Co., 109, 114, 116–17, 147, 159, 197, 202, 226, 279
British Standard Softness (B.S.S.), 253–4, 257–8, 302
Bronchitis, 478, 482–4
Brunner, Mond & Co., 15, 20, 21

Builders, 313–15, 392
Burroughs Wellcome, *see* Wellcome Foundation
Butyl rubber, 213, 282, 286

C.i.f. (cost, insurance and freight), 7, 70
Canada, 70, 71, 102, 120–2, 345, 395–6
Cancer, 343–4
   of the lung, 478–80
Capital, 53, 89–92
   to developing countries, 143–5
Capital-intensive industry, 17, 83–4, 94, 108–12, 125
Carothers, W. H., 30, 32
Cartels, 21, 115, 161
Cash flow, 91, 94, **95–100**
Catalysts
   in afterburners, 488
   Deacon process, 452
   enzyme, 513
   Friedel Crafts, 443
   heterogeneous, 409, 513
   hydrogenation, 438
   oxidation, 437
   polymerization, 454–7
   Redox, 247, 272, 285, **456**, 458
   Ziegler, 456
Causticization, 14, 385–6, 395, 465
Cellulose derivatives
   carboxymethyl cellulose (CMC), 314–15, 444–5
   cellophane, 294, 388
   celluloid, 29
   cellulose acetate, 29, 214, 216, 235, 253, **294–5**
   cellulose xanthate (viscose), 29, 293–4
   dyeing of, 331, 334–5
   ethyl cellulose, 445–6, 458
   fabrication of, 460
   nitrocellulose, 28, 29, 32, 252–3, **280**, 289, 448
   rayon, 29, 289–93, 293–4, 387–8
Cement, 365, 367, 377, **384–5**, 465
Centrally planned economies, 43, 144–5, 162
   see also individual countries; 'Eastern Europe'
Chain reactions, 453–7, 486–8
'Chemical age 200', 113, 125, 128–30
Chromophores, 326–8
Clean air act, 478, 481–2, 491
Coal, 33, 161–2, **184–91**, 226, 437–9, 480–2, 501–2, 507, 512
   see also 'Coke'; 'Smoke'; 'Coal tar'
Coal tar
   aromatics from, 217
   catalytic reduction, 191
   composition and distillation, **187–9**
   naphthalene from, 226
   phenol from, 218–19
   UK production, 22
Coke
   from coal, 187–8
   raw material for

     carbide, 189
     hydrogen, 437
     make-up gas, 373
     water gas, 190
Colgate-Palmolive, 305, 321
Companies
   limited liability, 86–7
   private, 87
   UK chemical, 113–20
   world chemical, 125–35
Comparative advantage, 68–9
Competition, 64–7
   in pharmaceutical industry, 348–50
Contact process, 17, 383–4
Cost-effectiveness, 45
Costings, costs
   conventions used, 7
   desalination, 152
   distillation columns, 424–6
   fixed and variable, 78–81
   for production of
     acetylene, 174
     ammonia, 374
     benzene, 170, 177
     ethyl alcohol, 193
     ethylene, 57, 170, 171
     gloss and emulsion paints, 273
     hydrogen, 179, 437
     phenol, 223
     polyethylene, 237
     polystyrene, 241
     polyvinyl chloride, 248
   pollution control, 491–2
   sewage treatment, 474
Cotton, 12, **289–93**
   cyanoethylation of, 301
   dirt removal from, 312
   dyeing of, 331, 334, 336
   linters, 293–4, 448
Cracking of petroleum
   acetylene production, 173
   catalytic, 167
   chemistry, 167–9
   economics, 57–8, 171–3
   history, 33
   products from *n*-hexadecane, 166
   technology, 170
   thermal, 34, 117
   unit process, 465
Crystallinity in polymers, 231–4, 456–7
Cuba, 72, 142
Cumene-phenol process, 117, 207–9, 221–3, 269
Customs duties, *see* Tariffs
Cyanamide process, 18, 130

DCF, *see* Discounted cash flow
Deacon process, 247, 390, 452
Dead Sea, 147, 396–8, 452
Dehydrosulphurization, *see* Sulphur, removal of
Demand, 50–3, 73–7
   cyclical, 505–7

for graduates, 505–7
for pharmaceuticals, 348
in developing countries, 139
Depreciation, 85, 91–100, 104
Detergents, **305–21**
  biodegradability, 212, 317–20
  future of, 510
  non-ionic, 200–1, 210–11
  propylene tetramer for, 212
  sodium hydroxide for, 388
  sodium perborate for, 394–5, 399
  sodium sesquicarbonate for, 390
  sodium tripolyphosphate for, 313–15, 391–2
  straight chain alcohols for, 255, 457
  sulphuric acid for, 379
  UK statistics, 105, 112
Diazotization, *see* Dyes, azo
Diminishing returns, 54–6
Discounted cash flow, **95–100**, 172
Distillation, 403–6
  columns, 427–9
  flash, 148–9, 417, 419, 433
  fractional, 417–29
  in vinyl acetate production, 431–4
Distillers Company Ltd., 32, 109, 116–17, 213, 221–3, 277, 279, 300, 406
Dow Chemical Co., **128–30**, 220–3
Drugs
  proprietary and ethical, 345, 348–9
  psychotropic, 344–7, 349, 354, 359–63, 501, 510
  sources of, 344–5
Du Pont (E. I. Du Pont de Nemours Inc.), 31, 32, 38, 43, 93, 113, **127**, 128, 130, 287, 298
Dumping, 70
Dyes, dyestuffs, 12, 16, 18, 22, 24, 105, 126, **324–39**, 446, 510
  acid and basic, 328–31
  azo, 215, 326–38, 353, 449, 465
  direct, 331–2
  disperse, 334
  fibre-reactive, 335–6
  sulphur, 331
  vat, 331, 333

Esso, Exxon, *see* Standard Oil of New Jersey
Eastern Europe, 72, 135, 160, 289, 306
Eastman Kodak, 129, 131
Economies of scale, 34–5, 56–8, 66, 76, 139–40, 171–2, 384, 403, 513
Effluents, 404–5, 434, **467–93**
  from electroplating, 475
  from petrochemical plant, 477
  from wool scouring, 476
  standards for, 471
  see also 'Pollution'
Elasticity of demand, 51–3, 320, 348
Elastomers, 231
  see also individual rubbers; Rubber
Electricity, non-electrolytic processes based on
  boron carbide, 395
  calcium carbide, 189

  electrodialysis, 150
  ozone, 437
Electrolysis, 465
  of brine, 16, 142, 221, **385–9**, 451
    diaphragm cell, 387–9
    mercury cell, 387–9
  electroplating wastes, 475
  production of
    hydrogen, 437, 179
    hydrogen peroxide, 399
    potassium hydroxide, 387
    sodium metal, 387
Employees
  in OECD chemical industries, 120
  in UK chemical industry, 105, 108
  in UK paint industry, 271
Emulsions, 310, 454, 457–8
Enzymes, 191–4, 315, 317, 320, 513
Epoxy resins, 31, 208–9, 259, **269–70**, 280, 429, 460
Esters, 440–1
Ethylene–propylene copolymer rubbers (EPM and EPDM), 282–3, **287–8**, 454
Evelyn, John, 477
Expectation of life, *see* Mortality
Explosives, 13, 18, 220, 225, 367, 369, 377, 407, **447–9**
Exports, 69–72, 105, 111, 159–60
Extenders
  for plasticizers, 257
  for paints, 272

f.o.b. (free on board), 7, 70
Fatty acids, 317, 439
Fermentation processes, 33, 35, **191–4**, 345, 356, 358, 407, 465, 512–13
  in sewage treatment, 472–6
  operated by Distiller's Co. Ltd., 116, 202
  Weizmann process, 19, 116, 192
Fertilizers
  basic slag, 392
  borax into, 394
  future of, 510, 512
  in nineteenth century, 13
  nitro-chalk, 377
  nitrogenous, 367–9, 377–8
  phosphatic, 368–9, 381, **391–3**
  potash, 368, 395–6
  production and consumption, 22, 26, 105, 112, 126, 137, 365–8
  sulphuric acid requirements, 379, 385
Fibres—synthetic, 26, 28, 231, **288–301**
  dyeing of, 334–5
  future of, 509, 511
  sulphuric acid requirements, 379
  see also under individual fibres
Fillers, 246, **258**, 262, 264
  in detergents, 313–14
  glass fibre, *see* 'Glass'
Fischer–Tropsch process, 190, 439
Fisons, 114, 118
Flow-sheets
  benzene–toluene–xylene plant, 176

526 *Subject index*

Flow-sheets–(*contd.*)
  domestic sewage plant, 470
  Haber process, 372
  'hydrostripping' oil refinery, 164
  lead chamber process, 382
  naphtha reformer, 375
  steam-naphtha cracker, 169
  vinyl acetate, 432
Fluid dynamics, 409–10
Fluorescent brighteners (optical bleaches), 314, 315, 332, 336
Fluorocarbons, 449–50
Foamed plastics, 211, 235, 259, 268–9, 450, 460, **463–4**
Forecasting, 502–7
France
  chemical
    and industrial production, 111
    companies, 128, 130, 132
    imports and exports, 70–1
    industry, 35, 120–4
    journals, 4
    statistical sources, 6
  production of
    carbide and acetylene, 189
    nylon 11, 298
    potash, 396
    sulphuric acid and plastics, 102
Friedel Crafts reaction, 216, 443
  in production of
    butyl rubber, 286
    cumene, 221
    detergent alkylate, 317
    nonyl phenols, 319
    styrene, 240
  reverse, 177
Fuel oil, 161–2, 411, 437, 488

Gas oil, 164
Gasoline
  'Alkylate', 443
  coal-based, 190–1, 439
  demand structure, 161–2, 165–6
  octane improvers, 224–5
    see also 'Anti-knock fluids'
  pollution by exhaust fumes, 484–8
  straight-run, 164
Geographical distribution of chemical industry, 103, 109–10, 121–4
Germany
  chemical
    and industrial production, 111
    companies, 128, 130, 132–3
    imports and exports, 70–1
    industry, 120–4
    journals, 4
    statistical sources, 6
  coal industry, 186
  history of chemical industry, 16–18, 31–2, 35–6, 191, 439
  pharmaceutical industry, 345
  production of
    acrylonitrile, 301
    carbide and acetylene, 189
    nylon 11, 298
    phenol from coal, 220
    potash, 395–6
    sulphuric acid and plastics, 102
    synthetic rubber, 283
Glass, 12, 390, 411, 465
  as material of construction, 429–30
  borosilicate, 393–4
  fibre, 267, 270, 394
Glaxo Group, 114, 117, 119
Goodwill, 89, 90
'Group of 77', 136
Guevara, Ché, 142, 155

Haber process, **369–72**, 438
  feedstock for, 178, 190, **372–6**
  history, 17, 18, 20, 367
Hardening of fats and oils, *see* Margarine
Heat transfer, 406, 409, **410–16**
  in polymerization, 457–8
Hitler, Adolf, 131–2, 283
Hoechst, Farbwerke AG, 17, 128, 130–2, 173, 335
Hoffman la Roche, F. & Cie, AG, 128, 130, 133
Holland, *see* Netherlands
'Hydrocracking', 165–6
'Hydrostripping' oil refinery, **164–5**, 184
Hypnotics, *see* Drugs, psychotropic

ICI, *see* Imperial Chemical Industries
IG Farben, 20, 31, 131–2, 283, 439
Imperial Chemical Industries, 20–1, 31–2, 38, 43, 109, **113–16**, 128, 130, 243, 245, 298
  balance sheet, 88–91
  current ratio, 93
  process for
    acetylene, 175
    acrylonitrile, 300
    alkyd resin paints, 271
    phthalic anhydride, 226
    Procion dyes, 335
    steam reforming, 373
  profit and loss account, 91–2
Imports, 69–72, 105, 111, 160–1
India, 4, 102
Inorganic chemicals, 22, 105, 112, 118, 126, **365–400**, 510–11
  see also under individual chemicals
Indifference curves, 46, 47, 48, 50
Inferior commodities, 51
Inflation, 66, 85, 94–5
Ingrain dyeing, 331, 334
Input-output table, 106, 107
Intermediate technology, 146
Investment
  and national income, 104
  by ICI, 91
  by individual chemical companies, 128–9
  DCF calculation of return on, 95–100
  in OECD chemical industries, 120
  in UK chemical industry, 105, 511

Subject index 527

Ireland (Eire and Northern Ireland), 69, 116, 120–1, 489
Israel, 102, 139–40, 147, 395–6, 452
Italy
  chemical
    and industrial production, 111
    companies, 134
    imports and exports, 70–1
    industry, 35, 120–4
  production of
    carbide and acetylene, 189
    sulphuric acid and plastics, 102

Japan
  acrylate process, 279
  carbide production, 189
  chemical
    companies, 128, 130–2
    industry, 26, 34–6, 120–4
    imports and exports, 70–2
    journals, 4
    statistical sources, 6
  coal industry, 186
  dyestuffs production, 324
  future growth, 498, 511
  nylon 11, 298
  petroleum consumption, 160–1
  pharmaceutical industry, 345
  pollution problems, 118
  soap and detergents, 306
  sulphuric acid and plastics, 101–2
  sulphur and sulphuric acid, 385
  synthetic fibres, 289
  synthetic rubber, 283

Kerosene, 164
Kolbe-Schmitt process, 350
Kuwait, 140, 151–2

LPG, *see* Liquid petroleum gas
Labour, 53, 140–1, 407
Labour-intensive industry, 83–4, 111, 140, 185
  future of, 507, 512
Land, 53, 152
Land-intensive projects, 83–4
Laporte Industries, 114, 119, 226, 282, 405
Leaded Petrol, *see* Anti-knock fluids
Leblanc process, 13–16, 395, 478
Liabilities, 89, 90, 93
Lilly, Eli & Co., 117, 129
Linseed oil, 256, 271, 278–80
Liquid petroleum gas, 164, 436
Loan stock, 85, 89, 115
Logistic curve, 503–5, 508
Lurgi process, 190

M/F resins, *see* Melamine-formaldehyde resins
Make-up gas, 178, 190, 372–6
Margarine, 436, 438–9
Marginal utility, profit etc., 46
  see also under Utility; profit; etc.

Mass transfer, 406, 409, 416–17
Melamine-formaldehyde resins, 31–2, 259, **264–6**, 280, 460, 464
Methylated spirits, 202
Micelles, 308–11, 458
Microbiological processes, *see* Fermentation processes
Middle distillates, 161–2, 175
Middle East, 159–62, 500–1
Molex process, 318
Money, 45, 48
Monsanto Co., 109, 128–30
Montecatini-Edison SpA, 31, 128, 130, 134
Mortality, 340–4
  bronchitis, 482–4
  lung cancer, 478–80
  maternal, 341, 352
  pneumonia, 342
  tuberculosis, 343
  trends – 1848–1965, 344

Naphtha
  acetylene and ethylene from, 247
  aromatics from, 175–6
  cracking of, 164–5, 169, 172–3
  light and heavy, 164
  steam reforming of, 373–6
National income, 101–6, 350, 507, 511
Natural gas, 161–2, 178, **182–4**, 186, 214, 373, 437–8, 447, 488, 508
Netherlands, 6, 35, 70–1, 102, **120–4**, 134, 184
Nitrocellulose, *see* Cellulose derivatives
Nitrile rubbers (Buna N), 31, 212, 282–3, **286–7**, 299
North Sea Gas, *see* Natural gas
Norway, 69, 102, 116, 120, 121, 189
Novolacs, 261
Nuclear power, 501, 512–13
Nylon, 30–2, 216, 257, **295–8**, 290–3, 299
  boric acid used as catalyst, 394
  dyeing of, 334
  n-heptanol by-product, 255, 298
  ICI overexpansion, 115
  market data, 235, 289
  relationship with fertilizers, 369

OECD, *see* Organization for Economic Cooperation and Development
Oil, *see* Petroleum
Organic chemicals, 22, 33, 105, 112, 125–6, **196–228**, 511
  see also under individual chemicals
Organization for Economic Cooperation and Development
  chemical industry in member countries, 120–35
  exports and imports, 70–2
  future growth, 511
  member countries and aims, 120
  petroleum usage, 160–1
  reports, 4–5
  sulphur and sulphuric acid, 385

Organization for Economic Cooperation and Development—(contd.)
  see also under individual countries
Over-capacity, 21, 64, 66, 127, 291, 508, 511
Oxo process, 178, 204, 206, 211, 255, **439–40**

P/F resins, see Phenol-formaldehyde resins
PFD Acetic acid process, 117, 213–14, 406
PTFE, see Polytetrafluoroethylene
PVC, see polyvinyl chloride
Paper, 12, 147, 292, 386, 388, 399, 508
Paints, 32, 105, 126, 257, **271–82**, 509
  for automobiles, 243, 265, 279–80
  strippers, 450
Penicillins, 191–3, 345–7, **355–7**
Perkin, W. H. Senior, 12, 16, 324
Pesticides, 386, 445, 451, 490
Petrochemical plant
  world distribution, 136
  wastes from, 476
Plasticizers, 245–6, 249–50, **251–7**, 441
  alcohols for, 440, 457
  low temperature, 256
  in plastisols, 464
  in PVA paints, 276
  polymeric, 210, 253, **256**
  secondary, 257
  softening power and volatility, 254
Petrol, see Gasoline
Petroleum, **159–81**, 418, 438–9, 500–1, 508, 512
  cracking, see Cracking of petroleum
Pfizer Inc., 128, 130
Pharmaceuticals, 22, 105, 112, **340–64**
  companies, 117–19, 129
  future of, 500, 510–12
  see also under individual companies; Drugs
Phenol-formaldehyde resins, 29, 31–2, **259–263**, 266, 460
Phenoplasts, 36, 264, 449
  see also Phenol-formaldehyde resins
Pigments, 271–3, 281–2, 324, 379, **336–8**, 510
Plastics and resins
  future of, 498, 508–11
  processing and fabrication of, 458–64
  production and consumption of, 26, 35–6, 102, 105, 112, 125–6
  see also under individual materials
Platforming, see Reforming, catalytic
Pneumonia, 342, 352
Poisonings, 354
Poland, 135, 186
Pollution
  automobile exhaust gases, 484–8
  cigarette smoke, 478–80
  coal burning, 480–4
  detergents, 212, 317, 319–20
  future levels of, 498, 508, 510
  lead, see Anti-knock fluids
  Leblanc process, 14, 478
  long-term, of biosphere, 488–91
  mercury, 387
  miscellaneous sources, 491

  pesticides, 451
  sulphur dioxide, 384, 480–6
Polyacetals, 235, **244**
Polyacrylates, see Acrylics
Polyacrylonitrile fibres, 31, 34, 289–93, **299–301**, 331, 454
Polyamide fibres, see Nylon
Poly-(butadiene-acrylonitrile), see Nitrile rubbers
Polybutadiene rubbers, 212, 282, 285, 454
Poly-(butadiene-styrene), see Styrene-butadiene rubber
Polycarbonate resins, 209, 234, 235, **244**
Poly-(2-chlorobutadiene), see Polychloroprene rubber
Polychloroprene rubber, 31, 282, **287**, 454
Polyester fibres, 30–4, 115, 201, 225, 234–5, 289–93, **298–9**, 334
Polyesters, 210, 218, 256, 259, **266–7**, 460
Polyethyl acrylate, 243, 292
Polyethylene
  crystallinity, 232–3, 237
  discovery, 31–3, 35–8
  end use patterns, 36–7, 238
  fabrication, 460
  future growth, 508
  manufacture (high and low density), **236–8**, 403
  market data, 36–7, 64, 198–9, 235
  polymerization, 454, 457
Polyethylene glycol terephthalate, see Polyester fibres
Polyethylene oxide, 201, 273
Polyisobutene, 213, 234, 454
Polyisoprene rubber (IR), 282–3, **284–5**, 454
Polymethyl methacrylate, 33, 209, 234–5, **242–3**, 259, 454, 457, 459–60
Polyolefines, 36, 289, 510
  see also polyethylene; polypropylene
Polypropoxy ethers, 210
Polypropylene
  crystallinity, 233
  discovery, 31, 34
  end use pattern, 238
  fabrication, 460
  fibres, 289, 291, 335
  manufacture, 238–9
  market data, 36, 205–6, 235
  polymerization, 454
Polypropylene glycols, 210
Polystyrene
  crystallinity, 234
  discovery, 31
  end-use pattern, 242
  fabrication, 460
  foamed, 463–4
  impact, 240–1
  manufacture, **239–41**
  market data, 36, 198–9, 235
  polymerization, 454
Polysulphide rubbers, 270, 288
Polytetrafluoroethylene, 235, **245**
Polyurethanes

Subject index 529

fibres, 291
foams, 460, 463–4
manufacture, **267–8**
market data, 259, 269–70
raw materials, 211, 225
rubbers, 282, 288
Polyvinyl acetate
discovery, 31
market data, 235
non-woven fabrics, 292
paints, 272–3, **274–6**
plasticization, 253
polymerization, 454, 458
Polyvinyl alcohol, 235, 274, 276
Polyvinyl butyral, 276
Polyvinyl chloride
additives, 245–6, 250–8
crystallinity, 234
discovery, 31, 33, 34
end-use pattern, 249–50
fabrication, 457, 460, 464
future uses, 508
manufacture, 70, 247–8
market data, 36, 235
polymerization, 454
Polyvinylidene chloride, 234–5, **244**
Population
of countries, 102, 136
rate of growth and density, 137–8
Preference shares, 85
Prescriptions
dispensed in UK and USA, 346–50
for psychotropic drugs, 362–3
Price, 48–9, 62–3, 65–7, 73, 82–3
Prices
α and γ picolines, 204
future, for oil and plastics, 501, 508, 512
phenol and acetone, 222
phenoplasts and aminoplasts, 266
secret, for ethylene, 197
sodium hydroxide and chlorine, 386–9
sources of, 7
UK chemical, 66
see also tables of market data throughout the book
Proctor and Gamble, 109, 120, 131, 305, 311, 316, 320–1
Producer gas, 190, 373
Production, 53–60
production figures, *see tables of market data throughout the book*
Productivity, 113, 121–3, 127, 507
Profit and loss account, 91–2
Profits
before and after tax, 92
in UK chemical industry, 112
of ICI, by sector, 115
optimization, 73–83
undistributed, 85, 91
Profitability
DCF calculation of, 95–100
measures of, 92–3
heavy inorganic chemical industry, 365

soap and detergent industry, 321
UK chemical industry, 111–13
Protein
in sewage, 469
semi-synthetic, including textured vegetable, protein (TVP), 152–3
via yeasts, 192, 194

Raschig process, 221, 223
Redox catalysts, *see* Catalysts
Rayon, *see* Cellulose derivatives
Reflux ratio, 421, 424–6
Reforming
catalytic, 166, 175–6, 439, 465
steam, 178, 373–6
Reppe synthesis, 277
Research and development, 16–21, 30, 32, 108, 128–9, 348–50, 512
Residual oil, 165
Resols, 260, 464
Resources, 53, 83–6, 499
Reverse yield gap, 86
Rhône-Poulenc SA, 128, 130, 132
Rio-Tinto Zinc Corporation, 114, 118
Rolls-Royce, 94
Royal Dutch Shell, 109, 114, **119**, 128, 130, 134, 159, 176
Rubber
chemicals for processing, 284, 429, 446
natural, 25, 27, 232, **282–4**
synthetic, 26–7, 33, **282–8**, 509
vulcanization, 25, 284–7
see also under individual rubbers

Searle's Lake, California, 393, 396
Sewage
biochemical oxygen demand (BOD), 470–2
suspended solids, 470–2
treatment, 386, 390, **468–74**
activated sludge method, 473–6
anaerobic digestion, 474
development of, 467–8
economics, 474
wool industry wastes, 476
Shale oil, 184, 501
Shareholders, 43, 86, 89–91, 93, 121
Shell, *see* Royal Dutch Shell
Shift reaction, 178, 190
Silicone rubbers, 282, 288
Slack wax, 165, 255
Smog, Los Angeles photochemical, 484–8
Smoke
cigarette, 478–80, 482–4, 510
coal, 476–8, 480–4, 488
Soap, 12, 23, 105, 112, **305–21**, 379, 388, 395, 441–2, 476
Social benefits, 45, 147, 142
Solids handling, 186, 226, 409–10
plastics processing and fabrication, 458–64
Solvay & Cie, 119, 128, 130, 133
Solvay process, 14–16, 385–7, 390
South Africa, 121, 182, 186, 190
Soya Beans, 152–3, 192, 256

Spain, 70, 71, 102, 120–4, 395
Spinning, 153, 298, 301
Square-cube law, 57, 76, 171
Stabilizers
   polypropylene, 239, 246
   PVC, 246, **250–1**
   foam, 315
Standard Oil of New Jersey (Exxon, formerly Esso), 31, 109, 159
Stanley, H. M., 39, 117
Starch
   fermentation, 191–3, 202
   hydrolysis, 442
Steam reforming, see Reforming
Stefan-Boltzmann Law, 412
Stereoregular (Ziegler) polymers, 454, 456–7
   crystallinity, 234
   ethylene-propylene, 287
   high-density polyethylene, 236–8
   polybutadiene, 286
   polyisoprene, 284–5
   polypropylene, 238–9
   styrene-butadiene, 285
Stimulants, see Drugs, psychotropic
Stoving lacquers, 261, 274, 279–80
Streptomycins, 345–7, **358**
Styrene-acrylonitrile resins (SAN), 199, 235, 241
Styrene-butadiene rubbers (Buna S, SBR, GR-S), 31, 32, 33, 212, 240, 282
Subsidies, 50, 63
Sulphonamides, 345–7, **352–3**
Sulphur, removal of, 165, 184, 186, 206, 373–4
Supply, 60–4
Surface tension, 305–10, 479
Surface coatings, see Paints
Surfactants, see Detergents
Sweden, 70, 71, 102, 120–4, 320, 345, 489
Switzerland, 70–1, 120–2, 131, **133**, 139, 189, 320, 345, 468
Synthesis gas, see Make-up gas

Tar sands, 184, 501
Tariffs, 1, 14, 19, 21, **67–70**, 127
Taxation, effect of, 48, 49, 62, 63, 98, 99
Terylene, see Polyester fibres
Tetracyclines, 345–7, 358
Thermal conductivities, 411
Thermodynamics
   contact process, 383
   cracking processes, 167
   Haber process, 369–72
   polymer crystallization, 231–4
   process design, 407
   sources of data, 434

Town gas, 374–5
Tranquilizers, see Drugs, psychotropic
Tuberculosis, 343–4, 352

U/F resins, see Urea-formaldehyde resins
UNCTAD, see United Nations Conference on Trade and Development
Udex process, 176
Ugine-Kuhlmann, 128, 130, 132
Unilever, 109, 114, **119–20**, 129, 134, 320–1
Union Carbide Co., 31, 38, 128–30
Union of Soviet Socialist Republics, 70, 72, 102, **134–5**, 160–2, 186, 220, 283–5, 395
United Alkali Co., 15, 21
United Kingdom, see note 2 on first page of this index
United Nations
   Conference on Trade and Development (UNCTAD), 135–6
   desalination work, 147
   Protein Advisory Group (PAG), 155
   Standard International Trade Classification (SITC), 1
   statistical sources, 5
United States of America, see note 2 on first page of this index
United States Tariff Commission, 5, 228
Urea-formaldehyde resins, 31–2, 259, **262–4**, 266, 458, 460
Utility, 45–50

Value added, 102, 108, 113, 120, 122–3
Viscose, see Cellulose derivatives

Wacker process
   acetaldehyde, 203, 301
   acetone, 207
   vinyl acetate, 115, 227, 275–6, 404–6, 431–4
Water
   consumption, 468
   desalination, 147–52
Water-gas, 190, 373
Weizmann process, see Fermentation processes
Wellcome Foundation, 114, 118
Wood, 33, 411, 430
   pulp, 293–4, 386, 399, 448
Wool, 28, **288–93**
   dyeing of, 329–31, 335
   grease, 476
World dynamics, 498–9, 514
Wulff process, 116, 173–5, 227

'Ziegler' alcohols, 255, 319, 457
'Ziegler' polymers, see Stereoregular polymers